# Elementary Differential Equations

# Elementary Differential Equations

## Sixth Edition

### Earl D. Rainville

Late Professor of Mathematics
University of Michigan

### Phillip E. Bedient

Professor of Mathematics
Franklin and Marshall College

*Macmillan Publishing Co., Inc.*
*New York*
*Collier Macmillan Publishers*
*London*

Macmillan Publishing Co., Inc.
866 Third Avenue, New York, New York 10022

Collier Macmillan Canada, Ltd.

Library of Congress Cataloging in Publication Data

Rainville, Earl David,
    Elementary differential equations.

    Includes index.
    1. Differential equations. I. Bedient, Philip
Edward,    –    joint author. II. Title.
QA371.R29   1980        515.3'5        80-12849
ISBN 0-02-397770-1

Printing:    ·5678    Year:    ·45678

ISBN    0-02-397770-1

# *Preface*

## to the Sixth Edition

This new edition of Professor Rainville's book maintains the simple and direct style of earlier editions and makes some modest changes. The balance between developing techniques for solving equations and the theory necessary to support those techniques is essentially unchanged. However, the variety and number of applications has been increased and placed as early in the text as is feasible.

The material is arranged to permit great flexibility in the choice of topics for a semester course. Except for Chapters 1, 2, 5, 16 through 18, and either 6 and 7 or 11 and 12, any chapter on ordinary differential equations can be omitted without interfering with the study of later chapters. Parts of chapters can be omitted in many instances.

For a course that aims at reaching power series as rapidly as is consistent with some treatment of more elementary methods, a reasonable syllabus should include Chapters 1 and 2, Chapters 5, 6, 7, 8, parts of Chapters 13 and 15, Chapters 17 and 18, and whatever applications the instructor cares to insert.

Chapters 1 through 16 of this book appear separately as *A Short Course in Differential Equations*, Sixth Edition. The shorter version is intended for courses that do not include discussion of infinite series methods.

The author wishes to thank those students and colleagues at Franklin and Marshall College whose suggestions and support have been most helpful.

Phillip E. Bedient

Lancaster, Pennsylvania

# Contents

## *12*   **Inverse Transforms**

## *13*   **Linear Systems of Equations**

## *14*   **Electric Circuits and Networks**

## *19*   Equations of Hypergeometric Type

## *20*   Numerical Methods

## *21*   Partial Differential Equations

# Definitions, Elimination of Arbitrary Constants

## 1. Examples of differential equations

In physics, engineering, and chemistry and, increasingly, in such subjects as biology, physiology, and economics it is necessary to build a mathematical model to represent certain problems. It is often the case that these mathematical models involve the search for an unknown function that satisfies an equation in which derivatives of the unknown function play an important role. Such equations are called differential equations. As in equation (3) below, a derivative may be involved implicitly through the presence of differentials. Our aim is to find methods for solving differential equations; that is, to find the unknown function or functions that satisfy the differential equation.

The following are examples of differential equations:

$$\frac{dy}{dx} = \cos x, \tag{1}$$

$$\frac{d^2y}{dx^2} + k^2 y = 0, \tag{2}$$

$$(x^2 + y^2)\,dx - 2xy\,dy = 0, \tag{3}$$

$$\frac{\partial u}{\partial t} = h^2\left(\frac{\partial^2 u}{\partial x^2} + \frac{\partial^2 u}{\partial y^2}\right), \tag{4}$$

$$L\frac{d^2 i}{dt^2} + R\frac{di}{dt} + \frac{1}{C}i = E\omega \cos \omega t, \tag{5}$$

$$\frac{\partial^2 V}{\partial x^2} + \frac{\partial^2 V}{\partial y^2} = 0, \tag{6}$$

$$\left(\frac{d^2 w}{dx^2}\right)^3 - xy\frac{dw}{dx} + w = 0, \tag{7}$$

$$\frac{d^3 x}{dy^3} + x\frac{dx}{dy} - 4xy = 0, \tag{8}$$

$$\frac{d^2 y}{dx^2} + 7\left(\frac{dy}{dx}\right)^3 - 8y = 0, \tag{9}$$

$$\frac{d^2 y}{dt^2} + \frac{d^2 x}{dt^2} = x, \tag{10}$$

$$x\frac{\partial f}{\partial x} + y\frac{\partial f}{\partial y} = nf. \tag{11}$$

When an equation involves one or more derivatives with respect to a particular variable, that variable is called an *independent* variable. A variable is called *dependent* if a derivative of that variable occurs.

In the equation

$$L\frac{d^2 i}{dt^2} + R\frac{di}{dt} + \frac{1}{C}i = E\omega \cos \omega t \tag{5}$$

$i$ is the dependent variable, $t$ the independent variable, and $L, R, C, E,$ and $\omega$ are called parameters. The equation

$$\frac{\partial^2 V}{\partial x^2} + \frac{\partial^2 V}{\partial y^2} = 0 \tag{6}$$

has one dependent variable $V$ and two independent variables.

Since the equation

$$(x^2 + y^2)\,dx - 2xy\,dy = 0 \tag{3}$$

may be written

$$x^2 + y^2 - 2xy\frac{dy}{dx} = 0$$

or

$$(x^2 + y^2)\frac{dx}{dy} - 2xy = 0,$$

we may consider either variable to be dependent, the other being the independent one.

### Exercise

Identify the independent variables, the dependent variables, and the parameters in the equations given as examples in this section.

## 2. Definitions

The *order* of a differential equation is the order of the highest-ordered derivative appearing in the equation. For instance,

$$\frac{d^2y}{dx^2} + 2b\left(\frac{dy}{dx}\right)^3 + y = 0 \tag{1}$$

is an equation of "order two." It is also referred to as a "second-order equation."

More generally, the equation

$$F(x, y, y', \ldots, y^{(n)}) = 0 \tag{2}$$

is called an "*n*th-order" ordinary differential equation. Under suitable restrictions on the function $F$, equation (2) can be solved explicitly for $y^{(n)}$ in terms of the other $n + 1$ variables $x, y, y', \ldots, y^{(n-1)}$, to obtain

$$y^{(n)} = f(x, y, y', \ldots, y^{(n-1)}). \tag{3}$$

For the purposes of this book we shall assume that this is always possible. Otherwise, an equation of the form of equation (2) may actually represent more than one equation of the form of equation (3).

For example, the equation

$$x(y')^2 + 4y' - 6x^2 = 0$$

actually represents the two different equations,

$$y' = \frac{-2 + \sqrt{4 + 6x^3}}{x} \quad \text{or} \quad y' = \frac{-2 - \sqrt{4 + 6x^3}}{x}.$$

A function $\phi$, defined on an interval $a < x < b$, is called a solution of the differential equation (3), provided that the $n$ derivatives of the function exist on the interval $a < x < b$ and

$$\phi^{(n)}(x) = f(x, \phi(x), \phi'(x), \ldots, \phi^{(n-1)}(x)),$$

for every $x$ in $a < x < b$.

For example, let us verify that

$$y = e^{2x}$$

is a solution of the equation

$$\frac{d^2y}{dx^2} + \frac{dy}{dx} - 6y = 0. \tag{4}$$

We substitute our tentative solution into the left member of equation (4) and find that

$$\frac{d^2y}{dx^2} + \frac{dy}{dx} - 6y = 4\,e^{2x} + 2\,e^{2x} - 6\,e^{2x} \equiv 0,$$

which completes the desired verification.

All of the equations we shall consider in Chapter 2 are of order one, and hence may be written

$$\frac{dy}{dx} = f(x, y).$$

For such equations it is sometimes convenient to use the definitions of elementary calculus to write the equation in the form

$$M(x, y)\,dx + N(x, y)\,dy = 0. \tag{5}$$

A very important concept in the study of differential equations is that of linearity. An ordinary differential equation of order $n$ is called linear if it may be written in the form

$$b_0(x)\frac{d^ny}{dx^n} + b_1(x)\frac{d^{n-1}y}{dx^{n-1}} + \cdots + b_{n-1}(x)\frac{dy}{dx} + b_n(x)y = R(x).$$

For example, equation (1) above is nonlinear, and equation (4) is linear. The equation

$$x^2y'' + xy' + (x^2 - n^2)y = 4x^3$$

is also linear.

The notion of linearity may be extended to partial differential equations. For example, the equation

$$b_0(x, y)\frac{\partial w}{\partial x} + b_1(x, y)\frac{\partial w}{\partial y} = R(x, y)$$

is the general first-order linear partial differential equation with two independent variables and

$$b_0(x, y)\frac{\partial^2 w}{\partial x^2} + b_1(x, y)\frac{\partial^2 w}{\partial x\, \partial y} + b_2(x, y)\frac{\partial^2 w}{\partial y^2}$$

$$+ b_3(x, y)\frac{\partial w}{\partial x} + b_4(x, y)\frac{\partial w}{\partial y} + b_5(x, y)w = R(x, y)$$

is the general second-order linear partial differential equation with two independent variables.

## Exercises

For each of the following, state whether the equation is ordinary or partial, linear or nonlinear, and give its order.

**1.** $\dfrac{d^2x}{dt^2} + k^2x = 0.$

**2.** $\dfrac{\partial^2 w}{\partial t^2} = a^2 \dfrac{\partial^2 w}{\partial x^2}.$

**3.** $(x^2 + y^2)\, dx + 2xy\, dy = 0.$

**4.** $y' + P(x)y = Q(x).$

**5.** $y''' - 3y' + 2y = 0.$

**6.** $yy'' = x.$

**7.** $\dfrac{\partial^2 u}{\partial x^2} + \dfrac{\partial^2 u}{\partial y^2} + \dfrac{\partial^2 u}{\partial z^2} = 0.$

**8.** $\dfrac{d^4y}{dx^4} = w(x).$

**9.** $x\dfrac{d^2y}{dt^2} - y\dfrac{d^2x}{dt^2} = c_1.$

**10.** $L\dfrac{di}{dt} + Ri = E.$

**11.** $(x + y)\, dx + (3x^2 - 1)\, dy = 0.$

**12.** $x(y'')^3 + (y')^4 - y = 0.$

**13.** $\left(\dfrac{d^3w}{dx^3}\right)^2 - 2\left(\dfrac{dw}{dx}\right)^4 + yw = 0.$

**14.** $\dfrac{dy}{dx} = 1 - xy + y^2.$

**15.** $y'' + 2y' - 8y = x^2 + \cos x.$

**16.** $a\, da + b\, db = 0.$

## 3. The elimination of arbitrary constants

In practice, differential equations arise in many ways, some of which we shall encounter later. There is one way of arriving at a differential equation, however, that is useful in that it gives us a feeling for the kinds of solutions to be expected. In this section we shall start with a relation involving arbitrary constants and, by elimination of those arbitrary constants, come to a differential equation consistent with the original relation. In a sense we start with the answer and find the problem.

Methods for the elimination of arbitrary constants vary with the way in which the constants enter the given relation. A method that is efficient for one problem may be poor for another. One fact persists throughout. Because each differentiation yields a new relation, the number of derivatives that need be used is the same as the number of arbitrary constants to be eliminated. We shall in each case determine the differential equation that is

(a) Of order equal to the number of arbitrary constants in the given relation.
(b) Consistent with that relation.
(c) Free from arbitrary constants.

EXAMPLE (a): Eliminate* the arbitrary constants $c_1$ and $c_2$ from the relation

$$y = c_1 e^{-2x} + c_2 e^{3x}. \tag{1}$$

Since two constants are to be eliminated, obtain the two derivatives,

$$y' = -2c_1 e^{-2x} + 3c_2 e^{3x}, \tag{2}$$

$$y'' = 4c_1 e^{-2x} + 9c_2 e^{3x}. \tag{3}$$

The elimination of $c_1$ from equations (2) and (3) yields

$$y'' + 2y' = 15c_2 e^{3x};$$

the elimination of $c_1$ from equations (1) and (2) yields

$$y' + 2y = 5c_2 e^{3x}.$$

Hence

$$y'' + 2y' = 3(y' + 2y),$$

or

$$y'' - y' - 6y = 0.$$

Another method for obtaining the differential equation in this example proceeds as follows: We know from a theorem in elementary algebra that the three equations (1), (2), and (3), considered as equations in the two unknowns $c_1$ and $c_2$, can have solutions only if

$$\begin{vmatrix} -y & e^{-2x} & e^{3x} \\ -y' & -2e^{-2x} & 3e^{3x} \\ -y'' & 4e^{-2x} & 9e^{3x} \end{vmatrix} = 0. \tag{4}$$

Since $e^{-2x}$ and $e^{3x}$ cannot be zero, equation (4) may be rewritten, with the factors $-1$, $e^{-2x}$, and $e^{3x}$ removed, as

* By differentiations and pertinent legitimate mathematical procedures. Elimination by erasure, for instance, is not permitted.

$$\begin{vmatrix} y & 1 & 1 \\ y' & -2 & 3 \\ y'' & 4 & 9 \end{vmatrix} = 0$$

from which the differential equation

$$y'' - y' - 6y = 0$$

follows immediately.

This latter method has the advantage of making it easy to see that the elimination of the constants $c_1, c_2, \ldots, c_n$ from a relation of the form

$$y = c_1 e^{m_1 x} + c_2 e^{m_2 x} + \cdots + c_n e^{m_n x}$$

will always lead to a linear differential equation

$$a_0 \frac{d^n y}{dx^n} + a_1 \frac{d^{n-1} y}{dx^{n-1}} + \cdots + a_{n-1} \frac{dy}{dx} + a_n y = 0,$$

in which the coefficients $a_0, a_1, \ldots, a_n$ are constants. The study of such differential equations will receive much of our attention.

EXAMPLE (b): Eliminate the constant $a$ from the equation

$$(x - a)^2 + y^2 = a^2.$$

Direct differentiation of the relation yields

$$2(x - a) + 2yy' = 0,$$

from which

$$a = x + yy'.$$

Therefore, using the original equation, we find that

$$(yy')^2 + y^2 = (x + yy')^2,$$

or

$$y^2 = x^2 + 2xyy',$$

which may be written in the form

$$(x^2 - y^2)\, dx + 2xy\, dy = 0.$$

Another method will be used in this example as an illustration of a device that is often helpful. The method is based upon the isolation of an arbitrary constant.

The equation

$$(x - a)^2 + y^2 = a^2$$

may be put in the form

$$x^2 + y^2 - 2ax = 0,$$

or

$$\frac{x^2 + y^2}{x} = 2a.$$

Then differentiation of both members leads to

$$\frac{x(2x\,dx + 2y\,dy) - (x^2 + y^2)\,dx}{x^2} = 0,$$

or

$$(x^2 - y^2)\,dx + 2xy\,dy = 0,$$

as desired.

It is interesting to speculate here about the significance of $x = 0$ upon the argument just used. The student should draw a few of the members of this family of circles and observe what is peculiar about their behavior at $x = 0$.

EXAMPLE (c):   Eliminate $B$ and $\alpha$ from the relation

$$x = B \cos(\omega t + \alpha), \tag{5}$$

in which $\omega$ is a parameter (not to be eliminated).

First we obtain two derivatives of $x$ with respect to $t$:

$$\frac{dx}{dt} = -\omega B \sin(\omega t + \alpha), \tag{6}$$

$$\frac{d^2x}{dt^2} = -\omega^2 B \cos(\omega t + \alpha). \tag{7}$$

Comparison of equations (5) and (7) shows at once that

$$\frac{d^2x}{dt^2} + \omega^2 x = 0.$$

EXAMPLE (d):   Eliminate $c$ from the equation

$$cxy + c^2x + 4 = 0.$$

At once we get

$$c(y + xy') + c^2 = 0.$$

Since $c \neq 0$,

$$c = -(y + xy')$$

and substitution into the original equation leads us to the result

$$x^3(y')^2 + x^2yy' + 4 = 0.$$

Our examples suggest that in a certain sense the totality of solutions of an $n$th-order equation depends on $n$ arbitrary constants. The sense in which this is true will be stated in Sections 5 and 12.

## Exercises

In each of the following eliminate the arbitrary constants.

1. $x \sin y + x^2 y = c$.    ANS.   $(\sin y + 2xy)\,dx + (x \cos y + x^2)\,dy = 0$.
2. $3x^2 - xy^2 = c$.    ANS.   $(6x - y^2)\,dx - 2xy\,dy = 0$.
3. $xy^2 - 1 = cy$.    ANS.   $y^3\,dx + (xy^2 + 1)\,dy = 0$.
4. $cx^2 + x + y^2 = 0$.    ANS.   $(x + 2y^2)\,dx - 2xy\,dy = 0$.

5. $x = A \sin (\omega t + \beta)$; $\omega$ a parameter, not to be eliminated.    ANS.   $\dfrac{d^2x}{dt^2} + \omega^2 x = 0$.

6. $x = c_1 \cos \omega t + c_2 \sin \omega t$; $\omega$ a parameter.    ANS.   $\dfrac{d^2x}{dt^2} + \omega^2 x = 0$.

7. $y = cx + c^2 + 1$.    ANS.   $y = xy' + (y')^2 + 1$.

8. $y = mx + \dfrac{h}{m}$; $h$ a parameter, $m$ to be eliminated.    ANS.   $y = xy' + \dfrac{h}{y'}$.

9. $y^2 = 4ax$.    ANS.   $2x\,dy - y\,dx = 0$.
10. $y = ax^2 + bx + c$.    ANS.   $y''' = 0$.
11. $y = c_1 + c_2 e^{3x}$.    ANS.   $y'' - 3y' = 0$.
12. $y = 4 + c_1 e^{3x}$.    ANS.   $y' - 3y = -12$.
13. $y = c_1 + c_2 e^{-4x}$.    ANS.   $y'' + 4y' = 0$.
14. $y = c_1 e^x + c_2 e^{-x}$.    ANS.   $y'' - y = 0$.
15. $y = x + c_1 e^x + c_2 e^{-x}$.    ANS.   $y'' - y = -x$.
16. $y = c_1 e^{2x} + c_2 e^{3x}$.    ANS.   $y'' - 5y' + 6y = 0$.
17. $y = x^2 + c_1 e^{2x} + c_2 e^{3x}$.    ANS.   $y'' - 5y' + 6y = 6x^2 - 10x + 2$.
18. $y = c_1 e^x + c_2 x e^x$.    ANS.   $y'' - 2y' + y = 0$.
19. $y = A e^{2x} + Bx e^{2x}$.    ANS.   $y'' - 4y' + 4y = 0$.
20. $y = c_1 e^{2x} \cos 3x + c_2 e^{2x} \sin 3x$.    ANS.   $y'' - 4y' + 13y = 0$.
21. $y = c_1 e^{ax} \cos bx + c_2 e^{ax} \sin bx$; $a$ and $b$ are parameters.    ANS.   $y'' - 2ay' + (a^2 + b^2)y = 0$.
22. $y = c_1 x + c_2 e^x$.    ANS.   $(x - 1)y'' - xy' + y = 0$.
23. $y = c_1 x^2 + c_2 e^{-x}$.    ANS.   $x(x + 2)y'' + (x^2 - 2)y' - 2(x + 1)y = 0$.
24. $y = x^2 + c_1 x + c_2 e^{-x}$.    ANS.   $(x + 1)y'' + xy' - y = x^2 + 2x + 2$.
25. $y = c_1 x^2 + c_2 e^{2x}$.    ANS.   $x(1 - x)y'' + (2x^2 - 1)y' - 2(2x - 1)y = 0$.

## 4. Families of curves

An equation involving a parameter, as well as one or both of the co-ordinates of a point in a plane, may represent a family of curves, one curve corresponding to each value of the parameter. For instance, the equation

$$(x - c)^2 + (y - c)^2 = 2c^2, \tag{1}$$

or

$$x^2 + y^2 - 2c(x + y) = 0, \tag{2}$$

may be interpreted as the equation of a family of circles, each having its center on the line $y = x$ and each passing through the origin. Figure 1 shows several elements, or members, of this family.

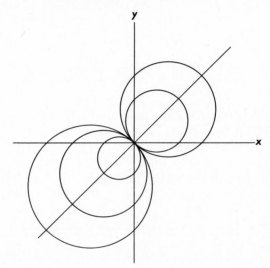

FIGURE 1

If the constant $c$ in equation (1) or in equation (2) is treated as an arbitrary constant and eliminated as in the preceding section, the result is called the *differential equation of the family represented by equation* (1). In this example, the elimination of $c$ is easily performed by isolating $c$, then differentiating throughout the equation with respect to $x$. Thus, from

$$\frac{x^2 + y^2}{x + y} = 2c$$

we find that

$$\frac{(x + y)(2x\,dx + 2y\,dy) - (x^2 + y^2)(dx + dy)}{(x + y)^2} = 0.$$

Therefore

$$(x^2 + 2xy - y^2)\,dx - (x^2 - 2xy - y^2)\,dy = 0 \qquad (3)$$

is the differential equation of the family of circles represented by equation (1).

Note that equation (3) associates a definite slope with each point $(x, y)$ in the plane

$$\frac{dy}{dx} = \frac{x^2 + 2xy - y^2}{x^2 - 2xy - y^2}, \qquad (4)$$

except where the denominator on the right in equation (4) vanishes. When the denominator vanishes, the curve passing through that point must have a vertical tangent. From

$$x^2 - 2xy - y^2 = 0$$

we see that

$$y = (-1 + \sqrt{2})x \qquad (5)$$

or

$$y = (-1 - \sqrt{2})x. \qquad (6)$$

In Figure 2, the straight lines (5) and (6) appear along with the family (1). It is seen that the lines (5) and (6) cut the members of the family of circles in precisely those points of vertical tangency.

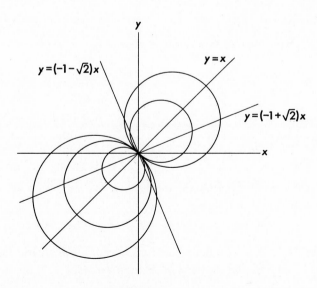

FIGURE 2

For a two-parameter family of curves, the differential equation will be of order two, and such a simple geometric interpretation is not available.

EXAMPLE (a):   Find the differential equation of the family of parabolas (Figure 3), having their vertices at the origin and their foci on the $y$-axis.

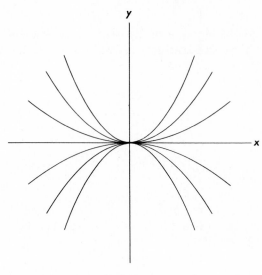

FIGURE 3

An equation of this family of parabolas is

$$y = ax^2, \qquad (7)$$

so that

$$y' = 2ax. \qquad (8)$$

It follows that

$$xy' - 2y = 0 \qquad (9)$$

is the differential equation of the family. We note that (9) is a first-order linear differential equation.

EXAMPLE (b):   Find the differential equation of the family of circles (Figure 4) having their centers on the $y$-axis.

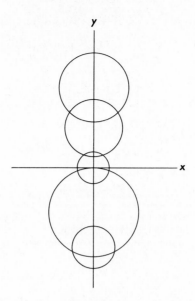

FIGURE 4

Because a member of the family of circles of this example may have its center anywhere on the $y$-axis and its radius of any magnitude, we are dealing with the two-parameter family

$$x^2 + (y - b)^2 = r^2. \tag{10}$$

We shall eliminate both $b$ and $r$ and obtain a second-order differential equation for the family (10).

At once

$$x + (y - b)y' = 0,$$

from which

$$\frac{x + yy'}{y'} = b.$$

Then

$$\frac{y'[1 + yy'' + (y')^2] - y''(x + yy')}{(y')^2} = 0,$$

so the desired differential equation is

$$xy'' - (y')^3 - y' = 0.$$

## *Exercises*

In each exercise, obtain the differential equation of the family of plane curves described and sketch several representative members of the family.

**1.** Straight lines through the origin.                    ANS.   $y\,dx - x\,dy = 0.$
**2.** Straight lines through the fixed point $(h, k)$; $h$ and $k$ not to be eliminated.
                                        ANS.   $(y - k)\,dx - (x - h)\,dy = 0.$
**3.** Straight lines with slope and $y$-intercept equal.    ANS.   $y\,dx - (x + 1)\,dy = 0.$
**4.** Straight lines with slope and $x$-intercept equal.        ANS.   $(y')^2 = xy' - y.$
**5.** Straight lines with algebraic sum of the intercepts fixed as $k$.
                                        ANS.   $(xy' - y)(y' - 1) + ky' = 0.$
**6.** Straight lines at a fixed distance $p$ from the origin.
                                        ANS.   $(xy' - y)^2 = p^2[1 + (y')^2].$
**7.** Circles with center at the origin.                    ANS.   $x\,dx + y\,dy = 0.$
**8.** Circles with center on the $x$-axis.                ANS.   $yy'' + (y')^2 + 1 = 0.$
**9.** Circles with fixed radius $r$ and tangent to the $x$-axis.
                                        ANS.   $(y \pm r)^2(y')^2 + y^2 \pm 2ry = 0.$
**10.** Circles tangent to the $x$-axis.        ANS.   $[1 + (y')^2]^3 = [yy'' + 1 + (y')^2]^2.$
**11.** Circles with center on the line $y = -x$, and passing through the origin.
        ANS.   $(x^2 - 2xy - y^2)\,dx + (x^2 + 2xy - y^2)\,dy = 0.$
**12.** Circles of radius unity. Use the fact that the radius of curvature is 1.
                                        ANS.   $(y'')^2 = [1 + (y')^2]^3.$
**13.** All circles. Use the curvature.        ANS.   $y'''[1 + (y')^2] = 3y'(y'')^2.$
**14.** Parabolas with vertex on the $x$-axis, with axis parallel to the $y$-axis, and with distance from focus to vertex fixed as $a$.                    ANS.   $a(y')^2 = y.$
**15.** Parabolas with vertex on the $y$-axis, with axis parallel to the $x$-axis, and with distance from focus to vertex fixed as $a$.                    ANS.   $x(y')^2 = a.$
**16.** Parabolas with axis parallel to the $y$-axis and with distance from vertex to focus fixed as $a$.                                        ANS.   $2ay'' = 1.$
**17.** Parabolas with axis parallel to the $x$-axis and with distance from vertex to focus fixed as $a$.                                    ANS.   $2ay'' + (y')^3 = 0.$

**18.** Work exercise 17, using differentiation with respect to $y$.    ANS.   $2a\dfrac{d^2x}{dy^2} = 1.$

**19.** Use the fact that

$$\frac{d^2x}{dy^2} = \frac{d}{dy}\left(\frac{dx}{dy}\right) = \frac{dx}{dy}\frac{d}{dx}\left(\frac{dx}{dy}\right) = \frac{dx}{dy}\frac{d}{dx}\left(\frac{dy}{dx}\right)^{-1} = \frac{-y''}{(y')^3}$$

to prove that the answers to exercises 17 and 18 are equivalent.

**20.** Parabolas with vertex and focus on the $x$-axis.     ANS.   $yy'' + (y')^2 = 0.$

**21.** Parabolas with axis parallel to the $x$-axis.     ANS.   $y'y''' - 3(y'')^2 = 0.$

**22.** Central conics with center at the origin and vertices on the coordinate axes.

ANS.   $xyy'' + x(y')^2 - yy' = 0.$

**23.** The confocal central conics

$$\frac{x^2}{a^2 + \lambda} + \frac{y^2}{b^2 + \lambda} = 1$$

with $a$ and $b$ held fixed.     ANS.   $(xy' - y)(yy' + x) = (a^2 - b^2)y'.$

**24.** The cubics $cy^2 = x^2(x - a)$ with $a$ held fixed.     ANS.   $2x(x - a)y' = y(3x - 2a).$

**25.** The cubics of exercise 24 with $c$ held fixed and $a$ to be eliminated.

ANS.   $2cy(xy' - y) = x^3.$

**26.** The quartics $c^2y^2 = x(x - a)^3$ with $a$ held fixed.     ANS.   $2x(x - a)y' = y(4x - a).$

**27.** The quartics of exercise 26 with $c$ held fixed and $a$ to be eliminated.

ANS.   $c^2(2xy' - y)^3 = 27x^4y.$

**28.** The strophoids $y^2 = \dfrac{x^2(a + x)}{a - x}.$     ANS.   $(x^4 - 4x^2y^2 - y^4)\,dx + 4x^3y\,dy = 0.$

**29.** The cissoids $y^2 = \dfrac{x^3}{a - x}.$     ANS.   $2x^3y' = y(y^2 + 3x^2).$

**30.** The trisectrices of Maclaurin $y^2(a + x) = x^2(3a - x).$

ANS.   $(3x^4 - 6x^2y^2 - y^4)\,dx + 8x^3y\,dy = 0.$

**31.** Circles through the intersections of the circle $x^2 + y^2 = 1$ and the line $y = x$. Use the "$u + kv$" form; that is, the equation

$$x^2 + y^2 - 1 + k(y - x) = 0.$$

ANS.   $(x^2 - 2xy - y^2 + 1)\,dx + (x^2 + 2xy - y^2 - 1)\,dy = 0.$

**32.** Circles through the fixed points $(a, 0)$ and $(-a, 0)$. Use the method of exercise 31.

ANS.   $2xy\,dx + (y^2 + a^2 - x^2)\,dy = 0.$

**33.** The circles $r = 2a(\sin \theta - \cos \theta)$.   ANS.   $(\cos \theta - \sin \theta)\,dr + r(\cos \theta + \sin \theta)\,d\theta = 0.$

**34.** The cardioids $r = a(1 - \sin \theta)$.     ANS.   $(1 - \sin \theta)\,dr + r \cos \theta\,d\theta = 0.$

**35.** The cissoids $r = a \sin \theta \tan \theta$. (See exercise 29.)

ANS.   $\sin \theta \cos \theta\,dr - r(1 + \cos^2 \theta)\,d\theta = 0.$

**36.** The strophoids $r = a(\sec \theta + \tan \theta)$.     ANS.   $\dfrac{dr}{d\theta} = r \sec \theta.$

**37.** The trisectrices of Maclaurin $r = a(4 \cos \theta - \sec \theta)$. (See exercise 30.)

ANS.   $\cos \theta(4 \cos^2 \theta - 1)\,dr + r \sin \theta(4 \cos^2 \theta + 1)\,d\theta = 0.$

# Equations of Order One

## 5. The isoclines of an equation

In this chapter we shall study several elementary methods for solving first-order differential equations. Before we begin that study, we shall take a brief look at some of the basic geometrical ideas that are involved.

Consider the equation of order one

$$\frac{dy}{dx} = f(x, y). \tag{1}$$

We can think of equation (1) as a machine that assigns to each point $(a, b)$ in the domain of $f$ some direction with slope $f(a, b)$. We can thus speak of the direction field of the differential equation. In a real sense any solution of equation (1) must have a graph, which at each point has the direction equation (1) requires.

One way to visualize this basic idea is to draw a short mark at each point to indicate the direction associated with that point. This can be done rather systematically by first drawing curves called isoclines, that is, curves along which the direction indicated by equation (1) is fixed.

EXAMPLE (a):   Consider the equation

$$\frac{dy}{dx} = y. \tag{2}$$

The isoclines are the straight lines $f(x, y) = y = c$. For each value of $c$ we obtain a line in which, at each point, the direction dictated by the differential equation is that number $c$. For example, at each point along the line $y = 1$, equation (2) determines a direction of slope 1. In Figure 5 we have drawn several of these isoclines, indicating the direction associated with each isocline by short markers. If one starts at any point in the plane and moves along a curve whose direction is always in the direction of the direction marks, then a solution curve is obtained. Several solution curves have been drawn in Figure 5.

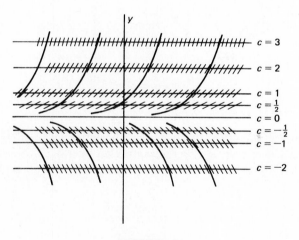

FIGURE 5

EXAMPLE (b):   Use the method of isoclines to sketch some of the solution curves for the equation

$$\frac{dy}{dx} = x^2 + y^2. \tag{3}$$

Here the isoclines will be the circles $x^2 + y^2 = c$, with $c > 0$. When $c = \frac{1}{4}$, the isocline has radius $\frac{1}{2}$; for $c = 1$, radius 1; for $c = 4$, radius 2. In Figure 6 we have drawn these isoclines, marking each of them with the appropriate direction indicator, and finally sketching several curves that represent solutions of equation (3).

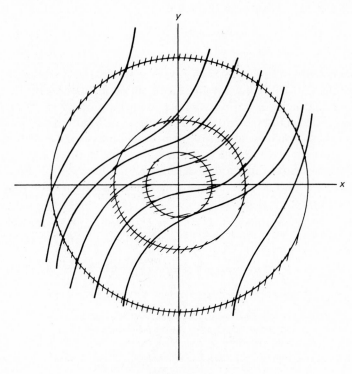

FIGURE 6

## Exercises

For each of the following differential equations, draw several isoclines with appropriate direction markers and sketch several solution curves for the equation.

1. $\dfrac{dy}{dx} = x.$

2. $\dfrac{dy}{dx} = \dfrac{y}{x}.$

3. $\dfrac{dy}{dx} = \dfrac{2y}{x}.$

4. $\dfrac{dy}{dx} = y - x.$

5. $\dfrac{dy}{dx} = x + y + 1.$

6. $\dfrac{dy}{dx} = x - y - 1.$

7. $\dfrac{dy}{dx} = 2x - y.$

8. $\dfrac{dy}{dx} = y - x^2.$

## 6. An existence theorem

It should be clear even to the casual reader that drawing isoclines is not a practical tool for finding solutions to any differential equations other than those involving the simplest of functions. Before discussing some of the analytic techniques for finding solutions, we will state an important theorem concerning the existence and uniqueness of solutions, a theorem which will be discussed in detail in Chapter 15.

Consider the equation of order one

$$\frac{dy}{dx} = f(x, y). \tag{1}$$

Let $T$ denote the rectangular region defined by

$$|x - x_0| \leqq a \qquad \text{and} \qquad |y - y_0| \leqq b,$$

a region with the point $(x_0, y_0)$ at its center. Suppose that $f$ and $\partial f/\partial y$ are continuous functions of $x$ and $y$ in $T$.

Under the conditions imposed on $f(x, y)$ above, an interval exists about $x_0, |x - x_0| \leqq h$, and a function $y(x)$ which has the properties:

(a) $y = y(x)$ is a solution of equation (1) on the interval $|x - x_0| \leqq h$;
(b) On the interval $|x - x_0| \leqq h$, $y(x)$ satisfies the inequality $|y(x) - y_0| \leqq b$;
(c) At $x = x_0$, $y = y(x_0) = y_0$;
(d) $y(x)$ is unique on the interval $|x - x_0| \leqq h$ in the sense that it is the only function that has all of the properties (a), (b), and (c).

The interval $|x - x_0| \leqq h$ may or may not need to be smaller than the interval $|x - x_0| \leqq a$ over which conditions were imposed upon $f(x, y)$.

In rough language, the theorem states that if $f(x, y)$ is sufficiently well behaved near the point $(x_0, y_0)$ then the differential equation

$$\frac{dy}{dx} = f(x, y) \tag{1}$$

has a solution that passes through the point $(x_0, y_0)$ and that solution is unique near $(x_0, y_0)$.

In Example (a) of Section 5 we can consider $(x_0, y_0)$ to be any point in the plane, since $f(x, y) = y$ and its partial derivative $\partial f/\partial y = 1$ are continuous in any rectangle. Therefore our existence theorem assures us that through any point $(x_0, y_0)$ there is exactly one solution, a situation that we assumed when we sketched the solution curves in Figure 5.

Again in Example (b), Section 5, the function $f(x, y) = x^2 + y^2$ and its partial derivative $\partial f/\partial y = 2y$ are continuous in any rectangle. It follows that through any point $(x_0, y_0)$ in the plane there is exactly one solution curve, a fact that is suggested by the solution curves in Figure 6.

## 7. Separation of variables

We begin our study of the methods for solving first-order equations by studying an equation of the form

$$M\,dx + N\,dy = 0,$$

where $M$ and $N$ may be functions of both $x$ and $y$. Some equations of this type are so simple that they can be put in the form

$$A(x)\,dx + B(y)\,dy = 0; \tag{1}$$

that is, the variables can be separated. Then a solution can be written at once. For it is only a matter of finding a function $F$ whose total differential is the left member of (1). Then $F = c$, where $c$ is an arbitrary constant, is the desired result.

EXAMPLE (a): Solve the equation

$$\frac{dy}{dx} = \frac{2y}{x}, \qquad \text{for } x > 0 \text{ and } y > 0. \tag{2}$$

We note that for the function in equation (2), the theorem of the previous section applies and assures the existence of a unique continuous solution through any point in the first quadrant. By separating the variables we can write

$$\frac{dy}{y} = \frac{2\,dx}{x}.$$

Hence we obtain a family of solutions

$$\ln|y| = 2\ln|x| + c \tag{3}$$

or, because we are in the first quadrant,

$$y = e^c x^2. \tag{4}$$

If we now put $c_1 = e^c$, we can write

$$y = c_1 x^2, \qquad c_1 > 0. \tag{5}$$

EXAMPLE (b): Solve the equation of the previous example for $x \neq 0$.

The argument now must be taken in two parts. First, if $y \neq 0$, we can proceed as before to equation (3). However, equation (5) must be written

$$|y| = c_1 x^2, \qquad c_1 > 0. \tag{6}$$

Second, if $y = 0$, we see immediately that since $x \neq 0$, $y = 0$ is a solution of the differential equation (2).

As a matter of convenience the solutions given by equation (6) are usually written

$$y = c_2 x^2, \tag{7}$$

where $c_2$ is taken to be an arbitrary real number. Indeed this form for the solutions incorporates the special case $y = 0$. Thus we say that the family of curves given in Figure 3, page 12, represents a family of solution curves for the differential equation (2).

We must be cautious, however. The function defined by

$$g(x) = x^2, \qquad x > 0$$
$$= -4x^2, \qquad x \leq 0,$$

obtained by piecing together two different parabolic arcs could also be considered a solution of the differential equation, even though this function is not included in the family of equation (7). The uniqueness statement in the theorem of Section 6 indicates that, as long as we restrict our attention to a point $(x_0, y_0)$ with $x_0 \neq 0$ and consider a rectangle with center at $(x_0, y_0)$ containing no points at which $x = 0$, then in that rectangle there is a unique solution that passes through $(x_0, y_0)$ and is continuous in the rectangle.

EXAMPLE (c):    Solve the equation

$$(1 + y^2)\, dx + (1 + x^2)\, dy = 0, \tag{8}$$

with the "initial condition" that when $x = 0$, $y = -1$.

If we write this equation in the form

$$\frac{dy}{dx} = \frac{-(1 + y^2)}{1 + x^2}$$

we observe that the right member and its partial derivative with respect to $y$ are continuous near $(0, -1)$. It follows that a unique solution exists for equation (8) that passes through the point $(0, -1)$.

From the differential equation we get

$$\frac{dx}{1 + x^2} + \frac{dy}{1 + y^2} = 0$$

from which it follows at once that

$$\text{arc tan } x + \text{arc tan } y = c. \tag{9}$$

In the set of solutions (9), each "arc tan" stands for the principal value of the inverse tangent and is subject to the restriction

$$-\tfrac{1}{2}\pi < \text{arc tan } x < \tfrac{1}{2}\pi.$$

The initial condition that $y = -1$ when $x = 0$ permits us to determine the value of $c$ that must be used to obtain the particular solution desired here. Since arc tan $0 = 0$ and arc tan $(-1) = -\tfrac{1}{4}\pi$, the solution of the initial value problem is

$$\text{arc tan } x + \text{arc tan } y = -\tfrac{1}{4}\pi. \tag{10}$$

Suppose next that we wish to sketch the graph of (10). Resorting to a device of trigonometry, we take the tangent of each side of (10). Because

$$\tan (\text{arc tan } x) = x$$

and

$$\tan (A + B) = \frac{\tan A + \tan B}{1 - \tan A \tan B},$$

we are led to the equation

$$\frac{x + y}{1 - xy} = -1,$$

or

$$xy - x - y - 1 = 0. \tag{11}$$

Now (11) is the equation of an equilateral hyperbola with asymptotes $x = 1$ and $y = 1$. But if we turn to (10), we see from

$$\text{arc tan } x = -\tfrac{1}{4}\pi - \text{arc tan } y$$

that, since $(-\text{arc tan } y) < \tfrac{1}{2}\pi$,

$$\text{arc tan } x < \tfrac{1}{4}\pi.$$

Hence $x < 1$, and equation (10) represents only one branch of the hyperbola (11). In Figure 7, the solid curve is the graph of equation (10); the solid curve and the dotted curve together are the graph of equation (11).

Each branch of the hyperbola (11) represents a solution of the differential equation, one branch for $x < 1$, the other for $x > 1$. In this example we were forced onto the left branch, equation (10), by the initial condition that $y = -1$ when $x = 0$.

FIGURE 7

One distinction between equations (10) and (11) can be seen by noting that a computing machine, given the differential equation (8) and seeking a solution that passes through the point $(0, -1)$, would draw only the left branch of the curve in Figure 7. The barrier (asymptote) at $x = 1$ would prevent the machine from learning of the existence of the other branch of the hyperbola (11).

EXAMPLE (d):   Solve the initial value problem

$$2x(y + 1)\,dx - y\,dy = 0, \tag{12}$$

where $x = 0$ and $y = -2$.

Separating the variables in equation (12), we obtain

$$2x\,dx = \left(1 - \frac{1}{y + 1}\right)dy, \qquad y \neq -1.$$

Integrating, we get a family of solutions given implicitly by

$$x^2 = y - \ln|y + 1| + c. \tag{13}$$

Since we seek a member of this family that passes through the point $(0, -2)$, we must have

$$0 = -2 - \ln|-1| + c,$$

or

$$c = 2.$$

Thus the solution to the problem is given implicitly by

$$x^2 = y - \ln|y + 1| + 2.$$

The reader should note how the theorem of Section 6 applies to this problem to indicate that we have found implicitly the unique solution to the initial value problem which is continuous for $y < -1$.

## *Exercises*

In exercises 1 through 24 obtain the general solution.

**1.** $(1 - x)y' = y^2$.   ANS.   $y \ln|c(1 - x)| = 1$.

**2.** $\sin x \sin y \, dx + \cos x \cos y \, dy = 0$.   ANS.   $\sin y = c \cos x$.

**3.** $xy^3 \, dx + e^{x^2} \, dy = 0$.   ANS.   $e^{-x^2} + y^{-2} = c$.

**4.** $2y \, dx = 3x \, dy$.   ANS.   $x^2 = cy^3$.

**5.** $my \, dx = nx \, dy$.   ANS.   $x^m = cy^n$.

**6.** $y' = xy^2$.   ANS.   $y(x^2 + c) + 2 = 0$.

**7.** $dV/dP = -V/P$.   ANS.   $PV = C$.

**8.** $y e^{2x} \, dx = (4 + e^{2x}) \, dy$.   ANS.   $c^2 y^2 = 4 + e^{2x}$.

**9.** $dr = b(\cos \theta \, dr + r \sin \theta \, d\theta)$.   ANS.   $r = c(1 - b \cos \theta)$.

**10.** $xy \, dx - (x + 2) \, dy = 0$.   ANS.   $e^x = cy(x + 2)^2$.

**11.** $x^2 \, dx + y(x - 1) \, dy = 0$.   ANS.   $(x + 1)^2 + y^2 + 2 \ln|c(x - 1)| = 0$.

**12.** $(xy + x) \, dx = (x^2 y^2 + x^2 + y^2 + 1) \, dy$.

  ANS.   $\ln(x^2 + 1) = y^2 - 2y + 4 \ln|c(y + 1)|$.

**13.** $x \cos^2 y \, dx + \tan y \, dy = 0$.   ANS.   $x^2 + \tan^2 y = c^2$.

**14.** $xy^3 \, dx + (y + 1) e^{-x} \, dy = 0$.   ANS.   $e^x(x - 1) = \dfrac{1}{y} + \dfrac{1}{2y^2} + c$.

**15.** $x^2 yy' = e^y$.   ANS.   $x(y + 1) = (1 + cx) e^y$.

**16.** $\tan^2 y \, dy = \sin^3 x \, dx$.   ANS.   $\cos^3 x - 3 \cos x = 3(\tan y - y + c)$.

**17.** $y' = \cos^2 x \cos y$.   ANS.   $4 \ln|\sec y + \tan y| = 2x + \sin 2x + c$.

**18.** $y' = y \sec x$.   ANS.   $y = c(\sec x + \tan x)$.

**19.** $dx = t(1 + t^2) \sec^2 x \, dt$.   ANS.   $2x + \sin 2x = c + (1 + t^2)^2$.

**20.** $(e^{2x} + 4)y' = y$.   ANS.   $y^8(1 + 4e^{-2x}) = c^2$.

**21.** $\alpha \, d\beta + \beta \, d\alpha + \alpha\beta(3 \, d\alpha + d\beta) = 0$.   ANS.   $c\alpha\beta = \exp(-3\alpha - \beta)$.

**22.** $(1 + \ln x) \, dx + (1 + \ln y) \, dy = 0$.   ANS.   $x \ln x + y \ln y = c$.

**23.** $x \, dx - \sqrt{a^2 - x^2} \, dy = 0$.

  ANS.   $y - c = -\sqrt{a^2 - x^2}$, the lower half of the circle $x^2 + (y - c)^2 = a^2$.

**24.** $x \, dx + \sqrt{a^2 - x^2} \, dy = 0$.

  ANS.   $y - c = \sqrt{a^2 - x^2}$, the upper half of the circle $x^2 + (y - c)^2 = a^2$.

**25.** $a^2 \, dx = x\sqrt{x^2 - a^2} \, dy$.   ANS.   $x = a \sec \dfrac{y + c}{a}$.

**26.** $y \ln x \ln y \, dx + dy = 0$.   ANS.   $x \ln x + \ln|\ln y| = x + c$.

In exercises 27 through 32, obtain the particular solution satisfying the initial condition indicated. In each exercise interpret your answer in the light of the existence theorem of Section 6 and draw a graph of the solution.

**27.** $dr/dt = -4rt$; when $t = 0, r = r_0$.               ANS.   $r = r_0 \exp(-2t^2)$.
**28.** $2xyy' = 1 + y^2$; when $x = 2, y = 3$.               ANS.   $y = \sqrt{5x - 1}$.
**29.** $xyy' = 1 + y^2$; when $x = 2, y = 3$.               ANS.   $y = \frac{1}{2}\sqrt{10x^2 - 4}$.
**30.** $2y\,dx = 3x\,dy$; when $x = 2, y = 1$.               ANS.   $y = (x/2)^{2/3}$.
**31.** $2y\,dx = 3x\,dy$; when $x = -2, y = 1$.              ANS.   $y = (x/2)^{2/3}$.
**32.** $2y\,dx = 3x\,dy$; when $x = 2, y = -1$.              ANS.   $y = -(x/2)^{2/3}$.

In exercises 33 through 36, obtain the particular solution satisfying the initial condition indicated.

**33.** $y' = x \exp(y - x^2)$; when $x = 0, y = 0$.          ANS.   $y = \ln\left[\dfrac{2}{1 + \exp(-x^2)}\right]$.

**34.** $xy^2\,dx + e^x\,dy = 0$; when $x \to \infty, y \to \frac{1}{2}$.   ANS.   $y = e^x/(2e^x - x - 1)$.
**35.** $(2a^2 - r^2)\,dr = r^3 \sin\theta\,d\theta$; when $\theta = 0, r = a$.
**36.** $v(dv/dx) = g$; when $x = x_0, v = v_0$.              ANS.   $v^2 - v_0^2 = 2g(x - x_0)$.

# 8. Homogeneous functions

Polynomials in which all terms are of the same degree, such as

$$x^2 - 3xy + 4y^2,$$
$$x^3 + y^3, \tag{1}$$
$$x^4y + 7y^5,$$

are called *homogeneous* polynomials. We wish now to extend the concept of homogeneity so it will apply to functions other than polynomials.

If we assign a physical dimension, say length, to each variable $x$ and $y$ in the polynomials in (1), then each polynomial itself also has a physical dimension, length to some power. This suggests the desired generalization. If, when certain variables are thought of as lengths, a function has the physical dimension length to the $k$th power, then we shall call that function homogeneous of degree $k$ in those variables. For example, the function

$$f(x, y) = 2y^3 \exp\left(\frac{y}{x}\right) - \frac{x^4}{x + 3y} \tag{2}$$

is of dimension (length)$^3$ when $x$ and $y$ are lengths. Therefore that function is said to be homogeneous of degree 3 in $x$ and $y$.

We permit the degree $k$ to be any number. The function $\sqrt{x + 4y}$ is called homogeneous of degree $\frac{1}{2}$ in $x$ and $y$. The function

$$\frac{x}{\sqrt{x^2 + y^2}}$$

is homogeneous of degree zero in $x$ and $y$.

A formal definition of homogeneity is: *The function $f(x, y)$ is said to be homogeneous of degree $k$ in $x$ and $y$ if, and only if,*

$$f(\lambda x, \lambda y) = \lambda^k f(x, y). \tag{3}$$

The definition is easily extended to functions of more than two variables.

For the function $f(x, y)$ of equation (2), the formal definition of homogeneity leads us to consider

$$f(\lambda x, \lambda y) = 2\lambda^3 y^3 \exp\left(\frac{\lambda y}{\lambda x}\right) - \frac{\lambda^4 x^4}{\lambda x + 3\lambda y}.$$

But we see at once that

$$f(\lambda x, \lambda y) = \lambda^3 f(x, y);$$

hence $f(x, y)$ is homogeneous of degree 3 in $x$ and $y$, as stated previously.

The following theorems prove useful in the next section.

**THEOREM 1**: *If $M(x, y)$ and $N(x, y)$ are both homogeneous and of the same degree, the function $M(x, y)/N(x, y)$ is homogeneous of degree zero.*

Proof of Theorem 1 is left to the student.

**THEOREM 2**: *If $f(x, y)$ is homogeneous of degree zero in $x$ and $y$, $f(x, y)$ is a function of $y/x$ alone.*

**PROOF.**   Let us put $y = vx$. Then Theorem 2 states that, if $f(x, y)$ is homogeneous of degree zero, $f(x, y)$ is a function of $v$ alone. Now

$$f(x, y) = f(x, vx) = x^0 f(1, v) = f(1, v), \tag{4}$$

in which the $x$ is now playing the role taken by $\lambda$ in the definition (3) above. By (4), $f(x, y)$ depends on $v$ alone as stated in Theorem 2.

## Exercises

Determine in each exercise whether or not the function is homogeneous. If it is homogeneous, state the degree of the function.

**1.** $4x^2 - 3xy + y^2$.

**2.** $x^3 - xy + y^3$.

**3.** $2y + \sqrt{x^2 + y^2}$.

**4.** $\sqrt{x - y}$.

**5.** $e^x$.

**6.** $\tan x$.

**7.** $\exp\left(\frac{x}{y}\right)$.

**8.** $\tan\dfrac{3y}{x}$.

**9.** $(x^2 + y^2)\exp\left(\dfrac{2x}{y}\right) + 4xy.$

**10.** $x\sin\dfrac{y}{x} - y\sin\dfrac{x}{y}.$

**11.** $\dfrac{x^2 + 3xy}{x - 2y}.$

**12.** $\dfrac{x^5}{x^2 + 2y^2}.$

**13.** $(u^2 + v^2)^{3/2}.$

**14.** $(u^2 - 4v^2)^{-1/2}.$

**15.** $y^2\tan\dfrac{x}{y}.$

**16.** $\dfrac{(x^2 + y^2)^{1/2}}{(x^2 - y^2)^{1/2}}.$

**17.** $\dfrac{a + 4b}{a - 4b}.$

**18.** $\ln\dfrac{x}{y}.$

**19.** $x\ln x - y\ln y.$

**20.** $x\ln x - x\ln y.$

ANS.   All functions are homogeneous except those of exercises 2, 5, 6, and 19.

## 9. Equations with homogeneous coefficients

Suppose that the coefficients $M$ and $N$ in an equation of order one,

$$M(x, y)\,dx + N(x, y)\,dy = 0, \tag{1}$$

are both homogeneous functions and are of the *same degree* in $x$ and $y$. By Theorems 1 and 2 of Section 8, the ratio $M/N$ is a function of $y/x$ alone. Hence equation (1) may be put in the form

$$\frac{dy}{dx} + g\!\left(\frac{y}{x}\right) = 0. \tag{2}$$

This suggests the introduction of a new variable $v$ by putting $y = vx$. Then (2) becomes

$$x\frac{dv}{dx} + v + g(v) = 0, \tag{3}$$

in which the variables are separable. We can obtain the solution of (3) by the method of Section 7, insert $y/x$ for $v$, and thus arrive at the solution of (1). We have shown that the substitution $y = vx$ will transform equation (1) into an equation in $v$ and $x$ in which the variables are separable.

The above method would have been equally successful had we used $x = vy$ to obtain from (1) an equation in $y$ and $v$. See Example (b) below.

EXAMPLE (a):   Solve the equation

$$(x^2 - xy + y^2)\,dx - xy\,dy = 0. \tag{4}$$

Since the coefficients in (4) are both homogeneous and of degree two in $x$ and $y$, let us put $y = vx$. Then (4) becomes

$$(x^2 - x^2 v + x^2 v^2)\, dx - x^2 v(v\, dx + x\, dv) = 0,$$

from which the factor $x^2$ should be removed at once. That done, we have to solve

$$(1 - v + v^2)\, dx - v(v\, dx + x\, dv) = 0,$$

or

$$(1 - v)\, dx - xv\, dv = 0.$$

Hence we separate variables to get

$$\frac{dx}{x} + \frac{v\, dv}{v - 1} = 0.$$

Then from

$$\frac{dx}{x} + \left[1 + \frac{1}{v - 1}\right] dv = 0$$

a family of solutions is seen to be

$$\ln |x| + v + \ln |v - 1| = \ln |c|,$$

or

$$x(v - 1)\, e^v = c.$$

In terms of the original variables, these solutions are given by

$$x\left(\frac{y}{x} - 1\right) \exp\left(\frac{y}{x}\right) = c,$$

or

$$(y - x) \exp\left(\frac{y}{x}\right) = c.$$

EXAMPLE (b):　Solve the equation

$$xy\, dx + (x^2 + y^2)\, dy = 0. \tag{5}$$

Again the coefficients in the equation are homogeneous and of degree two. We could use $y = vx$, but the relative simplicity of the $dx$ term in (5) suggests that we put $x = vy$. Then $dx = v\, dy + y\, dv$, and equation (5) is replaced by

$$vy^2(v\, dy + y\, dv) + (v^2 y^2 + y^2)\, dy = 0,$$

or

$$v(v\,dy + y\,dv) + (v^2 + 1)\,dy = 0.$$

Hence we need to solve

$$vy\,dv + (2v^2 + 1)\,dy = 0, \tag{6}$$

which leads at once to

$$\ln(2v^2 + 1) + 4\ln|y| = \ln c,$$

or

$$y^4(2v^2 + 1) = c.$$

Thus the desired solutions are given by

$$y^4\!\left(\frac{2x^2}{y^2} + 1\right) = c;$$

that is,

$$y^2(2x^2 + y^2) = c. \tag{7}$$

Since the left member of equation (7) cannot be negative, we may, for symmetry's sake, change the arbitrary constant to $c_1^4$, writing

$$y^2(2x^2 + y^2) = c_1^4.$$

It is worthwhile for the student to attack equation (5) using $y = vx$. That method leads directly to the equation

$$(v^3 + 2v)\,dx + x(v^2 + 1)\,dv = 0.$$

Frequently in equations with homogeneous coefficients, it is quite immaterial whether one uses $y = vx$ or $x = vy$. However, it is sometimes easier to substitute for the variable whose differential has the simpler coefficient.

### Exercises

In exercises 1 through 21, obtain a family of solutions.

1. $(x - 2y)\,dx + (2x + y)\,dy = 0.$      ANS.   $\ln(x^2 + y^2) + 4\arctan(y/x) = c.$
2. $2(2x^2 + y^2)\,dx - xy\,dy = 0.$      ANS.   $x^4 = c^2(4x^2 + y^2).$
3. $xy\,dx - (x^2 + 3y^2)\,dy = 0.$      ANS.   $x^2 = 6y^2 \ln|y/c|.$
4. $x^2 y' = 4x^2 + 7xy + 2y^2.$      ANS.   $x^2(y + 2x) = c(y + x).$
5. $3xy\,dx + (x^2 + y^2)\,dy = 0.$
6. $(x - y)(4x + y)\,dx + x(5x - y)\,dy = 0.$      ANS.   $x(y + x)^2 = c(y - 2x).$
7. $(5v - u)\,du + (3v - 7u)\,dv = 0.$      ANS.   $(3v + u)^2 = c(v - u).$
8. $(x^2 + 2xy - 4y^2)\,dx - (x^2 - 8xy - 4y^2)\,dy = 0.$      ANS.   $x^2 + 4y^2 = c(x + y).$
9. $(x^2 + y^2)\,dx - xy\,dy = 0.$      ANS.   $y^2 = 2x^2 \ln|x/c|.$

**10.** $x(x^2 + y^2)^2(y\,dx - x\,dy) + y^6\,dy = 0.$     ANS.   $(x^2 + y^2)^3 = 6y^6 \ln |c/y|.$

**11.** $(x^2 + y^2)\,dx + xy\,dy = 0.$     ANS.   $x^2(x^2 + 2y^2) = c^4.$

**12.** $xy\,dx - (x + 2y)^2\,dy = 0.$     ANS.   $y^3(x + y) = ce^{x/y}.$

**13.** $v^2\,dx + x(x + v)\,dv = 0.$     ANS.   $xv^2 = c(x + 2v).$

**14.** $[x\csc(y/x) - y]\,dx + x\,dy = 0.$     ANS.   $\ln |x/c| = \cos(y/x).$

**15.** $x\,dx + \sin^2(y/x)[y\,dx - x\,dy] = 0.$     ANS.   $4x \ln |x/c| - 2y + x\sin(2y/x) = 0.$

**16.** $(x - y \ln y + y \ln x)\,dx + x(\ln y - \ln x)\,dy = 0.$

ANS.   $(x - y)\ln x + y \ln y = cx + y.$

**17.** $[x - y \arctan(y/x)]\,dx + x \arctan(y/x)\,dy = 0.$

ANS.   $2y \arctan(y/x) = x \ln [c^2(x^2 + y^2)/x^4].$

**18.** $y^2\,dy = x(x\,dy - y\,dx)\,e^{x/y}.$     ANS.   $y \ln |y/c| = (y - x)\,e^{x/y}.$

**19.** $t(s^2 + t^2)\,ds - s(s^2 - t^2)\,dt = 0.$     ANS.   $s^2 = -2t^2 \ln |cst|.$

**20.** $y\,dx = (x + \sqrt{y^2 - x^2})\,dy.$     ANS.   $\arcsin(x/y) = \ln |y/c|.$

**21.** $(3x^2 - 2xy + 3y^2)\,dx = 4xy\,dy.$     ANS.   $(y - x)(y + 3x)^3 = cx^3.$

**22.** Prove that, with the aid of the substitution $y = vx$, you can solve any equation of the form

$$y^n f(x)\,dx + H(x, y)(y\,dx - x\,dy) = 0,$$

where $H(x, y)$ is homogeneous in $x$ and $y$.

In exercises 23 through 35, find the particular solution indicated.

**23.** $(x - y)\,dx + (3x + y)\,dy = 0;$ when $x = 2,\ y = -1.$

ANS.   $2(x + 2y) + (x + y)\ln(x + y) = 0.$

**24.** $(y - \sqrt{x^2 + y^2})\,dx - x\,dy = 0;$ when $x = \sqrt{3},\ y = 1.$     ANS.   $x^2 = 9 - 6y.$

**25.** $(y + \sqrt{x^2 + y^2})\,dx - x\,dy = 0;$ when $x = \sqrt{3},\ y = 1.$     ANS.   $x^2 = 2y + 1.$

**26.** $[x\cos^2(y/x) - y]\,dx + x\,dy = 0;$ when $x = 1,\ y = \pi/4.$

ANS.   $\tan(y/x) = \ln(e/x).$

**27.** $(y^2 + 7xy + 16x^2)\,dx + x^2\,dy = 0;$ when $x = 1,\ y = 1.$

ANS.   $x - y = 5(y + 4x)\ln x.$

**28.** $y^2\,dx + (x^2 + 3xy + 4y^2)\,dy = 0;$ when $x = 2,\ y = 1.$

ANS.   $4(2y + x)\ln y = 2y - x.$

**29.** $xy\,dx + 2(x^2 + 2y^2)\,dy = 0;$ when $x = 0,\ y = 1.$     ANS.   $y^4(3x^2 + 4y^2) = 4.$

**30.** $y(2x^2 - xy + y^2)\,dx - x^2(2x - y)\,dy = 0;$ when $x = 1,\ y = \frac{1}{2}.$

ANS.   $y^2 \ln x = 2y^2 + xy - x^2.$

**31.** $y(9x - 2y)\,dx - x(6x - y)\,dy = 0;$ when $x = 1,\ y = 1.$

ANS.   $3x^3 - x^2 y - 2y^2 = 0.$

**32.** $y(x^2 + y^2)\,dx + x(3x^2 - 5y^2)\,dy = 0;$ when $x = 2,\ y = 1.$

ANS.   $2y^5 - 2x^2 y^3 + 3x = 0.$

**33.** $(16x + 5y)\,dx + (3x + y)\,dy = 0;$ the curve to pass through the point $(1, -3).$

ANS.   $y + 3x = (y + 4x)\ln(y + 4x).$

**34.** $v(3x + 2v)\,dx - x^2\,dv = 0;$ when $x = 1,\ v = 2.$     ANS.   $2x^3 + 2x^2 v - 3v = 0.$

**35.** $(3x^2 - 2y^2)y' = 2xy;$ when $x = 0,\ y = -1.$     ANS.   $x^2 = 2y^2(y + 1).$

**36.** From Theorems 1 and 2, page 26, it follows that, if $F$ is homogeneous of degree $k$ in $x$ and $y$, $F$ can be written in the form

$$F = x^k \varphi\left(\frac{y}{x}\right).$$     (A)

Use (A) to prove Euler's theorem that, if $F$ is a homogeneous function of degree $k$ in $x$ and $y$,

$$x\frac{\partial F}{\partial x} + y\frac{\partial F}{\partial y} = kF.$$

## 10. Exact equations

In Section 7 it was noted that when an equation can be put in the form

$$A(x)\, dx + B(y)\, dy = 0,$$

a set of solutions can be determined by integration; that is, by finding a function whose differential is $A(x)\, dx + B(y)\, dy$.

That idea can be extended to some equations of the form

$$M(x, y)\, dx + N(x, y)\, dy = 0 \tag{1}$$

in which separation of variables may not be possible. Suppose that a function $F(x, y)$ can be found that has for its total differential the expression $M\, dx + N\, dy$; that is,

$$dF = M\, dx + N\, dy. \tag{2}$$

Then certainly

$$F(x, y) = c \tag{3}$$

defines implicitly a set of solutions of (1). For, from (3) it follows that

$$dF = 0,$$

or, in view of (2),

$$M\, dx + N\, dy = 0,$$

as desired.

Two things, then, are needed: first, to find out under what conditions on $M$ and $N$ a function $F$ exists such that its total differential is exactly $M\, dx + N\, dy$; second, if those conditions are satisfied, actually to determine the function $F$. If a function $F$ exists such that

$$M\, dx + N\, dy$$

is exactly the total differential of $F$, we call equation (1) an *exact equation.*

If the equation

$$M\, dx + N\, dy = 0 \tag{1}$$

is exact, then by definition $F$ exists such that

$$dF = M\, dx + N\, dy.$$

But, from calculus,

$$dF = \frac{\partial F}{\partial x} dx + \frac{\partial F}{\partial y} dy,$$

so

$$M = \frac{\partial F}{\partial x}, \qquad N = \frac{\partial F}{\partial y}.$$

These two equations lead to

$$\frac{\partial M}{\partial y} = \frac{\partial^2 F}{\partial y \, \partial x} \qquad \text{and} \qquad \frac{\partial N}{\partial x} = \frac{\partial^2 F}{\partial x \, \partial y}.$$

Again from calculus

$$\frac{\partial^2 F}{\partial y \, \partial x} = \frac{\partial^2 F}{\partial x \, \partial y},$$

provided that these partial derivatives are continuous. Therefore, if (1) is an exact equation, then

$$\frac{\partial M}{\partial y} = \frac{\partial N}{\partial x}. \qquad (4)$$

Thus, for (1) to be exact it is necessary that (4) be satisfied.

Let us now show that, if condition (4) is satisfied, (1) is an exact equation. Let $\phi(x, y)$ be a function for which

$$\frac{\partial \phi}{\partial x} = M.$$

The function $\phi$ is the result of integrating $M \, dx$ with respect to $x$ while holding $y$ constant. Now

$$\frac{\partial^2 \phi}{\partial y \, \partial x} = \frac{\partial M}{\partial y};$$

hence, if (4) is satisfied, then also

$$\frac{\partial^2 \phi}{\partial x \, \partial y} = \frac{\partial N}{\partial x}. \qquad (5)$$

Let us integrate both sides of this last equation with respect to $x$, holding $y$ fixed. In the integration with respect to $x$, the "arbitrary constant" may be any function of $y$. Let us call it $B'(y)$, for ease in indicating its integral. Then integration of (5) with respect to $x$ yields

$$\frac{\partial \phi}{\partial y} = N + B'(y). \qquad (6)$$

Now a function $F$ can be exhibited, namely,

$$F = \phi(x, y) - B(y),$$

for which

$$dF = \frac{\partial \phi}{\partial x} dx + \frac{\partial \phi}{\partial y} dy - B'(y) dy$$

$$= M\, dx + [N + B'(y)]\, dy - B'(y)\, dy$$

$$= M\, dx + N\, dy.$$

Hence, equation (1) is exact. We have completed a proof of the theorem stated below.

THEOREM 3: *If $M$, $N$, $\partial M/\partial y$, and $\partial N/\partial x$ are continuous functions of $x$ and $y$, then a necessary and sufficient condition that*

$$M\, dx + N\, dy = 0 \tag{1}$$

*be an exact equation is that*

$$\frac{\partial M}{\partial y} = \frac{\partial N}{\partial x}. \tag{4}$$

Furthermore, the proof contains the germ of a method for obtaining a set of solutions, a method used in Examples (a) and (b) below.

EXAMPLE (a): Solve the equation

$$3x(xy - 2)\, dx + (x^3 + 2y)\, dy = 0. \tag{7}$$

First, from the fact that

$$\frac{\partial M}{\partial y} = 3x^2 \quad \text{and} \quad \frac{\partial N}{\partial x} = 3x^2,$$

we conclude that equation (7) is exact. Therefore, its solution is $F = c$, where

$$\frac{\partial F}{\partial x} = M = 3x^2 y - 6x, \tag{8}$$

and

$$\frac{\partial F}{\partial y} = N = x^3 + 2y. \tag{9}$$

Let us attempt to determine $F$ from equation (8). Integration of both sides of (8) with respect to $x$, holding $y$ constant, yields

$$F = x^3 y - 3x^2 + T(y), \tag{10}$$

where the usual arbitrary constant in indefinite integration is now necessarily a function $T(y)$, as yet unknown. To determine $T(y)$, we use the fact that the function $F$ of equation (10) must also satisfy equation (9). Hence

$$x^3 + T'(y) = x^3 + 2y,$$

$$T'(y) = 2y.$$

No arbitrary constant is needed in obtaining $T(y)$, since one is being introduced on the right in the solution $F = c$. Then

$$T(y) = y^2,$$

and from (10)

$$F = x^3y - 3x^2 + y^2.$$

Finally, a set of solutions of equation (7) is defined by

$$x^3y - 3x^2 + y^2 = c.$$

**EXAMPLE (b):** Solve the equation

$$(2x^3 - xy^2 - 2y + 3)\,dx - (x^2y + 2x)\,dy = 0. \tag{11}$$

Here

$$\frac{\partial M}{\partial y} = -2xy - 2 = \frac{\partial N}{\partial x},$$

so equation (11) is exact.

A set of solutions of (11) is $F = c$, where

$$\frac{\partial F}{\partial x} = 2x^3 - xy^2 - 2y + 3 \tag{12}$$

and

$$\frac{\partial F}{\partial y} = -x^2y - 2x. \tag{13}$$

Because (13) is simpler than (12) and, for variety's sake, let us start the determination of $F$ from equation (13).

At once, from (13)

$$F = -\tfrac{1}{2}x^2y^2 - 2xy + Q(x),$$

where $Q(x)$ will be determined from (12). The latter yields

$$-xy^2 - 2y + Q'(x) = 2x^3 - xy^2 - 2y + 3,$$

$$Q'(x) = 2x^3 + 3.$$

Therefore

$$Q(x) = \tfrac{1}{2}x^4 + 3x,$$

and the desired set of solutions of (11) is defined implicitly by

$$-\tfrac{1}{2}x^2 y^2 - 2xy + \tfrac{1}{2}x^4 + 3x = \tfrac{1}{2}c,$$

or

$$x^4 - x^2 y^2 - 4xy + 6x = c.$$

## Exercises

Test each of the following equations for exactness and solve the equation. The equations that are not exact may, of course, be solved by methods discussed in the preceding sections.

**1.** $(x + 2y)\,dx + (2x + y)\,dy = 0.$        ANS.   $x^2 + 4xy + y^2 = c.$

**2.** $(2xy - 3x^2)\,dx + (x^2 + 2y)\,dy = 0.$       ANS.   $x^2 y - x^3 + y^2 = c.$

**3.** $(6x + y^2)\,dx + y(2x - 3y)\,dy = 0.$       ANS.   $3x^2 + xy^2 - y^3 = c.$

**4.** $(y^2 - 2xy + 6x)\,dx - (x^2 - 2xy + 2)\,dy = 0.$

                ANS.   $xy^2 - x^2 y + 3x^2 - 2y = c.$

**5.** $(2xy - y)\,dx + (x^2 + x)\,dy = 0.$         ANS.   $y(x + 1)^3 = cx.$

**6.** $v(2uv^2 - 3)\,du + (3u^2 v^2 - 3u + 4v)\,dv = 0.$    ANS.   $v(u^2 v^2 - 3u + 2v) = c.$

**7.** $(\cos 2y - 3x^2 y^2)\,dx + (\cos 2y - 2x \sin 2y - 2x^3 y)\,dy = 0.$

                ANS.   $\tfrac{1}{2}\sin 2y + x \cos 2y - x^3 y^2 = c.$

**8.** $(1 + y^2)\,dx + (x^2 y + y)\,dy = 0.$       ANS.   $2 \arctan x + \ln(1 + y^2) = c.$

**9.** $(1 + y^2 + xy^2)\,dx + (x^2 y + y + 2xy)\,dy = 0.$    ANS.   $2x + y^2(1 + x)^2 = c.$

**10.** $(w^3 + wz^2 - z)\,dw + (z^3 + w^2 z - w)\,dz = 0.$    ANS.   $(w^2 + z^2)^2 = 4wz + c.$

**11.** $(2xy - \tan y)\,dx + (x^2 - x \sec^2 y)\,dy = 0.$      ANS.   $x^2 y - x \tan y = c.$

**12.** $(\cos x \cos y - \cot x)\,dx - \sin x \sin y \, dy = 0.$    ANS.   $\sin x \cos y = \ln|c \sin x|.$

**13.** $(r + \sin \theta - \cos \theta)\,dr + r(\sin \theta + \cos \theta)\,d\theta = 0.$

                ANS.   $r^2 + 2r(\sin \theta - \cos \theta) = c.$

**14.** $x(3xy - 4y^3 + 6)\,dx + (x^3 - 6x^2 y^2 - 1)\,dy = 0.$

                ANS.   $x^3 y - 2x^2 y^3 + 3x^2 - y = c.$

**15.** $(\sin \theta - 2r \cos^2 \theta)\,dr + r \cos \theta(2r \sin \theta + 1)\,d\theta = 0.$

                ANS.   $r \sin \theta - r^2 \cos^2 \theta = c.$

**16.** $[2x + y \cos(xy)]\,dx + x \cos(xy)\,dy = 0.$     ANS.   $x^2 + \sin(xy) = c.$

**17.** $2xy\,dx + (y^2 + x^2)\,dy = 0.$          ANS.   $y(3x^2 + y^2) = c.$

**18.** $2xy\,dx + (y^2 - x^2)\,dy = 0.$           ANS.   $x^2 + y^2 = cy.$

**19.** $[2xy \cos(x^2) - 2xy + 1]\,dx + [\sin(x^2) - x^2]\,dy = 0.$

                ANS.   $y[\sin(x^2) - x^2] = c - x.$

**20.** $(2x - 3y)\,dx + (2y - 3x)\,dy = 0.$        ANS.   $x^2 + y^2 - 3xy = c.$

**21.** Do exercise 20 by a second method.

**22.** $(xy^2 + y - x)\,dx + x(xy + 1)\,dy = 0.$     ANS.   $x^2 y^2 + 2xy - x^2 = c.$

**23.** $3y(x^2 - 1)\,dx + (x^3 + 8y - 3x)\,dy = 0$; when $x = 0$, $y = 1$.

                ANS.   $xy(x^2 - 3) = 4(1 - y^2).$

**24.** $(1 - xy)^{-2} dx + [y^2 + x^2(1 - xy)^{-2}] dy = 0$; when $x = 4$, $y = 1$.

ANS. $xy^4 - y^3 + 3xy - 3x - 3 = 0$.

**25.** $(ye^{xy} - 2y^3) dx + (x e^{xy} - 6xy^2 - 2y) dy = 0$; when $x = 0$, $y = 2$.

ANS. $e^{xy} = 2xy^3 + y^2 - 3$.

**26.** $(3 + y + 2y^2 \sin^2 x) dx + (x + 2xy - y \sin 2x) dy = 0$.

ANS. $y^2 \sin 2x = c + 2x(3 + y + y^2)$.

**27.** $2x[3x + y - y \exp(-x^2)] dx + [x^2 + 3y^2 + \exp(-x^2)] dy = 0$.

ANS. $x^2y + y^3 + 2x^3 + y \exp(-x^2) = c$.

**28.** $(xy^2 + x - 2y + 3) dx + x^2y \, dy = 2(x + y) dy$; when $x = 1$, $y = 1$.

ANS. $(xy - 2)^2 + (x + 3)^2 = 2y^2 + 15$.

## 11. The linear equation of order one

In Section 10 we studied first-order differential equations that were exact. If an equation is not exact, it is natural to attempt to make it exact by the introduction of an appropriate factor, which is then called an integrating factor. Indeed, in Section 7 we multiplied by an integrating factor to separate the variables and thereby obtain an exact equation.

In general, very little can be said about the theory of integrating factors for first-order equations. In Chapter 4, we shall prove some theorems that will give some assistance in a few isolated situations. There is one important class of equations, however where the existence of an integrating factor can be demonstrated. This is the class of linear equations of order one.

An equation that is linear and of order one in the dependent variable $y$ must by definition (Section 2) be of the form

$$A(x)\frac{dy}{dx} + B(x)y = C(x). \tag{1}$$

By dividing each member of equation (1) by $A(x)$, we obtain

$$\frac{dy}{dx} + P(x)y = Q(x), \tag{2}$$

which we choose as the standard form for the linear equation of order one.

For the moment, suppose that there exists for equation (2) a positive integrating factor $v(x) > 0$, a function of $x$ alone. Then

$$v(x)\left[\frac{dy}{dx} + P(x)y\right] = v(x) \cdot Q(x) \tag{3}$$

must be an exact equation. But (3) is easily put into the form

$$M \, dx + N \, dy = 0$$

with

$$M = vPy - vQ,$$

and

$$N = v,$$

in which $v$, $P$, and $Q$ are functions of $x$ alone.

Therefore, if equation (3) is to be exact, it follows from the requirement

$$\frac{\partial M}{\partial y} = \frac{\partial N}{\partial x}$$

that $v$ must satisfy the equation

$$vP = \frac{dv}{dx}. \tag{4}$$

From (4), $v$ may be obtained readily, for

$$P \, dx = \frac{dv}{v},$$

so

$$\ln v = \int P \, dx,$$

or

$$v = \exp \left( \int P \, dx \right). \tag{5}$$

That is, if equation (2) has a positive integrating factor independent of $y$, then that factor must be as given by equation (5).

It remains to be shown that the $v$ given by equation (5) is actually an integrating factor of

$$\frac{dy}{dx} + P(x)y = Q(x). \tag{2}$$

Let us multiply (2) by the integrating factor, obtaining

$$\exp \left( \int P \, dx \right) \frac{dy}{dx} + P \exp \left( \int P \, dx \right) y = Q \exp \left( \int P \, dx \right). \tag{6}$$

The left member of (6) is the derivative of the product

$$y \exp \left( \int P \, dx \right);$$

the right member of (6) is a function of $x$ only. Hence equation (6) is exact, which is what we wanted to show.

Of course one integrating factor is sufficient. Hence we may use in the exponent $\left( \int P \, dx \right)$ any function whose derivative is $P$.

Because of the great importance of the ideas just discussed and the frequent occurrence of linear equations of first order, we summarize the steps involved in solving such equations:

(a)  Put the equation into standard form:

$$\frac{dy}{dx} + Py = Q.$$

(b)  Obtain the integrating factor exp $(\int P \, dx)$.
(c)  Apply the integrating factor to the equation in its standard form.
(d)  Solve the resultant exact equation.

Note in integrating the exact equation, that *the integral of the left member is always the product of the dependent variable and the integrating factor used.*

**EXAMPLE** (a):   Solve the equation

$$2(y - 4x^2) \, dx + x \, dy = 0.$$

The equation is linear in $y$. When put in standard form it becomes

$$\frac{dy}{dx} + \frac{2}{x} y = 8x, \qquad \text{when} \quad x \neq 0. \tag{7}$$

Then an integrating factor is

$$\exp\left(\int \frac{2 \, dx}{x}\right) = \exp\left(2 \ln |x|\right) = \exp\left(\ln x^2\right) = x^2.$$

Next apply the integrating factor to (7), thus obtaining the exact equation

$$x^2 \frac{dy}{dx} + 2xy = 8x^3, \tag{8}$$

which may be immediately rewritten as

$$\frac{d}{dx}(x^2 y) = 8x^3. \tag{9}$$

By integrating (9) we find that

$$x^2 y = 2x^4 + c. \tag{10}$$

This can be checked. From (10) we get (8) by differentiation. Then the original differential equation follows from (8) by a simple adjustment. Hence (10) defines a set of solutions of the original equation.

**EXAMPLE** (b):   Solve the equation

$$y \, dx + (3x - xy + 2) \, dy = 0.$$

Since the product $y\,dy$ occurs here, the equation is not linear in $y$. It is, however, linear in $x$. Therefore we arrange the terms as in

$$y\,dx + (3 - y)x\,dy = -2\,dy$$

and pass to the standard form,

$$\frac{dx}{dy} + \left(\frac{3}{y} - 1\right)x = \frac{-2}{y}, \qquad \text{for } y \neq 0. \tag{11}$$

Now

$$\int \left(\frac{3}{y} - 1\right) dy = 3 \ln |y| - y + c_1,$$

so that an integrating factor for equation (11) is

$$\exp(3 \ln |y| - y) = \exp(3 \ln |y|) \cdot e^{-y}$$
$$= \exp(\ln |y|^3) \cdot e^{-y}$$
$$= |y|^3\,e^{-y}.$$

It follows that for $y > 0$, $y^3\,e^{-y}$ is an integrating factor for equation (11) and for $y < 0$, $-y^3\,e^{-y}$ serves as an integrating factor. In either case we are led to the exact equation

$$y^3\,e^{-y}\,dx + y^2(3 - y)\,e^{-y}x\,dy = -2y^2\,e^{-y}\,dy$$

from which we get

$$xy^3\,e^{-y} = -2 \int y^2\,e^{-y}\,dy$$
$$= 2y^2\,e^{-y} + 4y\,e^{-y} + 4\,e^{-y} + c.$$

Thus a family of solutions is defined implicitly by

$$xy^3 = 2y^2 + 4y + 4 + c\,e^y.$$

## 12. The general solution of a linear equation

At the beginning of this chapter we stated an existence and uniqueness theorem for first-order differential equations. If the differential equation in that theorem happens to be a linear equation, we can prove a somewhat stronger statement.

Consider the linear differential equation

$$\frac{dy}{dx} + P(x)y = Q(x). \tag{1}$$

Suppose that $P$ and $Q$ are continuous functions on the interval $a < x < b$, and that $x = x_0$ is any number in that interval. If $y_0$ is an arbitrary real number, there exists a unique solution $y = y(x)$ of differential equation (1) that also satisfies the initial condition

$$y(x_0) = y_0.$$

Moreover, this solution satisfies equation (1) throughout the entire interval $a < x < b$.

The proof of this theorem has essentially been obtained in Section 11. Multiplication of equation (1) by the integrating factor $v = \exp(\int P\,dx)$ and integration gives

$$yv = \int vQ\,dx + c.$$

Since $v \neq 0$, we can write

$$y = v^{-1}\int vQ\,dx + cv^{-1}. \tag{2}$$

It is a simple matter to show that since $v \neq 0$ and $v$ is continuous on $a < x < b$, (2) is a family of solutions of equation (1).

It is also easy to see that given any $x_0$ on the interval $a < x < b$ together with any number $y_0$, we can choose the constant $c$ so that $y = y_0$ when $x = x_0$.

The effect of our argument is that every equation of the form of equation (1), for which $P$ and $Q$ have some common interval of continuity, will have a unique set of solutions containing one constant of integration that can be obtained by introducing the appropriate integrating factor. Because we are assured of the uniqueness of these solutions, we know that any other solution obtained by any other method must be one of the functions in our one-parameter family of solutions. It is for this reason that this set of solutions is called the *general solution* of equation (1). The word "general" is intended to mean that we have found all possible solutions that satisfy the differential equation on the interval $a < x < b$.

## Exercises

In exercises 1 through 25, find the general solution.

1. $(x^4 + 2y)\,dx - x\,dy = 0.$     ANS. $2y = x^4 + cx^2.$
2. $(3xy + 3y - 4)\,dx + (x + 1)^2\,dy = 0.$     ANS. $y = 2(x + 1)^{-1} + c(x + 1)^{-3}.$
3. $y' = \csc x - y\cot x.$     ANS. $y\sin x = x + c.$
4. $t(dx/dt) = 6te^{2t} + x(2t - 1).$     ANS. $xt = (3t^2 + c)e^{2t}.$
5. $dy = (x - 3y)\,dx.$     ANS. $9y = 3x - 1 + ce^{-3x}.$
6. $(3x - 1)y' = 6y - 10(3x - 1)^{1/3}.$     ANS. $y = 2(3x - 1)^{1/3} + c(3x - 1)^2.$

**7.** $(y - 2) dx + (3x - y) dy = 0.$    ANS.    $12x = 3y + 2 + c(y - 2)^{-3}.$

**8.** $(2xy + x^2 + x^4) dx - (1 + x^2) dy = 0.$    ANS.    $y = (1 + x^2)(c + x - \arctan x).$

**9.** $y' = x - 2xy.$ Solve by two methods.    ANS.    $2y = 1 + ce^{-x^2}.$

**10.** $(y - \cos^2 x) dx + \cos x \, dy = 0.$    ANS.    $y(\sec x + \tan x) = c + x - \cos x.$

**11.** $y' = x - 2y \cot 2x.$    ANS.    $4y \sin 2x = c + \sin 2x - 2x \cos 2x.$

**12.** $(y - x + xy \cot x) dx + x \, dy = 0.$    ANS.    $xy \sin x = c + \sin x - x \cos x.$

**13.** $\dfrac{dy}{dx} - my = c_1 e^{mx}$; where $c_1$ and $m$ are constants.    ANS.    $y = (c_1 x + c_2) e^{mx}.$

**14.** $\dfrac{dy}{dx} - m_2 y = c_1 e^{m_1 x}$; where $c_1, m_1, m_2$ are constants and $m_1 \neq m_2$.

ANS.    $y = c_3 e^{m_1 x} + c_2 e^{m_2 x}$; where $c_3 = c_1/(m_1 - m_2).$

**15.** $v \, dx + (2x + 1 - vx) dv = 0.$    ANS.    $xv^2 = v + 1 + ce^v.$

**16.** $x(x^2 + 1)y' + 2y = (x^2 + 1)^3.$    ANS.    $x^2 y = \frac{1}{4}(x^2 + 1)^3 + c(x^2 + 1).$

**17.** $2x(y - x^2) dx + dy = 0.$    ANS.    $y = x^2 - 1 + ce^{-x^2}.$

**18.** $(1 + xy) dx - (1 + x^2) dy = 0.$    ANS.    $y = x + c(1 + x^2)^{1/2}.$

**19.** $2y \, dx = (x^2 - 1)(dx - dy).$    ANS.    $(x - 1)y = (x + 1)(c + x - 2 \ln |x + 1|).$

**20.** $dx - (1 + 2x \tan y) dy = 0.$    ANS.    $2x \cos^2 y = y + c + \sin y \cos y.$

**21.** $(1 + \cos x)y' = \sin x(\sin x + \sin x \cos x - y).$

ANS.    $y = (1 + \cos x)(c + x - \sin x).$

**22.** $y' = 1 + 3y \tan x.$    ANS.    $3y \cos^3 x = c + 3 \sin x - \sin^3 x.$

**23.** $(x^2 + a^2) dy = 2x[(x^2 + a^2)^2 + 3y] dx$; $a$ is constant.

ANS.    $y = (x^2 + a^2)^2[c(x^2 + a^2) - 1].$

**24.** $(x + a)y' = bx - ny$; $a, b, n$ are constants with $n \neq 0, n \neq -1.$

ANS.    $n(n + 1)y = b(nx - a) + c(x + a)^{-n}.$

**25.** Solve the equation of exercise 24 for the exceptional cases $n = 0$ and $n = -1.$

ANS.    If $n = 0$, $y = bx + c - ab \ln |x + a|.$

If $n = -1$, $y = ab + c(x + a) + b(x + a) \ln |x + a|.$

**26.** In the standard form $dy + Py \, dx = Q \, dx$, put $y = vw$, thus obtaining

$$w(dv + Pv \, dx) + v \, dw = Q \, dx.$$

Then, by first choosing $v$ so that

$$dv + Pv \, dx = 0$$

and later determining $w$, show how to complete the solution of

$$dy + Py \, dx = Q \, dx.$$

In exercises 27 through 33, find the particular solution indicated.

**27.** $(2x + 3)y' = y + (2x + 3)^{1/2}$; when $x = -1, y = 0.$

ANS.    $2y = (2x + 3)^{1/2} \ln (2x + 3).$

**28.** $y' = x^3 - 2xy$; when $x = 1, y = 1.$    ANS.    $2y = x^2 - 1 + 2 \exp (1 - x^2).$

**29.** $L\dfrac{di}{dt} + Ri = E$; where $L, R,$ and $E$ are constants, when $t = 0, i = 0.$

ANS.    $i = \dfrac{E}{R}\left[1 - \exp\left(-\dfrac{Rt}{L}\right)\right].$

**30.** $L\dfrac{di}{dt} + Ri = E \sin \omega t$; when $t = 0$, $i = 0$.

ANS.   Let $Z^2 = R^2 + \omega^2 L^2$.
Then $i = EZ^{-2}[R \sin \omega t - \omega L \cos \omega t + \omega L \exp(-Rt/L)]$.

**31.** Find that solution of $y' = 2(2x - y)$ which passes through the point $(0, -1)$.

ANS.   $y = 2x - 1$.

**32.** Find that solution of $y' = 2(2x - y)$ which passes through the point $(0, 1)$.

ANS.   $y = 2x - 1 + 2e^{-2x}$.

**33.** $(1 + t^2)\,ds + 2t[st^2 - 3(1 + t^2)^2]\,dt = 0$; when $t = 0$, $s = 2$.

ANS.   $s = (1 + t^2)[3 - \exp(-t^2)]$.

## Miscellaneous Exercises

In each exercise, find a set of solutions, unless the statement of the exercise stipulates otherwise.

**1.** $y' = \exp(2x - y)$.                ANS.   $2e^y = e^{2x} + c$.

**2.** $(x + y)\,dx + x\,dy = 0$.                ANS.   $x(x + 2y) = c$.

**3.** $y^2\,dx - x(2x + 3y)\,dy = 0$.                ANS.   $y^2(x + y) = cx$.

**4.** $(x^2 + 1)\,dx + x^2 y^2\,dy = 0$.                ANS.   $xy^3 = 3(1 + cx - x^2)$.

**5.** $(x^3 + y^3)\,dx + y^2(3x + ky)\,dy = 0$; $k$ is constant.      ANS.   $ky^4 + 4xy^3 + x^4 = c$.

**6.** $y' = x^3 - 2xy$; when $x = 1$, $y = 2$.      ANS.   $2y = x^2 - 1 + 4\exp(1 - x^2)$.

**7.** $dy/dx - \cos x = \cos x \tan^2 y$.      ANS.   $2 \sin x = y + \sin y \cos y + c$.

**8.** $\cos x\,dy/dx = 1 - y - \sin x$.      ANS.   $y(1 + \sin x) = (x + c)\cos x$.

**9.** $\sin \theta\,dr/d\theta = -1 - 2r \cos \theta$.      ANS.   $r \sin^2 \theta = c + \cos \theta$.

**10.** $y(x + 3y)\,dx + x^2\,dy = 0$.                ANS.   $x^2 y = c(2x + 3y)$.

**11.** $dy/dx = \sec^2 x \sec^3 y$.      ANS.   $3 \tan x + c = 3 \sin y - \sin^3 y$.

**12.** $y(2x^3 - x^2 y + y^3)\,dx - x(2x^3 + y^3)\,dy = 0$.      ANS.   $2x^2 y \ln|cx| = 4x^3 - y^3$.

**13.** $(1 + x^2)y' = x^4 y^4$.      ANS.   $x^3 y^3 + 1 = y^3(c + 3x - 3 \arctan x)$.

**14.** $y(3 + 2xy^2)\,dx + 3(x^2 y^2 + x - 1)\,dy = 0$.      ANS.   $x^2 y^3 = 3(c + y - xy)$.

**15.** $(2x^2 - 2xy - y^2)\,dx + xy\,dy = 0$.      ANS.   $x^3 = c(y - x)\exp(y/x)$.

**16.** $y(x^2 + y^2)\,dx + x(3x^2 - 5y^2)\,dy = 0$; when $x = 2$, $y = 1$.

ANS.   $2y^5 - 2x^2 y^3 + 3x = 0$.

**17.** $y' + ay = b$; $a$ and $b$ constants. Solve by two methods.      ANS.   $y = b/a + ce^{-ax}$.

**18.** $(x - y)\,dx - (x + y)\,dy = 0$. Solve by two methods.      ANS.   $x^2 - 2xy - y^2 = c$.

**19.** $dx/dt = \cos x \cos^2 t$.      ANS.   $4 \ln|\sec x + \tan x| = 2t + \sin 2t + c$.

**20.** $(\sin y - y \sin x)\,dx + (\cos x + x \cos y)\,dy = 0$.      ANS.   $x \sin y + y \cos x = c$.

**21.** $(1 + 4xy - 4x^2 y)\,dx + (x^2 - x^3)\,dy = 0$; when $x = 2$, $y = \frac{1}{4}$.

ANS.   $2x^4 y = x^2 + 2x + 2\ln(x - 1)$.

**22.** $3x^3 y' = 2y(y - 3)$.                ANS.   $y = c(y - 3)\exp(x^{-2})$.

**23.** $(2y \cos x + \sin^4 x)\,dx = \sin x\,dy$; when $x = \frac{1}{2}\pi$, $y = 1$.

ANS.   $y = 2 \sin^2 x \sin^2 \frac{1}{2}x$.

**24.** $xy(dx - dy) = x^2\,dy + y^2\,dx$.      ANS.   $x = y \ln|cxy|$.

**25.** $a^2(dy - dx) = x^2\,dy + y^2\,dx$; $a$ constant.

ANS.   $2 \arctan(y/a) = \ln|c(x + a)/(x - a)|$.

**26.** $(y - \sin^2 x)\,dx + \sin x\,dy = 0$.      ANS.   $y(\csc x - \cot x) = x + c - \sin x$.

**27.** $(x - y)\,dx + (3x + y)\,dy = 0$; when $x = 2$, $y = -1$.

ANS.    $2(x + 2y) + (x + y)\ln(x + y) = 0$.

**28.** $y\,dx = (2x + 1)(dx - dy)$.    ANS.    $3y = (2x + 1) + c(2x + 1)^{-1/2}$.

In solving exercises 29 through 33, recall that the principal value arc sin $x$ of the inverse sine function is restricted as follows: $-\frac{1}{2}\pi \leq$ arc sin $x \leq \frac{1}{2}\pi$.

**29.** $\sqrt{1 - y^2}\,dx + \sqrt{1 - x^2}\,dy = 0$.

ANS.    arc sin $x$ + arc sin $y = c$, or a part of the ellipse
$x^2 + 2c_1xy + y^2 + c_1^2 - 1 = 0$; where $c_1 = \cos c$.

**30.** Solve the equation of exercise 29 with the added condition that, when $x = 0$, $y = \frac{1}{2}\sqrt{3}$.    ANS.    arc sin $x$ + arc sin $y = \frac{1}{3}\pi$, or that arc of the ellipse $x^2 + xy + y^2 = \frac{3}{4}$ that is indicated by a heavy solid line in Figure 8.

**31.** Solve the equation of exercise 29 with the added condition that, when $x = 0$, $y = -\frac{1}{2}\sqrt{3}$.    ANS.    arc sin $x$ + arc sin $y = -\frac{1}{3}\pi$, or that arc of the ellipse $x^2 + xy + y^2 = \frac{3}{4}$, that is indicated by a light solid line in Figure 8.

**32.** Show that after the answers to exercises 30 and 31 have been deleted, the remaining arcs of the ellipse

$$x^2 + xy + y^2 = \tfrac{3}{4}$$

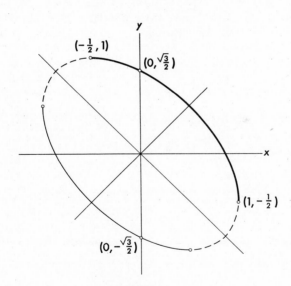

FIGURE 8

are not solutions of the differential equation

$$\sqrt{1 - y^2}\,dx + \sqrt{1 - x^2}\,dy = 0.$$

For this purpose consider the sign of the slope of the curve.

**33.** For the equation

$$\sqrt{1 - y^2}\, dx - \sqrt{1 - x^2}\, dy = 0$$

state and solve four problems analogous to exercises 29 through 32 above.

**34.** $v\, du = (e^v + 2uv - 2u)\, dv.$       ANS.   $v^2 u = c\, e^{2v} - (v + 1)\, e^v.$

**35.** $y\, dx = (3x + y^3 - y^2)\, dy$; when $x = 1, y = -1.$    ANS.   $x = y^2[1 + y \ln(-y)].$

**36.** $y^2\, dx - (xy + 2)\, dy = 0.$       ANS.   $xy = cy^2 - 1.$

**37.** $(x^2 - 2xy - y^2)\, dx - (x^2 + 2xy - y^2)\, dy = 0.$

ANS.   $(x + y)(x^2 - 4xy + y^2) = c^3.$

**38.** $y^2\, dx + (xy + y^2 - 1)\, dy = 0$; when $x = -1, y = 1.$

ANS.   $y^2 + 2xy + 1 = 2 \ln y.$

**39.** $y(y^2 - 3x^2)\, dx + x^3\, dy = 0.$       ANS.   $2x^6 = y^2(x^4 + c).$

**40.** $y' = \cos x - y \sec x$; when $x = 0, y = 1.$

ANS.   $y(1 + \sin x) = \cos x(x + 2 - \cos x).$

**41.** Find that solution of $y' = 3x + y$ which passes through the point $(-1, 0).$

ANS.   $y = -3(x + 1).$

**42.** Find that solution of $y' = 3x + y$ which passes through the point $(-1, 1).$

ANS.   $y = \exp(x + 1) - 3(x + 1).$

**43.** $y' = y \tan x + \cos x.$       ANS.   $2y = \sin x + (x + c) \sec x.$

**44.** $(x^2 - 1 + 2y)\, dx + (1 - x^2)\, dy = 0$; when $x = 2, y = 1.$

ANS.   $(x + 1)y = (x - 1)[x + 1 + 2 \ln(x - 1)].$

**45.** $(x^3 - 3xy^2)\, dx + (y^3 - 3x^2y)\, dy = 0.$       ANS.   $x^4 - 6x^2y^2 + y^4 = c.$

**46.** $(1 - x^2)y' = 1 - xy - 3x^2 + 2x^4.$       ANS.   $y = x - x^3 + c(1 - x^2)^{1/2}.$

**47.** $(y^2 + y)\, dx - (y^2 + 2xy + x)\, dy = 0$; when $x = 3, y = 1.$    ANS.   $2y^2 + y = x.$

**48.** $(y^3 - x^3)\, dx = xy(x\, dx + y\, dy).$       ANS.   $2x^2 \ln|x + y| = cx^2 + 2xy - y^2.$

**49.** $y' = \sec x - y \tan x.$       ANS.   $y = \sin x + c \cos x.$

**50.** $x^2 y' = y(1 - x).$       ANS.   $x \ln|cxy| = -1.$

**51.** $xy' = x - y + xy \tan x.$       ANS.   $xy \cos x = c + \cos x + x \sin x.$

**52.** $(3x^4 y - 1)\, dx + x^5\, dy = 0$; when $x = 1, y = 1.$       ANS.   $x^4 y = 2x - 1.$

**53.** $y^2\, dx + x^2\, dy = 2xy\, dy.$       ANS.   $y^2 = x(y + c).$

**54.** $(\sin x \sin y + \tan x)\, dx - \cos x \cos y\, dy = 0.$    ANS.   $\cos x \sin y = \ln|c \sec x|.$

**55.** $(3xy - 4y - 1)\, dx + x(x - 2)\, dy = 0$; when $x = 1, y = 2.$

ANS.   $2x^2(x - 2)y = x^2 - 5.$

# Elementary Applications

## 13. Velocity of escape from the earth

Many physical problems involve differential equations of order one.

Consider the problem of determining the velocity of a particle projected in a radial direction outward from the earth and acted upon by only one force, the gravitational attraction of the earth.

We shall assume an initial velocity in a radial direction so that the motion of the particle takes place entirely on a line through the center of the earth.

According to the Newtonian law of gravitation, the acceleration of the particle will be inversely proportional to the square of the distance from the particle to the center of the earth. Let $r$ be that variable distance, and let $R$ be the radius of the earth. If $t$ represents time, $v$ the velocity of the particle, $a$ its acceleration, and $k$ the constant of proportionality in the Newtonian law, then

$$a = \frac{dv}{dt} = \frac{k}{r^2}.$$

The acceleration is negative because the velocity is decreasing. Hence the constant $k$ is negative. When $r = R$, then $a = -g$, the acceleration of gravity at the surface of the earth. Thus

$$-g = \frac{k}{R^2},$$

from which

$$a = -\frac{gR^2}{r^2}.$$

We wish to express the acceleration in terms of the velocity and the distance. We have $a = dv/dt$ and $v = dr/dt$. Hence

$$a = \frac{dv}{dt} = \frac{dr}{dt}\frac{dv}{dr} = v\frac{dv}{dr},$$

so the differential equation for the velocity is now seen to be

$$v\frac{dv}{dr} = -\frac{gR^2}{r^2}. \tag{1}$$

The method of separation of variables applies to equation (1) and leads at once to the set of solutions

$$v^2 = \frac{2gR^2}{r} + C.$$

Suppose that the particle leaves the earth's surface with the velocity $v_0$. Then $v = v_0$ when $r = R$, from which the constant $C$ is easily determined to be

$$C = v_0^2 - 2gR.$$

Thus, a particle projected in a radial direction outward from the earth's surface with an initial velocity $v_0$ will travel with a velocity $v$ given by the equation

$$v^2 = \frac{2gR^2}{r} + v_0^2 - 2gR. \tag{2}$$

It is of considerable interest to determine whether the particle will escape from the earth. Now at the surface of the earth, at $r = R$, the velocity is positive, $v = v_0$. An examination of the right member of equation (2) shows that the velocity of the particle will remain positive if, and only if,

$$v_0^2 - 2gR \geqq 0. \tag{3}$$

If the inequality (3) is satisfied, the velocity given by equation (2) will remain positive because it cannot vanish, is continuous, and is positive at $r = R$. On the other hand, if (3) is not satisfied, then $v_0^2 - 2gR < 0$, and there will be a critical value of $r$ for which the right member of equation (2) is zero. That is, the particle would stop, the velocity would change from positive to negative, and the particle would return to the earth.

A particle projected from the earth with a velocity $v_0$ such that $v_0 \geqq \sqrt{2gR}$ will escape from the earth. Hence, the minimum such velocity of projection,

$$v_e = \sqrt{2gR}, \tag{4}$$

is called the *velocity of escape*.

The radius of the earth is approximately $R = 3960$ miles. The acceleration of gravity at the surface of the earth is approximately $g = 32.16$ feet per second per second (ft/sec$^2$), or $g = 6.09(10)^{-3}$ miles/sec$^2$. For the earth, the velocity of escape is easily found to be $v_e = 6.95$ miles/sec.

Of course, the gravitational pull of other celestial bodies, the moon, the sun, Mars, Venus, and so on has been neglected in the idealized problem treated here. It is not difficult to see that such approximations are justified, since we are interested in only the critical initial velocity $v_e$. Whether the particle actually recedes from the earth forever or becomes, for instance, a satellite of some heavenly body, is of no consequence in the present problem.

If in this study we happen to be thinking of the particle as an idealization of a ballistic-type rocket, then other elements must be considered. Air resistance in the first few miles may not be negligible. Methods for overcoming such difficulties are not suitable topics for discussion here.

It must be realized that the formula $v_e = \sqrt{2gR}$ applies equally well for the velocity of escape from the other members of the solar system, as long as $R$ and $g$ are given their appropriate values.

## 14. Newton's law of cooling

Experiment has shown that under certain conditions, a good approximation to the temperature of an object can be obtained by using Newton's law of cooling: The temperature of a body changes at a rate that is proportional to the difference in temperature between the outside medium and the body itself. We shall assume here that the constant of proportionality is the same whether the temperature is increasing or decreasing.

Suppose, for instance, that a thermometer, which has been at the reading 70°F inside a house, is placed outside where the air temperature is 10°F. Three minutes later it is found that the thermometer reading is 25°F. We wish to predict the thermometer reading at various later times.

Let $u$ (°F) represent the temperature of the thermometer at time $t$ (min), the time being measured from the instant the thermometer is placed outside. We are given that when $t = 0$, $u = 70$ and when $t = 3$, $u = 25$.

According to Newton's law, the time rate of change of temperature, $du/dt$, is proportional to the temperature difference $(u - 10)$. Since the thermometer temperature is decreasing, it is convenient to choose $(-k)$ as the constant of proportionality. Thus the $u$ is to be determined from the differential equation

$$\frac{du}{dt} = -k(u - 10), \tag{1}$$

and the conditions that

$$\text{when } t = 0, u = 70 \tag{2}$$

and

$$\text{when } t = 3, u = 25. \tag{3}$$

We need to know the thermometer reading at two different times because there are two constants to be determined, $k$ in equation (1) and the "arbitrary" constant that occurs in the solution of differential equation (1).

From equation (1) it follows at once that

$$u = 10 + C e^{-kt}.$$

Then condition (2) yields $70 = 10 + C$ from which $C = 60$, so we have

$$u = 10 + 60 e^{-kt}. \tag{4}$$

The value of $k$ will be determined now by using condition (3). Putting $t = 3$ and $u = 25$ into equation (4) we get

$$25 = 10 + 60 e^{-3k},$$

from which $e^{-3k} = \frac{1}{4}$, so $k = \frac{1}{3} \ln 4$.

Thus, the temperature is given by the equation

$$u = 10 + 60 \exp\left(-\tfrac{1}{3}t \ln 4\right). \tag{5}$$

Since $\ln 4 = 1.39$, equation (5) may be replaced by

$$u = 10 + 60 \exp\left(-0.46t\right), \tag{6}$$

which is convenient when a table of values of $e^{-x}$ is available.

## 15. Simple chemical conversion

It is known from the results of chemical experimentation that, in certain reactions in which a substance $A$ is being converted into another substance, the time rate of change of the amount $x$ of unconverted substance is proportional to $x$.

Let the amount of unconverted substance be known at some specified time; that is, let $x = x_0$ at $t = 0$. Then the amount $x$ at any time $t > 0$ is determined by the differential equation

$$\frac{dx}{dt} = -kx \tag{1}$$

and the condition that $x = x_0$ when $t = 0$. Since the amount $x$ is decreasing as time increases, the constant of proportionality in equation (1) is taken to be $(-k)$.

From equation (1) it follows that

$$x = C e^{-kt}.$$

But $x = x_0$ when $t = 0$. Hence $C = x_0$. Thus we have the result

$$x = x_0 e^{-kt}. \tag{2}$$

Let us now add another condition, which will enable us to determine $k$. Suppose it is known that at the end of half a minute, at $t = 30$ (sec), two-thirds of the original amount $x_0$ has already been converted. Let us determine how much unconverted substance remains at $t = 60$ (sec).

When two-thirds of the substance has been converted, one-third remains unconverted. Hence $x = \frac{1}{3}x_0$ when $t = 30$. Equation (2) now yields the relation

$$\tfrac{1}{3}x_0 = x_0 e^{-30k}$$

from which $k$ is easily found to be $\frac{1}{30} \ln 3$. Then with $t$ measured in seconds, the amount of unconverted substance is given by the equation

$$x = x_0 \exp\left(-\tfrac{1}{30}t \ln 3\right). \tag{3}$$

At $t = 60$,

$$x = x_0 \exp\left(-2 \ln 3\right) = x_0(3)^{-2} = \tfrac{1}{9}x_0.$$

### Exercises

1. The radius of the moon is roughly 1080 miles. The acceleration of gravity at the surface of the moon is about 0.165g, where $g$ is the acceleration of gravity at the surface of the earth. Determine the velocity of escape for the moon.

   ANS.   1.5 miles/sec.

2. Determine, to two significant figures, the velocity of escape for each of the celestial bodies listed below. The data given are rough and $g$ may be taken to be $6.1(10)^{-3}$ mile/sec$^2$.

| | Acceleration of gravity at surface | Radius (miles) | Answer (miles/sec) |
|---|---|---|---|
| Venus | 0.85g | 3,800 | 6.3 |
| Mars | 0.38g | 2,100 | 3.1 |
| Jupiter | 2.6g | 43,000 | 37 |
| Sun | 28g | 432,000 | 380 |
| Ganymede | 0.12g | 1,780 | 1.6 |

**3.** A thermometer reading 18°F is brought into a room where the temperature is 70°F; 1 min later the thermometer reading is 31°F. Determine the temperature reading as a function of time and, in particular, find the temperature reading 5 min after the thermometer is first brought into the room.

ANS. $u = 70 - 52 \exp(-0.29t)$; when $t = 5$, $u = 58$.

**4.** A thermometer reading 75°F is taken out where the temperature is 20°F. The reading is 30°F 4 min later. Find (a) the thermometer reading 7 min after the thermometer was brought outside, and (b) the time taken for the reading to drop from 75°F to within a half degree of the air temperature. ANS. (a) 23°F; (b) 11.5 min.

**5.** At 1:00 P.M., a thermometer reading 70°F is taken outside where the air temperature is $-10°F$ (ten below zero). At 1:02 P.M., the reading is 26°F. At 1:05 P.M., the thermometer is taken back indoors where the air is at 70°F. What is the thermometer reading at 1:09 P.M.? ANS. 56°F.

**6.** At 9 A.M., a thermometer reading 70°F is taken outdoors where the temperature is 15°F. At 9:05 A.M., the thermometer reading is 45°F. At 9:10 A.M., the thermometer is taken back indoors where the temperature is fixed at 70°F. Find (a) the reading at 9:20 A.M. and (b) when the reading, to the nearest degree, will show the correct (70°F) indoor temperature. ANS. (a) 58°F; (b) 9:46 A.M.

**7.** At 2:00 P.M., a thermometer reading 80°F is taken outside where the air temperature is 20°F. At 2:03 P.M., the temperature reading yielded by the thermometer is 42°F. Later, the thermometer is brought inside where the air is at 80°F. At 2:10 P.M., the reading is 71°F. When was the thermometer brought indoors? ANS. At 2:05 P.M.

**8.** Suppose that a chemical reaction proceeds according to the law given in Section 15 above. If half the substance $A$ has been converted at the end of 10 sec, find when nine-tenths of the substance will have been converted. ANS. 33 sec.

**9.** The conversion of a substance $B$ follows the law used in Section 15 above. If only a fourth of the substance has been converted at the end of 10 sec, find when nine-tenths of the substance will have been converted. ANS. 80 sec.

**10.** For a substance $C$, the time rate of conversion is proportional to the square of the amount $x$ of unconverted substance. Let $k$ be the numerical value of the constant of proportionality and let the amount of unconverted substance be $x_0$ at time $t = 0$. Determine $x$ for all $t \geq 0$. ANS. $x = x_0/(1 + x_0 kt)$.

**11.** Two substances, $A$ and $B$, are being converted into a single compound $C$. In the laboratory it has been shown that, for these substances, the following law of conversion holds: the time rate of change of the amount $x$ of compound $C$ is proportional to the product of the amounts of unconverted substances $A$ and $B$. Assume the units of measure so chosen that one unit of compound $C$ is formed from the combination of one unit of $A$ with one unit of $B$. If at time $t = 0$ there are $a$ units of substance $A$, $b$ units of substance $B$, and none of compound $C$ present, show that the law of conversion may be expressed by the equation

$$\frac{dx}{dt} = k(a - x)(b - x).$$

Solve this equation with the given initial condition.

ANS.    If $b \neq a$, $x = \dfrac{ab[\exp(b - a)kt - 1]}{b \exp(b - a)kt - a}$; if $b = a$, $x = \dfrac{a^2kt}{akt + 1}$.

12. In the solution of exercise 11 above, assume that $k > 0$ and investigate the behavior of $x$ as $t \rightarrow \infty$.      ANS.    If $b \geq a$, $x \rightarrow a$; if $b \leq a$, $x \rightarrow b$.
13. Radium decomposes at a rate proportional to the quantity of radium present. Suppose that it is found that in 25 years approximately 1.1 % of a certain quantity of radium has decomposed. Determine approximately how long it will take for one-half the original amount of radium to decompose.      ANS.   1600 years.
14. A certain radioactive substance has a half-life of 38 hr. Find how long it takes for 90 % of the radioactivity to be dissipated.      ANS.   126 hr.
15. A bacterial population $B$ is known to have a rate of growth proportional to $B$ itself. If between noon and 2 P.M. the population triples, at what time, no controls being exerted, should $B$ become 100 times what it was at noon?

                                   ANS.   About 8:22 P.M.

16. In the motion of an object through a certain medium (air at certain pressures is an example), the medium furnishes a resisting force proportional to the square of the velocity of the moving object. Suppose a body falls, due to the action of gravity, through such a medium. Let $t$ represent time, and $v$ represent velocity, positive downward. Let $g$ be the usual constant acceleration of gravity and let $w$ be the weight of the body. Use Newton's law, force equals mass times acceleration, to conclude that the differential equation of the motion is

$$\frac{w}{g}\frac{dv}{dt} = w - kv^2,$$

where $kv^2$ is the magnitude of the resisting force furnished by the medium.
17. Solve the differential equation of exercise 16 with the initial condition that $v = v_0$ when $t = 0$. Introduce the constant $a^2 = w/k$ to simplify the formulas.

                         ANS.   $\dfrac{a + v}{a - v} = \dfrac{a + v_0}{a - v_0} \exp\left(\dfrac{2gt}{a}\right).$

18. List a consistent set of units for the dimensions of the variables and parameters of exercises 16 and 17 above.      ANS.   $t$ in sec       $g$ in ft/sec$^2$

                                          $v$ in ft/sec     $k$ in (lb)(sec$^2$)/ft$^2$

                                          $w$ in lb        $a$ in ft/sec

19. There are mediums that resist motion through them with a force proportional to the first power of the velocity. For such a medium, state and solve problems analogous to exercises 16 through 18 above, except that for convenience a constant

$b = w/k$ may be introduced to replace the $a^2$ of exercise 17. Show that $b$ has the dimensions of a velocity.

ANS. $v = b + (v_0 - b) \exp\left(-\dfrac{gt}{b}\right).$

FIGURE 9

**20.** Figure 9 shows a weight, $w$ pounds (lb), sliding down an inclined plane which makes an angle $\alpha$ with the horizontal. Assume that no force other than gravity is acting on the weight; that is, there is no friction, no air resistance, and so on. At time $t = 0$, let $x = x_0$ and let the initial velocity be $v_0$. Determine $x$ for $t > 0$.

ANS. $x = \frac{1}{2}gt^2 \sin\alpha + v_0 t + x_0.$

**21.** A long, very smooth board is inclined at an angle of $10°$ with the horizontal. A weight starts from rest 10 ft from the bottom of the board and slides downward under the action of gravity alone. Find how long it will take the weight to reach the bottom of the board and determine the terminal speed.

ANS. 1.9 sec and 10.5 ft/sec.

**22.** Add to the conditions of exercise 20 above a retarding force of magnitude $kv$, where $v$ is the velocity. Determine $v$ and $x$ under the assumption that the weight starts from rest with $x = x_0$. Use the notation $a = kg/w$.

ANS. $v = a^{-1}g \sin\alpha(1 - e^{-at})$; $x = x_0 + a^{-2}g \sin\alpha(-1 + e^{-at} + at).$

FIGURE 10

**23.** A man, standing at $O$ in Figure 10, holds a rope of length $a$ to which a weight is attached, initially at $W_0$. The man walks to the right dragging the weight after him. When the man is at $M$, the weight is at $W$. Find the differential equation of the path (called the tractrix) of the weight and solve the equation.

$$\text{ANS.} \quad x = a\ln\frac{a + \sqrt{a^2 - y^2}}{y} - \sqrt{a^2 - y^2}.$$

**24.** A tank contains 80 gallons (gal) of pure water. A brine solution with 2 lb/gal of salt enters at 2 gal/min, and the well-stirred mixture leaves at the same rate. Find (a) the amount of salt in the tank at any time, and (b) the time at which the brine leaving will contain 1 lb/gal of salt.

ANS.   (a) $s = 160[1 - \exp(-t/40)]$; (b) $t = 40\ln 2$ min.

**25.** For the tank in the previous exercise, determine the limiting value for the amount of salt in the tank after a long time. How much time must pass before the amount of salt in the tank reaches 80 % of this limiting value?

ANS.   (a) $s = 160$ lb; (b) $t = 64$ min.

**26.** A certain sum of money $P$ draws interest compounded continuously. If at a certain time there is $P_0$ dollars in the account, determine the time when the principal attains the value $2P_0$ dollars, if the annual interest rate is (a) 2 %, or (b) 4 %.

ANS.   (a) 50 ln 2 years; (b) 25 ln 2 years.

**27.** A bank offers 5 % interest compounded continuously in a savings account. Determine (a) the amount of interest earned in 1 year on a deposit of $100 and (b) the equivalent rate if the compounding were done annually.

ANS.   (a) $5.13; (b) 5.13 %.

## 16. Logistic growth and the price of commodities

Numerous attempts have been made to develop models to study the growth of populations. One means of obtaining a simple model for that study is to assume that the average birthrate per individual is a positive constant and that the average death rate per individual is proportional to the population.

If we let $x(t)$ represent the population at time $t$, then the above assumptions lead to the differential equation

$$\frac{1}{x}\frac{dx}{dt} = b - ax, \tag{1}$$

where $b$ and $a$ are positive constants. This equation is commonly called the *logistic equation* and the growth of population determined by it is called *logistic growth.*

The variables in the logistic equation may be separated to obtain

$$\frac{dx}{x(b - ax)} = dt,$$

or

$$\left(\frac{1}{x} + \frac{a}{b - ax}\right) dx = b \, dt.$$

Integrating both sides gives us

$$\ln\left|\frac{x}{b - ax}\right| = bt + c,$$

or

$$\left|\frac{x}{b - ax}\right| = e^c e^{bt}. \tag{2}$$

To expedite the study of equation (2), let us further assume that at $t = 0$ the population is the positive number $x_0$. Then equation (2) may be written

$$\frac{x}{b - ax} = \frac{x_0}{b - ax_0} e^{bt},$$

and upon solving for $x$, we have

$$x(t) = \frac{b x_0 e^{bt}}{b - ax_0 + ax_0 e^{bt}}. \tag{3}$$

It is interesting to note that the population function obtained in equation (3) has a limiting value

$$\lim_{t \to \infty} x(t) = \lim_{t \to \infty} \frac{b x_0 e^{bt}}{b - ax_0 + ax_0 e^{bt}}$$

$$= \lim_{t \to \infty} \frac{b^2 x_0 e^{bt}}{abx_0 e^{bt}}$$

$$= \frac{b}{a},$$

where we have used l'Hospital's rule to evaluate the limit.

We should also note that the logistic equation (1) will dictate a growth or a decline in the population depending upon whether the initial population is less than or greater than $b/a$.

As a further example of an application in which a first-order differential equation occurs, we consider an economic model of a certain commodity market. We will assume that the price $P$, the supply $S$, and the demand $D$ of that commodity are functions of time and that the rate of change of the price is proportional to the difference between the demand and the supply. That is,

$$\frac{dP}{dt} = k(D - S). \tag{4}$$

We further assume that the constant $k$ is positive so that the price will increase if the demand exceeds the supply.

Different models of the commodity market will result depending upon the nature of the demand and supply functions that are indicated. If, for example, we assume that

$$D = c - dP \quad \text{and} \quad S = a + bP, \tag{5}$$

where $a, b, c,$ and $d$ are positive constants, we obtain a differential equation

$$\frac{dP}{dt} = k[(c - a) - (d + b)P] \tag{6}$$

that is linear in $P$. The assumptions (5) reflect the tendency for the demand to decrease as the price increases and the tendency for the supply to increase as the price increases, both reasonable assumptions for many commodities. We should also assume that $0 < P < c/d$, so that $D$ is not negative.

Equation (6) may be written

$$\frac{dP}{dt} + k(d + b)P = k(c - a), \tag{7}$$

and solved by multiplying by the integrating factor $e^{k(d+b)t}$ and integrating to obtain

$$P(t) = c_1 e^{-k(d+b)t} + \frac{c - a}{d + b}.$$

If the price at $t = 0$ is $P = P_0$, we have

$$c_1 = P_0 - (c - a)/(d + b),$$

so that

$$P(t) = \left(P_0 - \frac{c - a}{d + b}\right)e^{-k(d+b)t} + \frac{c - a}{d + b}. \tag{8}$$

Equation (8) shows that under the assumptions of (4) and (5) the price will stabilize at a value $(c - a)/(d + b)$ as $t$ becomes large.

## Exercises.

1. A certain population is known to be growing at a rate given by the logistic equation $dx/dt = x(b - ax)$. Show that the maximum rate of growth will occur when the population is equal to half its equilibrium size, that is, when the population is $b/2a$.

2. A bacterial population is known to have a logistic growth pattern with initial population 1000 and an equilibrium population of 10,000. A count shows that at the end of 1 hr there are 2000 bacteria present. Determine the population as a function of time.

$$\text{ANS.} \quad x(t) = \frac{10,000\,e^{bt}}{9 + e^{bt}}, \text{ where } b = \ln\left(\tfrac{9}{4}\right) \approx 0.81.$$

3. For the population of Exercise 2, determine the time at which the population is increasing most rapidly and draw a sketch of the logistic curve.

$$\text{ANS.} \quad t = \frac{\ln 9}{\ln 9 - \ln 4} \approx 2.7 \text{ hr.}$$

4. A college dormitory houses 100 students, each of whom is susceptible to a certain virus infection. A simple model of epidemics assumes that during the course of an epidemic the rate of change with respect to time of the number of infected students $I$ is proportional to the number of infected students and also proportional to the number of uninfected students, $100 - I$. (a) If at time $t = 0$ a single student becomes infected, show that the number of infected students at time $t$ is given by

$$I = \frac{100\,e^{100kt}}{99 + e^{100kt}}.$$

(b) If the constant of proportionality $k$ has value 0.01 when $t$ is measured in days, find the value of the rate of new cases $I'(t)$ at the end of each day for the first 9 days.

$$\text{ANS.} \quad 3, 6, 14, 23, 24, 16, 8, 3, 1.$$

5. Glucose is being fed intravenously into the bloodstream of a patient at a constant rate $c$ grams per minute. At the same time, the patient's body converts the glucose and removes it from the bloodstream at a rate proportional to the amount of glucose present. If the constant of proportionality is $k$, show that as time increases, the amount of glucose in the bloodstream approaches an equilibrium value of $c/k$.

6. The supply of food for a certain population is subject to a seasonal change that affects the growth rate of the population. The differential equation $dx/dt = cx(t)\cos t$, where $c$ is a positive constant, provides a simple model for the seasonal growth of the population. Solve the differential equation in terms of an initial population $x_0$ and the constant $c$. Determine the maximum and the minimum populations and the time interval between maxima. $\quad$ ANS. $\quad$ Max $x = x_0 e^c$; Min $x = x_0 e^{-c}$.

7. Suppose that the human body dissipates a drug at a rate proportional to the amount $y$ of drug present in the bloodstream at time $t$. At time $t = 0$ a first injection of $y_0$ grams of the drug is made into a body that was free from that drug prior to that time.

(a) Find the amount of residual drug in the bloodstream at the end of $T$ hours.
(b) If at time $T$ a second injection of $y_0$ grams is made, find the residual amount of drug at the end of $2T$ hours.
(c) If at the end of each time period of length $T$, an injection of $y_0$ grams is made, find the residual amount of drug at the end of $nT$ hours.
(d) Find the limiting value of the answer to part (c) as $n$ approaches infinity.

$$\text{ANS.} \quad \text{(d)} \frac{y_0 e^{-kT}}{1 - e^{-kT}}.$$

**8.** If the demand and supply functions for a commodity market are $D = c - dP$ and $S = a \sin \beta t$, determine $P(t)$ and analyze its behavior as $t$ increases.

ANS.   $P(t) = \dfrac{c}{d} + \dfrac{ka}{k^2 d^2 + \beta^2}(\beta \cos \beta t - kd \sin \beta t) + \left( P_0 - \dfrac{c}{d} - \dfrac{ka\beta}{k^2 d^2 + \beta^2} \right) e^{-kdt}.$

**9.** An analysis of a certain commodity market reveals that the demand and the supply functions are given by $D = c - dP$ and $S = a + bP + q \sin \beta t$, where $a, b, c, d, q$, and $\beta$ are positive constants. Determine $P(t)$ and analyze its behavior as $t$ increases.

ANS.   $P(t) = \dfrac{c - a}{d + b} + \dfrac{qk}{Q} \cos(\beta t + \alpha) + c_1 e^{-k(d+b)t},$

where $Q = \sqrt{k^2(d + b)^2 + \beta^2}$, $\alpha = \arccos \dfrac{\beta}{Q}$, and $c_1 = P_0 - \dfrac{c - a}{d + b} - \dfrac{\beta kq}{Q^2}.$

## 17. Orthogonal trajectories

Suppose that we have a family of curves given by

$$f(x, y, c) = 0, \tag{1}$$

one curve corresponding to each $c$ in some range of values of the parameter $c$. In certain applications it is found desirable to know what curves have the property of intersecting a curve of the family (1) at right angles whenever they do intersect.

That is, we wish to determine a family of curves with equations

$$g(x, y, k) = 0 \tag{2}$$

such that, at any intersection of a curve of the family (2) with a curve of the family (1), the tangents to the two curves are perpendicular. The families (1) and (2) are then said to be *orthogonal trajectories** of each other.

If two curves are to be orthogonal, then at each point of intersection the slopes of the curves must be negative reciprocals of each other. That fact leads us to a method for finding orthogonal trajectories of a given family of curves. First we find the differential equation of the given family. Then, replacing $dy/dx$ by $-dx/dy$ in that equation yields the differential equation of the orthogonal trajectories to the given curves. It remains only to solve the latter differential equation.

So far we have solved differential equations of only one form,

$$M \, dx + N \, dy = 0.$$

---

*The word orthogonal comes from the Greek ορθη (right) and γωνια (angle); the word trajectory comes from the Latin *trajectus* (cut across). Hence a curve that cuts across certain others at right angles is called an orthogonal trajectory of those others.

For such an equation

$$\frac{dy}{dx} = -\frac{M}{N},$$

so the differential equation of the orthogonal trajectories is

$$\frac{dy}{dx} = \frac{N}{M}$$

or

$$N\,dx - M\,dy = 0.$$

EXAMPLE: Find the orthogonal trajectories of all parabolas with vertices at the origin and foci on the $x$-axis.

   The algebraic equation of such parabolas is

$$y^2 = 4ax. \tag{3}$$

Hence, from

$$\frac{y^2}{x} = 4a,$$

we find the differential equation of the family (3) to be

$$2x\,dy - y\,dx = 0. \tag{4}$$

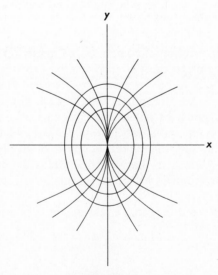

FIGURE 11

Therefore the orthogonal trajectories of the family (3) must satisfy the equation

$$2x\,dx + y\,dy = 0. \tag{5}$$

From (5) it follows that

$$2x^2 + y^2 = b^2, \tag{6}$$

where $b$ is the arbitrary constant. Thus the orthogonal trajectories of (3) are certain ellipses (6) with centers at the origin. See Figure 11.

## *Exercises*

In each exercise, find the orthogonal trajectories of the given family of curves. Draw a few representative curves of each family whenever a figure is requested.

1. $x - 4y = c$. Draw the figure.        ANS.   $4x + y = k$.
2. $x^2 + y^2 = c^2$. Draw the figure.        ANS.   $y = kx$.
3. $x^2 - y^2 = c_1$. Draw the figure.        ANS.   $xy = c_2$.
4. Circles through the origin with centers on the $x$-axis. Draw the figure.
       ANS.    Circles through the origin with centers on the $y$-axis.
5. Straight lines with slope and $y$-intercept equal. Draw the figure.
       ANS.   $(x + 1)^2 + y^2 = a^2$.
6. $y^2 = cx^3$. Draw the figure.        ANS.   $2x^2 + 3y^2 = k^2$.
7. $e^x + e^{-y} = c_1$.        ANS.   $e^y - e^{-x} = c_2$.
8. $y = c_1(\sec x + \tan x)$.        ANS.   $y^2 = 2(c_2 - \sin x)$.
9. $x^3 = 3(y - c)$. Draw the figure.        ANS.   $x(y - k) = 1$.
10. $x = c \exp(y^2)$.        ANS.   $y = c_1 \exp(-x^2)$.
11. $y = c\,e^{-mx}$; with $m$ held fixed. Draw the figure.        ANS.   $my^2 = 2(x - c_1)$.
12. Ellipses with centers at $(0, 0)$ and two vertices at $(1, 0)$ and $(-1, 0)$.
       ANS.   $x^2 + y^2 = 2 \ln |cx|$.
13. $x^2 - y^2 = cx$.        ANS.   $y(y^2 + 3x^2) = c_1$.
14. The cissoids, $y^2 = x^3/(a - x)$.        ANS.   $(x^2 + y^2)^2 = b(2x^2 + y^2)$.
15. The trisectrices of Maclaurin, $(a + x)y^2 = x^2(3a - x)$.
       ANS.   $(x^2 + y^2)^5 = cy^3(5x^2 + y^2)$.
16. $ax^2 + by^2 = c$; with $a$ and $b$ held fixed.        ANS.   $y^a = kx^b$.
17. $ax^2 + y^2 = 2acx$; with $a$ held fixed.        ANS.   If $a \neq 2$, $(2 - a)x^2 + y^2 = c_1 y^a$.
       If $a = 2$, $x^2 = -y^2 \ln(c_2 y)$.
18. $y(x^2 + c) + 2 = 0$.        ANS.   $y^3 = -3 \ln |kx|$.
19. $x^n + y^n = a^n$; with $n$ held fixed, $n \neq 2$.        ANS.   $x^{2-n} - y^{2-n} = c$.
20. $y^2 = x^2(1 - cx)$.        ANS.   $x^2 + 3y^2 = c_1 y$.
21. $y^2 = 4x^2(1 - cx)$.        ANS.   $2x^2 = 3y^2(1 - c_1 y^2)$.
22. $y^2 = ax^2(1 - cx)$; with $a$ held fixed.
       ANS.   If $a \neq 2$, $(a - 2)x^2 = 3y^2(1 - c_1 y^{a-2})$.
       If $a = 2$, $x^2 = -3y^2 \ln |c_2 y|$.
23. $y(x^2 + 1) = cx$.        ANS.   $y^2 = x^2 + 2 \ln |k(x^2 - 1)|$.
24. $y = 3x - 1 + ce^{-3x}$.        ANS.   $27x = 9y - 1 + k\,e^{-9y}$.

**25.** $y^2(2x^2 + y^2) = c^2$.                                    ANS.   $y^2 = 2x^2 \ln |kx|$.

**26.** $y^4 = c^2(x^2 + 4y^2)$.                                   ANS.   $x^8(2x^2 + 5y^2) = k^2$.

**27.** $x^4(4x^2 + 3y^2) = c^2$.                                  ANS.   $y^8 = k^2(3x^2 + 2y^2)$.

**28.** For the family $x^2 + 3y^2 = cy$, find that member of the orthogonal trajectories which passes through $(1, 2)$.                ANS.   $y^2 = x^2(3x + 1)$.

$$\frac{4}{}$$

# Additional Topics on
# Equations of Order One

## 18. Integrating factors found by inspection

In Section 11 we found that any linear equation of order one can be solved with the aid of an integrating factor. In Section 19 there is some discussion of tests for the determination of integrating factors.

At present we are concerned with equations that are simple enough to enable us to find integrating factors by inspection. The ability to do this depends largely upon recognition of certain common exact differentials and upon experience.

Below are four exact differentials that occur frequently:

$$d(xy) = x\,dy + y\,dx, \tag{1}$$

$$d\left(\frac{x}{y}\right) = \frac{y\,dx - x\,dy}{y^2}, \tag{2}$$

$$d\left(\frac{y}{x}\right) = \frac{x\,dy - y\,dx}{x^2}, \tag{3}$$

$$d\left(\arctan\frac{y}{x}\right) = \frac{x\,dy - y\,dx}{x^2 + y^2}. \tag{4}$$

Note the homogeneity of the coefficients of $dx$ and $dy$ in each of these differentials.

A differential involving only one variable, such as $x^{-2}\,dx$, is an exact differential.

EXAMPLE (a):   Solve the equation

$$y\,dx + (x + x^3y^2)\,dy = 0. \tag{5}$$

Let us group the terms of like degree, writing the equation in the form

$$(y\,dx + x\,dy) + x^3y^2\,dy = 0.$$

Now the combination $(y\,dx + x\,dy)$ attracts attention, so we rewrite the equation, obtaining

$$d(xy) + x^3y^2\,dy = 0. \tag{6}$$

Since the differential of $xy$ is present in equation (6), any factor that is a function of the product $xy$ will not disturb the integrability of that term. But the other term contains the differential $dy$, and hence should contain a function of $y$ alone. Therefore, let us divide by $(xy)^3$ and write

$$\frac{d(xy)}{(xy)^3} + \frac{dy}{y} = 0.$$

The equation above is integrable as it stands. A family of solutions is defined by

$$-\frac{1}{2x^2y^2} + \ln|y| = -\ln|c|,$$

or

$$2x^2y^2 \ln|cy| = 1.$$

EXAMPLE (b):   Solve the equation

$$y(x^3 - y)\,dx - x(x^3 + y)\,dy = 0. \tag{7}$$

Let us regroup the terms of (7) to obtain

$$x^3(y\,dx - x\,dy) - y(y\,dx + x\,dy) = 0. \tag{8}$$

Recalling that

$$d\left(\frac{x}{y}\right) = \frac{y\,dx - x\,dy}{y^2},$$

we divide the terms of equation (8) throughout by $y^2$ to get

$$x^3 \, d\left(\frac{x}{y}\right) - \frac{d(xy)}{y} = 0. \tag{9}$$

Equation (9) will be made exact by introducing a factor, if it can be found, to make the coefficient of $d(x/y)$ a function of $(x/y)$ and the coefficient of $d(xy)$ a function of $(xy)$. Some skill in obtaining such factors can be developed with a little practice.

There is a straightforward attack on equation (9) which has its good points. Assume that the integrating factor desired is $x^k y^n$, where $k$ and $n$ are to be determined. Applying that factor, we obtain

$$x^{k+3} y^n \, d\left(\frac{x}{y}\right) - x^k y^{n-1} \, d(xy) = 0. \tag{10}$$

Since the coefficient of $d(x/y)$ is to be a function of the ratio $(x/y)$, the exponents of $x$ and $y$ in that coefficient must be numerically equal but of opposite sign. That is,

$$k + 3 = -n. \tag{11}$$

In a similar manner, from the coefficient of $d(xy)$ it follows that we must put

$$k = n - 1. \tag{12}$$

From equations (11) and (12) we conclude that $k = -2$, $n = -1$. The desired integrating factor is $x^{-2} y^{-1}$ and (10) becomes

$$\frac{x}{y} \, d\left(\frac{x}{y}\right) - \frac{d(xy)}{x^2 y^2} = 0$$

of which a set of solutions is given by

$$\frac{1}{2}\left(\frac{x}{y}\right)^2 + \frac{1}{xy} = \frac{c}{2}.$$

Finally, we may write the desired solutions of equation (7) as

$$x^3 + 2y = cxy^2.$$

EXAMPLE (c):    Solve the equation

$$3x^2 y \, dx + (y^4 - x^3) \, dy = 0.$$

Two terms in the coefficients of $dx$ and $dy$ are of degree three, and the other coefficient is not of degree three. Let us regroup the terms to get

$$(3x^2 y \, dx - x^3 \, dy) + y^4 \, dy = 0,$$

or

$$y \, d(x^3) - x^3 \, dy + y^4 \, dy = 0.$$

The form of the first two terms now suggests the numerator in the differential of a quotient, as in

$$d\left(\frac{u}{v}\right) = \frac{v\,du - u\,dv}{v^2}.$$

Therefore we divide each term of our equation by $y^2$ and obtain

$$\frac{y\,d(x^3) - x^3\,dy}{y^2} + y^2\,dy = 0,$$

or

$$d\left(\frac{x^3}{y}\right) + y^2\,dy = 0.$$

Hence a solution set of the original equation is

$$\frac{x^3}{y} + \frac{y^3}{3} = \frac{c}{3}$$

or

$$3x^3 + y^4 = cy.$$

## Exercises

Except when the exercise indicates otherwise, find a set of solutions.

1. $y(2xy + 1)\,dx - x\,dy = 0.$     ANS.   $x(xy + 1) = cy.$
2. $y(y^3 - x)\,dx + x(y^3 + x)\,dy = 0.$     ANS.   $2xy^3 - x^2 = cy^2.$
3. $(x^3y^3 + 1)\,dx + x^4y^2\,dy = 0.$     ANS.   $x^3y^3 = -3\ln|cx|.$
4. $2t\,ds + s(2 + s^2t)\,dt = 0.$     ANS.   $1 + s^2t = cs^2t^2.$
5. $y(x^4 - y^2)\,dx + x(x^4 + y^2)\,dy = 0.$     ANS.   $y(3x^4 + y^2) = cx^3.$
6. $y(y^2 + 1)\,dx + x(y^2 - 1)\,dy = 0.$     ANS.   $x(y^2 + 1) = cy.$
7. Do exercise 6 by a second method.
8. $y(x^3 - y^5)\,dx - x(x^3 + y^5)\,dy = 0.$     ANS.   $x^4 = y^4(c + 4xy).$
9. $y(x^2 - y^2 + 1)\,dx - x(x^2 - y^2 - 1)\,dy = 0.$     ANS.   $x^2 + cxy + y^2 = 1.$
10. $(x^3 + xy^2 + y)\,dx + (y^3 + x^2y + x)\,dy = 0.$

    ANS.   $(x^2 + y^2)^2 = c - 4xy.$

11. $y(x^2 + y^2 - 1)\,dx + x(x^2 + y^2 + 1)\,dy = 0.$

    ANS.   $xy + \arctan(y/x) = c.$

12. $(x^3 + xy^2 - y)\,dx + (y^3 + x^2y + x)\,dy = 0.$

    ANS.   $2\arctan(y/x) = c - x^2 - y^2.$

13. $y(x^3e^{xy} - y)\,dx + x(y + x^3e^{xy})\,dy = 0.$     ANS.   $2x^2\,e^{xy} + y^2 = cx^2.$
14. $xy(y^2 + 1)\,dx + (x^2y^2 - 2)\,dy = 0$; when $x = 1, y = 1.$

    ANS.   $x^2(y^2 + 1) = 2 + 4\ln y.$

15. $y^2(1 - x^2)\,dx + x(x^2y + 2x + y)\,dy = 0.$     ANS.   $x^2y + x + y = cxy^2.$
16. $y(x^2y^2 - 1)\,dx + x(x^2y^2 + 1)\,dy = 0.$     ANS.   $x^2y^2 = 2\ln|cx/y|.$

17. $x^4 y' = -x^3 y - \csc(xy)$.     ANS.   $2x^2 \cos(xy) = cx^2 - 1$.
18. $[1 + y \tan(xy)]\,dx + x \tan(xy)\,dy = 0$.     ANS.   $\cos(xy) = c\,e^x$.
19. $y(x^2 y^2 - m)\,dx + x(x^2 y^2 + n)\,dy = 0$.     ANS.   $x^2 y^2 = 2 \ln|cx^m/y^n|$.
20. $x(x^2 - y^2 - x)\,dx - y(x^2 - y^2)\,dy = 0$; when $x = 2$, $y = 0$.

     ANS.   $3(x^2 - y^2)^2 = 4(x^3 + 4)$.

21. $y(x^2 + y)\,dx + x(x^2 - 2y)\,dy = 0$; when $x = 1$, $y = 2$.

     ANS.   $x^2 y - y^2 + 2x = 0$.

22. $y(x^3 y^3 + 2x^2 - y)\,dx + x^3(xy^3 - 2)\,dy = 0$; when $x = 1$, $y = 1$.

     ANS.   $x^3 y^3 + 4x^2 - 7xy + 2y = 0$.

23. $y(2 - 3xy)\,dx - x\,dy = 0$.     ANS.   $x^2(1 - xy) = cy$.
24. $y(2x + y^2)\,dx + x(y^2 - x)\,dy = 0$.     ANS.   $x(x + y^2) = cy$.
25. $y\,dx + 2(y^4 - x)\,dy = 0$.     ANS.   $y^4 + x = cy^2$.
26. $y(3x^3 - x + y)\,dx + x^2(1 - x^2)\,dy = 0$.     ANS.   $y \ln|cx| = x(1 - x^2)$.
27. $2x^5 y' = y(3x^4 + y^2)$.     ANS.   $x^4 = y^2(1 + cx)$.
28. $(x^n y^{n+1} + ay)\,dx + (x^{n+1} y^n + bx)\,dy = 0$.

     ANS.   If $n \neq 0$, $x^n y^n = n \ln|cx^{-a} y^{-b}|$.
          If $n = 0$, $xy = c_1 x^{-a} y^{-b}$.

29. $(x^{n+1} y^n + ay)\,dx + (x^n y^{n+1} + ax)\,dy = 0$.

     ANS.   If $n \neq 1$, $(n - 1)(xy)^{n-1}(x^2 + y^2 - c) = 2a$.
          If $n = 1$, $x^2 + y^2 - c = -2a \ln|xy|$.

## 19. The determination of integrating factors

Let us see what progress can be made on the problem of the determination of an integrating factor for the equation

$$M\,dx + N\,dy = 0. \tag{1}$$

Suppose that $u$, possibly a function of both $x$ and $y$, is to be an integrating factor of (1). Then the equation

$$uM\,dx + uN\,dy = 0 \tag{2}$$

must be exact. Therefore, by the result of Section 10,

$$\frac{\partial}{\partial y}(uM) = \frac{\partial}{\partial x}(uN).$$

Hence $u$ must satisfy the partial differential equation

$$u\frac{\partial M}{\partial y} + M\frac{\partial u}{\partial y} = u\frac{\partial N}{\partial x} + N\frac{\partial u}{\partial x},$$

or

$$u\left(\frac{\partial M}{\partial y} - \frac{\partial N}{\partial x}\right) = N\frac{\partial u}{\partial x} - M\frac{\partial u}{\partial y}. \tag{3}$$

Furthermore, by reversing the argument above, it can be seen that, if $u$ satisfies equation (3), $u$ is an integrating factor for equation (1). We have "reduced" the problem of solving the ordinary differential equation (1) to the problem of obtaining a particular solution of the partial differential equation (3).

Not much has been gained because we have developed no methods for attacking an equation such as (3). Therefore, we turn the problem back into the realm of ordinary differential equations by restricting $u$ to be a function of only one variable.

First let $u$ be a function of $x$ alone. Then $\partial u/\partial y = 0$ and $\partial u/\partial x$ becomes $du/dx$. Then (3) reduces to

$$u\left(\frac{\partial M}{\partial y} - \frac{\partial N}{\partial x}\right) = N\frac{du}{dx},$$

or

$$\frac{1}{N}\left(\frac{\partial M}{\partial y} - \frac{\partial N}{\partial x}\right) dx = \frac{du}{u}. \tag{4}$$

If the left member of the above equation is a function of $x$ alone, then we can determine $u$ at once. Indeed, if

$$\frac{1}{N}\left(\frac{\partial M}{\partial y} - \frac{\partial N}{\partial x}\right) = f(x), \tag{5}$$

then the desired integrating factor is $u = \exp\left(\int f(x)\,dx\right)$.

By a similar argument, assuming that $u$ is a function of $y$ alone, we are led to the conclusion that if

$$\frac{1}{M}\left(\frac{\partial M}{\partial y} - \frac{\partial N}{\partial x}\right) = g(y), \tag{6}$$

then an integrating factor for equation (1) is $u = \exp\left(-\int g(y)\,dy\right)$.

Our two results are expressed in the following rules:

(a) If $\dfrac{1}{N}\left(\dfrac{\partial M}{\partial y} - \dfrac{\partial N}{\partial x}\right) = f(x)$, a function of $x$ alone, then $\exp\left(\int f(x)\,dx\right)$ is an integrating factor for the equation

$$M\,dx + N\,dy = 0. \tag{1}$$

(b) If $\dfrac{1}{M}\left(\dfrac{\partial M}{\partial y} - \dfrac{\partial N}{\partial x}\right) = g(y)$, a function of $y$ alone, then $\exp\left(-\int g(y)\,dy\right)$

is an integrating factor for equation (1).

It should be emphasized that if neither of the preceding criteria is satisfied, we can say only that the equation does not have an integrating factor that is a function of $x$ or $y$ alone. For example, the student should show that the above criteria fail in the case of Example (a) of Section 18, even though $(xy)^{-3}$ is an integrating factor for the differential equation.

EXAMPLE (a):   Solve the equation

$$(4xy + 3y^2 - x)\,dx + x(x + 2y)\,dy = 0. \tag{7}$$

Here $M = 4xy + 3y^2 - x$, $N = x^2 + 2xy$, so

$$\frac{\partial M}{\partial y} - \frac{\partial N}{\partial x} = 4x + 6y - (2x + 2y) = 2x + 4y.$$

Hence

$$\frac{1}{N}\left(\frac{\partial M}{\partial y} - \frac{\partial N}{\partial x}\right) = \frac{2x + 4y}{x(x + 2y)} = \frac{2}{x}.$$

Therefore an integrating factor for equation (7) is

$$\exp\left(2\int \frac{dx}{x}\right) = \exp\left(2\ln|x|\right) = x^2.$$

Returning to the original equation (7), we insert the integrating factor and obtain

$$(4x^3y + 3x^2y^2 - x^3)\,dx + (x^4 + 2x^3y)\,dy = 0, \tag{8}$$

which we know must be an exact equation. The methods of Section 10 apply. We are then led to put equation (8) in the form

$$(4x^3y\,dx + x^4\,dy) + (3x^2y^2\,dx + 2x^3y\,dy) - x^3\,dx = 0,$$

from which the solution set

$$x^4y + x^3y^2 - \tfrac{1}{4}x^4 = \tfrac{1}{4}c,$$

or

$$x^3(4xy + 4y^2 - x) = c$$

follows at once.

EXAMPLE (b):   Solve the equation

$$y(x + y + 1)\,dx + x(x + 3y + 2)\,dy = 0. \tag{9}$$

First we form

$$\frac{\partial M}{\partial y} = x + 2y + 1, \qquad \frac{\partial N}{\partial x} = 2x + 3y + 2.$$

Then we see that

$$\frac{\partial M}{\partial y} - \frac{\partial N}{\partial x} = -x - y - 1,$$

so

$$\frac{1}{N}\left(\frac{\partial M}{\partial y} - \frac{\partial N}{\partial x}\right) = -\frac{x + y + 1}{x(x + 3y + 2)}$$

is not a function of $x$ alone. But

$$\frac{1}{M}\left(\frac{\partial M}{\partial y} - \frac{\partial N}{\partial x}\right) = -\frac{x + y + 1}{y(x + y + 1)} = -\frac{1}{y}.$$

Therefore $\exp(\ln|y|) = |y|$ is the desired integrating factor for (9).

It follows that for $y > 0$, $y$ itself is an integrating factor of equation (9) and for $y < 0$, $-y$ is an integrating factor. In either case (9) becomes

$$(xy^2 + y^3 + y^2)\,dx + (x^2 y + 3xy^2 + 2xy)\,dy = 0,$$

or

$$(xy^2\,dx + x^2 y\,dy) + (y^3\,dx + 3xy^2\,dy) + (y^2\,dx + 2xy\,dy) = 0.$$

Then a set of solutions of (9) is defined implicitly by

$$\tfrac{1}{2}x^2 y^2 + xy^3 + xy^2 = \tfrac{1}{2}c,$$

or

$$xy^2(x + 2y + 2) = c.$$

**EXAMPLE (c):**   Solve the equation

$$y(x + y)\,dx + (x + 2y - 1)\,dy = 0. \tag{10}$$

From $\partial M/\partial y = x + 2y$, $\partial N/\partial x = 1$, we conclude at once that

$$\frac{1}{N}\left(\frac{\partial M}{\partial y} - \frac{\partial N}{\partial x}\right) = \frac{x + 2y - 1}{x + 2y - 1} = 1.$$

Hence $e^x$ is an integrating factor for (10). Then

$$(xy\,e^x + y^2\,e^x)\,dx + (x\,e^x + 2y\,e^x - e^x)\,dy = 0$$

is an exact equation. Grouping the terms in the following manner,

$$[xy\,e^x\,dx + (x\,e^x - e^x)\,dy] + (y^2\,e^x\,dx + 2y\,e^x\,dy) = 0,$$

leads us at once to the family of solutions defined by

$$e^x(x - 1)y + y^2\,e^x = c,$$

or

$$y(x + y - 1) = c\,e^{-x}.$$

## Exercises

Solve each of the following equations.

1. $(x^2 + y^2 + 1) dx + x(x - 2y) dy = 0.$     ANS.   $x^2 - y^2 + xy - 1 = cx.$
2. $2y(x^2 - y + x) dx + (x^2 - 2y) dy = 0.$     ANS.   $y(x^2 - y) = c e^{-2x}.$
3. $y(2x - y + 1) dx + x(3x - 4y + 3) dy = 0.$     ANS.   $xy^3(x - y + 1) = c.$
4. $y(4x + y) dx - 2(x^2 - y) dy = 0.$     ANS.   $2x^2 + xy + 2y \ln |y| = cy.$
5. $(xy + 1) dx + x(x + 4y - 2) dy = 0.$     ANS.   $xy + \ln |x| + 2y^2 - 2y = c.$
6. $(2y^2 + 3xy - 2y + 6x) dx + x(x + 2y - 1) dy = 0.$
   ANS.   $x^2(y^2 + xy - y + 2x) = c.$
7. $y(y + 2x - 2) dx - 2(x + y) dy = 0.$     ANS.   $y(2x + y) = c e^x.$
8. $y^2 dx + (3xy + y^2 - 1) dy = 0.$     ANS.   $y^2(y^2 + 4xy - 2) = c.$
9. $2y(x + y + 2) dx + (y^2 - x^2 - 4x - 1) dy = 0.$
   ANS.   $x^2 + 2xy + y^2 + 4x + 1 = cy.$
10. $2(2y^2 + 5xy - 2y + 4) dx + x(2x + 2y - 1) dy = 0.$
   ANS.   $x^4(y^2 + 2xy - y + 2) = c.$
11. $3(x^2 + y^2) dx + x(x^2 + 3y^2 + 6y) dy = 0.$     ANS.   $x(x^2 + 3y^2) = c e^{-y}.$
12. $y(8x - 9y) dx + 2x(x - 3y) dy = 0.$     ANS.   $x^3y(2x - 3y) = c.$
13. Do exercise 12 by another method.
14. $y(2x^2 - xy + 1) dx + (x - y) dy = 0.$     ANS.   $y(2x - y) = c \exp(-x^2).$
15. Euler's theorem (exercise 36, page 30) on homogeneous functions states that, if $F$ is a homogeneous function of degree $k$ in $x$ and $y$, then

$$x\frac{\partial F}{\partial x} + y\frac{\partial F}{\partial y} = kF.$$

Use Euler's theorem to prove the result that, if $M$ and $N$ are homogeneous functions of the same degree, and if $Mx + Ny \neq 0$, then

$$\frac{1}{Mx + Ny}$$

is an integrating factor for the equation

$$M\,dx + N\,dy = 0. \tag{A}$$

16. In the result to be proved in exercise 15 above there is an exceptional case, namely, when $Mx + Ny = 0$. Solve equation (A) when $Mx + Ny = 0$.     ANS.   $y = cx.$

Use the integrating factor in the result of exercise 15 above to solve each of the equations in exercises 17 through 20.

17. $xy\,dx - (x^2 + 2y^2) dy = 0.$     ANS.   $x^2 = 4y^2 \ln |y/c|.$
18. $v^2 dx + x(x + v) dv = 0.$     ANS.   $xv^2 = c(x + 2v).$
19. $v(u^2 + v^2) du - u(u^2 + 2v^2) dv = 0.$     ANS.   $u^2 = 2v^2 \ln |cv^2/u|.$
20. $(x^2 + y^2) dx - xy\,dy = 0.$     (Exercise 9, page 29.)
21. Apply the method of this section to the general linear equation of order one.

## 20. Substitution suggested by the equation

An equation of the form

$$M\,dx + N\,dy = 0$$

may not yield at once (or at all) to the methods of Chapter 2. Even then the usefulness of those methods is not exhausted. It may be possible by some change of variables to transform the equation into a type that we know how to solve.

A natural source of suggestions for useful transformations is the differential equation itself. If a particular function of one or both variables stands out in the equation, then it is worthwhile to examine the equation after that function has been introduced as a new variable. For instance, in the equation

$$(x + 2y - 1)\,dx + 3(x + 2y)\,dy = 0 \tag{1}$$

the combination $(x + 2y)$ occurs twice and thus attracts attention. Hence we put

$$x + 2y = v,$$

and, because no other function of $x$ and $y$ stands out, we retain either $x$ or $y$ for the other variable. The solution is completed in Example (a) below.

In the equation

$$(1 + 3x \sin y)\,dx - x^2 \cos y\,dy = 0, \tag{2}$$

the presence of both $\sin y$ and its differential $\cos y\,dy$, and the fact that $y$ appears in the equation in no other manner, leads us to put $\sin y = w$ and to obtain the differential equation in $w$ and $x$. See Example (b) below.

EXAMPLE (a): Solve the equation

$$(x + 2y - 1)\,dx + 3(x + 2y)\,dy = 0. \tag{1}$$

As suggested above, put

$$x + 2y = v.$$

Then

$$dx = dv - 2\,dy$$

and equation (1) becomes

$$(v - 1)(dv - 2\,dy) + 3v\,dy = 0,$$

or

$$(v - 1)\,dv + (v + 2)\,dy = 0.$$

Now the variables can be separated. From the equation in the form

$$\frac{v-1}{v+2} dv + dy = 0,$$

we get

$$\left(1 - \frac{3}{v+2}\right) dv + dy = 0$$

and then

$$v - 3 \ln |v + 2| + y + c = 0.$$

But $v = x + 2y$, so our final result is

$$x + 3y + c = 3 \ln |x + 2y + 2|.$$

**EXAMPLE (b):** Solve the equation

$$(1 + 3x \sin y) dx - x^2 \cos y \, dy = 0. \qquad (2)$$

Put $\sin y = w$. Then $\cos y \, dy = dw$ and (2) becomes

$$(1 + 3xw) dx - x^2 \, dw = 0,$$

an equation linear in $w$. From the standard form

$$dw - \frac{3}{x} w \, dx = \frac{dx}{x^2}$$

an integrating factor is seen to be

$$\exp(-3 \ln |x|) = |x|^{-3}.$$

Application of the integrating factor yields the exact equation

$$x^{-3} \, dw - 3x^{-4} w \, dx = x^{-5} \, dx,$$

for either $x > 0$ or $x < 0$, from which we get

$$x^{-3} w = -\tfrac{1}{4} x^{-4} + \tfrac{1}{4} c,$$

or

$$4xw = cx^4 - 1.$$

Hence (2) has the solution set

$$4x \sin y = cx^4 - 1.$$

## 21. Bernoulli's equation

A well-known equation that fits into the category of Section 20 is Bernoulli's equation,

$$y' + P(x)y = Q(x)y^n. \tag{1}$$

If $n = 1$ in (1), the variables are separable, so we concentrate on the case $n \neq 1$. Equation (1) may be put in the form

$$y^{-n}\,dy + Py^{-n+1}\,dx = Q\,dx. \tag{2}$$

But the differential of $y^{-n+1}$ is $(1 - n)y^{-n}\,dy$, so equation (2) may be simplified by putting

$$y^{-n+1} = z,$$

from which

$$(1 - n)y^{-n}\,dy = dz.$$

Thus the equation in $z$ and $x$ is

$$dz + (1 - n)Pz\,dx = (1 - n)Q\,dx,$$

a linear equation in standard form. Hence any Bernoulli equation can be solved with the aid of the above change of dependent variable (unless $n = 1$, when no substitution is needed).

EXAMPLE (a):   Solve the equation

$$y(6y^2 - x - 1)\,dx + 2x\,dy = 0. \tag{3}$$

First let us group the terms according to powers of $y$, writing

$$2x\,dy - y(x + 1)\,dx + 6y^3\,dx = 0.$$

Now it can be seen that the equation is a Bernoulli equation, since it involves only terms containing respectively $dy$, $y$, and $y^n$ ($n = 3$ here). Therefore, we divide throughout by $y^3$, obtaining

$$2xy^{-3}\,dy - y^{-2}(x + 1)\,dx = -6\,dx.$$

This equation is linear in $y^{-2}$, so we put $y^{-2} = v$, obtain $dv = -2y^{-3}\,dy$, and need to solve the equation

$$x\,dv + v(x + 1)\,dx = 6\,dx,$$

or

$$dv + v(1 + x^{-1})\,dx = 6x^{-1}\,dx. \tag{4}$$

Since

$$\exp{(x + \ln|x|)} = |x|\, e^x$$

is an integrating factor for (4), the equation

$$x\, e^x\, dv + v\, e^x(x + 1)\, dx = 6\, e^x\, dx$$

is exact. Its solution set

$$xv\, e^x = 6\, e^x + c,$$

together with $v = y^{-2}$, leads us to the final result

$$y^2(6 + c\, e^{-x}) = x.$$

EXAMPLE (b):   Solve the equation

$$6y^2\, dx - x(2x^3 + y)\, dy = 0. \tag{5}$$

This is a Bernoulli equation with $x$ as the dependent variable, so it can be treated in the manner used in Example (a). That method of attack is left for the exercises.

Equation (5) can equally well be treated as follows. Note that if each member of (5) is multiplied by $x^2$, the equation becomes

$$6y^2x^2\, dx - x^3(2x^3 + y)\, dy = 0. \tag{6}$$

In (6), the variable $x$ appears only in the combinations $x^3$ and its differential $3x^2\, dx$. Hence a reasonable choice for a new variable is $w = x^3$. The equation in $w$ and $y$ is

$$2y^2\, dw - w(2w + y)\, dy = 0,$$

an equation with coefficients homogeneous of degree two in $y$ and $w$. The further change of variable $w = zy$ leads to the equation

$$2y\, dz - z(2z - 1)\, dy = 0,$$

$$\frac{4\, dz}{2z - 1} - \frac{2\, dz}{z} - \frac{dy}{y} = 0.$$

Therefore, we have

$$2\ln|2z - 1| - 2\ln|z| - \ln|y| = \ln|c|,$$

or

$$(2z - 1)^2 = cyz^2.$$

But $z = w/y = x^3/y$, so the solutions we seek are determined by

$$(2x^3 - y)^2 = cyx^6.$$

## Exercises

In exercises 1 through 21, solve the equation.

**1.** $(3x - 2y + 1) dx + (3x - 2y + 3) dy = 0$.

ANS.    $5(x + y + c) = 2 \ln |15x - 10y + 11|$.

**2.** $\sin y(x + \sin y) dx + 2x^2 \cos y \, dy = 0$.          ANS.    $x^3 \sin^2 y = c(3x + \sin y)^2$.

**3.** $dy/dx = (9x + 4y + 1)^2$.          ANS.    $3 \tan (6x + c) = 2(9x + 4y + 1)$.

**4.** $y' = y - xy^3 e^{-2x}$.          ANS.    $e^{2x} = y^2(x^2 + c)$.

**5.** $dy/dx = \sin (x + y)$.          ANS.    $x + c = \tan (x + y) - \sec (x + y)$.

**6.** $xy \, dx + (x^2 - 3y) dy = 0$.          ANS.    $x^2 y^2 = 2y^3 + c$.

**7.** $(3 \tan x - 2 \cos y) \sec^2 x \, dx + \tan x \sin y \, dy = 0$.

ANS.    $\cos y \tan^2 x = \tan^3 x + c$.

**8.** $(x + 2y - 1) dx + (2x + 4y - 3) dy = 0$. Solve by two methods.

ANS.    $(x + 2y - 1)^2 = 2y + c$.

**9.** Solve the equation $6y^2 \, dx - x(2x^3 + y) dy = 0$ of Example (b) above by treating it as a Bernoulli equation in the dependent variable $x$.

**10.** $2x^3 y' = y(y^2 + 3x^2)$. Solve by two methods.          ANS.    $y^2(c - x) = x^3$.

**11.** $(3 \sin y - 5x) dx + 2x^2 \cot y \, dy = 0$.          ANS.    $x^3(\sin y - x)^2 = c \sin^2 y$.

**12.** $y' = 1 + 6x \exp (x - y)$.          ANS.    $\exp (y - x) = 3x^2 + c$.

**13.** $dv/du = (u - v)^2 - 2(u - v) - 2$.          ANS.    $(u - v - 3) \exp (4u) = c(u - v + 1)$.

**14.** $2y \, dx + x(x^2 \ln y - 1) dy = 0$.          ANS.    $y(1 + x^2 - x^2 \ln y) = cx^2$.

**15.** $\cos y \sin 2x \, dx + (\cos^2 y - \cos^2 x) dy = 0$.

ANS.    $\cos^2 x(1 + \sin y) = \cos y(y + c - \cos y)$.

**16.** $(ke^{2v} - u) du = 2 e^{2v}(e^{2v} + ku) dv$.

ANS.    $2k \arctan (u e^{-2v}) = \ln |c(u^2 + e^{4v})|$.

**17.** $y' \tan x \sin 2y = \sin^2 x + \cos^2 y$.          ANS.    $(\sin^2 x + 3 \cos^2 y) \sin x = c$.

**18.** $(x + 2y - 1) dx - (x + 2y - 5) dy = 0$.

**19.** $y(x \tan x + \ln y) dx + \tan x \, dy = 0$.          ANS.    $\sin x \ln y = x \cos x - \sin x + c$.

**20.** $xy' - y = x^k y^n$, where $n \neq 1$ and $k + n \neq 1$.

ANS.    $(k + n - 1)y^{1-n} = (1 - n)x^k + cx^{1-n}$.

**21.** Solve the equation of exercise 20 for the values of $k$ and $n$ not included there.

ANS.    If $n = 1$ and $k \neq 0$, $x^k = k \ln |cy/x|$.

If $n = 1$ and $k = 0$, $y = cx^2$.

If $n \neq 1$ but $k + n = 1$, $y^{1-n} = (1 - n)x^{1-n} \ln |cx|$.

In exercises 22 through 27, find the particular solution required.

**22.** $4(3x + y - 2) dx - (3x + y) dy = 0$; when $x = 1$, $y = 0$.

ANS.    $7(4x - y - 4) = 8 \ln \dfrac{21x + 7y - 8}{13}$.

**23.** $y' = 2(3x + y)^2 - 1$; when $x = 0$, $y = 1$.          ANS.    $4 \arctan (3x + y) = 8x + \pi$.

**24.** $2xyy' = y^2 - 2x^3$. Find the solution that passes through the point $(1, 2)$.

ANS.    $y^2 = x(5 - x^2)$.

**25.** $(y^4 - 2xy) dx + 3x^2 \, dy = 0$; when $x = 2$, $y = 1$.          ANS.    $x^2 = y^3(x + 2)$.

**26.** $(2y^3 - x^3)\,dx + 3xy^2\,dy = 0$; when $x = 1$, $y = 1$. Solve by two methods.

ANS.   $5x^2y^3 = x^5 + 4.$

**27.** $(x^2 + 6y^2)\,dx - 4xy\,dy = 0$; when $x = 1$, $y = 1$. Solve by three methods.

ANS.   $2y^2 = x^2(3x - 1).$

##  22. Coefficients linear in the two variables

Consider the equation

$$(a_1 x + b_1 y + c_1)\,dx + (a_2 x + b_2 y + c_2)\,dy = 0, \tag{1}$$

in which the $a$'s, $b$'s, and $c$'s are constants. We know already how to solve the special case in which $c_1 = 0$ and $c_2 = 0$ because then the coefficients in (1) are each homogeneous and of degree one in $x$ and $y$. It is reasonable, therefore, to attempt to reduce equation (1) to that situation.

In connection with (1) consider the lines

$$a_1 x + b_1 y + c_1 = 0, \tag{2}$$

$$a_2 x + b_2 y + c_2 = 0.$$

They may be parallel or they may intersect. There will not be two lines if $a_1$ and $b_1$ are zero or if $a_2$ and $b_2$ are zero, but equation (1) will then be linear in one of its variables.

If the lines (2) intersect, let the point of intersection be $(h, k)$. Then the translation

$$x = u + h, \tag{3}$$

$$y = v + k$$

will change the equations (2) into equations of lines through the origin of the $uv$ coordinate system, namely,

$$a_1 u + b_1 v = 0, \tag{4}$$

$$a_2 u + b_2 v = 0.$$

Therefore, since $dx = du$ and $dy = dv$, the change of variables

$$x = u + h,$$

$$y = v + k,$$

where $(h, k)$ is the point of intersection of the lines (2), will transform the differential equation (1) into

$$(a_1 u + b_1 v)\,du + (a_2 u + b_2 v)\,dv = 0, \tag{5}$$

an equation that we know how to solve.

If the lines (2) do not intersect, a constant $k$ exists such that

$$a_2 x + b_2 y = k(a_1 x + b_1 y),$$

so that equation (1) appears in the form

$$(a_1 x + b_1 y + c_1) \, dx + [k(a_1 x + b_1 y) + c_2] \, dy = 0. \tag{6}$$

The recurrence of the expression $(a_1 x + b_1 y)$ in (6) suggests the introduction of a new variable $w = a_1 x + b_1 y$. Then the new equation, in $w$ and $x$ or in $w$ and $y$, is one with variables separable, since its coefficients contain only $w$ and constants.

EXAMPLE (a):   Solve the equation

$$(x + 2y - 4) \, dx - (2x + y - 5) \, dy = 0. \tag{7}$$

The lines

$$x + 2y - 4 = 0,$$
$$2x + y - 5 = 0$$

intersect at the point (2, 1). Hence put

$$x = u + 2,$$
$$y = v + 1.$$

Then equation (7) becomes

$$(u + 2v) \, du - (2u + v) \, dv = 0, \tag{8}$$

which has coefficients homogeneous and of degree one in $u$ and $v$. Therefore let $u = vz$, which transforms (8) into

$$(z + 2)(z \, dv + v \, dz) - (2z + 1) \, dv = 0,$$

or

$$(z^2 - 1) \, dv + v(z + 2) \, dz = 0.$$

Separation of the variables $v$ and $z$ leads us to the equation

$$\frac{dv}{v} + \frac{(z + 2) \, dz}{z^2 - 1} = 0.$$

With the aid of partial fractions, we can write the above equation in the form

$$\frac{2 \, dv}{v} + \frac{3 \, dz}{z - 1} - \frac{dz}{z + 1} = 0.$$

Hence we get

$$2 \ln |v| + 3 \ln |z - 1| - \ln |z + 1| = \ln |c|$$

from which it follows that

$$v^2(z - 1)^3 = c(z + 1),$$

or

$$(vz - v)^3 = c(vz + v).$$

Now $vz = u$, so a set of solutions appears as

$$(u - v)^3 = c(u + v).$$

But $u = x - 2$ and $v = y - 1$. Therefore the desired result in terms of $x$ and $y$ is

$$(x - y - 1)^3 = c(x + y - 3).$$

For other methods of solution of equation (7), see exercises 23 and 30 below.

EXAMPLE (b):    Solve the equation

$$(2x + 3y - 1)\, dx + (2x + 3y + 2)\, dy = 0, \tag{9}$$

with the condition that $y = 3$ when $x = 1$.

    The lines

$$2x + 3y - 1 = 0$$

and

$$2x + 3y + 2 = 0$$

are parallel. Therefore we proceed, as we should have upon first glancing at the equation, to put

$$2x + 3y = v.$$

Then $2\, dx = dv - 3\, dy$, and equation (9) is transformed into

$$(v - 1)(dv - 3\, dy) + 2(v + 2)\, dy = 0,$$

or

$$(v - 1)\, dv - (v - 7)\, dy = 0. \tag{10}$$

    Equation (10) is easily solved, leading us to the relation

$$v - y + c + 6 \ln |v - 7| = 0.$$

Therefore a solution set of (9) is

$$2x + 2y + c = -6 \ln |2x + 3y - 7|.$$

But $y = 3$ when $x = 1$, so $c = -8 - 6 \ln 4$. Hence the particular solution required is given by

$$x + y - 4 = -3 \ln \left[\tfrac{1}{4}(2x + 3y - 7)\right].$$

## *Exercises*

In exercises 1 through 18, solve the equations.

**1.** $(y - 2) dx - (x - y - 1) dy = 0.$        ANS.   $x - 3 = (2 - y) \ln |c(y - 2)|.$

**2.** $(x - 4y - 9) dx + (4x + y - 2) dy = 0.$

ANS.   $\ln [(x - 1)^2 + (y + 2)^2] - 8 \arc\tan \dfrac{x - 1}{y + 2} = c.$

**3.** $(2x - y) dx + (4x + y - 6) dy = 0.$        ANS.   $(x + y - 3)^3 = c(2x + y - 4)^2.$

**4.** $(x - 4y - 3) dx - (x - 6y - 5) dy = 0.$        ANS.   $(x - 2y - 1)^2 = c(x - 3y - 2).$

**5.** $(2x + 3y - 5) dx + (3x - y - 2) dy = 0.$ Solve by two methods.

**6.** $2 dx + (2x - y + 3) dy = 0.$ Use a change of variable.

ANS.   $y + c = -\ln |2x - y + 4|.$

**7.** Solve the equation of exercise 6 by using the fact that the equation is linear in $x$.

**8.** $(x - y + 2) dx + 3 dy = 0.$        ANS.   $x + c = 3 \ln |x - y + 5|.$

**9.** Solve exercise 8 by another method.

**10.** $(x + y - 1) dx + (2x + 2y + 1) dy = 0.$        ANS.   $x + 2y + c = 3 \ln |x + y + 2|.$

**11.** $(3x + 2y + 7) dx + (2x - y) dy = 0.$ Solve by two methods.

**12.** $(x - 2) dx + 4(x + y - 1) dy = 0.$        ANS.   $2(y + 1) = -(x + 2y) \ln |c(x + 2y)|.$

**13.** $(x - 3y + 2) dx + 3(x + 3y - 4) dy = 0.$

ANS.   $\ln [(x - 1)^2 + 9(y - 1)^2] - 2 \arc\tan \dfrac{x - 1}{3(y - 1)} = c.$

**14.** $(6x - 3y + 2) dx - (2x - y - 1) dy = 0.$        ANS.   $3x - y + c = 5 \ln |2x - y + 4|.$

**15.** $(9x - 4y + 4) dx - (2x - y + 1) dy = 0.$

ANS.   $y - 1 = 3(y - 3x - 1) \ln |c(3x - y + 1)|.$

**16.** $(x + 3y - 4) dx + (x + 4y - 5) dy = 0.$

ANS.   $y - 1 = (x + 2y - 3) \ln |c(x + 2y - 3)|.$

**17.** $(x + 2y - 1) dx - (2x + y - 5) dy = 0.$        ANS.   $(x - y - 4)^3 = c(x + y - 2).$

**18.** $(x - 1) dx - (3x - 2y - 5) dy = 0.$        ANS.   $(2y - x + 3)^2 = c(y - x + 2).$

In exercises 19 through 22 obtain the particular solution indicated.

**19.** $(2x - 3y + 4) dx + 3(x - 1) dy = 0$; when $x = 3, y = 2.$

ANS.   $3(y - 2) = -2(x - 1) \ln \dfrac{x - 1}{2}.$

**20.** Solve the equation of exercise 19 but with the condition that when $x = -1, y = 2.$

ANS.   $3(y - 2) = -2(x - 1) \ln \dfrac{1 - x}{2}.$

**21.** $(x + y - 4) dx - (3x - y - 4) dy = 0$; when $x = 4, y = 1.$

ANS.   $2(x + 2y - 6) = 3(x - y) \ln \dfrac{x - y}{3}.$

**22.** Solve the equation of exercise 21 but with the condition that when $x = 3, y = 7.$

ANS.   $y - 5x + 8 = 2(y - x) \ln \dfrac{y - x}{4}.$

**23.** Prove that the change of variables

$$x = \alpha_1 u + \alpha_2 v, \qquad y = u + v$$

will transform the equation

$$(a_1 x + b_1 y + c_1)\, dx + (a_2 x + b_2 y + c_2)\, dy = 0 \tag{A}$$

into an equation in which the variables $u$ and $v$ are separable, if $\alpha_1$ and $\alpha_2$ are roots of the equation

$$a_1 \alpha^2 + (a_2 + b_1)\alpha + b_2 = 0, \tag{B}$$

and if $\alpha_2 \neq \alpha_1$.

Note that this method of solution of (A) is not practical for us unless the roots of equation (B) are real and distinct.

Solve exercises 24 through 29 by the method indicated in exercise 23.

**24.** Do exercise 4 above. As a check, the equation (B) for this case is

$$\alpha^2 - 5\alpha + 6 = 0,$$

so we may choose $\alpha_1 = 2$ and $\alpha_2 = 3$.

The equation in $u$ and $v$ turns out to be

$$(v - 1)\, du - 2(u + 2)\, dv = 0.$$

**25.** Do exercise 3 above.              **26.** Do exercise 10 above.
**27.** Do exercise 17 above.              **28.** Do exercise 18 above.
**29.** Do Example (a) in the text of this section.
**30.** Prove that the change of variables

$$x = \alpha_1 u + \beta v, \qquad y = u + v$$

will transform the equation

$$(a_1 x + b_1 y + c_1)\, dx + (a_2 x + b_2 y + c_2)\, dy = 0 \tag{A}$$

into an equation that is linear in the variable $u$, if $\alpha_1$ is a root of the equation

$$a_1 \alpha^2 + (a_2 + b_1)\alpha + b_2 = 0, \tag{B}$$

and if $\beta$ is any number such that $\beta \neq \alpha_1$.

Note that this method is not practical for us unless the roots of equation (B) are real; however, they need not be distinct as they had to be in the theorem of exercise 23. The method of this exercise is particularly useful when the roots of (B) are equal.

Solve exercises 31 through 35 by the method indicated in exercise 30.

**31.** Do exercise 16 above. The only possible $\alpha_1$ is $(-2)$. Then $\beta$ may be chosen to be anything else.
**32.** Do exercise 12 above.        **33.** Do exercise 15 above.    **34.** Do exercise 18 above.
**35.** Do exercise 4 above. As seen in exercise 24, the roots of the "$\alpha$-equation" are 2 and 3. If you choose $\alpha_1 = 2$, for example, then you make $\beta$ anything except 2. Of course, if you choose $\alpha_1 = 2$ and $\beta = 3$, then you are reverting to the method of exercise 23.

## 23. Solutions involving nonelementary integrals

In solving differential equations we frequently are confronted with the need for integrating an expression that is not the differential of any elementary* function. Following is a short list of nonelementary integrals:

$$\int \exp(-x^2)\, dx \qquad \int \frac{e^{-x}}{x}\, dx \qquad \int x \tan x\, dx$$

$$\int \sin x^2\, dx \qquad \int \frac{\sin x}{x}\, dx \qquad \int \frac{dx}{\ln x}$$

$$\int \cos x^2\, dx \qquad \int \frac{\cos x}{x}\, dx \qquad \int \frac{dx}{\sqrt{1 - x^3}}.$$

Integrals involving the square root of a polynomial of degree greater than two are, in general, nonelementary. In special instances they may degenerate into elementary integrals.

The following example presents two ways of dealing with problems in which nonelementary integrals arise.

EXAMPLE: Solve the equation

$$y' - 2xy = 1$$

with the initial condition that when $x = 0$, $y = 1$.

The equation being linear in $y$, we write

$$dy - 2xy\, dx = dx,$$

obtain the integrating factor $\exp(-x^2)$, and prepare to solve

$$\exp(-x^2)\, dy - 2xy \exp(-x^2)\, dx = \exp(-x^2)\, dx. \tag{1}$$

The left member is, of course, the differential of $y \exp(-x^2)$. But the right member is not the differential of any elementary function; that is, $\int \exp(-x^2)\, dx$ is a nonelementary integral.

Let us turn to power series for help. From the series

$$\exp(-x^2) = \sum_{n=0}^{\infty} \frac{(-1)^n x^{2n}}{n!},$$

---

* By an elementary function we mean a function studied in the ordinary beginning calculus course. For example, polynomials, exponentials, logarithms, trigonometric, and inverse trigonometric functions are elementary. All functions obtained from them by a finite number of applications of the elementary operations of addition, subtraction, multiplication, division, extraction of roots, and raising to powers are elementary. Finally, we include such functions as $\sin(\sin x)$, in which the argument in a function previously classed as elementary is replaced by an elementary function.

obtained in calculus, it follows that

$$\int \exp(-x^2)\,dx = c + \sum_{n=0}^{\infty} \frac{(-1)^n x^{2n+1}}{n!(2n+1)}.$$

Thus the differential equation (1) has the general solution

$$y \exp(-x^2) = c + \sum_{n=0}^{\infty} \frac{(-1)^n x^{2n+1}}{n!(2n+1)}.$$

Since $y = 1$ when $x = 0$, $c$ may be found from

$$1 = c + 0.$$

Therefore, the particular solution desired is

$$y \exp(-x^2) = 1 + \sum_{n=0}^{\infty} \frac{(-1)^n x^{2n+1}}{n!(2n+1)}. \tag{2}$$

An alternative procedure is the introduction of a definite integral. In calculus, the error function defined by

$$\text{erf } x = \frac{2}{\sqrt{\pi}} \int_0^x \exp(-\beta^2)\,d\beta \tag{3}$$

is sometimes studied. Since, from (3),

$$\frac{d}{dx}\text{erf } x = \frac{2}{\sqrt{\pi}} \exp(-x^2),$$

we may integrate the exact equation (1) as follows:

$$y \exp(-x^2) = \tfrac{1}{2}\sqrt{\pi}\,\text{erf } x + c. \tag{4}$$

Since erf $0 = 0$, the condition that $y = 1$ when $x = 0$ yields $c = 1$. Hence, as an alternative to (2) we obtain

$$y \exp(-x^2) = 1 + \tfrac{1}{2}\sqrt{\pi}\,\text{erf } x. \tag{5}$$

Equation (5) means the same as

$$y \exp(-x^2) = 1 + \int_0^x \exp(-\beta^2)\,d\beta. \tag{6}$$

Writing a solution in the form of (6) implies that the definite integral is to be evaluated by power series, approximate integration such as Simpson's rule, mechanical quadrature, or any other available tool. If it happens, as in this case, that the definite integral is itself a tabulated function, that is a great convenience, but it is not vital. The essential thing is to reduce the solution to a computable form.

## Exercises

In each exercise, express the solution with the aid of power series or definite integrals.

**1.** $y' = y[1 - \exp(-x^2)]$.                ANS.   $\ln|cy| = x - \frac{1}{2}\sqrt{\pi}\,\text{erf}\,x$.

**2.** $(xy - \sin x)\,dx + x^2\,dy = 0$.                ANS.   $xy = c + \displaystyle\int_0^x \frac{\sin w}{w}\,dw$,

$$\text{or } y = cx^{-1} + \sum_{n=0}^{\infty} (-1)^n \frac{x^{2n}}{(2n+1)(2n+1)!}.$$

**3.** $y' = 1 - 4x^3 y$.                ANS.   $y = \exp(-x^4)\left[c + \displaystyle\int_0^x \exp(\beta^4)\,d\beta\right]$.

**4.** $(y\cos^2 x - x\sin x)\,dx + \sin x\cos x\,dy = 0$.

$$\text{ANS.}\quad y\sin x = c + \int_0^x \beta\tan\beta\,d\beta.$$

**5.** $(1 + xy)\,dx - x\,dy = 0$; when $x = 1$, $y = 0$.                ANS.   $y = e^x \displaystyle\int_1^x \frac{e^{-\beta}}{\beta}\,d\beta$.

**6.** $\left[x\exp\left(\dfrac{y^2}{x^2}\right) - y\right]dx + x\,dy = 0$; when $x = 1$, $y = 2$.

$$\text{ANS.}\quad \ln x = \int_{y/x}^{2} \exp(-\beta^2)\,d\beta.$$

**7.** $x(2y + x)\,dx - dy = 0$; when $x = 0$, $y = 1$.
                ANS.   $2y = 2\exp(x^2) - x + \frac{1}{2}\sqrt{\pi}\exp(x^2)\,\text{erf}\,x$.

## Miscellaneous Exercises

In each exercise, find a set of solutions unless the statement of the exercise stipulates otherwise.

**1.** $(y^2 - 3y - x)\,dx + (2y - 3)\,dy = 0$.                ANS.   $y^2 - 3y - x + 1 = c\,e^{-x}$.
**2.** $(y^3 + y + 1)\,dx + x(x - 3y^2 - 1)\,dy = 0$.                ANS.   $y^3 - xy + y + 1 = cx$.
**3.** $(x + 3y - 5)\,dx - (x - y - 1)\,dy = 0$.

                ANS.   $2(y - 1) = (x + y - 3)\ln|c(x + y - 3)|$.

**4.** $(x^5 - y^2)\,dx + 2xy\,dy = 0$.                ANS.   $x^5 + 4y^2 = cx$.
**5.** $(2x + y - 4)\,dx + (x - 3y + 12)\,dy = 0$.

                ANS.   $2x^2 + 2xy - 3y^2 - 8x + 24y = c$.

**6.** Solve in two ways the equation $y' = ax + by + c$; with $b \neq 0$.

                ANS.   $b^2 y = c_1 e^{bx} - abx - a - cb$.

**7.** $y^3 \sec^2 x\,dx - (1 - 2y^2\tan x)\,dy = 0$.                ANS.   $y^2\tan x = \ln|cy|$.
**8.** $x^3 y\,dx + (3x^4 - y^3)\,dy = 0$.                ANS.   $15x^4 y^{12} = 4y^{15} + c$.
**9.** $(a_1 x + ky + c_1)\,dx + (kx + b_2 y + c_2)\,dy = 0$.

                ANS.   $a_1 x^2 + 2kxy + b_2 y^2 + 2c_1 x + 2c_2 y = c$.

**10.** $(x - 4y + 7)\,dx + (x + 2y + 1)\,dy = 0$.

                ANS.   $(x - y + 4)^3 = c(x - 2y + 5)^2$.

**11.** $xy\,dx + (y^4 - 3x^2)\,dy = 0$.       ANS.   $x^2 = y^4(1 + cy^2)$.

**12.** $(x + 2y - 1)\,dx - (2x + y - 5)\,dy = 0$.

**13.** $(5x + 3\,e^y)\,dx + 2x\,e^y\,dy = 0$.       ANS.   $x^3(x + e^y)^2 = c$.

**14.** $(3x + y - 2)\,dx + (3x + y + 4)\,dy = 0$.    ANS.   $x + y + c = 3\ln|3x + y + 7|$.

**15.** $(x - 3y + 4)\,dx + 2(x - y - 2)\,dy = 0$.    ANS.   $(x + y - 8)^4 = c(x - 2y + 1)$.

**16.** $(x - 2)\,dx + 4(x + y - 1)\,dy = 0$.

      ANS.   $2(y + 1) = -(x + 2y)\ln|c(x + 2y)|$.

**17.** $y\,dx = x(1 + xy^4)\,dy$.       ANS.   $y(5 + xy^4) = cx$.

**18.** $2x\,dv + v(2 + v^2x)\,dx = 0$; when $x = 1, v = \frac{1}{2}$.       ANS.   $xv^2(5x - 1) = 1$.

**19.** $2(x - y)\,dx + (3x - y - 1)\,dy = 0$.    ANS.   $(x + y - 1)^4 = c(4x - 2y - 1)$.

**20.** $(2x - 5y + 12)\,dx + (7x - 4y + 15)\,dy = 0$.

      ANS.   $(x + 2y - 3)^3 = c(x - y + 3)$.

**21.** $y\,dx + x(x^2y - 1)\,dy = 0$.       ANS.   $y^2(2x^2y - 3) = cx^2$.

**22.** $dy/dx = \tan y \cot x - \sec y \cos x$.    ANS.   $\sin y + \sin x \ln|c \sin x| = 0$.

**23.** $[1 + (x + y)^2]\,dx + [1 + x(x + y)]\,dy = 0$.

      ANS.   $y^2 = (x + y)^2 + 2\ln|x + y| + c$.

**24.** $(x - 2y - 1)\,dx - (x - 3)\,dy = 0$. Solve by two methods.

      ANS.   $(x - 3)^2(x - 3y) = c$.

**25.** $(2x - 3y + 1)\,dx - (3x + 2y - 4)\,dy = 0$. Solve by two methods.

      ANS.   $x^2 + x - 3xy - y^2 + 4y = c$.

**26.** $(4x + 3y - 7)\,dx + (4x + 3y + 1)\,dy = 0$.

      ANS.   $x + y + c = 8\ln|4x + 3y + 25|$.

**27.** Find a change of variables that will reduce any equation of the form

$$xy' = yf(xy)$$

to an equation in which the variables are separable.

**28.** $(x + 4y + 3)\,dx - (2x - y - 3)\,dy = 0$.

      ANS.   $3(y + 1) = (x + y)\ln|c(x + y)|$.

**29.** $(3x - 3y - 2)\,dx - (x - y + 1)\,dy = 0$.

      ANS.   $2(y - 3x + c) = 5\ln|2x - 2y - 3|$.

**30.** $(x - 6y + 2)\,dx + 2(x + 2y + 2)\,dy = 0$.

      ANS.   $4y = -(x - 2y + 2)\ln|c(x - 2y + 2)|$.

**31.** $(x^4 - 4x^2y^2 - y^4)\,dx + 4x^3y\,dy = 0$; when $x = 1, y = 2$.

      ANS.   $y^2(5 - 3x) = x^2(5 + 3x)$.

**32.** $(x - y - 1)\,dx - 2(y - 2)\,dy = 0$.    ANS.   $(x + y - 5)^2(x - 2y + 1) = c$.

**33.** $(x - 3y + 3)\,dx + (3x + y + 9)\,dy = 0$.

      ANS.   $\ln[(x + 3)^2 + y^2] = c + 6\arctan[(x + 3)/y]$.

**34.** $(2x + 4y - 1)\,dx - (x + 2y - 3)\,dy = 0$.    ANS.   $\ln|x + 2y - 1| = y - 2x + c$.

**35.** $4y\,dx + 3(2x - 1)(dy + y^4\,dx) = 0$; when $x = 1, y = 1$.

      ANS.   $y^3(2x - 1)(5x - 4) = 1$.

**36.** $y(x - 1)\,dx - (x^2 - 2x - 2y)\,dy = 0$.    ANS.   $x^2 - 2x - 4y = cy^2$.

**37.** $(6xy - 3y^2 + 2y)\,dx + 2(x - y)\,dy = 0$.    ANS.   $y(2x - y) = c\,e^{-3x}$.

**38.** $y' = x - y + 2$. Solve by two methods.    ANS.   $\ln|x - y + 1| = c - x$.

**39.** $(x + y - 2)\,dx - (x - 4y - 2)\,dy = 0$.

      ANS.   $\ln[(x - 2)^2 + 4y^2] + \arctan[(x - 2)/2y] = c$.

**40.** $4\,dx + (x - y + 2)^2\,dy = 0$.    ANS.   $y + 2\arctan[(x - y + 2)/2] = c$.

5

# Linear Differential Equations

## 24. The general linear equation

The general linear differential equation of order $n$ is an equation that can be written

$$b_0(x)\frac{d^n y}{dx^n} + b_1(x)\frac{d^{n-1} y}{dx^{n-1}} + \cdots + b_{n-1}(x)\frac{dy}{dx} + b_n(x)y = R(x). \tag{1}$$

The functions $R(x)$ and $b_i(x)$; $i = 0, 1, \ldots, n$, are to be independent of the variable $y$. If $R(x)$ is identically zero, equation (1) is said to be linear and homogeneous*; if $R(x)$ is not identically zero, equation (1) is called linear and nonhomogeneous. In this chapter we shall obtain some fundamental and important properties of linear equations.

First we prove that if $y_1$ and $y_2$ are solutions of the homogeneous equation

$$b_0(x)y^{(n)} + b_1(x)y^{(n-1)} + \cdots + b_{n-1}(x)y' + b_n(x)y = 0, \tag{2}$$

* It is perhaps unfortunate that the word homogeneous as it is used here has a very different meaning from that in Sections 8 and 9.

and if $c_1$ and $c_2$ are constants, then

$$y = c_1 y_1 + c_2 y_2$$

is a solution of equation (2).

The statement that $y_1$ and $y_2$ are solutions of (2) means that

$$b_0 y_1^{(n)} + b_1 y_1^{(n-1)} + \cdots + b_{n-1} y_1' + b_n y_1 = 0 \qquad (3)$$

and

$$b_0 y_2^{(n)} + b_1 y_2^{(n-1)} + \cdots + b_{n-1} y_2' + b_n y_2 = 0. \qquad (4)$$

Now let us multiply each member of (3) by $c_1$, each member of (4) by $c_2$, and add the results. We get

$$b_0(c_1 y_1^{(n)} + c_2 y_2^{(n)}) + b_1(c_1 y_1^{(n-1)} + c_2 y_2^{(n-1)}) + \cdots$$
$$+ b_{n-1}(c_1 y_1' + c_2 y_2') + b_n(c_1 y_1 + c_2 y_2) = 0. \qquad (5)$$

Since $c_1 y_1' + c_2 y_2' = (c_1 y_1 + c_2 y_2)'$, and so on, equation (5) is neither more nor less than the statement that $c_1 y_1 + c_2 y_2$ is a solution of equation (2). The proof is completed. The special case $c_2 = 0$ is worth noting; that is, for a homogeneous linear equation any constant times a solution is also a solution.

In a similar manner, or by iteration of the above result, it can be seen that if $y_i$, with $i = 1, 2, \ldots, k$, are solutions of equation (2), and if $c_i$, with $i = 1, 2, \ldots, k$, are constants, then

$$y = c_1 y_1 + c_2 y_2 + \cdots + c_k y_k \qquad (6)$$

is a solution of equation (2).

The expression in equation (6) is called a linear combination of the functions $y_1, y_2, \ldots, y_k$. The theorem just proved can thus be stated as follows:

THEOREM 4: *Any linear combination of solutions of a linear homogeneous differential equation is also a solution.*

## 25. Linear independence

Given the functions $f_1, \ldots, f_n$, if constants $c_1, c_2, \ldots, c_n$, not all zero, exist such that

$$c_1 f_1(x) + c_2 f_2(x) + \cdots + c_n f_n(x) = 0 \qquad (1)$$

identically in some interval $a \leq x \leq b$, then the functions $f_1, f_2, \ldots, f_n$ are said to be *linearly dependent*. If no such relation exists, the functions are said to be *linearly independent*. That is, the functions $f_1, f_2, \ldots, f_n$ are linearly independent when equation (1) implies that $c_1 = c_2 = \cdots = c_n = 0$.

It should be clear that, if the functions of a set are linearly dependent, at least one of them is a linear combination of the others; if they are linearly independent, then none of them is a linear combination of the others.

## 26. An existence and uniqueness theorem

In Section 12 we stated an existence theorem for an initial value problem involving a first-order linear differential equation. The generalization of this theorem to $n$th order linear equations can be stated as follows:

THEOREM 5: *Let $P_1, P_2, \ldots, P_n$, and $R$ be functions that are continuous on an interval $a < x < b$. Suppose that $x_0$ is a real number in the interval and that $y_0, y_1, \ldots, y_{n-1}$ are $n$ arbitrary real numbers. Then a unique function $y = y(x)$ exists, defined on the interval $a < x < b$, which is a solution of the equation*

$$y^{(n)} + P_1 y^{(n-1)} + \cdots + P_n y = R \tag{1}$$

*on the interval and which satisfies the initial conditions*

$$y(x_0) = y_0, \qquad y'(x_0) = y_1, \ldots, y^{(n-1)}(x_0) = y_{n-1}.$$

EXAMPLE: Find the unique solution of the initial value problem

$$y'' + y = 0, y(0) = 0, y'(0) = 1. \tag{2}$$

We observe that $\sin x$ and $\cos x$ are solutions of (2), so that for arbitrary $c_1$ and $c_2$

$$y = c_1 \sin x + c_2 \cos x$$

is also a solution by the theorem of Section 24.

Because of the initial conditions in (2), we are led to choose $c_1$ and $c_2$ so that $c_1 \sin 0 + c_2 \cos 0 = 0$ and $c_1 \cos 0 - c_2 \sin 0 = 1$. This can be done in only one way, namely by choosing $c_1 = 1$ and $c_2 = 0$. We find that the function $\sin x$ is a solution of the initial value problem (2). Moreover, since the problem satisfies the conditions required in Theorem 5, no matter what interval we choose, $\sin x$ is the only solution to the problem given in (2).

## 27. The Wronskian

With the definitions of Section 25 in mind, we shall now obtain a sufficient condition that $n$ functions be linearly independent over an interval $a \leq x \leq b$.

Let us assume that each of the functions $f_1, f_2, \ldots, f_n$ is differentiable at least $(n - 1)$ times in the interval $a \leq x \leq b$. Then from the equation

$$c_1 f_1 + c_2 f_2 + \cdots + c_n f_n = 0, \tag{1}$$

it follows by successive differentiation that

$$c_1 f'_1 + c_2 f'_2 + \cdots + c_n f'_n = 0,$$
$$c_1 f''_1 + c_2 f''_2 + \cdots + c_n f''_n = 0,$$
$$\vdots$$
$$c_1 f_1^{(n-1)} + c_2 f_2^{(n-1)} + \cdots + c_n f_n^{(n-1)} = 0.$$

Considered as a system of equations in $c_1, c_2, \ldots, c_n$, the $n$ linear equations directly above will have no solution except the one with each of the $c$'s equal to zero, if the determinant of the system does not vanish. That is, if

$$\begin{vmatrix} f_1 & f_2 & \cdots & f_n \\ f'_1 & f'_2 & \cdots & f'_n \\ f''_1 & f''_2 & \cdots & f''_n \\ & & \vdots & \\ f_1^{(n-1)} & f_2^{(n-1)} & \cdots & f_n^{(n-1)} \end{vmatrix} \neq 0, \tag{2}$$

then the functions $f_1, f_2, \ldots, f_n$ are linearly independent. The determinant in (2) is called the *Wronskian** of the $n$ functions involved. We have shown that the nonvanishing of the Wronskian is a sufficient condition that the functions be linearly independent.

The nonvanishing of the Wronskian on an interval is not a necessary condition for linear independence. The Wronskian may vanish even when the functions are linearly independent, as exhibited in exercise 9 below.

If the $n$ functions involved are solutions of a homogeneous linear differential equation, the situation is simplified as is shown by Theorem 6. A proof of this theorem in the case $n = 2$, is suggested in exercises 10 through 13 below.

THEOREM 6: *If, on the interval $a < x < b$, $b_0(x) \neq 0$, $b_0, b_1, \ldots, b_n$ are continuous, and $y_1, y_2, \ldots, y_n$ are solutions of the equation*

$$b_0 y^{(n)} + b_1 y^{(n-1)} + \cdots + b_{n-1} y' + b_n y = 0, \tag{3}$$

*then a necessary and sufficient condition that $y_1, \ldots, y_n$ be linearly independent is the nonvanishing of the Wronskian of $y_1, \ldots, y_n$ on the interval $a < x < b$.*

---

* The Wronskian determinant is named after the Polish mathematician Hoëné Wronski (1778–1853).

The functions $\cos \omega t$, $\sin \omega t$, $\sin (\omega t + \alpha)$, in which $t$ is the variable and $\omega$ and $\alpha$ are constants, are linearly dependent because constants $c_1, c_2, c_3$ exist such that

$$c_1 \cos \omega t + c_2 \sin \omega t + c_3 \sin (\omega t + \alpha) = 0$$

for all $t$. Indeed, one set of such constants is $c_1 = \sin \alpha, c_2 = \cos \alpha, c_3 = -1$.

One of the best-known sets of $n$ linearly independent functions of $x$ is the set $1, x, x^2, \ldots, x^{n-1}$. The linear independence of the powers of $x$ follows at once from the fact that, if $c_1, c_2, \ldots, c_n$ are not all zero, the equation

$$c_1 + c_2 x + \cdots + c_n x^{n-1} = 0$$

can have, at most, $(n - 1)$ distinct roots and so cannot vanish identically in any interval. See also exercise 1 below.

### Exercises

**1.** Obtain the Wronskian of the functions

$$1, x, x^2, \ldots, x^{n-1} \text{ for } n > 1.$$

ANS.    $W = 0!1!2! \cdots (n - 1)!$.

**2.** Show that the functions $e^x$, $e^{2x}$, $e^{3x}$ are linearly independent.

ANS.    $W = 2 e^{6x} \neq 0$.

**3.** Show that the functions $e^x$, $\cos x$, $\sin x$ are linearly independent.

ANS.    $W = 2 e^x \neq 0$.

**4.** By determining constants $c_1$, $c_2$, $c_3$, $c_4$, which are not all zero and are such that $c_1 f_1 + c_2 f_2 + c_3 f_3 + c_4 f_4 = 0$ identically, show that the functions

$$f_1 = x, \quad f_2 = e^x, \quad f_3 = x e^x, \quad f_4 = (2 - 3x) e^x$$

are linearly dependent.

ANS.    One such set of $c$'s is: $c_1 = 0, c_2 = -2, c_3 = 3, c_4 = 1$.

**5.** Show that $\cos (\omega t - \beta)$, $\cos \omega t$, $\sin \omega t$ are linearly dependent functions of $t$.

**6.** Show that $1, \sin x, \cos x$ are linearly independent.

**7.** Show that $1, \sin^2 x, \cos^2 x$ are linearly dependent.

**8.** Show that two nonvanishing differentiable functions of $x$ are linearly dependent if, and only if, their Wronskian vanishes identically. This statement is not true for more than two functions.

**9.** Let $f_1(x) = 1 + x^3$ for $x \leq 0, f_1(x) = 1$ for $x \geq 0$;
$f_2(x) = 1$ for $x \leq 0, f_2(x) = 1 + x^3$ for $x \geq 0$;
$f_3(x) = 3 + x^3$ for all $x$.
Show that (a) $f, f', f''$ are continuous for all $x$ for each of $f_1, f_2, f_3$;
(b) the Wronskian of $f_1, f_2, f_3$ is zero for all $x$;
(c) $f_1, f_2, f_3$ are linearly independent over the interval $-1 \leq x \leq 1$.
In part (c) you must show that if $c_1 f_1(x) + c_2 f_2(x) + c_3 f_3(x) = 0$ for all $x$ in $-1 \leq x \leq 1$ then $c_1 = c_2 = c_3 = 0$. Use $x = -1, 0, 1$ successively to obtain three equations to solve for $c_1, c_2$, and $c_3$.

**10.** Given any interval $a < x < b$ with $x_0$ a fixed number in the interval and suppose $y$ is a solution of the homogeneous equation

$$y'' + Py' + Qy = 0. \tag{A}$$

Further, suppose that $y(x_0) = y'(x_0) = 0$. Use the existence and uniqueness theorem of Section 26 to prove that $y(x) = 0$ for every $x$ in the interval $a < x < b$.

**11.** Suppose that $y_1$ and $y_2$ are solutions of equation (A) of exercise 10 and suppose the Wronskian of $y_1$ and $y_2$ is identically zero on $a < x < b$. Show that for $x_0$ in the interval $a < x < b$, there must exist constants $\bar{c}_1$ and $\bar{c}_2$ not both zero such that

$$\bar{c}_1 y_1(x_0) + \bar{c}_2 y_2(x_0) = 0,$$

and

$$\bar{c}_1 y_1'(x_0) + \bar{c}_2 y_2'(x_0) = 0.$$

**12.** Consider the function defined by

$$y(x) = \bar{c}_1 y_1(x) + \bar{c}_2 y_2(x),$$

where $\bar{c}_1$ and $\bar{c}_2$ are the constants determined in exercise 11. Show that this function is a solution of equation (A) above and that it follows that $y(x) \equiv 0$ on $a < x < b$. (Use the results of exercise 10.)

**13.** Combine the results of exercises 10 to 12 to obtain a proof of the necessity condition of Theorem 6 in the case when $n = 2$. Also note that the sufficiency condition was established in the text of this section.

## 28. General solution of a homogeneous equation

One of the basic results of the subject of linear differential equations is contained in Theorem 7.

**THEOREM 7:** *Let* $\{y_1, y_2, \ldots, y_n\}$ *be a linearly independent set of solutions of the homogeneous linear equation*

$$b_0(x)y^{(n)} + b_1(x)y^{(n-1)} + \cdots + b_{n-1}(x)y' + b_n(x)y = 0, \tag{1}$$

*for $x$ on the interval $a < x < b$. Suppose further that $b_0(x), \ldots, b_n(x)$ are continuous functions and that $b_0(x) \neq 0$ on $a < x < b$.*

*If $\phi$ is any solution of equation (1), valid on $a < x < b$, there exist constants $\bar{c}_1, \bar{c}_2, \ldots, \bar{c}_n$ such that*

$$\phi = \bar{c}_1 y_1 + \bar{c}_2 y_2 + \cdots + \bar{c}_n y_n. \tag{2}$$

It is because of this theorem that we define the *general solution* of equation (1) to be

$$y = c_1 y_1 + c_2 y_2 + \cdots + c_n y_n, \tag{3}$$

where $c_1, c_2, \ldots, c_n$ are arbitrary constants.

In a sense each particular solution of the linear equation (1) is a special case (some choice of the $c$'s) of the general solution (3). The basic ideas needed for a proof of this important theorem are exhibited here for an equation of order two. No additional complications occur for equations of higher order.

PROOF.    Consider the equation

$$b_0(x)y'' + b_1(x)y' + b_2(x)y = 0. \tag{4}$$

Let $y_1$ and $y_2$ be linearly independent solutions of equation (4) on the interval $a < x < b$ and choose $x_0$ between $a$ and $b$. By Theorem 6, page 87, the Wronskian of $y_1$ and $y_2$ is not zero at $x_0$. That is,

$$W = \begin{vmatrix} y_1(x_0) & y_2(x_0) \\ y_1'(x_0) & y_2'(x_0) \end{vmatrix} \neq 0. \tag{5}$$

It follows that the system of equations

$$c_1 y_1(x_0) + c_2 y_2(x_0) = \phi(x_0),$$

$$c_1 y_1'(x_0) + c_2 y_2'(x_0) = \phi'(x_0),$$

has a unique solution $c_1 = \bar{c}_1, c_2 = \bar{c}_2$. That is,

$$\bar{c}_1 y_1(x_0) + \bar{c}_2 y_2(x_0) = \phi(x_0),$$

$$\bar{c}_1 y_1'(x_0) + \bar{c}_2 y_2'(x_0) = \phi'(x_0).$$

Now consider the function

$$f = \bar{c}_1 y_1 + \bar{c}_2 y_2. \tag{6}$$

Because $f$ is a linear combination of two solutions of equation (4) on the interval $a < x < b$, it is also a solution on that interval. Moreover,

$$f(x_0) = \bar{c}_1 y_1(x_0) + \bar{c}_2 y_2(x_0),$$

$$f'(x_0) = \bar{c}_1 y_1'(x_0) + \bar{c}_2 y_2'(x_0),$$

so that $f(x_0) = \phi(x_0)$ and $f'(x_0) = \phi'(x_0)$. It follows from the uniqueness theorem of Section 26 that $f$ and $\phi$ are the same solution. That is,

$$\phi = \bar{c}_1 y_1 + \bar{c}_2 y_2,$$

which completes the proof of the theorem.

It is necessary to keep in mind that the above discussion used the fact that $b_0(x) \neq 0$ in the interval $a < x < b$. It is easy to see that the linear equation

$$xy' - 2y = 0$$

has the general solution $y = cx^2$ and also such particular solutions as

$$y_1 = x^2, \qquad 0 \leqq x,$$
$$= -4x^2, \qquad x < 0.$$

The solution $y_1$ is not a special case of the general solution. But in any interval throughout which $b_0(x) = x \neq 0$, this particular solution is a special case of the general solution. It was, of course, made up by piecing together at $x = 0$ two parts, each drawn from the general solution.

## 29. General solution of a nonhomogeneous equation

Let $y_p$ be any particular solution (not necessarily involving any arbitrary constants) of the equation

$$b_0 y^{(n)} + b_1 y^{(n-1)} + \cdots + b_{n-1} y' + b_n y = R(x) \tag{1}$$

and let $y_c$ be a solution of the corresponding homogeneous equation

$$b_0 y^{(n)} + b_1 y^{(n-1)} + \cdots + b_{n-1} y' + b_n y = 0. \tag{2}$$

Then

$$y = y_c + y_p \tag{3}$$

is a solution of equation (1). For, using the $y$ of equation (3) we see that

$$b_0 y^{(n)} + \cdots + b_n y = (b_0 y_c^{(n)} + \cdots + b_n y_c)$$
$$+ (b_0 y_p^{(n)} + \cdots + b_n y_p) = 0 + R(x) = R(x).$$

If $y_1, y_2, \ldots, y_n$ are linearly independent solutions of equation (2), then

$$y_c = c_1 y_1 + c_2 y_2 + \cdots + c_n y_n, \tag{4}$$

in which the $c$'s are arbitrary constants, is the general solution of equation (2). The right member of equation (4) is called the *complementary function* for equation (1).

The general solution of the nonhomogeneous equation (1) is the sum of the complementary function and any particular solution. To justify this usage of the term "general solution" we must show that if $f$ is any solution of equation (1) then $f \equiv y_c + y_p$ for some particular choice of the $c_1, \ldots, c_n$. We note that since $f$ and $y_p$ are both solutions of the nonhomogeneous equation (1), $f - y_p$ is a solution of the homogeneous equation (2). Hence by Theorem 7 of Section 28

$$f - y_p \equiv c_1 y_1 + c_2 y_2 + \cdots + c_n y_n$$

for some particular choice of the $c_1, \ldots, c_n$. This establishes what we wished to show.

EXAMPLE (a):   Find the general solution of

$$y'' = 4. \tag{5}$$

We first observe that the functions 1 and $x$ are linearly independent on any interval and are solutions of the homogeneous equation $y'' = 0$. Hence, the complementary function for equation (5) is

$$y_c = c_1 + c_2 x.$$

On the other hand, the function $2x^2$ is a particular solution of equation (5). Hence, the general solution of equation (5) is

$$y = c_1 + c_2 x + 2x^2.$$

EXAMPLE (b):   Find the general solution of the equation

$$y'' - y = 4. \tag{6}$$

It is easily seen that $y = -4$ is a solution of equation (6). Therefore the $y_p$ in equation (3) may be taken to be $(-4)$. As we shall see later, the homogeneous equation

$$y'' - y = 0$$

has as its general solution

$$y_c = c_1 e^x + c_2 e^{-x}.$$

Thus the complementary function for equation (6) is $c_1 e^x + c_2 e^{-x}$ and a particular solution of (6) is $y_p = -4$. Hence the general solution of equation (6) is

$$y = c_1 e^x + c_2 e^{-x} - 4,$$

in which $c_1$ and $c_2$ are arbitrary constants.

## 30. Differential operators

Let $D$ denote differentiation with respect to $x$, $D^2$ differentiation twice with respect to $x$, and so on; that is, for positive integral $k$,

$$D^k y = \frac{d^k y}{dx^k}.$$

The expression

$$A = a_0 D^n + a_1 D^{n-1} + \cdots + a_{n-1} D + a_n \tag{1}$$

is called a differential operator of order $n$. It may be defined as that operator which, when applied to any function* $y$, yields the result

---

* The function $y$ is assumed to possess as many derivatives as may be encountered in whatever operations take place.

$$Ay = a_0 \frac{d^n y}{dx^n} + a_1 \frac{d^{n-1} y}{dx^{n-1}} + \cdots + a_{n-1} \frac{dy}{dx} + a_n y. \tag{2}$$

The coefficients $a_0, a_1, \ldots, a_n$ in the operator $A$ may be functions of $x$, but in this book the only operators used will be those with constant coefficients.

Two operators $A$ and $B$ are said to be equal if, and only if, the same result is produced when each acts upon the function $y$. That is, $A = B$ if, and only if, $Ay = By$ for all functions $y$ possessing the derivatives necessary for the operations involved.

The product $AB$ of two operators $A$ and $B$ is defined as that operator which produces the same result as is obtained by using the operator $B$ followed by the operator $A$. Thus $ABy = A(By)$. The product of two differential operators always exists and is a differential operator. For operators with *constant coefficients*, but not usually for those with variable coefficients, it is true that $AB = BA$.

EXAMPLE (a):  Let $A = D + 2$ and $B = 3D - 1$.
Then

$$By = (3D - 1)y = 3\frac{dy}{dx} - y$$

and

$$A(By) = (D + 2)\left(3\frac{dy}{dx} - y\right)$$

$$= 3\frac{d^2 y}{dx^2} - \frac{dy}{dx} + 6\frac{dy}{dx} - 2y$$

$$= 3\frac{d^2 y}{dx^2} + 5\frac{dy}{dx} - 2y$$

$$= (3D^2 + 5D - 2)y.$$

Hence $AB = (D + 2)(3D - 1) = 3D^2 + 5D - 2$.
Now consider $BA$. Acting upon $y$, the operator $BA$ yields

$$B(Ay) = (3D - 1)\left(\frac{dy}{dx} + 2y\right)$$

$$= 3\frac{d^2 y}{dx^2} + 6\frac{dy}{dx} - \frac{dy}{dx} - 2y$$

$$= 3\frac{d^2 y}{dx^2} + 5\frac{dy}{dx} - 2y.$$

Hence

$$BA = 3D^2 + 5D - 2 = AB.$$

EXAMPLE (b): Let $G = xD + 2$, and $H = D - 1$. Then

$$G(Hy) = (xD + 2)\left(\frac{dy}{dx} - y\right)$$

$$= x\frac{d^2y}{dx^2} - x\frac{dy}{dx} + 2\frac{dy}{dx} - 2y$$

$$= x\frac{d^2y}{dx^2} + (2 - x)\frac{dy}{dx} - 2y,$$

so

$$GH = xD^2 + (2 - x)D - 2.$$

On the other hand

$$H(Gy) = (D - 1)\left(x\frac{dy}{dx} + 2y\right)$$

$$= \frac{d}{dx}\left(x\frac{dy}{dx} + 2y\right) - \left(x\frac{dy}{dx} + 2y\right)$$

$$= x\frac{d^2y}{dx^2} + \frac{dy}{dx} + 2\frac{dy}{dx} - x\frac{dy}{dx} - 2y$$

$$= x\frac{d^2y}{dx^2} + (3 - x)\frac{dy}{dx} - 2y;$$

that is,

$$HG = xD^2 + (3 - x)D - 2.$$

It is worthy of notice that here we have two operators $G$ and $H$ (one of them with variable coefficients), whose product is dependent on the order of the factors. On this topic see also exercises 17 through 22 in the next section.

The sum of two differential operators is obtained by expressing each in the form

$$a_0D^n + a_1D^{n-1} + \cdots + a_{n-1}D + a_n$$

and adding corresponding coefficients. For instance, if

$$A = 3D^2 - D + x - 2$$

and

$$B = x^2D^2 + 4D + 7,$$

then

$$A + B = (3 + x^2)D^2 + 3D + x + 5.$$

Differential operators are linear operators; that is, if $A$ is any differential operator, $c_1$ and $c_2$ are constants, and $f_1$ and $f_2$ are any functions of $x$ each possessing the required number of derivatives, then

$$A(c_1 f_1 + c_2 f_2) = c_1 A f_1 + c_2 A f_2.$$

## 31.  The fundamental laws of operation

Let $A$, $B$, and $C$ be any differential operators as defined in Section 30. With the above definitions of addition and multiplication, it follows that differential operators satisfy the following:

(a)  The commutative law of addition:

$$A + B = B + A.$$

(b)  The associative law of addition:

$$(A + B) + C = A + (B + C).$$

(c)  The associative law of multiplication:

$$(AB)C = A(BC).$$

(d)  The distributive law of multiplication with respect to addition:

$$A(B + C) = AB + AC.$$

(e)  And if $A$ and $B$ are operators with *constant coefficients*, then they also satisfy the commutative law of multiplication:

$$AB = BA.$$

Therefore, differential operators with constant coefficients satisfy all the laws of the algebra of polynomials with respect to the operations of addition and multiplication.

If $m$ and $n$ are any two positive integers, then

$$D^m D^n = D^{m+n},$$

a useful result which follows immediately from the definitions.

Since for purposes of addition and multiplication the operators with constant coefficients behave just as algebraic polynomials behave, it is legitimate to use the tools of elementary algebra. In particular, synthetic division may be used to factor operators with constant coefficients.

### Exercises

Perform the indicated multiplications in exercises 1 through 4.

**1.** $(4D + 1)(D - 2)$.  
ANS. $4D^2 - 7D - 2$.

**2.** $(2D - 3)(2D + 3)$.  
ANS. $4D^2 - 9$.

**3.** $(D + 2)(D^2 - 2D + 5)$.  
ANS. $D^3 + D + 10$.

**4.** $(D - 2)(D + 1)^2$.  
ANS. $D^3 - 3D - 2$.

In exercises 5 through 16, factor each of the operators.

**5.** $2D^2 + 3D - 2$.  
ANS. $(D + 2)(2D - 1)$.

**6.** $2D^2 - 5D - 12$.

**7.** $D^3 - 2D^2 - 5D + 6$.  
ANS. $(D - 1)(D + 2)(D - 3)$.

**8.** $4D^3 - 4D^2 - 11D + 6$.

**9.** $D^4 - 4D^2$.  
ANS. $D^2(D - 2)(D + 2)$.

**10.** $D^3 - 3D^2 + 4$.

**11.** $D^3 - 21D + 20$.  
ANS. $(D - 1)(D - 4)(D + 5)$.

**12.** $2D^3 - D^2 - 13D - 6$.

**13.** $2D^4 + 11D^3 + 18D^2 + 4D - 8$.  
ANS. $(D + 2)^3(2D - 1)$.

**14.** $8D^4 + 36D^3 - 66D^2 + 35D - 6$.

**15.** $D^4 + D^3 - 2D^2 + 4D - 24$.  
ANS. $(D - 2)(D + 3)(D^2 + 4)$.

**16.** $D^3 - 11D - 20$.  
ANS. $(D - 4)(D^2 + 4D + 5)$.

Perform the indicated multiplications in exercises 17 through 22.

**17.** $(D - x)(D + x)$.  
ANS. $D^2 + 1 - x^2$.

**18.** $(D + x)(D - x)$.  
ANS. $D^2 - 1 - x^2$.

**19.** $D(xD - 1)$.  
ANS. $xD^2$.

**20.** $(xD - 1)D$.  
ANS. $xD^2 - D$.

**21.** $(xD + 2)(xD - 1)$.  
ANS. $x^2D^2 + 2xD - 2$.

**22.** $(xD - 1)(xD + 2)$.  
ANS. $x^2D^2 + 2xD - 2$.

## 32. Some properties of differential operators

Since for constant $m$ and positive integral $k$,

$$D^k e^{mx} = m^k e^{mx}, \tag{1}$$

it is easy to find the effect an operator has upon $e^{mx}$. Let $f(D)$ be a polynomial in $D$,

$$f(D) = a_0 D^n + a_1 D^{n-1} + \cdots + a_{n-1}D + a_n. \tag{2}$$

Then

$$f(D) e^{mx} = a_0 m^n e^{mx} + a_1 m^{n-1} e^{mx} + \cdots + a_{n-1} m e^{mx} + a_n e^{mx},$$

so

$$f(D) e^{mx} = e^{mx} f(m). \tag{3}$$

If $m$ is a root of the equation $f(m) = 0$, then in view of equation (3),

$$f(D) e^{mx} = 0.$$

Next consider the effect of the operator $D - a$ on the product of $e^{ax}$ and a function $y$. We have

$$(D - a)(e^{ax} y) = D(e^{ax} y) - a e^{ax} y$$
$$= e^{ax} D y,$$

and

$$(D - a)^2 (e^{ax} y) = (D - a)(e^{ax} D y)$$
$$= e^{ax} D^2 y.$$

Repeating the operation, we are led to

$$(D - a)^n (e^{ax} y) = e^{ax} D^n y. \tag{4}$$

Using the linearity of differential operators, we conclude that when $f(D)$ is a polynomial in $D$ with constant coefficients, then

$$e^{ax} f(D) y = f(D - a)[e^{ax} y]. \tag{5}$$

The relation (5) shows us how to shift an exponential factor from the left of a differential operator to the right of the operator. This relation has many uses, some of which we will examine in Chapter 6.

**EXAMPLE** (a):   Let $f(D) = 2D^2 + 5D - 12$. Then the equation $f(m) = 0$ is

$$2m^2 + 5m - 12 = 0,$$

or

$$(m + 4)(2m - 3) = 0,$$

of which the roots are $m_1 = -4$ and $m_2 = \frac{3}{2}$.

With the aid of equation (3) above it can be seen that

$$(2D^2 + 5D - 12) e^{-4x} = 0$$

and that

$$(2D^2 + 5D - 12) \exp\left(\tfrac{3}{2} x\right) = 0.$$

In other words, $y_1 = e^{-4x}$ and $y_2 = \exp(\tfrac{3}{2}x)$ are solutions of

$$(2D^2 + 5D - 12)y = 0.$$

EXAMPLE (b):   Show that

$$(D - m)^n(x^k\, e^{mx}) = 0 \qquad \text{for } k = 0, 1, \ldots, (n - 1). \tag{6}$$

In equation (5) we let $f(D) = (D - m)^n$ and $y = x^k$. Then using the exponential shift we obtain

$$(D - m)^n(x^k\, e^{mx}) = e^{mx}D^n x^k.$$

But $D^n x^k = 0$ for $k = 0, 1, 2, \ldots, n - 1$, which gives us equation (6) directly.

The results obtained in equations (3), (5), and (6) are of fundamental importance to the solving of linear differential equations with constant coefficients which we consider in Chapter 6.

EXAMPLE (c):   As an illustration of the use of the exponential shift we solve the differential equation

$$(D + 3)^4 y = 0. \tag{7}$$

First we multiply equation (7) by $e^{3x}$ to obtain

$$e^{3x}(D + 3)^4 y = 0.$$

Applying the exponential shift as in equation (5) leads to

$$D^4(e^{3x}y) = 0.$$

Integrating four times gives us

$$e^{3x}y = c_1 + c_2 x + c_3 x^2 + c_4 x^3,$$

and finally,

$$y = (c_1 + c_2 x + c_3 x^2 + c_4 x^3)\, e^{-3x}. \tag{8}$$

Note that each of the four functions $e^{-3x}$, $x\, e^{-3x}$, $x^2\, e^{-3x}$, and $x^3\, e^{-3x}$ is a solution of equation (7). This of course is assured by the theorem of equation (6) of Example (b).

If we now show that the four functions are linearly independent, equation (8) gives the general solution of equation (7). See exercise 5 below.

### Exercises

In exercises 1 through 4, use the exponential shift as in Example (c) above to find the general solution.

**1.** $(D - 2)^3 y = 0.$        ANS.  $y = (c_1 + c_2 x + c_3 x^2)\, e^{2x}.$
**2.** $(D + 1)^2 y = 0.$

**3.** $(2D - 1)^2 y = 0.$   ANS.   $y = (c_1 + c_2 x) \exp(\tfrac{1}{2}x).$

**4.** $(D + 7)^6 y = 0.$

**5.** To show that the four functions in Example (c) above are linearly independent on any interval, assume that they are linearly dependent and show that this leads to a contradiction of the results obtained in exercise 1 of Section 27.

**6.** Prove that the set of functions

$$e^{ax}, x\, e^{ax}, x^2\, e^{ax}, \dots, x^{n-1}\, e^{ax}$$

is a linearly independent set on any interval. See exercise 5.

# Linear Equations with Constant Coefficients

## 33. Introduction

Several methods for solving differential equations with constant coefficients are presented in this book. A classical technique is treated in this and the next chapter. Chapters 11 and 12 contain a development of the Laplace transform and its use in solving linear differential equations. Chapter 13 studies matrix techniques for solving linear equations with constant coefficients. Each method has its advantages and its disadvantages. Each is theoretically sufficient: all are necessary for maximum efficiency.

## 34. The auxiliary equation; distinct roots

Any linear homogeneous differential equation with constant coefficients,

$$a_0\frac{d^n y}{dx^n} + a_1\frac{d^{n-1}y}{dx^{n-1}} + \cdots + a_{n-1}\frac{dy}{dx} + a_n y = 0, \tag{1}$$

may be written in the form

$$f(D)y = 0, \tag{2}$$

where $f(D)$ is a linear differential operator. As we saw in the preceding chapter, if $m$ is any root of the algebraic equation $f(m) = 0$, then

$$f(D)\, e^{mx} = 0,$$

which means simply that $y = e^{mx}$ is a solution of equation (2). The equation

$$f(m) = 0 \tag{3}$$

is called the *auxiliary equation* associated with (1) or (2).

The auxiliary equation for (1) is of degree $n$. Let its roots be $m_1, m_2, \ldots, m_n$. If these roots are all real and distinct, then the $n$ solutions

$$y_1 = \exp{(m_1 x)}, y_2 = \exp{(m_2 x)}, \ldots, y_n = \exp{(m_n x)}$$

are linearly independent and the general solution of (1) can be written at once. It is

$$y = c_1 \exp{(m_1 x)} + c_2 \exp{(m_2 x)} + \cdots + c_n \exp{(m_n x)},$$

in which $c_1, c_2, \ldots, c_n$ are arbitrary constants.

Repeated roots of the auxiliary equation will be treated in the next section. Imaginary roots will be avoided until Section 37, where the corresponding solutions will be put into a desirable form.

EXAMPLE (a):   Solve the equation

$$\frac{d^3 y}{dx^3} - 4\frac{d^2 y}{dx^2} + \frac{dy}{dx} + 6y = 0.$$

First write the auxiliary equation

$$m^3 - 4m^2 + m + 6 = 0,$$

whose roots $m = -1, 2, 3$ may be obtained by synthetic division. Then the general solution is seen to be

$$y = c_1 e^{-x} + c_2 e^{2x} + c_3 e^{3x}.$$

EXAMPLE (b):   Solve the equation

$$(3D^3 + 5D^2 - 2D)y = 0.$$

The auxiliary equation is

$$3m^3 + 5m^2 - 2m = 0$$

and its roots are $m = 0, -2, \frac{1}{3}$. Using the fact that $e^{0x} = 1$, the desired solution may be written

$$y = c_1 + c_2 e^{-2x} + c_3 \exp(\tfrac{1}{3}x).$$

EXAMPLE (c):   Solve the equation

$$\frac{d^2 x}{dt^2} - 4x = 0$$

with the conditions that when $t = 0$, $x = 0$ and $dx/dt = 3$.
    The auxiliary equation is

$$m^2 - 4 = 0,$$

with roots $m = 2, -2$. Hence the general solution of the differential equation
is

$$x = c_1 e^{2t} + c_2 e^{-2t}.$$

It remains to enforce the conditions at $t = 0$. Now

$$\frac{dx}{dt} = 2c_1 e^{2t} - 2c_2 e^{-2t}.$$

Thus the condition that $x = 0$ when $t = 0$ requires that

$$0 = c_1 + c_2,$$

and the condition that $dx/dt = 3$ when $t = 0$ requires that

$$3 = 2c_1 - 2c_2.$$

From the simultaneous equations for $c_1$ and $c_2$ we conclude that $c_1 = \tfrac{3}{4}$ and
$c_2 = -\tfrac{3}{4}$. Therefore

$$x = \tfrac{3}{4}(e^{2t} - e^{-2t}),$$

which can also be put in the form

$$x = \tfrac{3}{2} \sinh(2t).$$

## Exercises

In exercises 1 through 22, find the general solution. When the operator $D$ is used, it
is implied that the independent variable is $x$.

1. $(D^2 - D - 2)y = 0$.                              ANS.  $y = c_1 e^{-x} + c_2 e^{2x}$.
2. $(D^2 + 3D)y = 0$.                                 ANS.  $y = c_1 + c_2 e^{-3x}$.
3. $(D^2 - D - 6)y = 0$.                              ANS.  $y = c_1 e^{-2x} + c_2 e^{3x}$.
4. $(D^2 + 5D + 6)y = 0$.                             ANS.  $y = c_1 e^{-2x} + c_2 e^{-3x}$.
5. $(D^3 + 2D^2 - 15D)y = 0$.                         ANS.  $y = c_1 + c_2 e^{3x} + c_3 e^{-5x}$.
6. $(D^3 + 2D^2 - 8D)y = 0$.                          ANS.  $y = c_1 + c_2 e^{2x} + c_3 e^{-4x}$.

7. $(D^3 - D^2 - 4D + 4)y = 0.$    ANS.  $y = c_1 e^{-2x} + c_2 e^x + c_3 e^{2x}.$
8. $(D^3 - 3D^2 - D + 3)y = 0.$    ANS.  $y = c_1 e^{3x} + c_2 e^x + c_3 e^{-x}.$
9. $(4D^3 - 13D + 6)y = 0.$    ANS.  $y = c_1 e^{x/2} + c_2 e^{3x/2} + c_3 e^{-2x}.$
10. $(4D^3 - 49D - 60)y = 0.$    ANS.  $y = c_1 e^{4x} + c_2 e^{-5x/2} + c_3 e^{-3x/2}.$

11. $\dfrac{d^3x}{dt^3} - 2\dfrac{d^2x}{dt^2} - 3\dfrac{dx}{dt} = 0.$

ANS.  $x = c_1 + c_2 e^{-t} + c_3 e^{3t}.$

12. $\dfrac{d^3x}{dt^3} - 7\dfrac{dx}{dt} + 6x = 0.$

ANS.  $x = c_1 e^t + c_2 e^{2t} + c_3 e^{-3t}.$

13. $(10D^3 + D^2 - 7D + 2)y = 0.$    ANS.  $y = c_1 e^{-x} + c_2 e^{x/2} + c_3 e^{2x/5}.$
14. $(4D^3 - 13D - 6)y = 0.$    ANS.  $y = c_1 e^{2x} + c_2 e^{-3x/2} + c_3 e^{-x/2}.$
15. $(D^3 - 5D - 2)y = 0.$    ANS.  $y = c_1 e^{-2x} + c_2 e^{(1+\sqrt{2})x} + c_3 e^{(1-\sqrt{2})x}.$
16. $(D^3 - 3D^2 - 3D + 1)y = 0.$    ANS.  $y = c_1 e^{-x} + c_2 e^{(2+\sqrt{3})x} + c_3 e^{(2-\sqrt{3})x}.$
17. $(4D^4 - 15D^2 + 5D + 6)y = 0.$    ANS.  $y = c_1 e^{-2x} + c_2 e^{-x/2} + c_3 e^{3x/2} + c_4 e^x.$
18. $(D^4 - 2D^3 - 13D^2 + 38D - 24)y = 0.$

ANS.  $y = c_1 e^x + c_2 e^{2x} + c_3 e^{3x} + c_4 e^{-4x}.$

19. $(6D^4 + 23D^3 + 28D^2 + 13D + 2)y = 0.$
20. $(4D^4 - 45D^2 - 70D - 24)y = 0.$

ANS.  $y = c_1 e^{4x} + c_2 e^{-2x} + c_3 e^{-x/2} + c_4 e^{-3x/2}.$

21. $(D^2 - 4aD + 3a^2)y = 0; a$ real $\neq 0.$
22. $[D^2 - (a + b)D + ab]y = 0; a$ and $b$ real and unequal.  ANS.  $y = c_1 e^{ax} + c_2 e^{bx}.$

In exercises 23 and 24, find the particular solution indicated.

23. $(D^2 - 2D - 3)y = 0;$ when $x = 0, y = 0, y' = -4.$    ANS.  $y = e^{-x} - e^{3x}.$
24. $(D^2 - D - 6)y = 0;$ when $x = 0, y = 0,$ and when $x = 1, y = e^3.$

ANS.  $y = (e^{3x} - e^{-2x})/(1 - e^{-5}).$

In exercises 25 through 29, find for $x = 1$ the $y$ value for the particular solution required.

25. $(D^2 - 2D - 3)y = 0;$ when $x = 0, y = 4, y' = 0.$

ANS.  When $x = 1, y = e^3 + 3 e^{-1} = 21.2.$

26. $(D^3 - 4D)y = 0;$ when $x = 0, y = 0, y' = 0, y'' = 2.$

ANS.  When $x = 1, y = \sinh^2 1.$

27. $(D^2 - D - 6)y = 0;$ when $x = 0, y = 3, y' = -1.$    ANS.  When $x = 1, y = 20.4.$
28. $(D^2 + 3D - 10)y = 0;$ when $x = 0, y = 0,$ and when $x = 2, y = 1.$

ANS.  When $x = 1, y = 0.135.$

29. $(D^3 - 2D^2 - 5D + 6)y = 0;$ when $x = 0, y = 1, y' = -7, y'' = -1.$

ANS.  When $x = 1, y = -19.8.$

## 35. The auxiliary equation; repeated roots

Suppose that in the equation

$$f(D)y = 0 \qquad\qquad (1)$$

the operator $f(D)$ has repeated factors; that is, the auxiliary equation $f(m) = 0$ has repeated roots. Then the method of the previous section does not yield the general solution. Let the auxiliary equation have three equal roots $m_1 = b$, $m_2 = b$, $m_3 = b$. The corresponding part of the solution yielded by the method of Section 34 is

$$y = c_1 e^{bx} + c_2 e^{bx} + c_3 e^{bx},$$

$$y = (c_1 + c_2 + c_3) e^{bx}. \tag{2}$$

Now (2) can be replaced by

$$y = c_4 e^{bx} \tag{3}$$

with $c_4 = c_1 + c_2 + c_3$. Thus, corresponding to the three roots under consideration, this method has yielded only the solution (3). The difficulty is present, of course, because the three solutions corresponding to the roots $m_1 = m_2 = m_3 = b$ are not linearly independent.

What is needed is a method for obtaining $n$ linearly independent solutions corresponding to $n$ equal roots of the auxiliary equation. Suppose that the auxiliary equation $f(m) = 0$ has the $n$ equal roots

$$m_1 = m_2 = \cdots = m_n = b.$$

Then the operator $f(D)$ must have a factor $(D - b)^n$. We wish to find $n$ linearly independent $y$'s for which

$$(D - b)^n y = 0. \tag{4}$$

Turning to the result (6) near the end of Section 32 and writing $m = b$, we find that

$$(D - b)^n (x^k e^{bx}) = 0 \qquad \text{for} \quad k = 0, 1, 2, \ldots, (n - 1). \tag{5}$$

The functions $y_k = x^k e^{bx}$ where $k = 0, 1, 2, \ldots, (n - 1)$ are linearly independent because, aside from the common factor $e^{bx}$, they contain only the respective powers $x^0, x^1, x^2, \ldots, x^{n-1}$. (See exercise 5, Section 32.)

The general solution of equation (4) is

$$y = c_1 e^{bx} + c_2 x e^{bx} + \cdots + c_n x^{n-1} e^{bx}. \tag{6}$$

Furthermore, if $f(D)$ contains the factor $(D - b)^n$, then the equation

$$f(D)y = 0 \tag{1}$$

can be written

$$g(D)(D - b)^n y = 0 \tag{7}$$

where $g(D)$ contains all the factors of $f(D)$ except $(D - b)^n$. Then any solution of

$$(D - b)^n y = 0 \tag{4}$$

is also a solution of (7) and therefore of (1).

Now we are in a position to write the solution of equation (1) whenever the auxiliary equation has only real roots. Each root of the auxiliary equation is either distinct from all the other roots or it is one of a set of equal roots. Corresponding to a root $m_i$ distinct from all others, there is the solution

$$y_i = c_i \exp (m_i x), \tag{8}$$

and corresponding to $n$ equal roots $m_1, m_2, \ldots, m_n$, each equal to $b$, there are the solutions

$$c_1 e^{bx}, c_2 x e^{bx}, \ldots, c_n x^{n-1} e^{bx}. \tag{9}$$

The collection of solutions (9) has the proper number of elements, a number equal to the order of the differential equation, because there is one solution corresponding to each root of the auxiliary equation. The solutions thus obtained can be proved to be linearly independent.

EXAMPLE (a):   Solve the equation

$$(D^4 - 7D^3 + 18D^2 - 20D + 8)y = 0. \tag{10}$$

With the aid of synthetic division, it is easily seen that the auxiliary equation

$$m^4 - 7m^3 + 18m^2 - 20m + 8 = 0$$

has the roots $m = 1, 2, 2, 2$. Then the general solution of equation (10) is

$$y = c_1 e^x + c_2 e^{2x} + c_3 x e^{2x} + c_4 x^2 e^{2x},$$

or

$$y = c_1 e^x + (c_2 + c_3 x + c_4 x^2) e^{2x}.$$

EXAMPLE (b):   Solve the equation

$$\frac{d^4 y}{dx^4} + 2\frac{d^3 y}{dx^3} + \frac{d^2 y}{dx^2} = 0.$$

The auxiliary equation is

$$m^4 + 2m^3 + m^2 = 0,$$

with roots $m = 0, 0, -1, -1$. Hence the desired solution is

$$y = c_1 + c_2 x + c_3 e^{-x} + c_4 x e^{-x}.$$

## Exercises

In exercises 1 through 20 find the general solution.

**1.** $(4D^2 - 4D + 1)y = 0.$         ANS.   $y = (c_1 + c_2 x) e^{x/2}.$

**2.** $(D^2 + 6D + 9)y = 0.$

**3.** $(D^3 - 4D^2 + 4D)y = 0.$ ANS. $y = c_1 + (c_2 + c_3 x) e^{2x}.$

**4.** $(9D^3 + 6D^2 + D)y = 0.$

**5.** $(2D^4 - 3D^3 - 2D^2)y = 0.$ ANS. $y = c_1 + c_2 x + c_3 e^{2x} + c_4 e^{-x/2}.$

**6.** $(2D^4 - 5D^3 - 3D^2)y = 0.$

**7.** $(D^3 + 3D^2 - 4)y = 0.$ ANS. $y = c_1 e^x + (c_2 + c_3 x) e^{-2x}.$

**8.** $(4D^3 - 27D + 27)y = 0.$

**9.** $(D^3 + 3D^2 + 3D + 1)y = 0.$ ANS. $y = (c_1 + c_2 x + c_3 x^2) e^{-x}.$

**10.** $(D^3 + 6D^2 + 12D + 8)y = 0.$

**11.** $(D^5 - D^3)y = 0.$ ANS. $y = c_1 + c_2 x + c_3 x^2 + c_4 e^x + c_5 e^{-x};$
or $y = c_1 + c_2 x + c_3 x^2 + c_6 \cosh x + c_7 \sinh x.$

**12.** $(D^5 - 16D^3)y = 0.$

**13.** $(4D^4 + 4D^3 - 3D^2 - 2D + 1)y = 0.$
ANS. $y = (c_1 + c_2 x) e^{x/2} + (c_3 + c_4 x) e^{-x}.$

**14.** $(4D^4 - 4D^3 - 23D^2 + 12D + 36)y = 0.$
ANS. $y = (c_1 + c_2 x) e^{-3x/2} + (c_3 + c_4 x) e^{2x}.$

**15.** $(D^4 + 3D^3 - 6D^2 - 28D - 24)y = 0.$
ANS. $y = c_1 e^{3x} + (c_2 + c_3 x + c_4 x^2) e^{-2x}.$

**16.** $(27D^4 - 18D^2 + 8D - 1)y = 0.$

**17.** $(4D^5 - 23D^3 - 33D^2 - 17D - 3)y = 0.$
ANS. $y = c_1 e^{3x} + (c_2 + c_3 x) e^{-x} + (c_4 + c_5 x) e^{-x/2}.$

**18.** $(4D^5 - 15D^3 - 5D^2 + 15D + 9)y = 0.$

**19.** $(D^4 - 5D^2 - 6D - 2)y = 0.$
ANS. $y = (c_1 + c_2 x) e^{-x} + c_3 e^{(1 + \sqrt{3})x} + c_4 e^{(1 - \sqrt{3})x}.$

**20.** $(D^5 - 5D^4 + 7D^3 + D^2 - 8D + 4)y = 0.$

In exercises 21 through 26, find the particular solution indicated.

**21.** $(D^2 + 4D + 4)y = 0$; when $x = 0$, $y = 1$, $y' = -1$. ANS. $y = (1 + x) e^{-2x}.$

**22.** The equation of exercise 21 with the condition that the graph of the solution pass through the points $(0, 2)$ and $(2, 0)$. ANS. $y = (2 - x) e^{-2x}.$

**23.** $(D^3 - 3D - 2)y = 0$; when $x = 0$, $y = 0$, $y' = 9$, $y'' = 0.$
ANS. $y = 2 e^{2x} + (3x - 2) e^{-x}.$

**24.** $(D^4 + 3D^3 + 2D^2)y = 0$; when $x = 0$, $y = 0$, $y' = 4$, $y'' = -6$, $y''' = 14.$
ANS. $y = 2(x + e^{-x} - e^{-2x}).$

**25.** The equation of exercise 24 with the conditions that when $x = 0$, $y = 0$, $y' = 3$, $y'' = -5$, $y''' = 9.$ ANS. $y = 2 - e^{-x} - e^{-2x}.$

**26.** $(D^3 + D^2 - D - 1)y = 0$; when $x = 0$, $y = 1$, when $x = 2$, $y = 0$, and also as $x \to \infty$, $y \to 0.$ ANS. $y = \frac{1}{2}(2 - x) e^{-x}.$

In exercises 27 through 29, find for $x = 2$ the $y$ value for the particular solution required.

**27.** $(4D^2 - 4D + 1)y = 0$; when $x = 0$, $y = -2$, $y' = 2.$
ANS. When $x = 2$, $y = 4 e.$

**28.** $(D^3 + 2D^2)y = 0$; when $x = 0$, $y = -3$, $y' = 0$, $y'' = 12.$
ANS. When $x = 2$, $y = 3 e^{-4} + 6.$

**29.** $(D^3 + 5D^2 + 3D - 9)y = 0$; when $x = 0$, $y = -1$, when $x = 1$, $y = 0$, and also as $x \to \infty$, $y \to 0.$ ANS. When $x = 2$, $y = e^{-6}.$

## 36.  A definition of exp z for imaginary z

Since the auxiliary equation may have imaginary roots, we need now to lay down a definition of exp z for imaginary z.

Let $z = \alpha + i\beta$ with $\alpha$ and $\beta$ real. Since it is desirable to have the ordinary laws of exponents remain valid, it is wise to require that

$$\exp(\alpha + i\beta) = e^\alpha \cdot e^{i\beta}. \tag{1}$$

To $e^\alpha$ with $\alpha$ real, we attach the usual meaning.

Now consider $e^{i\beta}$, $\beta$ real. In calculus it is shown that for all real $x$

$$e^x = 1 + \frac{x}{1!} + \frac{x^2}{2!} + \frac{x^3}{3!} + \cdots + \frac{x^n}{n!} + \cdots, \tag{2}$$

or

$$e^x = \sum_{n=0}^{\infty} \frac{x^n}{n!}. \tag{2}$$

If we now tentatively put $x = i\beta$ in (2) as a definition of $e^{i\beta}$, we get

$$e^{i\beta} = 1 + \frac{i\beta}{1!} + \frac{i^2\beta^2}{2!} + \frac{i^3\beta^3}{3!} + \frac{i^4\beta^4}{4!} + \cdots + \frac{i^n\beta^n}{n!} + \cdots. \tag{3}$$

Separating the even powers of $\beta$ from the odd powers of $\beta$ in (3) yields

$$e^{i\beta} = 1 + \frac{i^2\beta^2}{2!} + \frac{i^4\beta^4}{4!} + \cdots + \frac{i^{2k}\beta^{2k}}{(2k)!} + \cdots$$
$$+ \frac{i\beta}{1!} + \frac{i^3\beta^3}{3!} + \cdots + \frac{i^{2k+1}\beta^{2k+1}}{(2k+1)!} + \cdots, \tag{4}$$

or

$$e^{i\beta} = \sum_{k=0}^{\infty} \frac{i^{2k}\beta^{2k}}{(2k)!} + \sum_{k=0}^{\infty} \frac{i^{2k+1}\beta^{2k+1}}{(2k+1)!}. \tag{4}$$

Now $i^{2k} = (-1)^k$, so we may write

$$e^{i\beta} = 1 - \frac{\beta^2}{2!} + \frac{\beta^4}{4!} + \cdots + \frac{(-1)^k\beta^{2k}}{(2k)!} + \cdots$$
$$+ i\left[\frac{\beta}{1!} - \frac{\beta^3}{3!} + \cdots + \frac{(-1)^k\beta^{2k+1}}{(2k+1)!} + \cdots\right], \tag{5}$$

or

$$e^{i\beta} = \sum_{k=0}^{\infty} \frac{(-1)^k\beta^{2k}}{(2k)!} + i\sum_{k=0}^{\infty} \frac{(-1)^k\beta^{2k+1}}{(2k+1)!}. \tag{5}$$

But the series on the right in (5) are precisely those for $\cos \beta$ and $\sin \beta$ as developed in calculus. Hence we are led to the tentative result

$$e^{i\beta} = \cos \beta + i \sin \beta. \tag{6}$$

The student should realize that the manipulations above have no meaning in themselves at this stage (assuming that infinite series with complex terms are not a part of the content of elementary mathematics). What has been accomplished is this: the formal manipulations above have suggested the meaningful definition (6). Combining (6) with (1) we now put forward a reasonable *definition* of $\exp(\alpha + i\beta)$, namely,

$$\exp(\alpha + i\beta) = e^{\alpha}(\cos \beta + i \sin \beta), \qquad \text{when } \alpha \text{ and } \beta \text{ are real.} \tag{7}$$

Replacing $\beta$ by $(-\beta)$ in (7) yields a result that is of value to us in the next section,

$$\exp(\alpha - i\beta) = e^{\alpha}(\cos \beta - i \sin \beta).$$

It is interesting and important that, with the definition (7), the function $e^z$ for complex $z$ retains many of the properties possessed by the function $e^x$ for real $x$. Such matters are often studied in detail in books on complex variables.* Here we need in particular to know that if

$$y = \exp(a + ib)x,$$

with $a, b$, and $x$ real, then

$$(D - a - ib)y = 0.$$

The result desired follows at once by differentiation, with respect to $x$, of the function

$$y = e^{ax}(\cos bx + i \sin bx).$$

## 37. The auxiliary equation; imaginary roots

Consider a differential equation $f(D)y = 0$ for which the auxiliary equation $f(m) = 0$ has real coefficients. From elementary algebra we know that if the auxiliary equation has any imaginary roots those roots must occur in conjugate pairs. Thus if

$$m_1 = a + ib$$

is a root of the equation $f(m) = 0$, with $a$ and $b$ real and $b \neq 0$, then

$$m_2 = a - ib$$

*For example, R. V. Churchill, J. W. Brown, and R. F. Verhey, *Complex Variables and Applications*, 3rd ed. (New York: McGraw-Hill Book Company, 1976), pp. 52–56.

is also a root of $f(m) = 0$. It must be kept in mind that this result is a consequence of the reality of the coefficients in the equation $f(m) = 0$. Imaginary roots do not necessarily appear in pairs in an algebraic equation whose coefficients involve imaginaries.

We can now construct in usable form solutions of

$$f(D)y = 0 \tag{1}$$

corresponding to imaginary roots of $f(m) = 0$. For, since $f(m)$ is assumed to have real coefficients, any imaginary roots appear in conjugate pairs

$$m_1 = a + ib \qquad \text{and} \qquad m_2 = a - ib.$$

Then, according to the preceding section, equation (1) is satisfied by

$$y = c_1 \exp\left[(a + ib)x\right] + c_2 \exp\left[(a - ib)x\right]. \tag{2}$$

Taking $x$ to be real along with $a$ and $b$, we get from (2) the result

$$y = c_1 e^{ax}(\cos bx + i \sin bx) + c_2 e^{ax}(\cos bx - i \sin bx). \tag{3}$$

Now (3) may be written

$$y = (c_1 + c_2) e^{ax} \cos bx + i(c_1 - c_2) e^{ax} \sin bx.$$

Finally, let $c_1 + c_2 = c_3$, and $i(c_1 - c_2) = c_4$, where $c_3$ and $c_4$ are new arbitrary constants. Then equation (1) is seen to have the solutions

$$y = c_3 e^{ax} \cos bx + c_4 e^{ax} \sin bx, \tag{4}$$

corresponding to the two roots $m_1 = a + ib$ and $m_2 = a - ib$ $(b \neq 0)$ of the auxiliary equation.

The reduction of the solution (2) above to the desirable form (4) has been done once and that is enough. Whenever a pair of conjugate imaginary roots of the auxiliary equation appears, we write down at once, in the form given on the right in equation (4), the particular solution corresponding to those two roots.

EXAMPLE (a): Solve the equation

$$(D^3 - 3D^2 + 9D + 13)y = 0.$$

For the auxiliary equation

$$m^3 - 3m^2 + 9m + 13 = 0,$$

one root, $m_1 = -1$, is easily found. When the factor $(m + 1)$ is removed by synthetic division, it is seen that the other two roots are solutions of the quadratic equation

$$m^2 - 4m + 13 = 0.$$

Those roots are found to be $m_2 = 2 + 3i$ and $m_3 = 2 - 3i$. The auxiliary equation has the roots $m = -1, 2 \pm 3i$. Hence the general solution of the differential equation is

$$y = c_1 e^{-x} + c_2 e^{2x} \cos 3x + c_3 e^{2x} \sin 3x.$$

Repeated imaginary roots lead to solutions analogous to those brought in by repeated real roots. For instance, if the roots $m = a \pm ib$ occur three times, then the corresponding six linearly independent solutions of the differential equation are those appearing in the expression

$$(c_1 + c_2 x + c_3 x^2) e^{ax} \cos bx + (c_4 + c_5 x + c_6 x^2) e^{ax} \sin bx.$$

EXAMPLE (b):   Solve the equation

$$(D^4 + 8D^2 + 16)y = 0.$$

The auxiliary equation $m^4 + 8m^2 + 16 = 0$ may be written

$$(m^2 + 4)^2 = 0,$$

so its roots are seen to be $m = \pm 2i, \pm 2i$. The roots $m_1 = 2i$ and $m_2 = -2i$ occur twice each. Thinking of $2i$ as $0 + 2i$ and recalling that $e^{0x} = 1$, we write the solution of the differential equation as

$$y = (c_1 + c_2 x) \cos 2x + (c_3 + c_4 x) \sin 2x.$$

In such exercises as those below a fine check can be obtained by direct substitution of the result and its appropriate derivatives into the differential equation. The verification is particularly effective because the operations performed in the check are so different from those performed in obtaining the solution.

## 38. A note on hyperbolic functions

Two particular linear combinations of exponential functions appear with such frequency in both pure and applied mathematics that it has been worth-while to use special symbols for those combinations. The hyperbolic sine of $x$, written $\sinh x$, is defined by

$$\sinh x = \frac{e^x - e^{-x}}{2}; \tag{1}$$

the hyperbolic cosine of $x$, written $\cosh x$, is defined by

$$\cosh x = \frac{e^x + e^{-x}}{2}. \tag{2}$$

From the definitions of $\sinh x$ and $\cosh x$ it follows that

$$\sinh^2 x = \tfrac{1}{4}(e^{2x} - 2 + e^{-2x})$$

and

$$\cosh^2 x = \tfrac{1}{4}(e^{2x} + 2 + e^{-2x}),$$

so

$$\cosh^2 x - \sinh^2 x = 1, \tag{3}$$

an identity similar to the well-known identity $\cos^2 x + \sin^2 x = 1$ in trigonometry.

Directly from the definition we find that

$$y = \sinh u$$

is equivalent to

$$y = \tfrac{1}{2}(e^u - e^{-u}).$$

Hence, if $u$ is a function of $x$, then

$$\frac{dy}{dx} = \tfrac{1}{2}(e^u + e^{-u})\frac{du}{dx},$$

that is,

$$\frac{d}{dx}\sinh u = \cosh u \frac{du}{dx}. \tag{4}$$

The same method yields the result

$$\frac{d}{dx}\cosh u = \sinh u \frac{du}{dx}. \tag{5}$$

The graphs of $y = \cosh x$ and $y = \sinh x$ are exhibited in Figure 12. Note the important properties:

(a) $\cosh x \geq 1$ for all real $x$;
(b) the only real value of $x$ for which $\sinh x = 0$ is $x = 0$;
(c) $\cosh(-x) = \cosh x$; that is, $\cosh x$ is an even function of $x$;
(d) $\sinh(-x) = -\sinh x$; $\sinh x$ is an odd function of $x$.

The hyperbolic functions have no real period. Corresponding to the period $2\pi$ possessed by the circular functions, there is a period $2\pi i$ for the hyperbolic functions.

The hyperbolic cosine curve is that in which a transmission line, cable, piece of string, watch chain, etc., hangs between two points at which it is suspended. This result is obtained in Chapter 16.

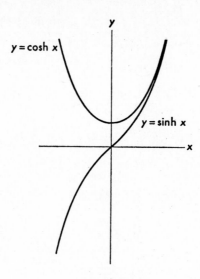

FIGURE 12

Since $D^2 \cosh ax = a^2 \cosh ax$ and $D^2 \sinh ax = a^2 \sinh ax$, it follows that both $\cosh ax$ and $\sinh ax$ are solutions of

$$(D^2 - a^2)y = 0, \qquad a \neq 0. \tag{6}$$

Furthermore the Wronskian of these two functions,

$$W(x) = \begin{vmatrix} \cosh ax & \sinh ax \\ a \sinh ax & a \cosh ax \end{vmatrix} = a,$$

is not zero, so that $\cosh ax$ and $\sinh ax$ are linearly independent solutions of equation (6). Hence the general solution of (6) may be written

$$y = c_1 \cosh ax + c_2 \sinh ax$$

instead of using the form

$$y = c_3 e^{ax} + c_4 e^{-ax}.$$

It is often very convenient to use this alternative form for representing the general solution of (6).

EXAMPLE: Find the solution of the problem

$$(D^2 - 4)y = 0; \qquad \text{when } x = 0, y = 0, y' = 2. \tag{7}$$

The general solution of the differential equation (7) may be written

$$y = c_1 \cosh 2x + c_2 \sinh 2x,$$

from which

$$y' = 2c_1 \sinh 2x + 2c_2 \cosh 2x.$$

The initial conditions now require that $0 = c_1$ and $2 = 2c_2$, so that finally

$$y = \sinh 2x.$$

Note that if we were to choose the alternative form

$$y = c_3 e^{2x} + c_4 e^{-2x}$$

for the general solution of (7), we would obtain the same result with a little more fuss in determining $c_3$ and $c_4$. Indeed, one major reason for using the hyperbolic functions is that $\cosh ax$ and $\sinh ax$ have values 1 and 0 when $x = 0$, a fact that is particularly useful in solving initial value problems.

## Exercises

Find the general solution except when the exercise stipulates otherwise.

1. Verify directly that the relation

$$y = c_3 e^{ax} \cos bx + c_4 e^{ax} \sin bx \qquad \text{(A)}$$

satisfies the equation

$$[(D - a)^2 + b^2]y = 0.$$

2. $(D^2 - 2D + 5)y = 0$. Verify your answer.     ANS.   $y = c_1 e^x \cos 2x + c_2 e^x \sin 2x$.
3. $(D^2 - 2D + 2)y = 0$.
4. $(D^2 + 9)y = 0$. Verify your answer.                    ANS.   $y = c_1 \cos 3x + c_2 \sin 3x$.
5. $(D^2 - 9)y = 0$.                                              ANS.   $y = c_1 \cosh 3x + c_2 \sinh 3x$.
6. $(D^2 + 6D + 13)y = 0$. Verify your answer.
7. $(D^2 - 4D + 7)y = 0$.          ANS.   $y = c_1 e^{2x} \cos \sqrt{3}x + c_2 e^{2x} \sin \sqrt{3}x$.
8. $(D^3 + 2D^2 + D + 2)y = 0$. Verify your answer.
9. $(D^2 - 1)y = 0$; when $x = 0$, $y = y_0$, $y' = 0$.          ANS.   $y = y_0 \cosh x$.
10. $(D^2 + 1)y = 0$; when $x = 0$, $y = y_0$, $y' = 0$.          ANS.   $y = y_0 \cos x$.
11. $(D^4 + 2D^3 + 10D^2)y = 0$.     ANS.   $y = c_1 + c_2 x + c_3 e^{-x} \cos 3x + c_4 e^{-x} \sin 3x$.
12. $(D^3 + 7D^2 + 19D + 13)y = 0$;  when  $x = 0$,  $y = 0$,  $y' = 2$,  $y'' = -12$.
                                                     ANS.   $y = e^{-3x} \sin 2x$.
13. $(D^5 + D^4 - 7D^3 - 11D^2 - 8D - 12)y = 0$.
                     ANS.   $y = c_1 \cos x + c_2 \sin x + c_3 e^{-2x} + c_4 x e^{-2x} + c_5 e^{3x}$.
14. $(D^4 - 2D^3 + 2D^2 - 2D + 1)y = 0$. Verify your answer.
15. $(D^4 + 18D^2 + 81)y = 0$.          ANS.   $y = (c_1 + c_2 x) \cos 3x + (c_3 + c_4 x) \sin 3x$.
16. $(2D^4 + 11D^3 - 4D^2 - 69D + 34)y = 0$. Verify your answer.
17. $(D^6 + 9D^4 + 24D^2 + 16)y = 0$.
             ANS.   $y = c_1 \cos x + c_2 \sin x + (c_3 + c_4 x) \cos 2x + (c_5 + c_6 x) \sin 2x$.
18. $(2D^3 - D^2 + 36D - 18)y = 0$.
                     ANS.   $y = c_1 e^{x/2} + c_2 \cos (3\sqrt{2}x) + c_3 \sin (3\sqrt{2}x)$.

**19.** $\dfrac{d^2x}{dt^2} + k^2 x = 0$, $k$ real; when $t = 0$, $x = 0$, $\dfrac{dx}{dt} = v_0$. Verify your result completely.

ANS.   $x = (v_0/k) \sin kt$.

**20.** $(D^3 + D^2 + 4D + 4)y = 0$; when $x = 0$, $y = 0$, $y' = -1$, $y'' = 5$.

ANS.   $y = e^{-x} - \cos 2x$.

**21.** $\dfrac{d^2x}{dt^2} + 2b\dfrac{dx}{dt} + k^2 x = 0$, $k > b > 0$; when $t = 0$, $x = 0$, $\dfrac{dx}{dt} = v_0$.

ANS.   $x = (v_0/a)\, e^{-bt} \sin at$; where $a = \sqrt{k^2 - b^2}$.

## Miscellaneous Exercises

Obtain the general solution unless otherwise instructed.

**1.** $(D^2 + 3D)y = 0$.  ANS.   $y = c_1 + c_2 e^{-3x}$.

**2.** $(9D^4 + 6D^3 + D^2)y = 0$.  ANS.   $y = c_1 + c_2 x + (c_3 + c_4 x)\exp\left(-\tfrac{1}{3}x\right)$.

**3.** $(D^2 + D - 6)y = 0$.  ANS.   $y = c_1 e^{2x} + c_2 e^{-3x}$.

**4.** $(D^3 + 2D^2 + D + 2)y = 0$.  ANS.   $y = c_1 e^{-2x} + c_2 \cos x + c_3 \sin x$.

**5.** $(D^3 - 3D^2 + 4)y = 0$.  ANS.   $y = c_1 e^{-x} + e^{2x}(c_2 + c_3 x)$.

**6.** $(D^3 - 2D^2 - 3D)y = 0$.  ANS.   $y = c_1 + c_2 e^{3x} + c_3 e^{-x}$.

**7.** $(4D^3 - 3D + 1)y = 0$.  ANS.   $y = c_1 e^{-x} + (c_2 + c_3 x)\exp\left(\tfrac{1}{2}x\right)$.

**8.** $(D^3 + 3D^2 - 4D - 12)y = 0$.  ANS.   $y = c_1 \cosh 2x + c_2 \sinh 2x + c_3 e^{-3x}$.

**9.** $(D^3 + 3D^2 + 3D + 1)y = 0$.  ANS.   $y = e^{-x}(c_1 + c_2 x + c_3 x^2)$.

**10.** $(4D^3 - 21D - 10)y = 0$.  ANS.   $y = c_1 e^{-2x} + c_2 \exp\left(\tfrac{5}{2}x\right) + c_3 \exp\left(-\tfrac{1}{2}x\right)$.

**11.** $(4D^3 - 7D + 3)y = 0$.  ANS.   $y = c_1 e^x + c_2 \exp\left(\tfrac{1}{2}x\right) + c_3 \exp\left(-\tfrac{3}{2}x\right)$.

**12.** $(D^2 - D - 6)y = 0$; when $x = 0$, $y = 2$, $y' = 1$.  ANS.   $y = e^{3x} + e^{-2x}$.

**13.** $(D^4 + 6D^3 + 9D^2)y = 0$; when $x = 0$, $y = 0$, $y' = 0$, $y'' = 6$, and as $x \to \infty$, $y' \to 1$. For this particular solution, find the value of $y$ when $x = 1$.   ANS.   $y = 1 - e^{-3}$.

**14.** $(D^3 + 6D^2 + 12D + 8)y = 0$; when $x = 0$, $y = 1$, $y' = -2$, $y'' = 2$.

ANS.   $y = e^{-2x}(1 - x^2)$.

**15.** $(D^3 - 14D + 8)y = 0$.

ANS.   $y = c_1 e^{-4x} + c_2 \exp\left[(2 + \sqrt{2})x\right] + c_3 \exp\left[(2 - \sqrt{2})x\right]$.

**16.** $(8D^3 - 4D^2 - 2D + 1)y = 0$.  ANS.   $y = (c_1 + c_2 x)\exp\left(\tfrac{1}{2}x\right) + c_3 \exp\left(-\tfrac{1}{2}x\right)$.

**17.** $(D^4 + D^3 - 4D^2 - 4D)y = 0$.

**18.** $(D^4 - 2D^3 + 5D^2 - 8D + 4)y = 0$.

ANS.   $y = e^x(c_1 + c_2 x) + c_3 \cos 2x + c_4 \sin 2x$.

**19.** $(D^4 + 2D^2 + 1)y = 0$.

**20.** $(D^4 + 5D^2 + 4)y = 0$.  ANS.   $y = c_1 \cos x + c_2 \sin x + c_3 \cos 2x + c_4 \sin 2x$.

**21.** $(D^4 + 3D^3 - 4D)y = 0$.

**22.** $(D^5 + D^4 - 9D^3 - 13D^2 + 8D + 12)y = 0$.

ANS.   $y = c_1 e^x + c_2 e^{3x} + c_3 e^{-x} + e^{-2x}(c_4 + c_5 x)$.

**23.** $(D^4 - 11D^3 + 36D^2 - 16D - 64)y = 0$.

**24.** $(D^2 + 2D + 5)y = 0$.  ANS.   $y = e^{-x}(c_1 \cos 2x + c_2 \sin 2x)$.

**25.** $(D^4 + 4D^3 + 2D^2 - 8D - 8)y = 0$.

**26.** $(4D^4 - 24D^3 + 35D^2 + 6D - 9)y = 0$.

ANS.   $y = e^{3x}(c_1 + c_2 x) + c_3 \cosh \tfrac{1}{2}x + c_4 \sinh \tfrac{1}{2}x$.

**27.** $(4D^4 + 20D^3 + 35D^2 + 25D + 6)y = 0$.

**28.** $(D^4 - 7D^3 + 11D^2 + 5D - 14)y = 0.$

**29.** $(D^3 + 5D^2 + 7D + 3)y = 0.$

**30.** $(D^3 - 2D^2 + D - 2)y = 0.$          ANS.   $y = c_1 e^{2x} + c_2 \cos x + c_3 \sin x.$

**31.** $(D^3 - D^2 + D - 1)y = 0.$

**32.** $(D^3 + 4D^2 + 5D)y = 0.$

**33.** $(D^4 - 13D^2 + 36)y = 0.$

**34.** $(D^4 - 5D^3 + 5D^2 + 5D - 6)y = 0.$

ANS.   $y = c_1 \cosh x + c_2 \sinh x + c_3 e^{2x} + c_4 e^{3x}.$

**35.** $(4D^3 + 8D^2 - 11D + 3)y = 0.$

**36.** $(D^3 + D^2 - 16D - 16)y = 0.$

**37.** $(D^4 - D^3 - 3D^2 + D + 2)y = 0.$          ANS.   $y = c_1 e^x + c_2 e^{2x} + e^{-x}(c_3 + c_4 x).$

**38.** $(D^3 - 2D^2 - 3D + 10)y = 0.$

**39.** $(D^5 + D^4 - 6D^3)y = 0.$

**40.** $(4D^3 + 28D^2 + 61D + 37)y = 0.$   ANS.   $y = c_1 e^{-x} + e^{-3x}(c_2 \cos \tfrac{1}{2}x + c_3 \sin \tfrac{1}{2}x).$

**41.** $(4D^3 + 12D^2 + 13D + 10)y = 0.$

**42.** $(18D^3 - 33D^2 + 20D - 4)y = 0.$

**43.** $(D^5 - 2D^3 - 2D^2 - 3D - 2)y = 0.$

ANS.   $y = e^{-x}(c_1 + c_2 x) + c_3 e^{2x} + c_4 \cos x + c_5 \sin x.$

**44.** $(D^4 - 2D^3 + 2D^2 - 2D + 1)y = 0.$

**45.** $(4D^5 + 4D^4 - 9D^3 - 11D^2 + D + 3)y = 0.$

**46.** $(D^5 - 15D^3 + 10D^2 + 60D - 72)y = 0.$

**47.** $(D^4 + 2D^3 - 6D^2 - 16D - 8)y = 0.$

ANS.   $y = e^{-2x}(c_1 + c_2 x) + e^x(c_3 \cos \sqrt{3}x + c_4 \sin \sqrt{3}x).$

# Nonhomogeneous Equations: Undetermined Coefficients

### 39. Construction of a homogeneous equation from a specified solution

In Section 29 we saw that the general solution of the equation

$$(b_0 D^n + b_1 D^{n-1} + \cdots + b_{n-1} D + b_n)y = R(x) \tag{1}$$

is

$$y = y_c + y_p,$$

where $y_c$, the complementary function, is the general solution of the homogeneous equation

$$(b_0 D^n + b_1 D^{n-1} + \cdots + b_{n-1} D + b_n)y = 0 \tag{2}$$

and $y_p$ is any particular solution of the original equation (1).

Various methods for getting a particular solution of (1) when the $b_0$, $b_1, \ldots, b_n$ are constants will be presented. In preparation for the method of undetermined coefficients it is wise to obtain proficiency in writing a homogeneous differential equation of which a given function of proper form is a solution.

Recall that in solving homogeneous equations with constant coefficients, a term such as $c_1 e^{ax}$ occurred only when the auxiliary equation $f(m) = 0$ had a root $m = a$, and then the operator $f(D)$ had a factor $(D - a)$. In like manner, $c_2 x e^{ax}$ appeared only when $f(D)$ contained the factor $(D - a)^2$, $c_3 x^2 e^{ax}$ only when $f(D)$ contained $(D - a)^3$, and so on.    Such terms as $c e^{ax} \cos bx$ or $c e^{ax} \sin bx$ correspond to roots $m = a \pm ib$, or to a factor $[(D - a)^2 + b^2]$.

EXAMPLE (a):    Find a homogeneous linear equation, with constant coefficients, that has as a particular solution

$$y = 7 e^{3x} + 2x.$$

First note that the coefficients (7 and 2) are quite irrelevant for the present problem, so long as they are not zero. We shall obtain an equation satisfied by $y = c_1 e^{3x} + c_2 x$, no matter what the constants $c_1$ and $c_2$ may be.

A term $c_1 e^{3x}$ occurs along with a root $m = 3$ of the auxiliary equation. The term $c_2 x$ will appear if the auxiliary equation has $m = 0, 0$; that is, a double root $m = 0$. We have recognized that the equation

$$D^2(D - 3)y = 0,$$

or

$$(D^3 - 3D^2)y = 0,$$

has $y = c_1 e^{3x} + c_2 x + c_3$ as its general solution, and therefore that it also has $y = 7 e^{3x} + 2x$ as a particular solution.

EXAMPLE (b):    Find a homogeneous linear equation with real, constant coefficients that is satisfied by

$$y = 6 + 3x e^x - \cos x. \tag{3}$$

The term 6 is associated with $m = 0$, the term $3x e^x$ with a double root $m = 1, 1$, and the term $(-\cos x)$ with the pair of imaginary roots $m = 0 \pm i$. Hence the auxiliary equation is

$$m(m - 1)^2(m^2 + 1) = 0,$$

or

$$m^5 - 2m^4 + 2m^3 - 2m^2 + m = 0.$$

Therefore the function in (3) is a solution of the differential equation

$$(D^5 - 2D^4 + 2D^3 - 2D^2 + D)y = 0. \tag{4}$$

That is, from the general solution

$$y = c_1 + (c_2 + c_3 x) e^x + c_4 \cos x + c_5 \sin x$$

of equation (4), the relation (3) follows by an appropriate choice of the constants: $c_1 = 6$, $c_2 = 0$, $c_3 = 3$, $c_4 = -1$, $c_5 = 0$.

EXAMPLE (c):   Find a homogeneous linear equation with real, constant coefficients that is satisfied by

$$y = 4x\, e^x \sin 2x.$$

The desired equation must have its auxiliary equation with roots $m = 1 \pm 2i, 1 \pm 2i$. The roots $m = 1 \pm 2i$ correspond to factors $(m - 1)^2 + 4$, so the auxiliary equation must be

$$[(m - 1)^2 + 4]^2 = 0,$$

or

$$m^4 - 4m^3 + 14m^2 - 20m + 25 = 0.$$

Hence the desired equation is

$$(D^4 - 4D^3 + 14D^2 - 20D + 25)y = 0.$$

Note that in all such problems, a correct (but undesirable) solution may be obtained by inserting additional roots of the auxiliary equation.

### Exercises

In exercises 1 through 14, obtain in factored form a linear differential equation with real, constant coefficients that is satisfied by the given function.

1. $y = 4\, e^{2x} + 3\, e^{-x}$.       ANS.   $(D - 2)(D + 1)y = 0$.
2. $y = 7 - 2x + \frac{1}{2} e^{4x}$.       ANS.   $D^2(D - 4)y = 0$.
3. $y = -2x + \frac{1}{2}\, e^{4x}$.       ANS.   $D^2(D - 4)y = 0$.
4. $y = x^2 - 5 \sin 3x$.       ANS.   $D^3(D^2 + 9)y = 0$.
5. $y = 2\, e^x \cos 3x$.       ANS.   $(D - 1 - 3i)(D - 1 + 3i)y = 0$;
   or $[(D - 1)^2 + 9]y = 0$; or $(D^2 - 2D + 10)y = 0$.
6. $y = 3\, e^{2x} \sin 3x$.       ANS.   $(D^2 - 4D + 13)y = 0$.
7. $y = -2\, e^{3x} \cos x$.       ANS.   $(D^2 - 6D + 10)y = 0$.
8. $y = e^{-x} \sin 2x$.       ANS.   $(D^2 + 2D + 5)y = 0$.
9. $y = xe^{-x} \sin 2x + 3\, e^{-x} \cos 2x$.       ANS.   $(D^2 + 2D + 5)^2 y = 0$.
10. $y = \sin 2x + 3 \cos 2x$.       ANS.   $(D^2 + 4)y = 0$.
11. $y = \cos kx$.       ANS.   $(D^2 + k^2)y = 0$.
12. $y = x \sin 2x$.       ANS.   $(D^2 + 4)^2 y = 0$.
13. $y = 4 \sinh x$.       ANS.   $(D^2 - 1)y = 0$.
14. $y = 2 \cosh 2x - \sinh 2x$.       ANS.   $(D^2 - 4)y = 0$.

In exercises 15 through 34, list the roots of the auxiliary equation for a homogeneous linear equation with real, constant coefficients that has the given function as a particular solution.

15. $y = 3x\, e^{2x}$.       ANS.   $m = 2, 2$.
16. $y = x^2\, e^{-x} + 4\, e^x$.       ANS.   $m = -1, -1, -1, 1$.

**17.** $y = e^{-x} \cos 4x$.    ANS.  $m = -1 \pm 4i$.
**18.** $y = 3 e^{-x} \cos 4x + 15 e^{-x} \sin 4x$.    ANS.  $m = -1 \pm 4i$.
**19.** $y = x(e^{2x} + 4)$.    ANS.  $m = 0, 0, 2, 2$.
**20.** $y = 4 + 2x^2 - e^{-3x}$.    ANS.  $m = 0, 0, 0, -3$.
**21.** $y = x e^x$.    ANS.  $m = 1, 1$.
**22.** $y = x e^x + 5 e^x$.    ANS.  $m = 1, 1$.
**23.** $y = 4 \cos 2x$.    ANS.  $m = \pm 2i$.
**24.** $y = 4 \cos 2x - 3 \sin 2x$.    ANS.  $m = \pm 2i$.
**25.** $y = x \cos 2x$.    ANS.  $m = 2i, 2i, -2i, -2i$.
**26.** $y = e^{-2x} \cos 3x$.    ANS.  $m = -2 \pm 3i$.
**27.** $y = x \cos 2x - 3 \sin 2x$.    ANS.  $m = 2i, 2i, -2i, -2i$.
**28.** $y = e^{-2x}(\cos 3x + \sin 3x)$.    ANS.  $m = -2 \pm 3i$.
**29.** $y = \sin^3 x$. Use the fact that $\sin^3 x = \frac{1}{4}(3 \sin x - \sin 3x)$.    ANS.  $m = \pm i, \pm 3i$.
**30.** $y = \cos^2 x$.    ANS.  $m = 0, \pm 2i$.
**31.** $y = x^2 - x + e^{-x}(x + \cos x)$.
**32.** $y = x^2 \sin x$.
**33.** $y = x^2 \sin x + x \cos x$.
**34.** $y = 8 \cos 4x + \sin 3x$.

## 40. Solution of a nonhomogeneous equation

Before proceeding to the theoretical basis and the actual working technique of the useful method of undetermined coefficients, let us examine the underlying ideas as applied to a simple numerical example.

Consider the equation

$$D^2(D - 1)y = 3 e^x + \sin x. \tag{1}$$

The complementary function may be determined at once from the roots

$$m = 0, 0, 1 \tag{2}$$

of the auxiliary equation. The complementary function is

$$y_c = c_1 + c_2 x + c_3 e^x. \tag{3}$$

Since the general solution of (1) is

$$y = y_c + y_p,$$

where $y_c$ is as given in (3) and $y_p$ is any particular solution of (1), all that remains for us to do is to find a particular solution of (1).

The right-hand member of (1),

$$R(x) = 3 e^x + \sin x, \tag{4}$$

is a particular solution of a homogeneous linear differential equation whose auxiliary equation has the roots

$$m' = 1, \pm i. \tag{5}$$

Therefore the function $R$ is a particular solution of the equation

$$(D - 1)(D^2 + 1)R = 0. \tag{6}$$

We wish to convert (1) into a homogeneous linear differential equation with constant coefficients, because we know how to solve any such equation. But, by (6), the operator $(D - 1)(D^2 + 1)$ will annihilate the right member of (1). Therefore, we apply that operator to both sides of equation (1) and get

$$(D - 1)(D^2 + 1)D^2(D - 1)y = 0. \tag{7}$$

Any solution of (1) must be a particular solution of (7). The general solution of (7) can be written at once from the roots of its auxiliary equation, those roots being the values $m = 0, 0, 1$ from (2) and the values $m' = 1, \pm i$ from (5). Thus the general solution of (7) is

$$y = c_1 + c_2 x + c_3 e^x + c_4 x e^x + c_5 \cos x + c_6 \sin x. \tag{8}$$

But the desired general solution of (1) is

$$y = y_c + y_p, \tag{9}$$

where

$$y_c = c_1 + c_2 x + c_3 e^x,$$

the $c_1, c_2, c_3$ being arbitrary constants as in (8). Thus there must exist a particular solution of (1) containing at most the remaining terms in (8). Using different letters as coefficients to emphasize that they are not arbitrary, we conclude that (1) has a particular solution

$$y_p = A x e^x + B \cos x + C \sin x. \tag{10}$$

We now have only to determine the numerical coefficients $A, B, C$ by direct use of the original equation

$$D^2(D - 1)y = 3 e^x + \sin x. \tag{1}$$

From (10) it follows that

$$Dy_p = A(x e^x + e^x) - B \sin x + C \cos x,$$

$$D^2 y_p = A(x e^x + 2 e^x) - B \cos x - C \sin x,$$

$$D^3 y_p = A(x e^x + 3 e^x) + B \sin x - C \cos x.$$

Substitution of $y_p$ into (1) then yields

$$A e^x + (B + C) \sin x + (B - C) \cos x = 3 e^x + \sin x. \tag{11}$$

Because (11) is to be an identity and because $e^x$, $\sin x$, and $\cos x$ are linearly independent, the corresponding coefficients in the two members of (11) must be equal; that is,

$$A = 3$$

$$B + C = 1$$

$$B - C = 0.$$

Therefore $A = 3$, $B = \frac{1}{2}$, $C = \frac{1}{2}$. Returning to (10), we find that a particular solution of equation (1) is

$$y_p = 3x\,e^x + \tfrac{1}{2}\cos x + \tfrac{1}{2}\sin x.$$

The general solution of the original equation

$$D^2(D - 1)y = 3\,e^x + \sin x \tag{1}$$

is therefore obtained by adding to the complementary function the $y_p$ found above:

$$y = c_1 + c_2 x + c_3\,e^x + 3x\,e^x + \tfrac{1}{2}\cos x + \tfrac{1}{2}\sin x. \tag{12}$$

A careful analysis of the ideas behind the process used shows that to arrive at the solution (12), we need perform only the following steps:

(a) From (1) find the values of $m$ and $m'$ as exhibited in (2) and (5).
(b) From the values of $m$ and $m'$ write $y_c$ and $y_p$ as in (3) and (10).
(c) Substitute $y_p$ into (1), equate corresponding coefficients, and obtain the numerical values of the coefficients in $y_p$.
(d) Write the general solution of (1).

## 41. The method of undetermined coefficients

Let us examine the general problem of the type treated in the preceding section. Let $f(D)$ be a polynomial in the operator $D$. Consider the equation

$$f(D)y = R(x). \tag{1}$$

Let the roots of the auxiliary equation $f(m) = 0$ be

$$m = m_1, m_2, \ldots, m_n. \tag{2}$$

The general solution of (1) is

$$y = y_c + y_p \tag{3}$$

where $y_c$ can be obtained at once from the values of $m$ in (2) and where $y = y_p$ is any particular solution (yet to be obtained) of (1).

Now suppose that the right member $R(x)$ of (1) is *itself a particular solution of some homogeneous linear differential equation with constant coefficients,*

$$g(D)R = 0, \tag{4}$$

whose auxiliary equation has the roots

$$m' = m'_1, m'_2, \ldots, m'_k. \tag{5}$$

Recall that the values of $m'$ in (5) can be obtained by inspection from $R(x)$.
The differential equation

$$g(D)f(D)y = 0 \tag{6}$$

has as the roots of its auxiliary equation the values of $m$ from (2) and $m'$ from (5). Hence the general solution of (6) contains the $y_c$ of (3) and so is of the form

$$y = y_c + y_q.$$

But also any particular solution of (1) must satisfy (6). Now, if

$$f(D)(y_c + y_q) = R(x),$$

then $f(D)y_q = R(x)$ because $f(D)y_c = 0$. Then deleting the $y_c$ from the general solution of (6) leaves a function $y_q$ that for some numerical values of its coefficients must satisfy (1); that is, the coefficients in $y_q$ can be determined so that $y_q = y_p$. The determination of those numerical coefficients may be accomplished as in the examples below.

*It must be kept in mind that the method of this section is applicable when, and only when, the right member of the equation is itself a particular solution of some homogeneous linear differential equation with constant coefficients.*

EXAMPLE (a):   Solve the equation

$$(D^2 + D - 2)y = 2x - 40 \cos 2x. \tag{7}$$

Here we have

$$m = 1, -2$$

and

$$m' = 0, 0, \pm 2i.$$

Therefore we may write

$$y_c = c_1 e^x + c_2 e^{-2x},$$

$$y_p = A + Bx + C \cos 2x + E \sin 2x,$$

in which $c_1$ and $c_2$ are arbitrary constants, whereas $A, B, C,$ and $E$ are to be determined numerically so that $y_p$ will satisfy the equation (7).

Since

$$Dy_p = B - 2C \sin 2x + 2E \cos 2x$$

and

$$D^2 y_p = -4C \cos 2x - 4E \sin 2x,$$

direct substitution of $y_p$ into (7) yields

$$-4C \cos 2x - 4E \sin 2x + B - 2C \sin 2x + 2E \cos 2x - 2A$$
$$-2Bx - 2C \cos 2x - 2E \sin 2x = 2x - 40 \cos 2x. \qquad (8)$$

But (8) is to be an identity in $x$, so we must equate coefficients of each of the set of linearly independent functions $\cos 2x$, $\sin 2x$, $x$, and 1 appearing in the identity. Thus it follows that

$$-6C + 2E = -40,$$
$$-6E - 2C = 0,$$
$$-2B = 2,$$
$$B - 2A = 0.$$

The above equations determine $A$, $B$, $C$, and $E$. Indeed, they lead to

$$A = -\tfrac{1}{2}, \qquad C = 6,$$
$$B = -1, \qquad E = -2.$$

Since the general solution of (7) is $y = y_c + y_p$, we can now write the desired result,

$$y = c_1 e^x + c_2 e^{-2x} - \tfrac{1}{2} - x + 6 \cos 2x - 2 \sin 2x.$$

EXAMPLE (b): Solve the equation

$$(D^2 + 1)y = \sin x. \qquad (9)$$

At once $m = \pm i$ and $m' = \pm i$. Therefore

$$y_c = c_1 \cos x + c_2 \sin x,$$
$$y_p = Ax \cos x + Bx \sin x.$$

Now

$$y_p'' = A(-x \cos x - 2 \sin x) + B(-x \sin x + 2 \cos x),$$

so the requirement that $y_p$ satisfy equation (9) yields

$$-2A \sin x + 2B \cos x = \sin x,$$

from which $A = -\tfrac{1}{2}$ and $B = 0$.

The general solution of (9) is

$$y = c_1 \cos x + c_2 \sin x - \tfrac{1}{2}x \cos x.$$

EXAMPLE (c):   Determine $y$ so that it will satisfy the equation

$$y''' - y' = 4 e^{-x} + 3 e^{2x} \tag{10}$$

with the conditions that when $x = 0$, $y = 0$, $y' = -1$, and $y'' = 2$.
First we note that $m = 0, 1, -1$, and $m' = -1, 2$. Thus

$$y_c = c_1 + c_2 e^x + c_3 e^{-x},$$

$$y_p = Ax e^{-x} + B e^{2x}.$$

Now

$$y_p' = A(-x e^{-x} + e^{-x}) + 2B e^{2x},$$

$$y_p'' = A(x e^{-x} - 2 e^{-x}) + 4B e^{2x},$$

$$y_p''' = A(-x e^{-x} + 3 e^{-x}) + 8B e^{2x}.$$

Then

$$y_p''' - y_p' = 2A e^{-x} + 6B e^{2x},$$

so that from (10) we may conclude that $A = 2$ and $B = \tfrac{1}{2}$.
The general solution of (10) is therefore

$$y = c_1 + c_2 e^x + c_3 e^{-x} + 2x e^{-x} + \tfrac{1}{2} e^{2x}. \tag{11}$$

We must determine $c_1$, $c_2$, $c_3$ so (11) will satisfy the conditions that when $x = 0$, $y = 0$, $y' = -1$, and $y'' = 2$.
From (11) it follows that

$$y' = c_2 e^x - c_3 e^{-x} - 2x e^{-x} + 2 e^{-x} + e^{2x} \tag{12}$$

and

$$y'' = c_2 e^x + c_3 e^{-x} + 2x e^{-x} - 4 e^{-x} + 2 e^{2x}. \tag{13}$$

We put $x = 0$ in each of (11), (12), and (13) to get the equations for the determination of $c_1, c_2$, and $c_3$. These are

$$0 = c_1 + c_2 + c_3 + \tfrac{1}{2},$$

$$-1 = c_2 - c_3 + 3,$$

$$2 = c_2 + c_3 - 2,$$

from which $c_1 = -\tfrac{9}{2}$, $c_2 = 0$, $c_3 = 4$. Therefore, the final result is

$$y = -\tfrac{9}{2} + 4 e^{-x} + 2x e^{-x} + \tfrac{1}{2} e^{2x}.$$

An important point, sometimes overlooked by students, is that it is the general solution, the $y$ of (11), that must be made to satisfy the initial conditions.

## Exercises

In exercises 1 through 35, obtain the general solution.

**1.** $(D^2 - 3D + 2)y = 2x^3 - 9x^2 + 6x.$
   ANS.  $y = c_1 e^x + c_2 e^{2x} + x^3.$

**2.** $(D^2 + 4)y = 5e^x - 4x.$
   ANS.  $y = c_1 \sin 2x + c_2 \cos 2x + e^x - x.$

**3.** $(D^2 + 4)y = 5e^x - 4x^2.$
   ANS.  $y = c_1 \sin 2x + c_2 \cos 2x + e^x - x^2 + \frac{1}{2}.$

**4.** $(D^2 + D)y = \sin x.$
   ANS.  $y = c_1 + c_2 e^{-x} - \frac{1}{2} \sin x - \frac{1}{2} \cos x.$

**5.** $(D^2 - 4D + 4)y = e^x.$
   ANS.  $y = (c_1 + c_2 x) e^{2x} + e^x.$

**6.** $(D^2 - 3D + 2)y = 2x^2 + 1.$
   ANS.  $y = c_1 e^x + c_2 e^{2x} + x^2 + 3x + 4.$

**7.** $y'' - 3y' - 4y = 6e^x.$
   ANS.  $y = c_1 e^{4x} + c_2 e^{-x} - e^x.$

**8.** $y'' - 3y' - 4y = 5e^{4x}.$
   ANS.  $y = (c_1 + x) e^{4x} + c_2 e^{-x}.$

**9.** $(D^2 - 4)y = 8e^{2x} - 12.$
   ANS.  $y = (c_1 + 2x) e^{2x} + c_2 e^{-2x} + 3.$

**10.** $(D^2 - D - 2)y = 1 - 2x - 9e^{-x}.$
   ANS.  $y = (c_1 + 3x) e^{-x} + c_2 e^{2x} + x - 1.$

**11.** $y'' - 4y' + 3y = 20 \cos x.$
   ANS.  $y = c_1 e^x + c_2 e^{3x} + 2 \cos x - 4 \sin x.$

**12.** $y'' - 4y' + 3y = 2 \cos x + 4 \sin x.$
   ANS.  $y = c_1 e^x + c_2 e^{3x} + \cos x.$

**13.** $y'' + 2y' + y = 7 + 75 \sin 2x.$
   ANS.  $y = e^{-x}(c_1 + c_2 x) + 7 - 12 \cos 2x - 9 \sin 2x.$

**14.** $(D^2 + 4D + 5)y = 50x + 13e^{3x}.$
   ANS.  $y = e^{-2x}(c_1 \cos x + c_2 \sin x) + 10x - 8 + \frac{1}{2} e^{3x}.$

**15.** $(D^2 + 1)y = \cos x.$
   ANS.  $y = c_1 \cos x + c_2 \sin x + \frac{1}{2} x \sin x.$

**16.** $(D^2 - 4D + 4)y = e^{2x}.$
   ANS.  $y = e^{2x}(c_1 + c_2 x + \frac{1}{2} x^2).$

**17.** $(D^2 - 1)y = e^{-x}(2 \sin x + 4 \cos x).$

**18.** $(D^2 - 1)y = 8xe^x.$
   ANS.  $y = c_1 e^{-x} + e^x(c_2 - 2x + 2x^2).$

**19.** $(D^3 - D)y = x.$
   ANS.  $y = c_1 + c_2 e^x + c_3 e^{-x} - \frac{1}{2} x^2.$

**20.** $(D^3 - D^2 + D - 1)y = 4 \sin x.$
   ANS.  $y = c_1 e^x + (c_2 + x) \cos x + (c_3 - x) \sin x.$

**21.** $(D^3 + D^2 - 4D - 4)y = 3e^{-x} - 4x - 6.$
   ANS.  $y = c_1 e^{2x} + c_2 e^{-2x} + (c_3 - x) e^{-x} + x + \frac{1}{2}.$

**22.** $(D^4 - 1)y = 7x^2.$

**23.** $(D^4 - 1)y = e^{-x}.$
   ANS.  $y = c_1 e^x + (c_2 - \frac{1}{4} x) e^{-x} + c_3 \cos x + c_4 \sin x.$

**24.** $(D^2 - 1)y = 10 \sin^2 x.$ Use the identity $\sin^2 x = \frac{1}{2}(1 - \cos 2x).$
   ANS.  $y = c_1 e^x + c_2 e^{-x} - 5 + \cos 2x.$

**25.** $(D^2 + 1)y = 12 \cos^2 x.$
   ANS.  $y = c_1 \cos x + c_2 \sin x + 6 - 2 \cos 2x.$

**26.** $(D^2 + 4)y = 4 \sin^2 x.$
   ANS.  $y = c_1 \cos 2x + c_2 \sin 2x + \frac{1}{2}(1 - x \sin 2x).$

**27.** $y'' - 3y' - 4y = 16x - 50 \cos 2x.$
   ANS.  $y = c_1 e^{4x} + c_2 e^{-x} + 3 - 4x + 4 \cos 2x + 3 \sin 2x.$

**28.** $(D^3 - 3D - 2)y = 100 \sin 2x.$

**29.** $y'' + 4y' + 3y = 15e^{2x} + e^{-x}.$
   ANS.  $y = c_1 e^{-x} + c_2 e^{-3x} + e^{2x} + \frac{1}{2} x e^{-x}.$

**30.** $y'' - y = e^x - 4.$

**31.** $y'' - y' - 2y = 6x + 6e^{-x}.$
   ANS.  $y = c_1 e^{2x} + c_2 e^{-x} + \frac{3}{2} - 3x - 2x e^{-x}.$

**32.** $y'' + 6y' + 13y = 60 \cos x + 26.$

**33.** $(D^3 - 3D^2 + 4)y = 6 + 80 \cos 2x$.

           ANS.    $y = c_1 e^{-x} + e^{2x}(c_2 + c_3 x) + \frac{3}{2} + 4 \cos 2x - 2 \sin 2x$.

**34.** $(D^3 + D - 10)y = 29 e^{4x}$.

**35.** $(D^3 + D^2 - 4D - 4)y = 8x + 8 + 6e^{-x}$.

           ANS.    $y = c_1 \cosh 2x + c_2 \sinh 2x + c_3 e^{-x} - 2x - 2xe^{-x}$.

In exercises 36 through 44, find the particular solution indicated.

**36.** $(D^2 + 1)y = 10 e^{2x}$; when $x = 0$, $y = 0$, $y' = 0$.

           ANS.    $y = 2(e^{2x} - \cos x - 2 \sin x)$.

**37.** $(D^2 - 4)y = 2 - 8x$; when $x = 0$, $y = 0$, $y' = 5$.

           ANS.    $y = e^{2x} - \frac{1}{2}e^{-2x} + 2x - \frac{1}{2}$.

**38.** $(D^2 + 3D)y = -18x$; when $x = 0$, $y = 0$, $y' = 5$.

           ANS.    $y = 1 + 2x - 3x^2 - e^{-3x}$.

**39.** $(D^2 + 4D + 5)y = 10 e^{-3x}$; when $x = 0$, $y = 4$, $y' = 0$.

**40.** $\dfrac{d^2x}{dt^2} + 4\dfrac{dx}{dt} + 5x = 10$; when $t = 0$, $x = 0$, $\dfrac{dx}{dt} = 0$.

           ANS.    $x = 2(1 - e^{-2t} \cos t - 2 e^{-2t} \sin t)$.

**41.** $\ddot{x} + 4\dot{x} + 5x = 8 \sin t$; when $t = 0$, $x = 0$, $\dot{x} = 0$. Note that $\dot{x} = dx/dt$, $\ddot{x} = d^2x/dt^2$ is a common notation when the independent variable is time.

           ANS.    $x = (1 + e^{-2t}) \sin t - (1 - e^{-2t}) \cos t$.

**42.** $y'' + 9y = 81x^2 + 14 \cos 4x$; when $x = 0$, $y = 0$, $y' = 3$.

**43.** $(D^3 + 4D^2 + 9D + 10)y = -24 e^x$; when $x = 0$, $y = 0$, $y' = -4$, $y'' = 10$.

**44.** $y'' + 2y' + 5y = 8e^{-x}$; when $x = 0$, $y = 0$, $y' = 8$.

           ANS.    $y = e^{-x}(2 + 4 \sin 2x - 2 \cos 2x)$.

In exercises 45 through 48 obtain, from the particular solution indicated, the value of $y$ and the value of $y'$ at $x = 2$.

**45.** $y'' + 2y' + y = x$; at $x = 0$, $y = -3$, and at $x = 1$, $y = -1$.

           ANS.    At $x = 2$, $y = e^{-2}$, $y' = 1$.

**46.** $y'' + 2y' + y = x$; at $x = 0$, $y = -2$, $y' = 2$.

           ANS.    At $x = 2$, $y = 2 e^{-2}$, $y' = 1 - e^{-2}$.

**47.** $4y'' + y = 2$; at $x = \pi$, $y = 0$, $y' = 1$.

           ANS.    At $x = 2$, $y = -0.7635$, $y' = +0.3012$.

**48.** $2y'' - 5y' - 3y = -9x^2 - 1$; at $x = 0$, $y = 1$, $y' = 0$.

           ANS.    At $x = 2$, $y = 5.64$, $y' = 5.68$.

**49.** $(D^2 + D)y = x + 1$; when $x = 0$, $y = 1$, and when $x = 1$, $y = \frac{1}{2}$. Compute the value of $y$ at $x = 4$.       ANS.    At $x = 4$, $y = 8 - e^{-1} - e^{-2} - e^{-3}$.

**50.** $(D^2 + 1)y = x^3$; when $x = 0$, $y = 0$, and when $x = \pi$, $y = 0$. Show that this boundary value problem has no solution.

**51.** $(D^2 + 1)y = 2 \cos x$; when $x = 0$, $y = 0$, and when $x = \pi$, $y = 0$. Show that this boundary value problem has an unlimited number of solutions and obtain them.

           ANS.    $y = (c + x) \sin x$.

**52.** For the equation $(D^3 + D^2)y = 4$, find the solution whose graph has at the origin a point of inflection with a horizontal tangent line.

           ANS.    $y = 4 - 4x + 2x^2 - 4 e^{-x}$.

**53.** For the equation $(D^2 - D)y = 2 - 2x$, find a particular solution that has at some point (to be determined) on the $x$-axis an inflection point with a horizontal tangent line.          ANS.   The point is $(1, 0)$; the solution is $y = x^2 + 1 - 2 \exp(x - 1)$.

## 42. Solution by inspection

It is frequently easy to obtain a particular solution of a nonhomogeneous equation

$$(b_0 D^n + b_1 D^{n-1} + \cdots + b_{n-1}D + b_n)y = R(x) \tag{1}$$

by inspection.

For example, if $R(x)$ is a constant $R_0$ and if $b_n \neq 0$,

$$y_p = \frac{R_0}{b_n} \tag{2}$$

is a solution of

$$(b_0 D^n + b_1 D^{n-1} + \cdots + b_n)y = R_0, \qquad b_n \neq 0, R_0 \text{ constant}, \tag{3}$$

because all derivatives of $y_p$ are zero, so

$$(b_0 D^n + b_1 D^{n-1} + \cdots + b_n)y_p = b_n R_0 / b_n = R_0.$$

Suppose that $b_n = 0$ in equation (3). Let $D^k y$ be the lowest-ordered derivative that actually appears in the differential equation. Then the equation may be written

$$(b_0 D^n + \cdots + b_{n-k}D^k)y = R_0, \qquad b_{n-k} \neq 0, R_0 \text{ constant}. \tag{4}$$

Now $D^k x^k = k!$, a constant, so that all higher derivatives of $x^k$ are zero. Thus it becomes evident that (4) has a solution

$$y_p = \frac{R_0 x^k}{k! b_{n-k}}, \tag{5}$$

for then $(b_0 D^n + \cdots + b_{n-k}D^k)y_p = b_{n-k}R_0 k!/k! b_{n-k} = R_0$.

EXAMPLE (a):   Solve the equation

$$(D^2 - 3D + 2)y = 16. \tag{6}$$

By the methods of Chapter 6 we obtain the complementary function,

$$y_c = c_1 e^x + c_2 e^{2x}.$$

By inspection a particular solution of the original equation is

$$y_p = \tfrac{16}{2} = 8.$$

Hence the general solution of (6) is

$$y = c_1 e^x + c_2 e^{2x} + 8.$$

EXAMPLE (b):  Solve the equation

$$\frac{d^5y}{dx^5} + 4\frac{d^3y}{dx^3} = 7. \tag{7}$$

From the auxiliary equation $m^5 + 4m^3 = 0$ we get $m = 0, 0, 0, \pm 2i$. Hence

$$y_c = c_1 + c_2 x + c_3 x^2 + c_4 \cos 2x + c_5 \sin 2x.$$

A particular solution of (7) is

$$y_p = \frac{7x^3}{3! \cdot 4} = \frac{7x^3}{24}.$$

As a check, note that

$$(D^5 + 4D^3)\frac{7x^3}{24} = 0 + 4 \cdot \frac{7 \cdot 6}{24} = 7.$$

The general solution of equation (7) is

$$y = c_1 + c_2 x + c_3 x^2 + \tfrac{7}{24}x^3 + c_4 \cos 2x + c_5 \sin 2x,$$

in which the $c_1, \ldots, c_5$ are arbitrary constants.

Examination of

$$(D^2 + 4)y = \sin 3x \tag{8}$$

leads us to search for a solution proportional to $\sin 3x$ because, if $y$ is proportional to $\sin 3x$, so is $D^2y$. Indeed, from

$$y = A \sin 3x \tag{9}$$

we get

$$D^2y = -9A \sin 3x,$$

so (9) is a solution of (8) if

$$(-9 + 4)A = 1$$

$$A = -\tfrac{1}{5}.$$

Thus (8) has the general solution

$$y = c_1 \cos 2x + c_2 \sin 2x - \tfrac{1}{5}\sin 3x,$$

a result easily obtained mentally.

For equation (8), the general method of undetermined coefficients leads us to write

$$m = \pm 2i, \qquad m' = \pm 3i,$$

and so to write

$$y_p = A \sin 3x + B \cos 3x. \tag{10}$$

When the $y_p$ of (10) is substituted into (8), it is found, of course, that

$$A = -\tfrac{1}{5} \qquad B = 0.$$

In contrast, consider the equation

$$(D^2 + 4D + 4)y = \sin 3x. \tag{11}$$

Here any attempt to find a solution proportional to $\sin 3x$ is doomed to failure because, although $D^2 y$ will also be proportional to $\sin 3x$, the term $Dy$ will involve $\cos 3x$. There is no other term on either side of (11) to compensate for this cosine term, so no solution of the form $y = A \sin 3x$ is possible. For this equation, $m = -2, -2, m' = \pm 3i$, and, in the particular solution

$$y_p = A \sin 3x + B \cos 3x,$$

it must turn out that $B \neq 0$. No labor has been saved by the inspection.

In more complicated situations such as

$$(D^2 + 4)y = x \sin 3x - 2 \cos 3x,$$

the method of inspection will save no work.

For the equation

$$(D^2 + 4)y = e^{5x}, \tag{12}$$

we see, since $(D^2 + 4) e^{5x} = 29 e^{5x}$, that

$$y_p = \tfrac{1}{29} e^{5x}$$

is a solution.

Finally, note that if $y_1$ is a solution of

$$f(D)y = R_1(x)$$

and $y_2$ is a solution of

$$f(D)y = R_2(x),$$

then

$$y_p = y_1 + y_2$$

is a solution of

$$f(D)y = R_1(x) + R_2(x).$$

It follows readily that the task of obtaining a particular solution of

$$f(D)y = R(x)$$

may be split into parts by treating separate terms of $R(x)$ independently, if convenient. See the examples below. This is the basis of the "method of superposition," which plays a useful role in applied mathematics.

EXAMPLE (c):   Find a particular solution of

$$(D^2 - 9)y = 3 e^x + x - \sin 4x. \tag{13}$$

Since $(D^2 - 9) e^x = -8 e^x$, we see by inspection that

$$y_1 = -\tfrac{3}{8} e^x$$

is a particular solution of

$$(D^2 - 9)y_1 = 3 e^x.$$

In a similar manner, we see that $y_2 = -\tfrac{1}{9}x$ satisfies

$$(D^2 - 9)y_2 = x$$

and that

$$y_3 = \tfrac{1}{25} \sin 4x$$

satisfies

$$(D^2 - 9)y_3 = -\sin 4x.$$

Hence

$$y_p = -\tfrac{3}{8} e^x - \tfrac{1}{9}x + \tfrac{1}{25} \sin 4x$$

is a solution of equation (13).

EXAMPLE (d):   Find a particular solution of

$$(D^2 + 4)y = \sin x + \sin 2x. \tag{14}$$

At once we see that $y_1 = \tfrac{1}{3} \sin x$ is a solution of

$$(D^2 + 4)y_1 = \sin x.$$

Then we seek a solution of

$$(D^2 + 4)y_2 = \sin 2x \tag{15}$$

by the method of undetermined coefficients. Because $m = \pm 2i$ and $m' = \pm 2i$, we put

$$y_2 = Ax \sin 2x + Bx \cos 2x$$

into (15) and easily determine that

$$4A \cos 2x - 4B \sin 2x = \sin 2x,$$

from which $A = 0, B = -\frac{1}{4}$.

Thus a particular solution of (14) is

$$y_p = \tfrac{1}{3} \sin x - \tfrac{1}{4} x \cos 2x.$$

EXAMPLE (e):    Find a particular solution of

$$(D^2 + a^2)y = \cos bx. \tag{16}$$

If $b \neq a$, then a particular solution of the form $y = A \cos bx$ will exist. It follows from (16) that

$$(-b^2 A + a^2 A) \cos bx = \cos bx$$

and $A = (a^2 - b^2)^{-1}$. A particular solution of (16) is

$$y = (a^2 - b^2)^{-1} \cos bx.$$

If $b = a$, then equation (16) becomes

$$(D^2 + a^2)y = \cos ax, \tag{17}$$

and no function of the form $A \cos ax$ is a particular solution since the operator $D^2 + a^2$ will annihilate $A \cos ax$. However, a solution of the form $Ax \cos ax + Bx \sin ax$ exists. Upon substitution into (17) we require that

$$-2aA \sin ax + 2aB \cos ax = \cos ax,$$

an equation that is satisfied only if $A = 0$ and $B = 1/2a$. Therefore

$$y = \frac{x}{2a} \sin ax \tag{18}$$

is a particular solution of (17).

We have seen in this example an important distinction between the cases $b \neq a$ and $b = a$. In a physical application considered in Chapter 10, the presence of a solution of the form given in (18) results in a phenomenon called resonance. At this point we need only notice that the solution in (18) will be oscillatory in character, but the amplitudes of the oscillation will become increasingly large as $x$ increases.

### Exercises

**1.** Show that if $b \neq a$, then

$$(D^2 + a^2)y = \sin bx$$

has the particular solution $y = (a^2 - b^2)^{-1} \sin bx$.

2. Show that the equation

$$(D^2 + a^2)y = \sin ax$$

has no solution of the form $y = A \sin ax$, with $A$ constant. Find a particular solution of the equation.

ANS. $y = -\dfrac{x}{2a} \cos ax.$

In exercises 3 through 50 find a particular solution by inspection. Verify your solution.

3. $(D^2 + 4)y = 12.$     ANS. $y = 3.$
4. $(D^2 + 9)y = 18.$     ANS. $y = 2.$
5. $(D^2 + 4D + 4)y = 8.$     ANS. $y = 2.$
6. $(D^2 + 2D - 3)y = 6.$     ANS. $y = -2.$
7. $(D^3 - 3D + 2)y = -7.$     8. $(D^4 + 4D^2 + 4)y = -20.$
9. $(D^2 + 4D)y = 12.$     10. $(D^3 - 9D)y = 27.$
11. $(D^3 + 5D)y = 15.$     ANS. $y = 3x.$
12. $(D^3 + D)y = -8.$     ANS. $y = -8x.$
13. $(D^4 - 4D^2)y = 24.$     ANS. $y = -3x^2.$
14. $(D^4 + D^2)y = -12.$     ANS. $y = -6x^2.$
15. $(D^5 - D^3)y = 24.$     ANS. $y = -4x^3.$
16. $(D^5 - 9D^3)y = 27.$     ANS. $y = \tfrac{1}{2}x^3.$
17. $(D^2 + 4)y = 6 \sin x.$     ANS. $y = 2 \sin x.$
18. $(D^2 + 4)y = 10 \cos 3x.$     ANS. $y = -2 \cos 3x.$
19. $(D^2 + 4)y = 8x + 1 - 15 e^x.$     ANS. $y = 2x + \tfrac{1}{4} - 3 e^x.$
20. $(D^2 + D)y = 6 + 3 e^{2x}.$     ANS. $y = 6x + \tfrac{1}{2}e^{2x}.$
21. $(D^2 + 3D - 4)y = 18 e^{2x}.$     ANS. $y = 3 e^{2x}.$
22. $(D^2 + 2D + 5)y = 4 e^x - 10.$     ANS. $y = \tfrac{1}{2}e^x - 2.$
23. $(D^2 - 1)y = 2 e^{3x}.$     24. $(D^2 - 1)y = 2x + 3.$
25. $(D^2 - 1)y = \cos 2x.$     26. $(D^2 - 1)y = \sin 2x.$
27. $(D^2 + 1)y = e^x + 3x.$     28. $(D^2 + 1)y = 5 e^{-3x}.$
29. $(D^2 + 1)y = -2x + \cos 2x.$     30. $(D^2 + 1)y = 4 e^{-2x}.$
31. $(D^2 + 1)y = 10 \sin 4x.$     32. $(D^2 + 1)y = -6 e^{-3x}.$
33. $(D^2 + 2D + 1)y = 12 e^x.$     ANS. $y = 3 e^x.$
34. $(D^2 + 2D + 1)y = 7e^{-2x}.$     ANS. $y = 7e^{-2x}.$
35. $(D^2 - 2D + 1)y = 12 e^{-x}.$     ANS. $y = 3e^{-x}.$
36. $(D^2 - 2D + 1)y = 6 e^{-2x}.$     ANS. $y = \tfrac{2}{3}e^{-2x}.$
37. $(D^2 - 2D - 3)y = e^x.$     ANS. $y = -\tfrac{1}{4}e^x.$
38. $(D^2 - 2D - 3)y = e^{2x}.$     ANS. $y = -\tfrac{1}{3}e^{2x}.$
39. $(4D^2 + 1)y = 12 \sin x.$     ANS. $y = -4 \sin x.$
40. $(4D^2 + 1)y = -12 \cos x.$     ANS. $y = 4 \cos x.$
41. $(4D^2 + 4D + 1)y = 18 e^x - 5.$     42. $(4D^2 + 4D + 1)y = 7 e^{-x} + 2.$
43. $(D^3 - 1)y = e^{-x}.$     44. $(D^3 - 1)y = 4 - 3x^2.$
45. $(D^3 - D)y = e^{2x}.$     46. $(D^4 + 4)y = 5 e^{2x}.$
47. $(D^4 + 4)y = 6 \sin 2x.$     48. $(D^4 + 4)y = \cos 2x.$
49. $(D^3 - D)y = 5 \sin 2x.$     50. $(D^3 - D)y = 5 \cos 2x.$

# Variation of Parameters

## 43. Introduction

In Chapter 7 we solved the nonhomogeneous linear equation with constant coefficients

$$(b_0 D^n + b_1 D^{n-1} + \cdots + b_{n-1}D + b_n)y = R(x) \tag{1}$$

by the method of undetermined coefficients. We saw that this method would be applicable only for a certain class of differential equations: those for which $R(x)$ itself was a solution of a homogeneous linear equation with constant coefficients.

In this chapter we shall study two methods that carry no such restrictions. In fact, much of what we do will be applicable to linear equations with variable coefficients.

We begin with a procedure by D'Alembert that is often called the method of reduction of order.

## 44. Reduction of order

Consider the general second-order linear equation

$$y'' + py' + qy = R. \tag{1}$$

Suppose that we know a solution $y = y_1$ of the corresponding homogeneous equation

$$y'' + py' + qy = 0. \tag{2}$$

Then the introduction of a new dependent variable $v$ by the substitution

$$y = y_1 v \tag{3}$$

will lead to a solution of equation (1) in the following way.

From (3) it follows that

$$y' = y_1 v' + y_1' v,$$
$$y'' = y_1 v'' + 2y_1' v' + y_1'' v,$$

so substitution of (3) into (1) yields

$$y_1 v'' + 2y_1' v' + y_1'' v + py_1 v' + py_1' v + qy_1 v = R,$$

or

$$y_1 v'' + (2y_1' + py_1)v' + (y_1'' + py_1' + qy_1)v = R. \tag{4}$$

But $y = y_1$ is a solution of (2). That is,

$$y_1'' + py_1' + qy_1 = 0$$

and equation (4) reduces to

$$y_1 v'' + (2y_1' + py_1)v' = R. \tag{5}$$

Now let $v' = w$ so equation (5) becomes

$$y_1 w' + (2y_1' + py_1)w = R, \tag{6}$$

a linear equation of first order in $w$.

By the usual method (integrating factor) we can find $w$ from (6). Then we can get $v$ from $v' = w$ by an integration. Finally $y = y_1 v$.

*Note that the method is not restricted to equations with constant coefficients. It depends only upon our knowing a particular solution of equation (2); that is, upon our knowledge of the complementary function. For practical purposes, the method depends also upon our being able to effect the integrations.*

EXAMPLE (a):   Solve the equation

$$y'' - y = e^x. \tag{7}$$

The complementary function of (7) is

$$y_c = c_1 e^x + c_2 e^{-x}.$$

We shall take the particular solution $e^x$ and use the method of reduction of order by setting

$$y = v e^x.$$

Then

$$y' = v e^x + v' e^x,$$

and

$$y'' = v e^x + 2v' e^x + v'' e^x.$$

Substituting into equation (7) gives

$$v'' + 2v' = 1. \tag{8}$$

Equation (8) is a first-order linear equation in the variable $v'$. Applying the integrating factor $e^{2x}$ yields

$$e^{2x}(v'' + 2v') = e^{2x}.$$

Thus

$$e^{2x}v' = \tfrac{1}{2} e^{2x} + c, \tag{9}$$

where $c$ is an arbitrary constant. Equation (9) readily gives

$$v' = \tfrac{1}{2} + c e^{-2x},$$

and hence

$$v = c_1 e^{-2x} + c_2 + \tfrac{1}{2}x,$$

where $c_1$ and $c_2$ are arbitrary constants.

Remembering that $y = v e^x$, we finally have

$$y = c_1 e^{-x} + c_2 e^x + \tfrac{1}{2}x e^x.$$

Of course, the solution to equation (7) could have been obtained by the method of undetermined coefficients. Let us now solve a problem not solvable by that method.

EXAMPLE (b):   Solve the equation

$$(D^2 + 1)y = \csc x. \tag{10}$$

The complementary function is

$$y_c = c_1 \cos x + c_2 \sin x. \tag{11}$$

We may use any special case of (11) as the $y_1$ in the theory above. Let us then put

$$y = v \sin x.$$

We find that

$$y' = v' \sin x + v \cos x$$

and

$$y'' = v'' \sin x + 2v' \cos x - v \sin x.$$

The equation for $v$ is

$$v'' \sin x + 2v' \cos x = \csc x,$$

or

$$v'' + 2v' \cot x = \csc^2 x. \tag{12}$$

Put $v' = w$; then equation (12) becomes

$$w' + 2w \cot x = \csc^2 x,$$

for which an integrating factor is $\sin^2 x$. Thus

$$\sin^2 x \, dw + 2w \sin x \cos x \, dx = dx \tag{13}$$

is exact. From (13) we get

$$w \sin^2 x = x,$$

and if we seek only a particular solution, we have

$$w = x \csc^2 x$$

or

$$v' = x \csc^2 x.$$

Hence

$$v = \int x \csc^2 x \, dx,$$

or

$$v = -x \cot x + \ln |\sin x|,$$

a result easily obtained using integration by parts.
    Now

$$y = v \sin x,$$

so the particular solution which we sought is

$$y_p = -x \cos x + \sin x \ln |\sin x|.$$

Finally, the complete solution of (10) is seen to be

$$y = c_1 \cos x + c_2 \sin x - x \cos x + \sin x \ln |\sin x|.$$

## Exercises

Use the method of reduction of order to solve the following equations.

1. $(D^2 - 1)y = x - 1.$          ANS.    $y = c_1 e^x + c_2 e^{-x} - x + 1.$
2. $(D^2 - 5D + 6)y = 2 e^x.$        ANS.    $y = c_1 e^{2x} + c_2 e^{3x} + e^x.$
3. $(D^2 - 4D + 4)y = e^x.$        ANS.   See exercise 5, Section 41.
4. $(D^2 + 4)y = \sin x.$       ANS.    $y = c_1 \sin 2x + c_2 \cos 2x + \frac{1}{3} \sin x.$
5. Use the substitution $y = v \cos x$ to solve the equation of Example (b) above.
6. Use $y = v e^{-x}$ to solve the equation of Example (a) above.
7. $(D^2 + 1)y = \sec x.$
8. $(D^2 + 1)y = \sec^3 x.$ Use $y = v \sin x.$
9. $(D^2 + 1)y = \csc^3 x.$ Take a hint from exercise 8.
10. $(D^2 + 2D + 1)y = (e^x - 1)^{-2}.$      ANS.    $y = e^{-x}(c_1 + c_2 x - \ln |1 - e^{-x}|).$
11. $(D^2 - 3D + 2)y = (1 + e^{2x})^{-1/2}.$
12. Verify that $y = e^x$ is a solution of the equation

$$(x - 1)y'' - xy' + y = 0.$$

Use this fact to find the general solution of

$$(x - 1)y'' - xy' + y = 1.$$

ANS.    $y = c_1 x + c_2 e^x + 1.$

13. Observe that $y = x$ is a particular solution of the equation

$$2x^2 y'' + xy' - y = 0$$

and find the general solution. For what values of $x$ is the solution valid?

ANS.    $y = c_1 x + c_2 x^{-1/2}.$

14. In Chapter 19 we shall study Bessel's differential equation of index zero

$$xy'' + y' + xy = 0.$$

Suppose that one solution of this equation is given the name $J_0(x)$. Show that a second solution takes the form

$$J_0(x) \int \frac{dx}{x[J_0(x)]^2}.$$

15. One solution of the Legendre differential equation

$$(1 - x^2)y'' - 2xy' + 2y = 0$$

is $y = x$. Find a second solution.       ANS.    $y = -2 + x \ln \left| \dfrac{1 + x}{1 - x} \right|.$

## 45. Variation of parameters

In the previous section we saw that if $y_1$ is a solution of the homogeneous equation

$$y'' + p(x)y' + q(x)y = 0 \tag{1}$$

then we can use it to determine the general solution of the nonhomogeneous equation

$$y'' + p(x)y' + q(x)y = R(x). \tag{2}$$

In using the method of reduction of order we proceeded as follows. Because $y_1$ is a solution of (1), the function $c_1y_1$ is also a solution for an arbitrary constant $c_1$. We replaced the constant $c_1$ by a function $v(x)$ and considered the possibility of the existence of a solution of equation (2) of the form $v \cdot y_1$. This led us to a first-order linear equation in the variable $v'$ that we were able to solve.

Suppose now that we know the general solution of the homogeneous equation (1). That is, suppose

$$y_c = c_1y_1 + c_2y_2 \tag{3}$$

is a solution of (1), where $y_1$ and $y_2$ are linearly independent on an interval $a < x < b$. Let us see what happens if we replace both of the constants in (3) with functions of $x$. That is, we consider

$$y = Ay_1 + By_2 \tag{4}$$

and try to determine $A(x)$ and $B(x)$ so that $Ay_1 + By_2$ is a solution of equation (2).

Note that we are involved with two unknown functions $A(x)$ and $B(x)$ and that we have only insisted that these functions satisfy one condition: the function in (4) is to be a solution of equation (2). We may therefore expect to impose a second condition on $A(x)$ and $B(x)$ in some way which would be to our advantage. Indeed, if we simply impose the condition $B(x) \equiv 0$, then we will be dealing with the method of reduction of order. Actually we impose a somewhat different condition on $A$ and $B$.

From (4) it follows that

$$y' = Ay_1' + By_2' + A'y_1 + B'y_2. \tag{5}$$

Rather than become involved with derivatives of $A$ and $B$ of higher order than the first, we now choose some particular function for the expression

$$A'y_1 + B'y_2.$$

Technically, we could let this function be $\sin x$, $e^x$, or any other suitable function. For simplicity we choose

$$A'y_1 + B'y_2 = 0. \tag{6}$$

It then follows from (5) that

$$y'' = Ay_1'' + By_2'' + A'y_1' + B'y_2'. \tag{7}$$

Because $y$ was to be a solution of (2), we substitute from (4), (5), and (7) into equation (2) to obtain

$$A(y_1'' + py_1' + qy_1) + B(y_2'' + py_2' + qy_2) + A'y_1' + B'y_2' = R(x).$$

But $y_1$ and $y_2$ are solutions of the homogeneous equation (1), so that finally

$$A'y_1' + B'y_2' = R(x). \tag{8}$$

Equations (6) and (8) now give us two equations that we wish to solve for $A'$ and $B'$. This solution exists providing the determinant

$$\begin{vmatrix} y_1 & y_2 \\ y_1' & y_2' \end{vmatrix}$$

does not vanish. But this determinant is precisely the Wronskian of the functions $y_1$ and $y_2$, which were presumed to be linearly independent on the interval $a < x < b$. Therefore, the Wronskian does not vanish on that interval and we can find $A'$ and $B'$. By integration we can now find $A$ and $B$. Once $A$ and $B$ are known, equation (4) gives us the desired $y$.

*This argument can easily be extended to equations of order higher than two, but no essentially new ideas appear. Moreover, there is nothing in the method that prohibits the linear differential equation involved from having variable coefficients.*

**EXAMPLE (a):** Solve the equation

$$(D^2 + 1)y = \sec x \tan x. \tag{9}$$

Of course,

$$y_c = c_1 \cos x + c_2 \sin x.$$

Let us seek a particular solution by variation of parameters. Put

$$y = A \cos x + B \sin x, \tag{10}$$

from which

$$y' = -A \sin x + B \cos x + A' \cos x + B' \sin x.$$

Next set

$$A' \cos x + B' \sin x = 0, \tag{11}$$

so that

$$y' = -A \sin x + B \cos x.$$

Then

$$y'' = -A \cos x - B \sin x - A' \sin x + B' \cos x. \tag{12}$$

Next we eliminate $y$ by combining equations (10) and (12) with the original equation (9). Thus we get the relation

$$-A' \sin x + B' \cos x = \sec x \tan x. \tag{13}$$

From (13) and (11), $A'$ is easily eliminated. The result is

$$B' = \tan x,$$

so that

$$B = \ln |\sec x|, \tag{14}$$

in which the arbitrary constant has been disregarded because we are seeking only a particular solution to add to our previously determined complementary function $y_c$.

From equations (13) and (11) it also follows easily that

$$A' = -\sin x \sec x \tan x,$$

or

$$A' = -\tan^2 x.$$

Then

$$A = -\int \tan^2 x \, dx = \int (1 - \sec^2 x) \, dx,$$

so that

$$A = x - \tan x, \tag{15}$$

again disregarding the arbitrary constant.

Returning to equation (10) with the known $A$ from (15) and the known $B$ from (14), we write the particular solution

$$y_p = (x - \tan x) \cos x + \sin x \ln |\sec x|,$$

or

$$y_p = x \cos x - \sin x + \sin x \ln |\sec x|.$$

Then the general solution of (9) is

$$y = c_1 \cos x + c_3 \sin x + x \cos x + \sin x \ln |\sec x|, \tag{16}$$

where the term $(-\sin x)$ in $y_p$ has been absorbed in the complementary function term $c_3 \sin x$, since $c_3$ is an arbitrary constant.

The solution (16) can, as usual, be verified by direct substitution into the original differential equation.

EXAMPLE (b): Solve the equation

$$(D^2 - 3D + 2)y = \frac{1}{1 + e^{-x}}. \tag{17}$$

Here

$$y_c = c_1 e^x + c_2 e^{2x},$$

so we put

$$y = A e^x + B e^{2x}. \tag{18}$$

Because

$$y' = A e^x + 2B e^{2x} + A' e^x + B' e^{2x},$$

we impose the condition

$$A' e^x + B' e^{2x} = 0. \tag{19}$$

Then

$$y' = A e^x + 2B e^{2x}, \tag{20}$$

from which it follows that

$$y'' = A e^x + 4B e^{2x} + A' e^x + 2B' e^{2x}. \tag{21}$$

Combining (18), (20), (21), and the original equation (17), we find that

$$A' e^x + 2B' e^{2x} = \frac{1}{1 + e^{-x}}. \tag{22}$$

Elimination of $B'$ from equations (19) and (22) yields

$$A' e^x = -\frac{1}{1 + e^{-x}},$$

$$A' = -\frac{e^{-x}}{1 + e^{-x}}.$$

Then

$$A = \ln(1 + e^{-x}).$$

Similarly,

$$B' e^{2x} = \frac{1}{1 + e^{-x}}$$

so that

$$B = \int \frac{e^{-2x}}{1 + e^{-x}} dx = \int \left( e^{-x} - \frac{e^{-x}}{1 + e^{-x}} \right) dx,$$

or

$$B = -e^{-x} + \ln (1 + e^{-x}).$$

Then, from (18),

$$y_p = e^x \ln (1 + e^{-x}) - e^x + e^{2x} \ln (1 + e^{-x}).$$

The term $(-e^x)$ in $y_p$ can be absorbed into the complementary function. The general solution of equation (17) is

$$y = c_3 e^x + c_2 e^{2x} + (e^x + e^{2x}) \ln (1 + e^{-x}).$$

## 46. Solution of $y'' + y = f(x)$

Consider next the equation

$$(D^2 + 1)y = f(x), \tag{1}$$

in which all that we require of $f(x)$ is that it be integrable in the interval on which we seek a solution. For instance, $f(x)$ may be any continuous function or any function with only a finite number of finite discontinuities on the interval $a \le x \le b$.

The method of variation of parameters will now be applied to the solution of (1). Put

$$y = A \cos x + B \sin x. \tag{2}$$

Then

$$y' = -A \sin x + B \cos x + A' \cos x + B' \sin x,$$

and if we choose

$$A' \cos x + B' \sin x = 0, \tag{3}$$

we obtain

$$y'' = -A \cos x - B \sin x - A' \sin x + B' \cos x. \tag{4}$$

From (1), (2), and (4) it follows that

$$-A' \sin x + B' \cos x = f(x). \tag{5}$$

Equations (3) and (5) may be solved for $A'$ and $B'$, yielding

$$A' = -f(x) \sin x \quad \text{and} \quad B' = f(x) \cos x.$$

We may now write

$$A = -\int_a^x f(\beta) \sin \beta \, d\beta, \tag{6}$$

$$B = \int_a^x f(\beta) \cos \beta \, d\beta, \tag{7}$$

for any $x$ in $a \le x \le b$. It is here that we use the integrability of $f(x)$ on the interval $a \le x \le b$.

The $A$ and $B$ of (6) and (7) may be inserted in (2) to give us the particular solution

$$y_p = -\cos x \int_a^x f(\beta) \sin \beta \, d\beta + \sin x \int_a^x f(\beta) \cos \beta \, d\beta$$

$$= \int_a^x f(\beta)(\sin x \cos \beta - \cos x \sin \beta) \, d\beta. \tag{8}$$

Hence we have

$$y_p = \int_a^x f(\beta) \sin(x - \beta) \, d\beta, \tag{9}$$

and we can now write the general solution of equation (1):

$$y = c_1 \cos x + c_2 \sin x + \int_a^x f(\beta) \sin(x - \beta) \, d\beta. \tag{10}$$

## Exercises

In exercises 1 through 18 use variation of parameters.

1. $(D^2 - 1)y = e^x + 1$.  ANS.  $y = c_1 e^x + c_2 e^{-x} + \frac{1}{2}x e^x - \frac{1}{4}e^x - 1$.

2. $(D^2 + 1)y = \csc x \cot x$.

  ANS.  $y = c_1 \cos x + c_2 \sin x - x \sin x - \cos x \ln |\sin x|$.

3. $(D^2 + 1)y = \csc x$.  ANS.  $y = c_1 \sin x + c_2 \cos x - x \cos x + \sin x \ln |\sin x|$.

4. $(D^2 + 2D + 2)y = e^{-x} \csc x$.  ANS.  $y_p = -xe^{-x} \cos x + e^{-x} \sin x \ln |\sin x|$.

5. $(D^2 + 1)y = \sec^3 x$.  ANS.  $y = y_c + \frac{1}{2} \sec x$.

6. $(D^2 + 1)y = \sec^4 x$.  ANS.  $y = y_c - \frac{1}{2} + \frac{1}{6} \sec^2 x + \frac{1}{2} \sin x \ln |\sec x + \tan x|$.

7. $(D^2 + 1)y = \tan x$.  ANS.  $y = y_c - \cos x \ln |\sec x + \tan x|$.

8. $(D^2 + 1)y = \tan^2 x$.  ANS.  $y = y_c - 2 + \sin x \ln |\sec x + \tan x|$.

9. $(D^2 + 1)y = \sec x \csc x$.

  ANS.  $y = y_c - \cos x \ln |\sec x + \tan x| - \sin x \ln |\csc x + \cot x|$.

10. $(D^2 + 1)y = \sec^2 x \csc x$.  ANS.  $y = y_c - \sin x \ln |\csc 2x + \cot 2x|$.

11. $(D^2 - 2D + 1)y = e^{2x}(e^x + 1)^{-2}$.  ANS.  $y = y_c + e^x \ln(1 + e^x)$.

12. $(D^2 - 3D + 2)y = e^{2x}/(1 + e^{2x})$.

  ANS.  $y = y_c + e^x \arctan(e^{-x}) - \frac{1}{2}e^{2x} \ln(1 + e^{-2x})$.

13. $(D^2 - 3D + 2)y = \cos(e^{-x})$.  ANS.  $y = y_c - e^{2x} \cos(e^{-x})$.

14. $(D^2 - 1)y = 2(1 - e^{-2x})^{-1/2}$.

    ANS. $\quad y = c_1 e^x + c_2 e^{-x} - e^x \arcsin(e^{-x}) - (1 - e^{-2x})^{1/2}$.

15. $(D^2 - 1)y = e^{-2x} \sin e^{-x}$.

    ANS. $\quad y = y_c - \sin e^{-x} - e^x \cos e^{-x}$.

16. $(D - 1)(D - 2)(D - 3)y = e^x$.

    ANS. $\quad y = y_c + \tfrac{1}{2}x\, e^x$.

17. $y''' - y' = x$.

18. $y''' + y' = \tan x$.

19. Observe that $x$ and $e^x$ are solutions of the homogeneous equation associated with

$$(1 - x)y'' + xy' - y = 2(x - 1)^2 e^{-x}.$$

Use this fact to solve the nonhomogeneous equation.

20. Solve the equation

$$y'' - y = e^x$$

by the method of variation of parameters, but instead of setting $A'y_1 + B'y_2 = 0$ as in equation (6) Section 45, choose $A'y_1 + B'y_2 = k$, for constant $k$.

21. Apply the suggestion of exercise 20 to exercise 5 above.

22. Let $y_1$ and $y_2$ be solutions of the homogeneous equation associated with

$$y'' + p(x)y' + q(x)y = f(x). \tag{A}$$

Let $W(x)$ be the Wronskian of $y_1$ and $y_2$, and assume $W(x) \neq 0$ on the interval $a < x < b$. Show that a particular solution of equation (A) is given by

$$y_p = \int_a^x \frac{f(\beta)[y_1(\beta)y_2(x) - y_1(x)y_2(\beta)]\, d\beta}{W(\beta)}. \tag{B}$$

23. The conditions of exercise 22 imply that

$$y_1'' + py_1' + qy_1 = 0 \tag{C}$$

and

$$y_2'' + py_2' + qy_2 = 0. \tag{D}$$

If we multiply equation (C) by $y_2$ and equation (D) by $y_1$ and then subtract the two equations, we obtain

$$(y_2 y_1'' - y_1 y_2'') + p(y_2 y_1' - y_1 y_2') = 0.$$

From this equation show that the Wronskian of $y_1$ and $y_2$ can be written

$$W(x) = c \exp\left(-\int p\, dx\right), \tag{E}$$

where $c$ is constant. Equation (E) is known as Abel's formula.

24. Conclude from exercise 23 that if $W(x_0) = 0$ for some $x_0$ on the interval $a < x < b$, then $W(x) \equiv 0$ for all $a < x < b$.

25. Solve the initial value problem

$$y'' + y = f(x); \qquad \text{when } x = x_0, y = y_0, y' = y_0'.$$

*Hint:* Show that the constant $a$ in equations (6) and (7) of Section 46 could have been chosen to be $x_0$. Determine the $c_1$ and $c_2$ of equation (10) by using the form of $y_p$ in equation (8).

ANS.   $y = y_0 \cos(x - x_0) + y_0' \sin(x - x_0) + \displaystyle\int_{x_0}^{x} f(\beta) \sin(x - \beta)\, d\beta.$

## Miscellaneous Exercises

**1.** $(D^2 - 1)y = 2e^{-x}(1 + e^{-2x})^{-2}.$     ANS.   $y = y_c - x e^{-x} - \frac{1}{2}e^{-x}\ln(1 + e^{-2x}).$

**2.** $(D^2 - 1)y = (1 - e^{2x})^{-3/2}.$     ANS.   $y = y_c - (1 - e^{2x})^{1/2}.$

**3.** $(D^2 - 1)y = e^{2x}(3\tan e^x + e^x \sec^2 e^x).$     ANS.   $y = y_c + e^x \ln|\sec e^x|.$

**4.** $(D^2 + 1)y = \sec^2 x \tan x.$     ANS.   $y = y_c + \frac{1}{2}\tan x + \frac{1}{2}\cos x \ln|\sec x + \tan x|.$

**5.** Do exercise 4 by another method.

**6.** $(D^2 + 1)y = \cot x.$     ANS.   $y = y_c - \sin x \ln|\csc x + \cot x|.$

**7.** $(D^2 + 1)y = \sec x.$     ANS.   $y = c_1 \cos x + c_2 \sin x + x \sin x + \cos x \ln|\cos x|.$

**8.** Do exercise 7 by another method.

**9.** $(D^2 - 1)y = 2/(1 + e^x).$     ANS.   $y = y_c - 1 - x e^x + (e^x - e^{-x})\ln(1 + e^x).$

**10.** $(D^3 + D)y = \sec^2 x.$ *Hint:* integrate once first.

ANS.   $y = c_1 + c_2 \cos x + c_3 \sin x - \cos x \ln|\sec x + \tan x|.$

**11.** $(D^2 - 1)y = 2/(e^x - e^{-x}).$     ANS.   $y = y_c - x e^{-x} + \frac{1}{2}(e^x - e^{-x})\ln|1 - e^{-2x}|.$

**12.** $(D^2 - 3D + 2)y = \sin e^{-x}.$     ANS.   $y = y_c - e^{2x}\sin e^{-x}.$

**13.** $(D^2 - 1)y = 1/(e^{2x} + 1).$     ANS.   $y = y_c - \frac{1}{2} - \cosh x \arctan e^{-x}.$

**14.** $y'' + y = \sec^3 x \tan x.$     ANS.   $y = y_c + \frac{1}{6}\sec x \tan x.$

**15.** $y'' + y = \sec x \tan^2 x.$ Verify your answer.

**16.** $y'' + 4y' + 3y = \sin e^x.$     ANS.   $y = y_c - e^{-2x}\sin e^x - e^{-3x}\cos e^x.$

**17.** $y'' + y = \csc^3 x \cot x.$     ANS.   $y = y_c + \frac{1}{6}\cot x \csc x.$

**9**

# Inverse Differential Operators

## 47. The exponential shift

In Chapter 5 we studied some of the properties of the algebra of linear differential operators with constant coefficients. We found this algebra useful in finding solutions of homogeneous linear equations. In this chapter we show briefly how differential operators may be used to find particular solutions for nonhomogeneous linear equations.

As a first illustration we make use of the exponential shift theorem that was derived in Section 32:

$$e^{ax}f(D)y = f(D - a)[e^{ax}y], \tag{1}$$

where $f(D)$ is a linear differential operator with constant coefficients.

EXAMPLE (a):  Solve the equation

$$(D^2 - 2D + 5)y = 16x^3 e^{3x}. \tag{2}$$

Note that the complementary function is

$$y_c = c_1 e^x \cos 2x + c_2 e^x \sin 2x. \tag{3}$$

We can conclude also that there is a particular solution,

$$y_p = Ax^3 e^{3x} + Bx^2 e^{3x} + Cx e^{3x} + E e^{3x}, \tag{4}$$

which can be obtained by the method of Chapter 7. But the task of obtaining the derivatives of $y_p$ and finding the numerical values of $A$, $B$, $C$, and $E$ is a little tedious. It can be made easier by using the exponential shift (1).

Let us write (2) in the form

$$e^{-3x}(D^2 - 2D + 5)y = 16x^3,$$

and then apply the relation (1), with $a = -3$. In shifting the exponential $e^{-3x}$ from the left to the right of the differential operator, we must replace $D$ by $(D + 3)$ throughout, thus obtaining

$$[(D + 3)^2 - 2(D + 3) + 5](e^{-3x}y) = 16x^3,$$

or

$$(D^2 + 4D + 8)(e^{-3x}y) = 16x^3. \tag{5}$$

In equation (5), the dependent variable is $(e^{-3x}y)$. We know at once that (5) has a particular solution of the form

$$e^{-3x}y_p = Ax^3 + Bx^2 + Cx + E. \tag{6}$$

Successive differentiations of (6) are simple. Indeed,

$$D(e^{-3x}y_p) = 3Ax^2 + 2Bx + C,$$
$$D^2(e^{-3x}y_p) = 6Ax + 2B,$$

thus, from (5) we get

$$6Ax + 2B + 12Ax^2 + 8Bx + 4C + 8Ax^3 + 8Bx^2 + 8Cx + 8E = 16x^3.$$

Hence

$$8A = 16,$$
$$12A + 8B = 0,$$
$$6A + 8B + 8C = 0,$$
$$2B + 4C + 8E = 0,$$

from which $A = 2$, $B = -3$, $C = \frac{3}{2}$, $E = 0$.

Therefore

$$e^{-3x}y_p = 2x^3 - 3x^2 + \tfrac{3}{2}x,$$

or

$$y_p = (2x^3 - 3x^2 + \tfrac{3}{2}x)\, e^{3x},$$

and the general solution of the original equation (2) is

$$y = c_1\, e^x \cos 2x + c_2\, e^x \sin 2x + (2x^3 - 3x^2 + \tfrac{3}{2}x)\, e^{3x}.$$

EXAMPLE (b):   Solve the equation

$$(D^2 - 2D + 1)y = x\, e^x + 7x - 2. \tag{7}$$

Here the immediate use of the exponential shift would do no good, because removing the $e^x$ factor from the first term on the right would only insert a factor $e^{-x}$ in the second and third terms on the right. The terms $7x - 2$ on the right give us no trouble as they stand. Therefore we break (7) into two problems, obtaining a particular solution for each of the equations

$$(D^2 - 2D + 1)y_1 = x\, e^x \tag{8}$$

and

$$(D^2 - 2D + 1)y_2 = 7x - 2. \tag{9}$$

On (8) we use the exponential shift, passing from

$$e^{-x}(D - 1)^2 y_1 = x$$

to

$$D^2(e^{-x}y_1) = x. \tag{10}$$

A particular solution of (10) is easily obtained:

$$e^{-x}y_1 = \tfrac{1}{6}x^3,$$

so

$$y_1 = \tfrac{1}{6}x^3\, e^x. \tag{11}$$

Equation (9) is treated as in Chapter 7. Put

$$y_2 = Ax + B.$$

Then $Dy_2 = A$, and from (9) it is easily found that $A = 7$, $B = 12$. Thus a particular solution of (9) is

$$y_2 = 7x + 12. \tag{12}$$

Using (11), (12), and the roots of the auxiliary equation for (7), the general solution of (7) can now be written. It is

$$y = (c_1 + c_2 x)\, e^x + \tfrac{1}{6}x^3\, e^x + 7x + 12.$$

EXAMPLE (c):   Solve the equation

$$D^2(D + 4)^2 y = 96\, e^{-4x}. \tag{13}$$

At once we have $m = 0, 0, -4, -4$ and $m' = -4$. We seek first a particular solution. Therefore we integrate each member of (13) twice before using the exponential shift. From (13) it follows that

$$(D + 4)^2 y_p = 6\, e^{-4x}, \tag{14}$$

the constants of integration being disregarded because only a particular solution is sought. Equation (14) yields

$$e^{4x}(D + 4)^2 y_p = 6,$$

$$D^2(y_p\, e^{4x}) = 6,$$

$$y_p\, e^{4x} = 3x^2,$$

$$y_p = 3x^2\, e^{-4x}.$$

Thus the general solution of (13) is seen to be

$$y = c_1 + c_2 x + (c_3 + c_4 x + 3x^2)\, e^{-4x}.$$

The exponential shift is particularly helpful when applied in connection with terms for which the values of $m'$ (using the notations of Chapter 7) are repetitions of values of $m$.

### Exercises

In exercises 1 through 12 use the exponential shift to find a particular solution.

1. $(D - 3)^2 y = e^{3x}$.          ANS.   $y = \tfrac{1}{2}x^2 e^{3x}$.
2. $(D - 1)^2 y = e^x$.           ANS.   $y = \tfrac{1}{2}x^2 e^x$.
3. $(D + 2)^2 y = 12x e^{-2x}$.      ANS.   $y = 2x^3 e^{-2x}$.
4. $(D + 1)^2 y = 3x e^{-x}$.        ANS.   $y = \tfrac{1}{2}x^3 e^{-x}$.
5. $(D - 2)^3 y = 6x e^{2x}$.        ANS.   $y = \tfrac{1}{4}x^4 e^{2x}$.
6. $(D + 4)^3 y = 8x e^{-4x}$.      ANS.   $y = \tfrac{1}{3}x^4 e^{-4x}$.
7. $(D + 3)^3 y = 15x^2 e^{-3x}$.    ANS.   $y = \tfrac{1}{4}x^5 e^{-3x}$.
8. $(D - 4)^3 y = 15x^2 e^{4x}$.     ANS.   $y = \tfrac{1}{4}x^5 e^{4x}$.
9. $D^2(D - 2)^2 y = 16 e^{2x}$.     ANS.   $y = 2x^2 e^{2x}$.
10. $D^2(D + 3)^2 y = 9 e^{-3x}$.    ANS.   $y = \tfrac{1}{2}x^2 e^{-3x}$.
11. $(D^2 - D - 2)y = 18x e^{-x}$.    ANS.   $y = -(3x^2 + 2x)e^{-x}$.
12. $(D^2 - D - 2)y = 36x e^{2x}$.    ANS.   $y = e^{2x}(6x^2 - 4x)$.

In exercises 13 through 15, find a particular solution, using the exponential shift in part of your work, as in Example (b) above.

13. $(D - 2)^2 y = 20 - 3x e^{2x}$.    ANS.   $y = 5 - \tfrac{1}{2}x^3 e^{2x}$.
14. $(D - 2)^2 y = 4 - 8x + 6x e^{2x}$.    ANS.   $y = x^3 e^{2x} - 2x - 1$.

**15.** $y'' - 9y = 9(2x - 3 + 4xe^{3x})$.     ANS.   $y = 3 - 2x + (3x^2 - x)e^{3x}$.

**16.** $y'' + 4y' + 4y = 4x - 6e^{-2x} + 3e^x$.     ANS.   $y = \frac{1}{3}e^x - 3x^2 e^{-2x} + x - 1$.

**17.** $(D + 1)^2 y = e^{-x} + 3x$.     ANS.   $y = \frac{1}{2}x^2 e^{-x} + 3x - 6$.

**18.** $(D^2 - 4)y = 16x e^{-2x} + 8x + 4$.     ANS.   $y = -(2x + 1)(x e^{-2x} + 1)$.

In exercises 19 through 28 find the general solution.

**19.** $y'' - 4y = 8x e^{2x}$.     ANS.   $y = c_1 e^{-2x} + (c_2 - \frac{1}{2}x + x^2) e^{2x}$.

**20.** $y'' - 9y = -72x e^{-3x}$.

**21.** $D(D + 1)^2 y = e^{-x}$.     ANS.   $y = c_1 + (c_2 + c_3 x - \frac{1}{2}x^2) e^{-x}$.

**22.** $D^2(D - 2)^2 y = 2e^{2x}$.

**23.** $y'' + 2y' + y = 48 e^{-x} \cos 4x$.     ANS.   $y = (c_1 + c_2 x - 3 \cos 4x) e^{-x}$.

**24.** $y'' + 4y' + 4y = 18 e^{-2x} \cos 3x$.

**25.** $(D - 1)^2 y = e^x \sec^2 x \tan x$.     ANS.   $y = e^x(c_1 + c_2 x + \frac{1}{2}\tan x)$.

**26.** $(D^2 + 4D + 4)y = -x^{-2} e^{-2x}$.     ANS.   $y = e^{-2x}(c_1 + c_2 x + \ln|x|)$.

**27.** $(D - a)^2 y = e^{ax} f''(x)$.     ANS.   $y = e^{ax}[c_1 + c_2 x + f(x)]$.

**28.** $(D^2 + 7D + 12)y = e^{-3x} \sec^2 x(1 + 2 \tan x)$.

            ANS.   $y = c_1 e^{-4x} + e^{-3x}(c_2 + \tan x)$.

## 48. The operator $1/f(D)$

In seeking a particular solution of

$$f(D)y = R(x), \tag{1}$$

it is natural to write

$$y = \frac{1}{f(D)} R(x) \tag{2}$$

and to try to define an operator $1/f(D)$ so that the function $y$ of (2) will have meaning and will satisfy equation (1).

Instead of building a theory of such inverse differential operators, we shall adopt the following method of attack. Purely formal (unjustified) manipulations of the symbols will be performed, thus leading to a tentative evaluation of

$$\frac{1}{f(D)} R(x).$$

After all, the only thing that we require of our evaluation is that

$$f(D) \cdot \frac{1}{f(D)} R(x) = R(x). \tag{3}$$

Hence the burden of proof will be placed on a direct verification of the condition (3) in each instance.

## 49. Evaluation of $[1/f(D)]\,e^{ax}$

We proved (Section 32) with slightly different notation that

$$f(D)\,e^{ax} = e^{ax}f(a) \tag{1}$$

and

$$(D-a)^n(x^n\,e^{ax}) = n!\,e^{ax}. \tag{2}$$

Equation (1) suggests

$$\frac{1}{f(D)}\,e^{ax} = \frac{e^{ax}}{f(a)}, \qquad f(a) \neq 0. \tag{3}$$

Now from (1) it follows that

$$f(D)\frac{e^{ax}}{f(a)} = \frac{f(a)\,e^{ax}}{f(a)} = e^{ax}.$$

Hence (3) is verified.

Now suppose that $f(a) = 0$. Then $f(D)$ contains the factor $(D-a)$. Suppose that the factor occurs precisely $n$ times in $f(D)$; that is, let

$$f(D) = \phi(D)(D-a)^n, \qquad \phi(a) \neq 0.$$

With the aid of (2) we obtain

$$\phi(D)(D-a)^n(x^n\,e^{ax}) = \phi(D)n!\,e^{ax},$$

from which, by (1), it follows that

$$\phi(D)(D-a)^n(x^n\,e^{ax}) = n!\phi(a)\,e^{ax}. \tag{4}$$

Therefore we write

$$\frac{1}{\phi(D)(D-a)^n}\,e^{ax} = \frac{x^n\,e^{ax}}{n!\phi(a)}, \qquad \phi(a) \neq 0, \tag{5}$$

which is easily verified. Indeed,

$$\phi(D)(D-a)^n\frac{x^n\,e^{ax}}{n!\phi(a)} = \frac{n!\phi(a)\,e^{ax}}{n!\phi(a)} = e^{ax}.$$

Note that formula (3) is included in formula (5) as the special case, $n = 0$. See also exercise 35, page 154.

EXAMPLE (a):   Solve the equation

$$(D^2 + 1)y = e^{2x}. \tag{6}$$

Here the roots of the auxiliary equation are $m = \pm i$. Further,

$$f(D) = (D^2 + 1)$$

and

$$f(2) \neq 0.$$

Hence, using (3),

$$y_p = \frac{1}{D^2 + 1} e^{2x} = \frac{e^{2x}}{2^2 + 1} = \frac{1}{5} e^{2x},$$

so the solution of (6) is

$$y = c_1 \cos x + c_2 \sin x + \tfrac{1}{5} e^{2x}.$$

EXAMPLE (b):   Solve the equation

$$D^2(D - 1)^3(D + 1)y = e^x. \tag{7}$$

Here $m = 0, 0, 1, 1, 1, -1$. To get a particular solution of (7), we use formula (5) with $a = 1, n = 3, \phi(D) = D^2(D + 1)$. Then

$$\phi(1) = 1^2 \cdot 2$$

and a particular solution of (7) is given by

$$y_p = \frac{1}{(D - 1)^3 D^2(D + 1)} e^x = \frac{x^3 e^x}{3!1^2 \cdot 2} = \frac{1}{12} x^3 e^x.$$

Then the general solution of (7) is

$$y = c_1 + c_2 x + c_3 e^{-x} + (c_4 + c_5 x + c_6 x^2 + \tfrac{1}{12} x^3) e^x.$$

## 50. Evaluation of $(D^2 + a^2)^{-1} \sin ax$ and $(D^2 + a^2)^{-1} \cos ax$

No special device is needed for the evaluation of

$$(D^2 + a^2)^{-1} \sin bx$$

when $b \neq a$. In fact, it is easy to show that

$$\frac{1}{D^2 + a^2} \sin bx = \frac{\sin bx}{a^2 - b^2}, \qquad b \neq a;$$

and a similar result holds for the expression $(D^2 + a^2)^{-1} \cos bx$, when $b \neq a$.
Consider next the evaluation of

$$\frac{1}{D^2 + a^2} \sin ax. \tag{1}$$

The formulas of the preceding section can be put to good use here, since

$$\sin ax = \frac{e^{aix} - e^{-aix}}{2i}.$$

Then

$$\frac{1}{D^2 + a^2} \sin ax = \frac{1}{2i} \frac{1}{(D - ai)(D + ai)}(e^{aix} - e^{-aix})$$

$$= \frac{1}{2i}\left(\frac{x\,e^{aix}}{1!\,2ai} - \frac{x\,e^{-aix}}{1!(-2ai)}\right)$$

$$= -\frac{x}{2a} \frac{e^{aix} + e^{-aix}}{2},$$

so

$$\frac{1}{D^2 + a^2} \sin ax = -\frac{x}{2a} \cos ax. \tag{2}$$

The verification of (2) is left as an exercise.

Another useful result,

$$\frac{1}{D^2 + a^2} \cos ax = \frac{x}{2a} \sin ax, \tag{3}$$

can be obtained in the same way.

## Exercises

1. Verify formula (2) of Section 50.
2. Obtain and verify formula (3) of Section 50.

In exercises 3 through 34, find the general solution.

3. $(D^2 - 1)y = e^{2x}$.    ANS.   $y = c_1 e^x + c_2 e^{-x} + \frac{1}{3}e^{2x}$.
4. $(D^2 - 1)y = e^x$.    ANS.   $y = c_1 e^x + c_2 e^{-x} + \frac{1}{2}xe^x$.
5. $(D^2 + 1)y = \sin x$.    ANS.   $y = c_1 \cos x + c_2 \sin x - \frac{1}{2}x \cos x$.
6. $(D^2 + 4)y = \cos 2x$.    ANS.   $y = c_1 \cos 2x + c_2 \sin 2x + \frac{1}{4}x \sin 2x$.
7. $(D^2 + 9)y = e^{2x}$.    8. $(D^2 + 4)y = e^{3x}$.
9. $(4D^2 + 1)y = e^{-2x}$.    10. $D(D - 2)y = e^{-x}$.
11. $D(D - 2)^2 y = e^{2x}$.    12. $D(D + 3)^2 y = e^{-3x}$.
13. $(D^2 + 4)y = \cos 3x$.    14. $(D^2 + 9)y = \cos 3x$.
15. $(D^2 + 4)y = \sin 2x$.    16. $(D^2 + 36)y = \sin 6x$.
17. $(D^2 + 9)y = \sin 3x$.    18. $(D^2 + 36)y = \cos 6x$.
19. $(D^2 + 3D - 4)y = 12 e^{2x}$.    ANS.   $y = c_1 e^{-4x} + c_2 e^x + 2e^{2x}$.
20. $(D^2 + 3D - 4)y = 21 e^{3x}$.    ANS.   $y = c_1 e^x + c_2 e^{-4x} + \frac{3}{2}e^{3x}$.
21. $(D^2 + 3D - 4)y = 15 e^x$.    ANS.   $y = c_1 e^{-4x} + (c_2 + 3x)e^x$.

**22.** $(D^2 + 3D - 4)y = 20e^{-4x}$.    ANS.   $y = c_1 e^x + (c_2 - 4x)e^{-4x}$.

**23.** $(D^2 - 3D + 2)y = e^x + e^{2x}$.    ANS.   $y = (c_1 - x)e^x + (c_2 + x)e^{2x}$.

**24.** $(4D^2 - 1)y = e^{x/2} + 12e^x$.    ANS.   $y = c_1 e^{-x/2} + (c_2 + \frac{1}{4}x)e^{x/2} + \frac{1}{3}e^x$.

**25.** $D^2(D - 2)^3 y = 48e^{2x}$.    ANS.   $y = c_1 + c_2 x + (c_3 + c_4 x + c_5 x^2 + 2x^3)e^{2x}$.

**26.** $(D^4 - 18D^2 + 81)y = 36e^{3x}$.    ANS.   $y = (c_1 + c_2 x + \frac{1}{2}x^2)e^{3x} + (c_3 + c_4 x)e^{-3x}$.

**27.** $(D^2 + 16)y = 14\cos 3x$.    ANS.   $y = c_1 \cos 4x + c_2 \sin 4x + 2\cos 3x$.

**28.** $(4D^2 + 1)y = 35\sin 3x$.    ANS.   $y = c_1 \cos\frac{1}{2}x + c_2 \sin\frac{1}{2}x - \sin 3x$.

**29.** $y'' + 16y = 24\sin 4x$.    ANS.   $y = (c_1 - 3x)\cos 4x + c_2 \sin 4x$.

**30.** $y'' + 16y = 48\cos 4x$.    ANS.   $y = c_1 \cos 4x + (c_2 + 6x)\sin 4x$.

**31.** $y'' + y = 12\cos 2x - \sin x$.    ANS.   $y = (c_1 + \frac{1}{2}x)\cos x + c_2 \sin x - 4\cos 2x$.

**32.** $y'' + y = \sin 3x + 4\cos x$.    ANS.   $y = c_1 \cos x + (c_2 + 2x)\sin x - \frac{1}{8}\sin 3x$.

**33.** $(D^2 - 2D + 5)y = e^x \cos 2x$. *Hint:* use the exponential shift followed by formula (3) of Section 50.    ANS.   $y = e^x(c_1 \cos 2x + c_2 \sin 2x) + \frac{1}{4}x\, e^x \sin 2x$.

**34.** $(D^2 + 2D + 5)y = e^{-x}\sin 2x$.

ANS.   $y = e^{-x}(c_1 \cos 2x + c_2 \sin 2x) - \frac{1}{4}x\, e^{-x}\cos 2x$.

**35.** Prove that if $f(x) = (x - a)^n \phi(x)$, then $f^{(n)}(a) = n!\,\phi(a)$. Then use equation (5) of Section 49 above to prove that

$$\frac{1}{f(D)} e^{ax} = \frac{x^n e^{ax}}{f^{(n)}(a)},$$

where $n$ is the smallest nonnegative integer for which $f^{(n)}(a) \neq 0$.*

In exercises 36 through 39 use the formula of exercise 35 above.

**36.** Exercise 21.                **37.** Exercise 22.

**38.** Exercise 25.                **39.** Exercise 26.

In exercises 40 to 45 verify the formulas† stated.

**40.** $\dfrac{1}{f(D)}\sin ax = \dfrac{f(-D)\sin ax}{f(ai)f(-ai)};\ f(ai)f(-ai) \neq 0$.

**41.** $\dfrac{1}{f(D)}\cos ax = \dfrac{f(-D)\cos ax}{f(ai)f(-ai)};\ f(ai)f(-ai) \neq 0$.

**42.** $\dfrac{1}{f(D)}\sinh ax = \dfrac{f(-D)\sinh ax}{f(a)f(-a)};\ f(a)f(-a) \neq 0$.

**43.** $\dfrac{1}{f(D)}\cosh ax = \dfrac{f(-D)\cosh ax}{f(a)f(-a)};\ f(a)f(-a) \neq 0$.

**44.** $\dfrac{1}{(D^2 + a^2)^n}\sin ax = \dfrac{x^n}{(2a)^n n!}\sin(ax - \frac{1}{2}n\pi)$.

**45.** $\dfrac{1}{(D^2 + a^2)^n}\cos ax = \dfrac{x^n}{(2a)^n n!}\cos(ax - \frac{1}{2}n\pi)$.

* See C. A. Hutchinson: Another note on linear operators. *Amer. Math. Mon.*, **46**:161 (1939).

† The formulas of exercises 40 through 43 were obtained by C. A. Hutchinson: An operational formula. *Amer. Math. Mon.*, **40**:482–483 (1933); those of exercises 44 and 45 were given by C. A. Hutchinson, Note on an operational formula, *Amer. Math. Mon.*, **44**:371–372 (1937).

In exercises 46 to 49 use the formulas of exercises 40 and 41 above.

**46.** Exercise 4, page 125.          **47.** Exercise 11, page 125.
**48.** Exercise 12, page 125.         **49.** Exercise 32, page 125.

# Applications

## 51. Vibration of a spring

Consider a steel spring attached to a support and hanging downward. Within certain elastic limits the spring will obey Hooke's law: if the spring is stretched or compressed, its change in length will be proportional to the force exerted upon it and, when that force is removed, the spring will return to its original position with its length and other physical properties unchanged. There is, therefore, associated with each spring a numerical constant, the ratio of the force exerted to the displacement produced by that force. If a force of magnitude $Q$ pounds (lb) stretches the spring $c$ feet (ft), the relation

$$Q = kc \tag{1}$$

defines the spring constant $k$ in units of pounds per foot (lb/ft).

Let a body $B$ weighing $w$ lb be attached to the lower end of a spring (Figure 13) and brought to the point of equilibrium where it can remain at rest. Once the weight $B$ is moved from the point of equilibrium $E$ in Figure 14, the motion of $B$ will be determined by a differential equation and associated initial conditions.

Let $t$ be time measured in seconds after some initial moment when the motion begins. Let $x$, in feet, be distance measured positive downward (negative upward) from the point of equilibrium, as in Figure 14. We assume that the motion of $B$ takes place entirely in a vertical line, so the velocity and acceleration are given by the first and second derivatives of $x$ with respect to $t$.

FIGURE 13              FIGURE 14

In addition to the force proportional to displacement (Hooke's law), there will in general be a retarding force caused by resistance of the medium in which the motion takes place or by friction. We are interested here only in such retarding forces as can be well approximated by a term proportional to the velocity because we restrict our study to problems involving linear differential equations. Such a retarding force will contribute to the total force acting on $B$ a term $bx'(t)$, in which $b$ is a constant to be determined experimentally for the medium in which the motion takes place. Some common retarding forces, such as one proportional to the cube of the velocity, lead to nonlinear differential equations.

The weight of the spring is usually negligible compared to the weight $B$, so we use for the mass of our system the weight of $B$ divided by $g$, the constant acceleration of gravity. If no forces other than those described above act upon the weight, the displacement $x$ must satisfy the equation

$$\frac{w}{g}x''(t) + bx'(t) + kx(t) = 0. \tag{2}$$

Suppose that an additional vertical force, due to the motion of the support or to presence of a magnetic field, and so on, is imposed upon the system. The new, impressed force, will depend upon time and we may use $F(t)$ to denote the acceleration that it alone would impart to the weight $B$. Then the impressed force is $(w/g)F(t)$ and equation (2) is replaced by

$$\frac{w}{g}x''(t) + bx'(t) + kx(t) = \frac{w}{g}F(t). \tag{3}$$

At time zero, let the weight be displaced by an amount $x_0$ from the equilibrium point and let the weight be given an initial velocity $v_0$. Either or both of $x_0$ and $v_0$ may be zero in specific instances. The problem of determining the position of the weight at any time $t$ becomes that of solving the initial value problem consisting of the differential equation

$$\frac{w}{g}x''(t) + bx'(t) + kx(t) = \frac{w}{g}F(t), \qquad \text{for } t > 0, \tag{4}$$

and the initial conditions

$$x(0) = x_0, \qquad x'(0) = v_0. \tag{5}$$

It is convenient to rewrite equation (4) in the form

$$x''(t) + 2\gamma x'(t) + \beta^2 x(t) = F(t), \tag{6}$$

in which we have put

$$\frac{bg}{w} = 2\gamma, \qquad \frac{kg}{w} = \beta^2.$$

We may choose $\beta > 0$ and we know $\gamma \geqq 0$. Note that $\gamma = 0$ corresponds to a negligible retarding force.

A number of special cases of the initial value problem contained in equations (5) and (6) will now be studied.

## 52. Undamped vibrations

If $\gamma = 0$ in the problem of Section 51, the differential equation becomes

$$x''(t) + \beta^2 x(t) = F(t), \tag{1}$$

a second-order linear equation with constant coefficients in which $\beta^2 = kg/w$. The complementary function associated with the homogeneous equation $x''(t) + \beta^2 x(t) = 0$ is

$$x_c = c_1 \sin \beta t + c_2 \cos \beta t,$$

and the general solution of equation (1) will be of the form

$$x = c_1 \sin \beta t + c_2 \cos \beta t + x_p, \tag{2}$$

where $x_p$ is any particular solution of the nonhomogeneous equation.

We now look at a number of examples of the motion described by equation (2) for different functions $F(t)$ in equation (1).

EXAMPLE (a): Solve the spring problem with no damping but with $F(t) = A \sin \omega t$, where $\beta \neq \omega$. The case $\beta = \omega$ leads to resonance, which will be discussed in the next section.

The differential equation of motion is

$$\frac{w}{g} x''(t) + kx(t) = \frac{w}{g} A \sin \omega t$$

and may be written

$$x''(t) + \beta^2 x(t) = A \sin \omega t, \tag{3}$$

with the introduction of $\beta^2 = kg/w$. We shall assume initial conditions

$$x(0) = x_0, \qquad x'(0) = v_0. \tag{4}$$

A particular solution of equation (3) will be of the form

$$x_p = E \sin \omega t,$$

and we may obtain $E$ by direct substitution into equation (3). We have

$$-E\omega^2 \sin \omega t + \beta^2 E \sin \omega t = A \sin \omega t,$$

an equation that is satisfied for all $t$ only if we choose

$$E = \frac{A}{\beta^2 - \omega^2}.$$

The general solution of (3) now becomes

$$x(t) = c_1 \sin \beta t + c_2 \cos \beta t + \frac{A}{\beta^2 - \omega^2} \sin \omega t \tag{5}$$

with derivative

$$x'(t) = c_1 \beta \cos \beta t - c_2 \beta \sin \beta t + \frac{A\omega}{\beta^2 - \omega^2} \cos \omega t.$$

The initial conditions (4) now require

$$x_0 = c_2 \qquad \text{and} \qquad v_0 = c_1 \beta + \frac{A\omega}{\beta^2 - \omega^2}$$

and force us to choose

$$c_1 = \frac{v_0}{\beta} - \frac{A\omega}{\beta(\beta^2 - \omega^2)} \qquad \text{and} \qquad c_2 = x_0.$$

From (5) it follows at once that

$$x(t) = \frac{v_0}{\beta}\sin \beta t + x_0 \cos \beta t - \frac{A\omega}{\beta(\beta^2 - \omega^2)}\sin \beta t + \frac{A}{\beta^2 - \omega^2}\sin \omega t. \quad (6)$$

The $x$ of (6) has two parts. The first two terms represent the natural simple harmonic component of the motion, a motion that would be present if $A$ were zero. The last two terms in (6) are caused by the presence of the external force $(w/g)A \sin \omega t$.

EXAMPLE (b):  A spring is such that it would be stretched 6 inches (in.) by a 12-lb weight. Let the weight be attached to the spring and pulled down 4 in. below the equilibrium point. If the weight is started with an upward velocity of 2 ft/sec, describe the motion. No damping or impressed force is present.

We know that the acceleration of gravity enters our work in the expression for the mass. We wish to use the value $g = 32$ feet per second per second (ft/sec$^2$) and we must use consistent units, so we put all lengths into feet.

First we determine the spring constant $k$ from the fact that the 12-lb weight stretches the spring 6 in., $\frac{1}{2}$ ft. Thus $12 = \frac{1}{2}k$ so that $k = 24$ lb/ft.

The differential equation of the motion is therefore

$$\tfrac{12}{32}x''(t) + 24x(t) = 0. \quad (7)$$

At time zero the weight is 4 in. ($\frac{1}{3}$ ft) below the equilibrium point, so $x(0) = \frac{1}{3}$. The initial velocity is negative (upward), so $x'(0) = -2$. Thus our problem is that of solving

$$x''(t) + 64x(t) = 0; \qquad x(0) = \tfrac{1}{3}, x'(0) = -2. \quad (8)$$

The general solution of equation (8) is

$$x(t) = c_1 \sin 8t + c_2 \cos 8t,$$

from which

$$x'(t) = 8c_1 \cos 8t - 8c_2 \sin 8t.$$

The initial conditions now require that

$$\tfrac{1}{3} = c_2 \qquad \text{and} \qquad -2 = 8c_1,$$

so that finally

$$x(t) = -\tfrac{1}{4}\sin 8t + \tfrac{1}{3}\cos 8t. \quad (9)$$

A detailed study of the motion is straightforward once (9) has been obtained. The amplitude of the motion is

$$\sqrt{(\tfrac{1}{3})^2 + (\tfrac{1}{4})^2} = \tfrac{5}{12};$$

that is, the weight oscillates between points 5 in. above and below $E$. The period is $\frac{1}{4}\pi$ sec.

## 53. Resonance

In Example (a) of the previous section we postponed the study of the special case, $\beta = \omega$. In that case, the differential equation to be solved is

$$x''(t) + \beta^2 x(t) = A \sin \beta t, \tag{1}$$

where we had let $\beta^2 = kg/w$.

The complementary function associated with the homogeneous equation $x''(t) + \beta^2 x(t) = 0$ will be the same as it was before, but the previous particular solution $x_p$ will not exist because $\beta = \omega$.

The method of undetermined coefficients may be applied here to seek a particular solution of the form

$$x_p = Pt \sin \beta t + Qt \cos \beta t, \tag{2}$$

where $P$ and $Q$ are constants to be determined. Direct substitution of the $x_p$ of (2) into equation (1) yields

$$2P\beta \cos \beta t - 2Q\beta \sin \beta t = A \sin \beta t,$$

an equation that can be satisfied for all $t$ only if $P = 0$ and $Q = -A/2\beta$. Thus

$$x_p = \frac{-At}{2\beta} \cos \beta t, \tag{3}$$

and the general solution of (1) is

$$x(t) = c_1 \sin \beta t + c_2 \cos \beta t - \frac{At}{2\beta} \cos \beta t, \tag{4}$$

from which we obtain

$$x'(t) = c_1 \beta \cos \beta t - c_2 \beta \sin \beta t + \frac{At}{2} \sin \beta t - \frac{A}{2\beta} \cos \beta t.$$

The initial conditions $x(0) = x_0$ and $x'(0) = v_0$ now force us to take

$$c_2 = x_0 \quad \text{and} \quad c_1 = \frac{v_0}{\beta} + \frac{A}{2\beta^2}.$$

The final solution may now be written

$$x(t) = x_0 \cos \beta t + \frac{v_0}{\beta} \sin \beta t + \frac{A}{2\beta^2}(\sin \beta t - \beta t \cos \beta t). \tag{5}$$

That (5) satisfies the initial value problem is readily verified.

In the solution (5) the terms proportional to $\cos \beta t$ and $\sin \beta t$ are bounded, but the term with $\beta t \cos \beta t$ can be made as large as we wish by proper choice of $t$. This building up of large amplitudes in the vibration is called *resonance*.

## Exercises

1. A spring is such that a 5-lb weight stretches it 6 in. The 5-lb weight is attached, the spring reaches equilibrium, then the weight is pulled down 3 in. below the equilibrium point and started off with an upward velocity of 6 ft/sec. Find an equation giving the position of the weight at all subsequent times.
   ANS.   $x = \frac{1}{4}(\cos 8t - 3 \sin 8t)$.

2. A spring is stretched 1.5 in. by a 2-lb weight. Let the weight be pushed up 3 in. above $E$ and then released. Describe the motion.        ANS.   $x = -\frac{1}{4} \cos 16t$.

3. For the spring and weight of exercise 2, let the weight be pulled down 4 in. below $E$ and given a downward initial velocity of 8 ft/sec. Describe the motion.
   ANS.   $x = \frac{1}{3} \cos 16t + \frac{1}{2} \sin 16t$.

4. Show that the answer to exercise 3 can be written $x = 0.60 \sin (16t + \phi)$ where $\phi = \arctan \frac{2}{3}$.

5. A spring is such that a 4-lb weight stretches it 6 in. An impressed force $\frac{1}{2} \cos 8t$ is acting on the spring. If the 4-lb weight is started from the equilibrium point with an imparted upward velocity of 4 ft/sec, determine the position of the weight as a function of time.        ANS.   $x = \frac{1}{4}(t - 2) \sin 8t$.

6. A spring is such that it is stretched 6 in. by a 12-lb weight. The 12-lb weight is pulled down 3 in. below the equilibrium point and then released. If there is an impressed force of magnitude $9 \sin 4t$ lb, describe the motion. Assume that the impressed force acts downward for very small $t$.
   ANS.   $x = \frac{1}{4} \cos 8t - \frac{1}{4} \sin 8t + \frac{1}{2} \sin 4t$.

7. Show that the answer to exercise 6 can be written

$$x = \frac{1}{4}\sqrt{2} \cos (8t + \pi/4) + \frac{1}{2} \sin 4t.$$

8. A spring is such that a 2-lb weight stretches it $\frac{1}{2}$ ft. An impressed force $\frac{1}{4} \sin 8t$ is acting upon the spring. If the 2-lb weight is released from a point 3 in. below the equilibrium point, determine the equation of motion.
   ANS.   $x = \frac{1}{4}(1 - t) \cos 8t + \frac{1}{32} \sin 8t$ (ft).

9. For the motion of exercise 8, find the first four times at which stops occur and find the position at each stop.        ANS.   $t = \pi/8, \pi/4, 1, 3\pi/8$ (sec)
   and $x = -0.15, +0.05, +0.03, +0.04$ (ft), respectively.

10. Determine the approximate position to be expected, if nothing such as breakage interferes, at the time of the 65th stop, when $t = 8\pi$ (sec), in exercise 8.
   ANS.   $x = -6.0$ (ft).

11. A spring is such that a 16-lb weight stretches it 1.5 in. The weight is pulled down to a point 4 in. below the equilibrium point and given an initial downward velocity of 4 ft/sec. An impressed force of $360 \cos 4t$ lb is applied. Find the position and velocity of the weight at time $t = \pi/8$ sec.
   ANS.   At $t = \pi/8$ (sec), $x = -\frac{8}{3}$ (ft), $v = -8$ (ft/sec).

12. A spring is stretched 3 in. by a 5-lb weight. Let the weight be started from $E$ with an upward velocity of 12 ft/sec. Describe the motion.    ANS.    $x = -1.06 \sin 11.3t$

13. For the spring and weight of exercise 12, let the weight be pulled down 4 in. below $E$ and then given an upward velocity of 8 ft/sec. Describe the motion.
$$\text{ANS.} \quad x = 0.33 \cos 11.3t - 0.71 \sin 11.3t.$$

14. Find the amplitude of the motion in exercise 13.    ANS.    0.78 ft.

15. A 20-lb weight stretches a certain spring 10 in. Let the spring first be compressed 4 in., and then the 20-lb weight attached and given an initial downward velocity of 8 ft/sec. Find how far the weight would drop.    ANS.    35 in.

16. A spring is such that an 8-lb weight would stretch it 6 in. Let a 4-lb weight be attached to the spring, which is then pushed up 2 in. above its equilibrium point and released. Describe the motion.    ANS.    $x = -\frac{1}{6} \cos 11.3t.$

17. If the 4-lb weight of exercise 16 starts at the same point, 2 in. above $E$, but with an upward velocity of 15 ft/sec, when will the weight reach its lowest point?
$$\text{ANS.} \quad \text{At } t = \text{approximately 0.4 sec.}$$

18. A spring is such that it is stretched 4 in. by a 10-lb weight. Suppose the 10-lb weight to be pulled down 5 in. below $E$ and then given a downward velocity of 15 ft/sec. Describe the motion.
$$\text{ANS.} \quad x = 0.42 \cos 9.8t + 1.53 \sin 9.8t$$
$$= 1.59 \cos (9.8t - \phi), \text{ where } \phi = \text{arc tan } 3.64.$$

19. A spring is such that it is stretched 4 in. by an 8-lb weight. Suppose the weight to be pulled down 6 in. below $E$ and then given an upward velocity of 8 ft/sec. Describe the motion.    ANS.    $x = 0.50 \cos 9.8t - 0.82 \sin 9.8t.$

20. Show that the answer to exercise 19 can be written $x = 0.96 \cos (9.8t + \phi)$ where $\phi = \text{arc tan } 1.64.$

21. A spring is such that a 4 lb weight stretches it 6 in. The 4-lb weight is attached to the vertical spring and reaches its equilibrium point. The weight is then $(t = 0)$ drawn downward 3 in. and released. There is a simple harmonic exterior force equal to $\sin 8t$ impressed upon the whole system. Find the time for each of the first four stops following $t = 0$. Put the stops in chronological order.
$$\text{ANS.} \quad t = \pi/8, \tfrac{1}{2}, \pi/4, 3\pi/8 \text{ (sec).}$$

22. A spring is stretched 1.5 in. by a 4-lb weight. Let the weight be pulled down 3 in. below equilibrium and released. If there is an impressed force $8 \sin 16t$ acting upon the spring, describe the motion.    ANS.    $x = \tfrac{1}{4}(1 - 8t) \cos 16t + \tfrac{1}{8} \sin 16t.$

23. For the motion of exercise 22, find the first four times at which stops occur and find the position at each stop.    ANS.    $t = \tfrac{1}{8}, \pi/16, \pi/8, 3\pi/16$ (sec) and
$$x = +0.11, +0.14, -0.54, +0.93 \text{ (ft), respectively.}$$

## 54. Damped vibrations

In the general linear spring problem of Section 51, we were confronted with

$$x''(t) + 2\gamma x'(t) + \beta^2 x(t) = F(t); \qquad x(0) = x_0, x'(0) = v_0, \qquad (1)$$

in which $2\gamma = bg/w$ and $\beta^2 = kg/w$, $\beta > 0$. The auxiliary equation $m^2 + 2\gamma m + \beta^2 = 0$ has roots $-\gamma \pm \sqrt{\gamma^2 - \beta^2}$ and we see that the nature of the complementary function depends upon whether $\beta > \gamma$, $\beta = \gamma$, or $\beta < \gamma$.

If $\beta > \gamma$, $\beta^2 - \gamma^2 > 0$, so let us put

$$\beta^2 - \gamma^2 = \delta^2. \tag{2}$$

Then the general solution of (1) will be

$$x(t) = e^{-\gamma t}(c_1 \cos \delta t + c_2 \sin \delta t) + \psi_1(t), \tag{3}$$

in which $\psi_1(t)$ is any particular solution of equation (1). The presence of the function $e^{-\gamma t}$, called a damping factor, will cause the natural part of the solution, that is, the part independent of the external force $(w/g)F(t)$, to approach zero as $t \to \infty$.

If in (1) we have $\beta = \gamma$, the two roots of the auxiliary equation are equal and the general solution becomes

$$x(t) = e^{-\gamma t}(c_1 + c_2 t) + \psi_2(t), \tag{4}$$

in which $\psi_2(t)$ is a particular solution of (1). Again the natural component has the damping factor $e^{-\gamma t}$ in it.

If in (1) we have $\beta < \gamma$ and $\gamma^2 - \beta^2 > 0$, then we can set

$$\gamma^2 - \beta^2 = \sigma^2, \qquad \sigma > 0. \tag{5}$$

Since $\sigma < \gamma$, the two roots of the auxiliary equation are both real and negative, and we have

$$x(t) = c_1 e^{(-\gamma+\sigma)t} + c_2 e^{(-\gamma-\sigma)t} + \psi_3(t). \tag{6}$$

Again $\psi_3(t)$ is a particular solution of (1), and we see that the damping factor $e^{-\gamma t}$ causes the natural component of (6) to approach zero as $t \to \infty$.

Suppose for the moment that we have $F(t) \equiv 0$, so the natural component of the motion is all that is under consideration. If $\beta > \gamma$, equation (3) holds and the motion is a *damped oscillatory* one. If $\beta = \gamma$, equation (4) holds and the motion is not oscillatory; it is called *critically damped* motion. If $\beta < \gamma$, (6) holds and the motion is said to be *overdamped*; the parameter $\gamma$ is larger than it needs to be to remove the oscillations. Figure 15 shows a representative graph of each type of motion mentioned in this paragraph, a damped oscillatory motion, a critically damped motion, and an overdamped motion.

EXAMPLE: Solve the problem of Example (b), Section 52, with an added damping force of magnitude $0.6|v|$. Such a damping force can be realized by immersing the weight $B$ in a thick liquid.

The initial value problem to be solved is

$$\tfrac{12}{32}x''(t) + 0.6x'(t) + 24x(t) = 0; \qquad x(0) = \tfrac{1}{3}, x'(0) = -2. \tag{7}$$

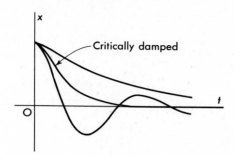

FIGURE 15

The auxiliary equation of (7) may be written

$$m^2 + 1.6m + 64 = 0,$$

an equation that has roots $-0.8 \pm \sqrt{63.36}\,i$. Therefore, the general solution of (7) is

$$x(t) = e^{-0.8t}(c_1 \cos 8.0t + c_2 \sin 8.0t)$$

and

$$x'(t) = e^{-0.8t}[(-8c_1 - 0.8c_2)\sin 8.0t + (8c_2 - 0.8c_1)\cos 8.0t].$$

The initial conditions in (7) now give us

$$\tfrac{1}{3} = c_1 \qquad \text{and} \qquad -2 = 8c_2 - 0.8c_1,$$

so that $c_1 = 0.33$ and $c_2 = -0.22$.

Therefore the desired solution is

$$x(t) = \exp(-0.8t)(0.33 \cos 8.0t - 0.22 \sin 8.0t), \tag{8}$$

a portion of its graph being shown in Figure 16.

FIGURE 16

## *Exercises*

1. A certain straight-line motion is determined by the differential equation

$$\frac{d^2x}{dt^2} + 2\gamma\frac{dx}{dt} + 169x = 0$$

and the conditions that when $t = 0$, $x = 0$, and $v = 8$ ft/sec.
   (a) Find the value of $\gamma$ that leads to critical damping, determine $x$ in terms of $t$, and draw the graph for $0 \leqq t \leqq 0.2$.     ANS.   $\gamma = 13(1/\text{sec})$, $x = 8t\,e^{-13t}$.
   (b) Use $\gamma = 12$. Find $x$ in terms of $t$ and draw the graph.
                                ANS.   $x = 1.6\,e^{-12t}\sin 5t$.
   (c) Use $\gamma = 14$. Find $x$ in terms of $t$ and draw the graph.
                             ANS.   $x = 0.77(e^{-8.8t} - e^{-19.2t})$.

2. A spring is such that a 2-lb weight stretches it $\frac{1}{2}$ ft. An impressed force $\frac{1}{4}\sin 8t$ and a damping force of magnitude $|v|$ are both acting on the spring. The weight starts $\frac{1}{4}$ ft below the equilibrium point with an imparted upward velocity of 3 ft/sec. Find a formula for the position of the weight at time $t$.
           ANS.   $x = \frac{3}{32}e^{-8t}(3 - 8t) - \frac{1}{32}\cos 8t$.

3. A spring is such that a 4-lb weight stretches it 0.64 ft. The 4-lb weight is pushed up $\frac{1}{3}$ ft above the point of equilibrium and then started with a downward velocity of 5 ft/sec. The motion takes place in a medium which furnishes a damping force of magnitude $\frac{1}{4}|v|$ at all times. Find the equation describing the position of the weight at time $t$.     ANS.   $x = \frac{1}{3}e^{-t}(2\sin 7t - \cos 7t)$.

4. A spring is such that a 4-lb weight stretches it 0.32 ft. The weight is attached to the spring and moves in a medium which furnishes a damping force of magnitude $\frac{3}{2}|v|$. The weight is drawn down $\frac{1}{2}$ ft below the equilibrium point and given an initial upward velocity of 4 ft/sec. Find the position of the weight thereafter.
           ANS.   $x = \frac{1}{8}e^{-6t}(4\cos 8t - \sin 8t)$.

5. A spring is such that a 4-lb weight stretches the spring 0.4 ft. The 4-lb weight is attached to the spring (suspended from a fixed support) and the system is allowed to reach equilibrium. Then the weight is started from equilibrium position with an imparted upward velocity of 2 ft/sec. Assume that the motion takes place in a medium that furnishes a retarding force of magnitude numerically equal to the speed, in feet per second, of the moving weight. Determine the position of the weight as a function of time.     ANS.   $x = -\frac{1}{4}e^{-4t}\sin 8t$.

6. A spring is stretched 6 in. by a 3-lb weight. The 3-lb weight is attached to the spring and then started from equilibrium with an imparted upward velocity of 12 ft/sec. Air resistance furnishes a retarding force equal in magnitude to $0.03|v|$. Find the equation of motion.     ANS.   $x = -1.5\,e^{-0.16t}\sin 8t$.

7. A spring is such that a 2-lb weight stretches it 6 in. There is a damping force present, with magnitude the same as the magnitude of the velocity. An impressed force $(2\sin 8t)$ is acting on the spring. If, at $t = 0$, the weight is released from a point 3 in. below the equilibrium point, find its position for $t > 0$.
           ANS.   $x = (\frac{1}{2} + 4t)e^{-8t} - \frac{1}{4}\cos 8t$.

8. A spring is stretched 10 in. by a 4-lb weight. The weight is started 6 in. below the equilibrium point with an upward velocity of 8 ft/sec. If a resisting medium furnishes

a retarding force of magnitude $\frac{1}{4}|v|$, describe the motion.

ANS.   $x = e^{-t}[0.50 \cos 6.1t - 1.23 \sin 6.1t]$.

9. For exercise 8, find the times of the first three stops and the position (to the nearest inch) of the weight at each stop.     ANS.   $t_1 = 0.3$ sec, $x_1 = -12$ in.; $t_2 = 0.8$ sec,
$$x_2 = +6 \text{ in.}; t_3 = 1.3 \text{ sec}, x_3 = -4 \text{ in.}$$

10. A spring is stretched 4 in. by a 2-lb weight. The 2-lb weight is started from the equilibrium point with a downward velocity of 12 ft/sec. If air resistance furnishes a retarding force of magnitude 0.02 of the velocity, describe the motion.

ANS.   $x = 1.22 \, e^{-0.16t} \sin 9.8t$.

11. For exercise 10, find how long it takes the damping factor to drop to one-tenth its initial value.     ANS.   14.4 sec.

12. For exercise 10, find the position of the weight at: (a) the first stop; (b) the second stop.     ANS.   (a) $x = 1.2$ ft; (b) $x = -1.1$ ft.

13. Let the motion of exercise 8, page 162, be retarded by a damping force of magnitude $0.6|v|$. Find the equation of motion.

ANS.   $x = 0.30 \, e^{-4.8t} \cos 6.4t + 0.22 \, e^{-4.8t} \sin 6.4t - 0.05 \cos 8t$ (ft).

14. Show that whenever $t > 1$ (sec), the solution of exercise 13 can be replaced (to the nearest 0.01 ft) by $x = -0.05 \cos 8t$.

15. Let the motion of exercise 8, page 162, be retarded by a damping force of magnitude $|v|$. Find the equation of motion and also determine its form (to the nearest 0.01 ft) for $t > 1$ (sec).

ANS.   $x = \frac{9}{32}(8t + 1) \, e^{-8t} - \frac{1}{32} \cos 8t$ (ft); for $t > 1$, $x = -\frac{1}{32} \cos 8t$.

16. Let the motion of exercise 8, page 162, be retarded by a damping force of magnitude $\frac{5}{3}|v|$. Find the equation of motion.

ANS.   $x = 0.30 \, e^{-(8/3)t} - 0.03 \, e^{-24t} - 0.02 \cos 8t$.

17. Alter exercise 6, page 162, by inserting a damping force of magnitude one-half that of the velocity and then determine $x$.

ANS.   $x = \exp\left(-\frac{2}{3}t\right)(0.30 \cos 8.0t - 0.22 \sin 8.0t) - 0.05 \cos 4t + 0.49 \sin 4t$.

18. A spring is stretched 6 in. by a 4 lb weight. Let the weight be pulled down 6 in. below equilibrium and given an initial upward velocity of 7 ft/sec. Assuming a damping force twice the magnitude of the velocity, describe the motion and sketch the graph at intervals of 0.05 sec for $0 \le t \le 0.3$ (sec).     ANS.   $x = \frac{1}{2} e^{-8t}(1 - 6t)$.

19. An object weighing $w$ lb is dropped from a height $h$ ft above the earth. At time $t$ (sec) after the object is dropped, let its distance from the starting point be $x$ (ft), measured positive downward. Assuming air resistance to be negligible, show that $x$ must satisfy the equation

$$\frac{w}{g} \frac{d^2x}{dt^2} = w$$

as long as $x < h$. Find $x$.     ANS.   $x = \frac{1}{2}gt^2$.

20. Let the weight of exercise 19 be given an initial velocity $v_0$. Let $v$ be the velocity at time $t$. Determine $v$ and $x$.     ANS.   $v = gt + v_0, x = \frac{1}{2}gt^2 + v_0 t$.

21. From the results in exercise 20, find a relation that does not contain $t$ explicitly.

ANS.   $v^2 = v_0^2 + 2gx$.

22. If air resistance furnishes an additional force proportional to the velocity in the motion studied in exercises 19 and 20, show that the equation of motion becomes

$$\frac{w}{g} \frac{d^2x}{dt^2} + b\frac{dx}{dt} = w. \tag{A}$$

Solve equation (A) given the conditions $t = 0$, $x = 0$, and $v = v_0$. Use $a = bg/w$.

ANS. $x = a^{-1}gt + a^{-2}(av_0 - g)(1 - e^{-at})$.

**23.** To compare the results of exercises 20 and 22 when $a = bg/w$ is small, use the power series for $e^{-at}$ in the answer for exercise 22 and discard all terms involving $a^n$ for $n \geq 3$. ANS. $x = \frac{1}{2}gt^2 + v_0 t - \frac{1}{6}at^2(3v_0 + gt) + \frac{1}{24}a^2t^3(4v_0 + gt)$.

**24.** The equation of motion of the vertical fall of a man with a parachute may be roughly approximated by equation (A) of exercise 22. Suppose a 180-lb man drops from a great height and attains a velocity of 20 miles per hour (mph) after a long time. Determine the implied coefficient $b$ of equation (A). ANS. 6.1 (lb)(sec)/ft.

**25.** A particle is moving along the $x$-axis according to the law

$$\frac{d^2x}{dt^2} + 6\frac{dx}{dt} + 25x = 0.$$

If the particle started at $x = 0$ with an initial velocity of 12 ft/sec to the left, determine: (a) $x$ in terms of $t$; (b) the times at which stops occur; and (c) the ratio between the numerical values of $x$ at successive stops.

(a) $x = -3\,e^{-3t}\sin 4t$.

ANS. (b) $t = 0.23 + \frac{1}{4}n\pi, n = 0, 1, 2, 3, \ldots$.

(c) 0.095.

## 55. The simple pendulum

A rod of length $C$ ft is suspended by one end so it can swing freely in a vertical plane. Let a weight $B$ (the bob) of $w$ lb be attached to the free end of the rod, and let the weight of the rod be negligible compared to the weight of the bob.

Let $\theta$ (radians) be the angular displacement from the vertical, as shown in Figure 17, of the rod at time $t$ (sec). The tangential component of the force

FIGURE 17

$w$ (lb) is $w \sin \theta$ and it tends to decrease $\theta$. Then, neglecting the weight of the rod and using $S = C\theta$ as a measure of arc length from the vertical position, we may conclude that

$$\frac{w}{g}\frac{d^2 S}{dt^2} = -w \sin \theta. \tag{1}$$

Since $S = C\theta$ and $C$ is constant, (1) becomes

$$\frac{d^2 \theta}{dt^2} + \frac{g}{C}\sin \theta = 0. \tag{2}$$

The solution of equation (2) is not elementary; it involves an elliptic integral. If $\theta$ is small, however, $\sin \theta$ and $\theta$ are nearly equal and (2) is closely approximated by the much simpler equation

$$\frac{d^2 \theta}{dt^2} + \beta^2 \theta = 0; \qquad \beta^2 = \frac{g}{C}. \tag{3}$$

The solution of (3) with pertinent initial conditions gives usable results whenever those conditions are such that $\theta$ remains small, say $|\theta| < 0.3$ (radians).

### Exercises

1. A clock has a 6-in. pendulum. The clock ticks once for each time that the pendulum completes a swing, returning to its original position. How many times does the clock tick in 30 sec?  ANS.  38 times.
2. A 6-in. pendulum is released from rest at an angle one-tenth of a radian from the vertical. Using $g = 32$ (ft/sec²), describe the motion.

    ANS.  $\theta = 0.1 \cos 8t$ (radians).
3. For the pendulum of exercise 2, find the maximum angular speed and its first time of occurrence.  ANS.  0.8 (radians/sec) at 0.2 sec.
4. A 6-in. pendulum is started with a velocity of 1 radian/sec, toward the vertical, from a position one-tenth radian from the vertical. Describe the motion.

    ANS.  $\theta = \frac{1}{10} \cos 8t - \frac{1}{8} \sin 8t$ (radians).
5. For exercise 4, find to the nearest degree the maximum angular displacement from the vertical.  ANS.  9°.
6. Interpret as a pendulum problem and solve:

$$\frac{d^2 \theta}{dt^2} + \beta^2 \theta = 0 : \beta^2 = \frac{g}{C}, \qquad \text{when } t = 0, \theta = \theta_0, \omega = \frac{d\theta}{dt} = \omega_0.$$

    ANS.  $\theta = \theta_0 \cos \beta t + \beta^{-1}\omega_0 \sin \beta t$ (radians).
7. Find the maximum angular displacement from the vertical for the pendulum of exercise 6.  ANS.  $\theta_{max} = (\theta_0^2 + \beta^{-2}\omega_0^2)^{1/2}$.

# The Laplace Transform

### 56. The transform concept

The reader is already familiar with some operators that transform functions into functions. An outstanding example is the differential operator $D$, which transforms each function of a large class (those possessing a derivative) into another function.

We have already found that the operator $D$ is useful in the treatment of linear differential equations with constant coefficients. In this chapter we study another transformation (a mapping of functions onto functions) which has played an increasingly important role in both pure and applied mathematics in the past few decades. The operator $L$, to be introduced in Section 57, is particularly effective in the study of initial value problems involving linear differential equations with constant coefficients.

One class of transformations, which are called integral transforms, may be defined by

$$T\{F(t)\} = \int_{-\infty}^{\infty} K(s, t)F(t)\, dt = f(s). \tag{1}$$

Given a function $K(s, t)$, called the kernel of the transformation, equation (1) associates with each $F(t)$ of the class of functions for which the above integral exists a function $f(s)$ defined by (1). Generalizations and abstractions of (1), as well as studies of special cases, are to be found in profusion in mathematical literature.

Various particular choices of $K(s, t)$ in (1) have led to special transforms, each with its own properties to make it useful in specific circumstances. The transform defined by choosing

$$K(s, t) = 0, \qquad \text{for } t < 0,$$
$$= e^{-st}, \qquad \text{for } t \geq 0,$$

is the one to which this chapter is devoted.

## 57. Definition of the Laplace transform

Let $F(t)$ be any function such that the integrations encountered may be legitimately performed on $F(t)$. The *Laplace transform* of $F(t)$ is denoted by $L\{F(t)\}$ and is defined by

$$L\{F(t)\} = \int_0^\infty e^{-st} F(t)\, dt. \tag{1}$$

The integral in (1) is a function of the parameter $s$; call that function $f(s)$. We may write

$$L\{F(t)\} = \int_0^\infty e^{-st} F(t)\, dt = f(s). \tag{2}$$

It is customary to refer to $f(s)$, as well as to the symbol $L\{F(t)\}$, as the transform, or the Laplace transform, of $F(t)$.

We may also look upon (2) as a definition of a Laplace operator $L$, which transforms each function $F(t)$ of a certain set of functions into some function $f(s)$.

It is easy to show that if the integral in (2) does converge, it will do so for all $s$ greater than* some fixed value $s_0$. That is, equation (2) will define $f(s)$ for $s > s_0$. In extreme cases the integral may converge for all finite $s$.

It is important that the operator $L$, like the differential operator $D$, is a linear operator. If $F_1(t)$ and $F_2(t)$ have Laplace transforms and if $c_1$ and $c_2$ are any constants,

$$L\{c_1 F_1(t) + c_2 F_2(t)\} = c_1 L\{F_1(t)\} + c_2 L\{F_2(t)\}. \tag{3}$$

---

* If $s$ is not to be restricted to real values, the convergence takes place for all $s$ with real part greater than some fixed value.

Using elementary properties of definite integrals, the student can easily show the validity of equation (3).

We shall hereafter employ the relation (3) without restating the fact that the operator $L$ is a linear one.

## 58. Transforms of elementary functions

The transforms of certain exponential and trigonometric functions and of polynomials will now be obtained. These results enter our work frequently.

EXAMPLE (a):   Find $L\{e^{kt}\}$.
We proceed as follows:

$$L\{e^{kt}\} = \int_0^\infty e^{-st} \cdot e^{kt} \, dt = \int_0^\infty e^{-(s-k)t} \, dt.$$

For $s \le k$, the exponent on $e$ is positive or zero and the integral diverges. For $s > k$, the integral converges.
Indeed, for $s > k$,

$$L\{e^{kt}\} = \int_0^\infty e^{-(s-k)t} \, dt$$

$$= \left[ \frac{-e^{-(s-k)t}}{s-k} \right]_0^\infty$$

$$= 0 + \frac{1}{s-k}.$$

Thus we find that

$$L\{e^{kt}\} = \frac{1}{s-k}, \qquad s > k. \tag{1}$$

Note the special case $k = 0$:

$$L\{1\} = \frac{1}{s}, \qquad s > 0. \tag{2}$$

EXAMPLE (b):   Obtain $L\{\sin kt\}$.
From elementary calculus we obtain

$$\int e^{ax} \sin mx \, dx = \frac{e^{ax}(a \sin mx - m \cos mx)}{a^2 + m^2} + C.$$

Since

$$L\{\sin kt\} = \int_0^\infty e^{-st} \sin kt \, dt,$$

it follows that

$$L\{\sin kt\} = \left[ \frac{e^{-st}(-s \sin kt - k \cos kt)}{s^2 + k^2} \right]_0^\infty . \tag{3}$$

For positive $s$, $e^{-st} \to 0$ as $t \to \infty$. Furthermore, $\sin kt$ and $\cos kt$ are bounded as $t \to \infty$. Therefore (3) yields

$$L\{\sin kt\} = 0 - \frac{1(0 - k)}{s^2 + k^2},$$

or

$$L\{\sin kt\} = \frac{k}{s^2 + k^2}, \qquad s > 0. \tag{4}$$

The result

$$L\{\cos kt\} = \frac{s}{s^2 + k^2}, \qquad s > 0, \tag{5}$$

can be obtained in a similar manner.

**EXAMPLE (c):**  Obtain $L\{t^n\}$ for $n$ a positive integer.

By definition

$$L\{t^n\} = \int_0^\infty e^{-st} t^n \, dt.$$

Let us attack the integral using integration by parts with the choice exhibited in the table.

| $t^n$ | $e^{-st} \, dt$ |
|---|---|
| $nt^{n-1} \, dt$ | $-\dfrac{1}{s} e^{-st}$ |

We thus obtain

$$\int_0^\infty e^{-st} t^n \, dt = \left[ \frac{-t^n e^{-st}}{s} \right]_0^\infty + \frac{n}{s} \int_0^\infty e^{-st} t^{n-1} \, dt. \tag{6}$$

For $s > 0$ and $n > 0$, the first term on the right in (6) is zero, and we are left with

$$\int_0^\infty e^{-st} t^n \, dt = \frac{n}{s} \int_0^\infty e^{-st} t^{n-1} \, dt, \qquad s > 0,$$

or

$$L\{t^n\} = \frac{n}{s} L\{t^{n-1}\}, \qquad s > 0. \tag{7}$$

From (7) we may conclude that, for $n > 1$,

$$L\{t^{n-1}\} = \frac{n-1}{s} L\{t^{n-2}\}$$

so

$$L\{t^n\} = \frac{n(n-1)}{s^2} L\{t^{n-2}\}. \tag{8}$$

Iteration of this process yields

$$L\{t^n\} = \frac{n(n-1)(n-2)\cdots 2\cdot 1}{s^n} L\{t^0\}.$$

From Example (a) above, we have

$$L\{t^0\} = L\{1\} = s^{-1}.$$

Hence, for $n$ a positive integer,

$$L\{t^n\} = \frac{n!}{s^{n+1}}, \qquad s > 0. \tag{9}$$

The Laplace transform of $F(t)$ will exist even if the object function $F(t)$ is discontinuous, provided the integral in the definition of $L\{F(t)\}$ exists. Little will be done at this time with specific discontinuous $F(t)$, because more efficient methods for obtaining such transforms are to be developed later.

EXAMPLE (d):   Find the Laplace transform of $H(t)$ where

$$H(t) = t, \qquad 0 < t < 4,$$

$$= 5, \qquad\quad t > 4.$$

Note that the fact that $H(t)$ is not defined at $t = 0$ and $t = 4$ has no bearing whatever on the existence, or the value, of $L\{H(t)\}$. We turn to the definition of $L\{H(t)\}$ to obtain

$$L\{H(t)\} = \int_0^\infty e^{-st} H(t)\, dt$$

$$= \int_0^4 e^{-st} t\, dt + \int_4^\infty e^{-st} 5\, dt.$$

Using integration by parts on the next-to-last integral above, we soon arrive, for $s > 0$, at

$$L\{H(t)\} = \left[ -\frac{t}{s}e^{-st} - \frac{1}{s^2}e^{-st} \right]_0^4 + \left[ -\frac{5}{s}e^{-st} \right]_4^\infty.$$

Thus

$$L\{H(t)\} = -\frac{4\,e^{-4s}}{s} - \frac{e^{-4s}}{s^2} + 0 + \frac{1}{s^2} - 0 + \frac{5\,e^{-4s}}{s}$$

$$= \frac{1}{s^2} + \frac{e^{-4s}}{s} - \frac{e^{-4s}}{s^2}.$$

## Exercises

1. Show that $L\{\cos kt\} = \dfrac{s}{s^2 + k^2}$; for $s > 0$.

2. Euler's formula $e^{ikt} = \cos kt + i \sin kt$ can be used to obtain the formula $\cos kt = \frac{1}{2}(e^{ikt} + e^{-ikt})$. Show that the result of exercise 1 can now be obtained with a formal application of the Laplace transform.

3. Obtain the transform for $\sin kt$ by an argument similar to the one suggested in exercise 2.

4. Obtain $L\{t^2 + 4t - 5\}$.        ANS.   $\dfrac{2}{s^3} + \dfrac{4}{s^2} - \dfrac{5}{s}, s > 0.$

5. Obtain $L\{t^3 - t^2 + 4t\}$.        ANS.   $\dfrac{6}{s^4} - \dfrac{2}{s^3} + \dfrac{4}{s^2}, s > 0.$

6. Evaluate $L\{e^{-2t} + 4\,e^{-3t}\}$.        ANS.   $\dfrac{5s + 11}{(s + 2)(s + 3)}, s > -2.$

7. Evaluate $L\{3\,e^{4t} - e^{-2t}\}$.        ANS.   $\dfrac{2s + 10}{(s - 4)(s + 2)}, s > 4.$

8. Show that $L\{\cosh kt\} = \dfrac{s}{s^2 - k^2}$; for $s > |k|$.

9. Show that $L\{\sinh kt\} = \dfrac{k}{s^2 - k^2}$; for $s > |k|$.

10. Use the trigonometric identity $\cos^2 A = \frac{1}{2}(1 + \cos 2A)$ and equation (5), Section 58, to evaluate $L\{\cos^2 kt\}$.        ANS.   $\dfrac{s^2 + 2k^2}{s(s^2 + 4k^2)}, s > 0.$

11. Parallel the method suggested in exercise 8 to obtain $L\{\sin^2 kt\}$.

                                         ANS.   $\dfrac{2k^2}{s(s^2 + 4k^2)}, s > 0.$

12. Obtain $L\{\sin^2 kt\}$ directly from the answer to exercise 10.

13. Evaluate $L\{\sin kt \cos kt\}$ with the aid of a trigonometric identity.

$$\text{ANS.} \quad \frac{k}{s^2 + 4k^2}, s > 0.$$

14. Evaluate $L\{e^{-at} - e^{-bt}\}$.

$$\text{ANS.} \quad \frac{b - a}{(s + a)(s + b)}, s > \max(-a, -b).$$

15. Find $L\{\psi(t)\}$ where

$$\psi(t) = 4, \quad 0 < t < 1,$$
$$= 3, \quad t > 1.$$

$$\text{ANS.} \quad \frac{1}{s}(4 - e^{-s}), s > 0.$$

16. Find $L\{\phi(t)\}$ where

$$\phi(t) = 1, \quad 0 < t < 2,$$
$$= t, \quad t > 2.$$

$$\text{ANS.} \quad \frac{1}{s} + \frac{e^{-2s}}{s} + \frac{e^{-2s}}{s^2}, s > 0.$$

17. Find $L\{A(t)\}$ where

$$A(t) = 0, \quad 0 < t < 1,$$
$$= t, \quad 1 < t < 2,$$
$$= 0, \quad t > 2.$$

$$\text{ANS.} \quad \left(\frac{1}{s^2} + \frac{1}{s}\right)e^{-s} - \left(\frac{1}{s^2} + \frac{2}{s}\right)e^{-2s}, s > 0.$$

18. Find $L\{B(t)\}$ where

$$B(t) = \sin 2t, \quad 0 < t < \pi,$$
$$= 0, \quad t > \pi.$$

$$\text{ANS.} \quad \frac{2(1 - e^{-\pi s})}{s^2 + 4}, s > 0.$$

## 59. Sectionally continuous functions

It should be apparent that, if we are to find problems for which the Laplace transform method is useful, we must learn a good deal more about the transforms of more complicated functions than those we considered in the previous sections.

Our approach will be to prove a number of useful properties of the Laplace transform and then consider initial value problems in which we can make use of those properties.

In Section 58 we began this study by actually determining the transforms of some simple functions. However, it soon becomes tiresome to test each $F(t)$ we encounter to determine whether the integral

$$\int_0^\infty e^{-st} F(t)\, dt \tag{1}$$

exists for some range of values of $s$. We therefore seek a fairly large class of functions for which we can prove once and for all that the integral (1) exists.

One of our avowed interests in the Laplace transform is in its usefulness as a tool in solving problems in more or less elementary applications, particularly initial value problems in differential equations. Therefore we do not hesitate to restrict our study to functions $F(t)$ that are continuous or even differentiable, except possibly at a discrete set of points, in the semi-infinite range $t \geq 0$.

For such functions, the existence of the integral (1) can be endangered only at points of discontinuity of $F(t)$ or by divergence due to behavior of the integrand as $t \to \infty$.

In elementary calculus we found that finite discontinuities, or finite jumps, of the integrand did not interfere with the existence of the integral. We therefore introduce a term to describe functions that are continuous except for such jumps.

DEFINITION: *The function $F(t)$ is said to be sectionally continuous over the closed interval $a \leq t \leq b$ if that interval can be divided into a finite number of subintervals $c \leq t \leq d$ such that in each subinterval:*

(a) *$F(t)$ is continuous in the open interval $c < t < d$,*
(b) *$F(t)$ approaches a limit as $t$ approaches each endpoint from within the interval; that is, $\lim_{t \to c^+} F(t)$ and $\lim_{t \to d^-} F(t)$ exist.*

FIGURE 18

Figure 18 shows the graph of a function $F(t)$ that is sectionally continuous over the interval

$$0 \leq t \leq 6.$$

The student should realize that there is no implication that $F(t)$ must be sectionally continuous for $L\{F(t)\}$ to exist. Indeed, we shall meet several counterexamples to any such notion. The concept of sectionally continuous functions will, in Section 61, play a role in a set of conditions sufficient for the existence of the transform.

## 60. Functions of exponential order

If the integral of $e^{-st}F(t)$ between the limits 0 and $t_0$ exists for every finite positive $t_0$, the only remaining threat to the existence of the transform

$$\int_0^\infty e^{-st}F(t)\,dt \tag{1}$$

is the behavior of the integrand as $t \to \infty$.

We know that

$$\int_0^\infty e^{-ct}\,dt \tag{2}$$

converges for $c > 0$. This arouses our interest in functions $F(t)$ that are, for large $t$ ($t \geq t_0$), essentially bounded by some exponential $e^{bt}$ so that the integrand in (1) will behave like the integrand in (2) for $s$ large enough.

DEFINITION: *The function $F(t)$ is said to be of exponential order as $t \to \infty$ if constants M and b and a fixed t-value $t_0$ exist such that*

$$|F(t)| < M\,e^{bt}, \qquad for\ t \geq t_0. \tag{3}$$

If $b$ is to be emphasized, we say that $F(t)$ is of the order of $e^{bt}$ as $t \to \infty$. We also write

$$F(t) = O(e^{bt}), \qquad t \to \infty, \tag{4}$$

to mean that $F(t)$ is of exponential order, the exponential being $e^{bt}$, as $t \to \infty$. That is, (4) is another way of expressing (3).

The integral in (1) may be split into parts as follows:

$$\int_0^\infty e^{-st}F(t)\,dt = \int_0^{t_0} e^{-st}F(t)\,dt + \int_{t_0}^\infty e^{-st}F(t)\,dt. \tag{5}$$

If $F(t)$ is of exponential order, $F(t) = O(e^{bt})$, the last integral in equation (5) exists because from the inequality (3) it follows that for $s > b$,

$$\int_{t_0}^{\infty} |e^{-st}F(t)| \, dt < M \int_{t_0}^{\infty} e^{-st} \cdot e^{bt} \, dt = \frac{M \exp\left[-t_0(s - b)\right]}{s - b}. \tag{6}$$

For $s > b$, the last member of (6) approaches zero as $t_0 \to \infty$. Therefore the last integral in (5) is absolutely convergent* for $s > b$. We have proved the following result.

THEOREM 8: *If the integral of $e^{-st}F(t)$ between the limits 0 and $t_0$ exists for every finite positive $t_0$, and if $F(t)$ is of exponential order, $F(t) = O(e^{bt})$ as $t \to \infty$, the Laplace transform*

$$L\{F(t)\} = \int_0^{\infty} e^{-st}F(t) \, dt = f(s) \tag{7}$$

*exists for $s > b$.*

We know that a function that is sectionally continuous over an interval is integrable over that interval. This leads us to the following useful special case of Theorem 8.

THEOREM 9: *If $F(t)$ is sectionally continuous over every finite interval in the range $t \geqq 0$, and if $F(t)$ is of exponential order, $F(t) = O(e^{bt})$ as $t \to \infty$, the Laplace transform $L\{F(t)\}$ exists for $s > b$.*

Functions of exponential order play a dominant role throughout our work. It is therefore wise to develop proficiency in determining whether or not a specified function is of exponential order.

Surely if a constant $b$ exists such that

$$\lim_{t \to \infty} \left[e^{-bt}|F(t)|\right] \tag{8}$$

exists, the function $F(t)$ is of exponential order, indeed of the order of $e^{bt}$. To see this, let the value of the limit (8) be $K \neq 0$. Then, for $t$ large enough, $|e^{-bt}F(t)|$ can be made as close to $K$ as is desired, so certainly

$$|e^{-bt}F(t)| < 2K.$$

Therefore, for $t$ sufficiently large,

$$|F(t)| < M e^{bt}, \tag{9}$$

with $M = 2K$. If the limit in (8) is zero, we may write (9) with $M = 1$.

---

* If complex $s$ is to be used, the integral converges for $\mathrm{Re}(s) > b$.

On the other hand if, for every fixed $c$,

$$\lim_{t \to \infty} [e^{-ct}|F(t)|] = \infty, \tag{10}$$

the function $F(t)$ is not of exponential order. For, assume that $b$ exists such that

$$|F(t)| < M e^{bt}, \qquad t \geq t_0; \tag{11}$$

then the choice $c = 2b$ would yield, by (11),

$$|e^{-2bt}F(t)| < M e^{-bt},$$

so $e^{-2bt}F(t) \to 0$ as $t \to \infty$, which disagrees with (10).

**EXAMPLE (a):**   Show that $t^3$ is of exponential order as $t \to \infty$.
   We consider, with $b$ as yet unspecified,

$$\lim_{t \to \infty} (e^{-bt}t^3) = \lim_{t \to \infty} \frac{t^3}{e^{bt}}. \tag{12}$$

If $b > 0$, the limit in (12) is of a type treated in calculus. In fact,

$$\lim_{t \to \infty} \frac{t^3}{e^{bt}} = \lim_{t \to \infty} \frac{3t^2}{b\, e^{bt}} = \lim_{t \to \infty} \frac{6t}{b^2\, e^{bt}} = \lim_{t \to \infty} \frac{6}{b^3\, e^{bt}} = 0.$$

Therefore $t^3$ is of exponential order,

$$t^3 = O(e^{bt}), \qquad t \to \infty,$$

for any fixed positive $b$.

**EXAMPLE (b):**   Show that $\exp(t^2)$ is not of exponential order as $t \to \infty$.
   Consider

$$\lim_{t \to \infty} \frac{\exp(t^2)}{\exp(bt)}. \tag{13}$$

If $b \leq 0$, the limit in (13) is infinite. If $b > 0$,

$$\lim_{t \to \infty} \frac{\exp(t^2)}{\exp(bt)} = \lim_{t \to \infty} \exp[t(t - b)] = \infty.$$

Thus, no matter what fixed $b$ we use, the limit in (13) is infinite and $\exp(t^2)$ cannot be of exponential order.
   The exercises at the end of the next section give additional opportunities for practice in determining whether or not a function is of exponential order.

## 61. Functions of class A

For brevity we shall hereafter use the term "a function of class A" for any function that

(a) is sectionally continuous over every finite interval in the range $t \geq 0$ and
(b) is of exponential order as $t \to \infty$.

We may then reword Theorem 9 as follows.

THEOREM 10:    *If $F(t)$ is a function of class A, $L\{F(t)\}$ exists.*

It is important to realize that Theorem 10 states only that, for $L\{F(t)\}$ to exist, it is sufficient that $F(t)$ be of class A. The condition is not necessary. A classic example showing that functions other than those of class A do have Laplace transforms is

$$F(t) = t^{-1/2}.$$

This function is not sectionally continuous in every finite interval in the range $t \geq 0$, because $F(t) \to \infty$ as $t \to 0^+$. But $t^{-1/2}$ is integrable from 0 to any positive $t_0$. Also $t^{-1/2} \to 0$ as $t \to \infty$, so $t^{-1/2}$ is of exponential order, with $M = 1$ and $b = 0$ in the inequality (3), page 178. Hence, by Theorem 8, page 179, $L\{t^{-1/2}\}$ exists.

Indeed, for $s > 0$,

$$L\{t^{-1/2}\} = \int_0^\infty e^{-st} t^{-1/2} \, dt,$$

in which the change of variable $st = y^2$ leads to

$$L\{t^{-1/2}\} = 2s^{-1/2} \int_0^\infty \exp(-y^2) \, dy, \qquad s > 0.$$

In elementary calculus we found that $\displaystyle\int_0^\infty \exp(-y^2) \, dy = \tfrac{1}{2}\sqrt{\pi}.$

Therefore

$$L\{t^{-1/2}\} = 2s^{-1/2} \cdot \tfrac{1}{2}\sqrt{\pi}$$
$$= \left(\frac{\pi}{s}\right)^{1/2}, \qquad s > 0, \tag{1}$$

even though $t^{-1/2} \to \infty$ as $t \to 0^+$. Additional examples are easily constructed and we shall meet some of them later in the book.

If $F(t)$ is of class A, $F(t)$ is bounded over the range $0 \leq t \leq t_0$,

$$|F(t)| < M_1, \qquad 0 \leq t \leq t_0. \tag{2}$$

But $F(t)$ is also of exponential order,

$$|F(t)| < M_2 e^{bt}, \qquad t \geq t_0. \tag{3}$$

If we choose $M$ as the larger of $M_1$ and $M_2$ and $c$ as the larger of $b$ and zero, we may write

$$|F(t)| < M e^{ct}, \qquad t \geq 0. \tag{4}$$

Therefore, for any function $F(t)$ of class A,

$$\left| \int_0^\infty e^{-st} F(t)\, dt \right| < M \int_0^\infty e^{-st} \cdot e^{ct}\, dt = \frac{M}{s - c}, \qquad s > c. \tag{5}$$

Since the right member of (5) approaches zero as $s \to \infty$, we have proved the following useful result.

THEOREM 11: *If $F(t)$ is of class A and if $L\{F(t)\} = f(s)$,*

$$\lim_{s \to \infty} f(s) = 0.$$

From (5) we may also conclude the stronger result that the transform $f(s)$ of a function $F(t)$ of class A must be such that $sf(s)$ is bounded as $s \to \infty$.

## Exercises

1. Prove that if $F_1(t)$ and $F_2(t)$ are each of exponential order as $t \to \infty$, then $F_1(t) \cdot F_2(t)$ and $F_1(t) + F_2(t)$ are also of exponential order as $t \to \infty$.
2. Prove that if $F_1(t)$ and $F_2(t)$ are of class A (see page 181), then $F_1(t) + F_2(t)$ and $F_1(t) \cdot F_2(t)$ are also of class A.
3. Show that $t^x$ is of exponential order as $t \to \infty$ for all real $x$.

In exercises 4 through 17, show that the given function is of class A. In these exercises, $n$ denotes a nonnegative integer, $k$ any real number.

4. $\sin kt$.
5. $\cos kt$.
6. $\cosh kt$.
7. $\sinh kt$.
8. $t^n$.
9. $t^n e^{kt}$.
10. $t^n \sin kt$.
11. $t^n \cos kt$.
12. $t^n \sinh kt$.
13. $t^n \cosh kt$.

14. $\dfrac{\sin kt}{t}$.
15. $\dfrac{1 - \exp(-t)}{t}$.

16. $\dfrac{1 - \cos kt}{t}$.
17. $\dfrac{\cos t - \cosh t}{t}$.

## 62. Transforms of derivatives

Any function of class A (see page 181) has a Laplace transform, but the derivative of such a function may or may not be of class A. For the function

$$F_1(t) = \sin [\exp (t)]$$

with derivative

$$F'_1(t) = \exp (t) \cos [\exp (t)],$$

both $F_1$ and $F'_1$ are of exponential order as $t \to \infty$. Here $F_1$ is bounded so it is of the order of $\exp (0 \cdot t)$; $F'_1$ is of the order of $\exp (t)$. On the other hand, the function

$$F_2(t) = \sin [\exp (t^2)]$$

with derivative

$$F'_2(t) = 2t \exp (t^2) \cos [\exp (t^2)]$$

is such that $F_2$ is of the order of $\exp (0 \cdot t)$, but $F'_2$ is not of exponential order. From Example (b), page 180,

$$\lim_{t \to \infty} \frac{\exp (t^2)}{\exp (bt)} = \infty$$

for any real $b$. Since the factors $2t \cos [\exp (t^2)]$ do not even approach zero as $t \to \infty$, the product $F'_2 \exp (-ct)$ cannot be bounded as $t \to \infty$ no matter how large a fixed $c$ is chosen.

Therefore, in studying the transforms of derivatives, we shall stipulate that the derivatives themselves be of class A.

If $F(t)$ is continuous for $t \geq 0$ and of exponential order as $t \to \infty$, and if $F'(t)$ is of class A, the integral in

$$L\{F'(t)\} = \int_0^\infty e^{-st} F'(t) \, dt \tag{1}$$

may be simplified by integration by parts with the choice exhibited in the table.

| $e^{-st}$ | $F'(t) \, dt$ |
|---|---|
| $-s \, e^{-st} \, dt$ | $F(t)$ |

We thus obtain, for $s$ greater than some fixed $s_0$,

$$\int_0^\infty e^{-st} F'(t) \, dt = \left[ e^{-st} F(t) \right]_0^\infty + s \int_0^\infty e^{-st} F(t) \, dt,$$

or

$$L\{F'(t)\} = -F(0) + sL\{F(t)\}.$$  (2)

THEOREM 12: *If $F(t)$ is continuous for $t \geq 0$ and of exponential order as $t \to \infty$, and if $F'(t)$ is of class A (see page 181), it follows from $L\{F(t)\} = f(s)$ that*

$$L\{F'(t)\} = sf(s) - F(0).$$  (3)

In treating a differential equation of order $n$, we seek solutions for which the highest-ordered derivative present is reasonably well behaved, say sectionally continuous. The integral of a sectionally continuous function is continuous. Hence, we lose nothing by requiring continuity for all derivatives of order lower than $n$. The requirement that the various derivatives be of exponential order is forced upon us by our desire to use the Laplace transform as a tool. For our purposes, iteration of Theorem 12 to obtain transforms of higher derivatives makes sense,

From (3) we obtain, if $F$, $F'$, $F''$ are suitably restricted,

$$L\{F''(t)\} = sL\{F'(t)\} - F'(0),$$

or

$$L\{F''(t)\} = s^2 f(s) - sF(0) - F'(0),$$  (4)

and the process can be repeated as many times as we wish.

THEOREM 13: *If $F(t)$, $F'(t), \ldots, F^{(n-1)}(t)$ are continuous for $t \geq 0$ and of exponential order as $t \to \infty$, and if $F^{(n)}(t)$ is of class A, then from*

$$L\{F(t)\} = f(s)$$

*it follows that*

$$L\{F^{(n)}(t)\} = s^n f(s) - \sum_{k=0}^{n-1} s^{n-1-k} F^{(k)}(0).$$  (5)

Thus

$$L\{F^{(3)}(t)\} = s^3 f(s) - s^2 F(0) - sF'(0) - F''(0),$$

$$L\{F^{(4)}(t)\} = s^4 f(s) - s^3 F(0) - s^2 F'(0) - sF''(0) - F^{(3)}(0), \text{ etc.}$$

Theorem 13 is basic in employing the Laplace transform to solve linear differential equations with constant coefficients. The theorem permits us to transform such differential equations into algebraic ones.

The restriction that $F(t)$ be continuous can be relaxed, but discontinuities in $F(t)$ bring in additional terms in the transform of $F'(t)$. As an example, consider an $F(t)$ that is continuous for $t \geq 0$ except for a finite jump at

$t = t_1$, as in Figure 19. If $F(t)$ is also of exponential order as $t \to \infty$, and if $F'(t)$ is of class A, we may write

$$L\{F'(t)\} = \int_0^\infty e^{-st}F'(t)\,dt$$

$$= \int_0^{t_1} e^{-st}F'(t)\,dt + \int_{t_1}^\infty e^{-st}F'(t)\,dt.$$

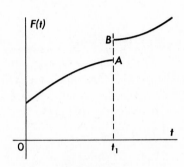

**FIGURE 19**

Then, integration by parts applied to the last two integrals yields

$$L\{F'(t)\} = \left[ e^{-st}F(t) \right]_0^{t_1} + s\int_0^{t_1} e^{-st}F(t)\,dt + \left[ e^{-st}F(t) \right]_{t_1}^\infty + s\int_{t_1}^\infty e^{-st}F(t)\,dt$$

$$= s\int_0^\infty e^{-st}F(t)\,dt + \exp(-st_1)F(t_1^-) - F(0) + 0 - \exp(-st_1)F(t_1^+)$$

$$= sL\{F(t)\} - F(0) - \exp(-st_1)[F(t_1^+) - F(t_1^-)].$$

In Figure 19 the directed distance $AB$ is of length $[F(t_1^+) - F(t_1^-)]$.

THEOREM 14: *If $F(t)$ is of exponential order as $t \to \infty$ and $F(t)$ is continuous for $t \geq 0$ except for a finite jump at $t = t_1$, and if $F'(t)$ is of class A, then from*

$$L\{F(t)\} = f(s)$$

*it follows that*

$$L\{F'(t)\} = sf(s) - F(0) - \exp(-st_1)[F(t_1^+) - F(t_1^-)]. \tag{6}$$

If $F(t)$ has more than one finite discontinuity, additional terms, similar to the last term in (6), enter the formula for $L\{F'(t)\}$.

## 63. Derivatives of transforms

For functions of class A, the theorems of advanced calculus show that it is legitimate to differentiate the Laplace transform integral. That is, if $F(t)$ is of class A, from

$$f(s) = \int_0^\infty e^{-st}F(t)\,dt \tag{1}$$

it follows that

$$f'(s) = \int_0^\infty (-t)\,e^{-st}F(t)\,dt. \tag{2}$$

The integral on the right in (2) is the transform of the function $(-t)F(t)$.

THEOREM 15:   *If $F(t)$ is a function of class A, it follows from*

$$L\{F(t)\} = f(s)$$

*that*

$$f'(s) = L\{-tF(t)\}. \tag{3}$$

When $F(t)$ is of class A, $(-t)^k F(t)$ is also of class A for any positive integer $k$.

THEOREM 16:   *If $F(t)$ is of class A, it follows from $L\{F(t)\} = f(s)$ that for any positive integer n*

$$\frac{d^n}{ds^n} f(s) = L\{(-t)^n F(t)\}. \tag{4}$$

These theorems are useful in several ways. One immediate application is to add to our list of transforms with very little labor. We know that

$$\frac{k}{s^2 + k^2} = L\{\sin kt\}, \tag{5}$$

and therefore, by Theorem 15,

$$\frac{-2ks}{(s^2 + k^2)^2} = L\{-t \sin kt\}.$$

Thus we obtain

$$\frac{s}{(s^2 + k^2)^2} = L\left\{\frac{t}{2k} \sin kt\right\}. \tag{6}$$

From the known formula

$$\frac{s}{s^2 + k^2} = L\{\cos kt\}$$

we obtain, by differentiation with respect to $s$,

$$\frac{k^2 - s^2}{(s^2 + k^2)^2} = L\{-t \cos kt\}. \tag{7}$$

Let us add to each side of (7) the corresponding member of

$$\frac{1}{s^2 + k^2} = L\left\{\frac{1}{k} \sin kt\right\}$$

to get

$$\frac{s^2 + k^2 + k^2 - s^2}{(s^2 + k^2)^2} = L\left\{\frac{1}{k} \sin kt - t \cos kt\right\},$$

from which it follows that

$$\frac{1}{(s^2 + k^2)^2} = L\left\{\frac{1}{2k^3}(\sin kt - kt \cos kt)\right\}. \tag{8}$$

## 64. The gamma function

For obtaining the Laplace transform of nonintegral powers of $t$, we need a function not usually discussed in elementary mathematics.

The gamma function $\Gamma(x)$ is defined by

$$\Gamma(x) = \int_0^\infty e^{-\beta}\beta^{x-1}\, d\beta, \qquad x > 0. \tag{1}$$

Substitution of $(x + 1)$ for $x$ in (1) gives

$$\Gamma(x + 1) = \int_0^\infty e^{-\beta}\beta^x\, d\beta. \tag{2}$$

An integration by parts, integrating $e^{-\beta}\, d\beta$ and differentiating $\beta^x$, yields

$$\Gamma(x + 1) = \left[-e^{-\beta}\beta^x\right]_0^\infty + x \int_0^\infty e^{-\beta}\beta^{x-1}\, d\beta. \tag{3}$$

Because $x > 0$, $\beta^x \to 0$ as $\beta \to 0$, and, because $x$ is fixed, $e^{-\beta}\beta^x \to 0$ as $\beta \to \infty$. Thus

$$\Gamma(x + 1) = x \int_0^\infty e^{-\beta}\beta^{x-1}\, d\beta = x\Gamma(x). \tag{4}$$

THEOREM 17:   *For $x > 0$, $\Gamma(x + 1) = x\Gamma(x)$.*

Suppose that $n$ is a positive integer. Iteration of Theorem 17 gives us

$$\Gamma(n + 1) = n\Gamma(n)$$
$$= n(n - 1)\Gamma(n - 1)$$
$$\vdots$$
$$= n(n - 1)(n - 2)\cdots 2 \cdot 1 \cdot \Gamma(1)$$
$$= n!\,\Gamma(1).$$

But, by definition,

$$\Gamma(1) = \int_0^\infty e^{-\beta}\beta^0 \, d\beta = \left[ -e^{-\beta} \right]_0^\infty = 1.$$

THEOREM 18:   *For positive integral $n$, $\Gamma(n + 1) = n!$.*

In the integral for $\Gamma(x + 1)$ in (2), let us put $\beta = st$ with $s > 0$ and $t$ as the new variable of integration. This yields, since $t \to 0$ as $\beta \to 0$ and $t \to \infty$ as $\beta \to \infty$,

$$\Gamma(x + 1) = \int_0^\infty e^{-st}s^x t^x s \, dt = s^{x+1} \int_0^\infty e^{-st}t^x \, dt, \qquad (5)$$

which is valid for $x + 1 > 0$. We thus obtain

$$\frac{\Gamma(x + 1)}{s^{x+1}} = \int_0^\infty e^{-st}t^x \, dt, \qquad s > 0, x > -1,$$

which in our Laplace transform notation says that

$$L\{t^x\} = \frac{\Gamma(x + 1)}{s^{x+1}}, \qquad s > 0, x > -1. \qquad (6)$$

If in (6) we put $x = -\frac{1}{2}$, we get

$$L\{t^{-1/2}\} = \frac{\Gamma(\frac{1}{2})}{s^{1/2}}.$$

But we already know that $L\{t^{-1/2}\} = (\pi/s)^{1/2}$. Hence

$$\Gamma(\tfrac{1}{2}) = \sqrt{\pi}. \qquad (7)$$

## 65. Periodic functions

Suppose that the function $F(t)$ is periodic with period $\omega$:

$$F(t + \omega) = F(t). \qquad (1)$$

The function is completely determined by (1) once the nature of $F(t)$ throughout one period, $0 \leqq t < \omega$, is given. If $F(t)$ has a transform,

$$L\{F(t)\} = \int_0^\infty e^{-st} F(t) \, dt, \tag{2}$$

the integral can be written as a sum of integrals,

$$L\{F(t)\} = \sum_{n=0}^\infty \int_{n\omega}^{(n+1)\omega} e^{-st} F(t) \, dt. \tag{3}$$

Let us put $t = n\omega + \beta$. Then (3) becomes

$$L\{F(t)\} = \sum_{n=0}^\infty \int_0^\omega \exp(-sn\omega - s\beta) F(\beta + n\omega) \, d\beta.$$

But $F(\beta + n\omega) = F(\beta)$, by iteration of (1). Hence

$$L\{F(t)\} = \sum_{n=0}^\infty \exp(-sn\omega) \int_0^\omega \exp(-s\beta) F(\beta) \, d\beta. \tag{4}$$

The integral on the right in (4) is independent of $n$ and we can sum the series on the right;

$$\sum_{n=0}^\infty \exp(-sn\omega) = \sum_{n=0}^\infty [\exp(-s\omega)]^n = \frac{1}{1 - e^{-s\omega}}.$$

THEOREM 19: *If $F(t)$ has a Laplace transform and if $F(t + \omega) = F(t)$,*

$$L\{F(t)\} = \frac{\displaystyle\int_0^\omega e^{-s\beta} F(\beta) \, d\beta}{1 - e^{-s\omega}}. \tag{5}$$

Next suppose that a function $H(t)$ has a period $2c$ and that we demand that $H(t)$ be zero throughout the right half of each period. That is,

$$H(t + 2c) = H(t), \tag{6}$$

$$H(t) = g(t), \qquad 0 \leqq t < c, \tag{7}$$

$$= 0, \qquad c \leqq t < 2c.$$

Then we say that $H(t)$ is a half-wave rectification of $g(t)$. Using (5) we may conclude that for the $H(t)$ defined by (6) and (7),

$$L\{H(t)\} = \frac{\displaystyle\int_0^c \exp(-s\beta) g(\beta) \, d\beta}{1 - \exp(-2cs)}. \tag{8}$$

EXAMPLE (a):  Find the transform of the function $\psi(t, c)$ shown in Figure 20 and defined by

$$\psi(t, c) = 1, \qquad 0 < t < c, \tag{9}$$

$$= 0, \qquad c < t < 2c;$$

$$\psi(t + 2c, c) = \psi(t, c). \tag{10}$$

FIGURE 20

We may use equation (8) and the fact that

$$\int_0^c \exp(-s\beta)\, d\beta = \frac{1 - \exp(-sc)}{s}$$

to conclude that

$$L\{\psi(t, c)\} = \frac{1}{s} \cdot \frac{1 - \exp(-sc)}{1 - \exp(-2sc)} = \frac{1}{s} \cdot \frac{1}{1 + \exp(-sc)}. \tag{11}$$

EXAMPLE (b):  Find the transform of the square-wave function $Q(t, c)$ shown in Figure 21 and defined by

$$Q(t, c) = 1, \qquad 0 < t < c, \tag{12}$$

$$= -1, \qquad c < t < 2c;$$

$$Q(t + 2c, c) = Q(t, c). \tag{13}$$

This transform can be obtained by using Theorem 19, but also

$$Q(t, c) = 2\psi(t, c) - 1; \tag{14}$$

**FIGURE 21**

hence, from (11),

$$L\{Q(t, c)\} = \frac{1}{s}\left[\frac{2}{1 + \exp(-sc)} - 1\right] = \frac{1}{s} \cdot \frac{1 - \exp(-sc)}{1 + \exp(-sc)}. \tag{15}$$

By multiplying numerator and denominator of the last fraction above by $\exp(\tfrac{1}{2}sc)$, we may put (15) in the form

$$L\{Q(t, c)\} = \frac{1}{s}\tanh\frac{cs}{2}. \tag{16}$$

### *Exercises*

1. Show that $L\{t^{1/2}\} = \dfrac{1}{2s}\left(\dfrac{\pi}{s}\right)^{1/2}, \quad s > 0.$

2. Show that $L\{t^{5/2}\} = \dfrac{15}{8s^3}\left(\dfrac{\pi}{s}\right)^{1/2}, \quad s > 0.$

3. Use equation (4), page 184, to derive $L\{\sin kt\}$.

4. Use equation (4), page 184, to derive $L\{\cos kt\}$.

5. Check the known transforms of $\sin kt$ and $\cos kt$ against one another by using Theorem 12, page 184.

6. If $n$ is a positive integer, obtain $L\{t^n e^{kt}\}$ from the known $L\{e^{kt}\}$ by using Theorem 16, page 186.      ANS. $\dfrac{n!}{(s - k)^{n+1}}, \quad s > k.$

7. Find $L\{t^2 \sin kt\}$.      ANS. $\dfrac{2k(3s^2 - k^2)}{(s^2 + k^2)^3}, \quad s > 0.$

**8.** Find $L\{t^2 \cos kt\}$.                    ANS. $\dfrac{2s(s^2 - 3k^2)}{(s^2 + k^2)^3}$, $s > 0$.

**9.** For the function

$$F(t) = t + 1, \qquad 0 \leq t \leq 2,$$
$$= 3, \qquad\qquad t > 2,$$

graph $F(t)$ and $F'(t)$. Find $L\{F(t)\}$. Find $L\{F'(t)\}$ in two ways.

ANS. $L\{F'(t)\} = s^{-1}(1 - e^{-2s})$, $s > 0$.

**10.** For the function

$$H(t) = t + 1, \qquad 0 \leq t \leq 2,$$
$$= 6, \qquad\qquad t > 2,$$

parallel exercise 9 above.

**11.** Define a triangular-wave function $T(t, c)$ by

$$T(t, c) = t, \qquad\qquad 0 \leq t \leq c,$$
$$= 2c - t, \qquad c < t < 2c;$$
$$T(t + 2c, c) = T(t, c).$$

Sketch $T(t, c)$ and find its Laplace transform.        ANS. $\dfrac{1}{s^2} \tanh \dfrac{cs}{2}$.

**12.** Show that the derivative of the function $T(t, c)$ of exercise 11 is, except at certain points, the function $Q(t, c)$ of Example (b), Section 65. Obtain $L\{T(t, c)\}$ from $L\{Q(t, c)\}$.

**13.** Find $L\{|\sin kt|\}$.                    ANS. $\dfrac{k}{s^2 + k^2} \cdot \coth \pi \dfrac{s}{2k}$.

**14.** Find $L\{|\cos kt|\}$.

**15.** Define the function $G(t)$ by

$$G(t) = e^t, \qquad 0 \leq t < c,$$
$$G(t + c) = G(t), \qquad t \geq 0.$$

Sketch the graph of $G(t)$ and find its Laplace transform.

ANS. $\dfrac{-1}{s - 1} \cdot \dfrac{1 - \exp c(1 - s)}{1 - \exp(-cs)}$, $s > 1$.

**16.** Define the function $S(t)$ by

$$S(t) = 1 - t, \qquad 0 \leq t < 1,$$
$$S(t + 1) = S(t), \qquad t \geq 0.$$

Sketch the graph of $S(t)$ and find its Laplace transform.

17. Sketch a half-wave rectification of the function $\sin \omega t$, as described below, and find its transform.

$$F(t) = \sin \omega t, \qquad 0 \leq t \leq \frac{\pi}{\omega},$$

$$= 0, \qquad \frac{\pi}{\omega} < t < \frac{2\pi}{\omega};$$

$$F\left(t + \frac{2\pi}{\omega}\right) = F(t).$$

ANS.  $\dfrac{\omega}{s^2 + \omega^2} \cdot \dfrac{1}{1 - \exp(-s\pi/\omega)}.$

18. Find $L\{F(t)\}$ where $F(t) = t$ for $0 < t < \omega$ and $F(t + \omega) = F(t)$.

ANS.  $\dfrac{1}{s^2} - \dfrac{\omega}{s} \dfrac{\exp(-s\omega)}{1 - \exp(-s\omega)} = \dfrac{1}{s^2} + \dfrac{\omega}{2s}\left(1 - \coth \dfrac{\omega s}{2}\right).$

19. Prove that if $L\{F(t)\} = f(s)$ and if $F(t)/t$ is of class A,

$$L\{F(t)/t\} = \int_s^\infty f(\beta)\,d\beta.$$

*Hint:* Use Theorem 15, page 186.

# Inverse Transforms

### 66. Definition of an inverse transform

Suppose that the function $F(t)$ is to be determined from a differential equation with initial conditions. The Laplace operator $L$ is used to transform the original problem into a new problem from which the transform $f(s)$ is to be found. If the Laplace transformation is to be effective, the new problem must be simpler than the original problem. We first find $f(s)$ and then must obtain $F(t)$ from $f(s)$. It is therefore desirable to develop methods for finding the object function $F(t)$ when its transform $f(s)$ is known.

If

$$L\{F(t)\} = f(s), \tag{1}$$

we say that $F(t)$ is an *inverse Laplace transform*, or an inverse transform, of $f(s)$ and we write

$$F(t) = L^{-1}\{f(s)\}. \tag{2}$$

Since (1) means that

$$\int_0^\infty e^{-st} F(t)\,dt = f(s), \tag{3}$$

it follows at once that an inverse transform is not unique. For example, if $F_1(t)$ and $F_2(t)$ are identical except at a discrete set of points and differ at these points, the value of the integral in (3) is the same for the two functions; their transforms are identical.

Let us employ the term *null function* for any function $N(t)$ for which

$$\int_0^{t_0} N(t)\, dt = 0 \tag{4}$$

for every positive $t_0$. Lerch's theorem (not proved here) states that if $L\{F_1(t)\} = L\{F_2(t)\}$, then $F_1(t) - F_2(t) = N(t)$. That is, an inverse Laplace transform is unique except for the addition of an arbitrary null function.

The only continuous null function is the zero function. If an $f(s)$ has a continuous inverse $F(t)$, then $F(t)$ is the only continuous inverse of $f(s)$. If $f(s)$ has an inverse $F_1(t)$ continuous over a specified closed interval, every inverse that is also continuous over that interval is identical with $F_1(t)$ on that interval. Essentially, inverses of the same $f(s)$ differ at most at their points of discontinuity.

In applications, failure of uniqueness caused by addition of a null function is not vital, because the effect of that null function on physical properties of the solution is null. In the problems we treat, the inverse $F(t)$ is required either to be continuous for $t \geq 0$ or to be sectionally continuous with the values of $F(t)$ at the points of discontinuity specified by each problem. The $F(t)$ is then unique.

A crude, but sometimes effective, method for finding inverse Laplace transforms is to construct a table of transforms (page 231) and then to use it in reverse to find inverse transforms.

We know from exercise 1, page 175, that

$$L\{\cos kt\} = \frac{s}{s^2 + k^2}. \tag{5}$$

Therefore

$$L^{-1}\left\{\frac{s}{s^2 + k^2}\right\} = \cos kt. \tag{6}$$

We shall refine the above method, and actually make it quite powerful, by developing theorems by which a given $f(s)$ may be expanded into component parts whose inverses are known (found in the table). Other theorems will permit us to write $f(s)$ in alternative forms that yield the desired inverse. The most fundamental of such theorems is one that states that the inverse transformation is a linear operation.

**THEOREM 20:**    *If $c_1$ and $c_2$ are constants,*

$$L^{-1}\{c_1 f_1(s) + c_2 f_2(s)\} = c_1 L^{-1}\{f_1(s)\} + c_2 L^{-1}\{f_2(s)\}.$$

Next let us prove a simple, but extremely useful, theorem on the manipulation of inverse transforms. From

$$f(s) = \int_0^\infty e^{-st} F(t)\, dt, \tag{7}$$

we obtain

$$f(s - a) = \int_0^\infty e^{-(s-a)t} F(t)\, dt$$

$$= \int_0^\infty e^{-st} [e^{at} F(t)]\, dt.$$

Thus, from $L^{-1}\{f(s)\} = F(t)$ it follows that

$$L^{-1}\{f(s - a)\} = e^{at} F(t),$$

or

$$L^{-1}\{f(s - a)\} = e^{at} L^{-1}\{f(s)\}. \tag{8}$$

Equation (8) may be rewritten with the exponential transferred to the other side of the equation. We thus obtain the following result.

THEOREM 21: $\quad L^{-1}\{f(s)\} = e^{-at} L^{-1}\{f(s - a)\}.$

EXAMPLE (a): Find $L^{-1}\left\{\dfrac{15}{s^2 + 4s + 13}\right\}.$

First complete the square in the denominator.

$$L^{-1}\left\{\frac{15}{s^2 + 4s + 13}\right\} = L^{-1}\left\{\frac{15}{(s + 2)^2 + 9}\right\}.$$

Since we know that $L^{-1}\left\{\dfrac{k}{s^2 + k^2}\right\} = \sin kt$, we proceed as follows:

$$L^{-1}\left\{\frac{15}{s^2 + 4s + 13}\right\} = 5L^{-1}\left\{\frac{3}{(s + 2)^2 + 9}\right\} = 5\,e^{-2t} L^{-1}\left\{\frac{3}{s^2 + 9}\right\}$$

$$= 5\,e^{-2t} \sin 3t,$$

in which we have used Theorem 21.

EXAMPLE (b): Evaluate $L^{-1}\left\{\dfrac{s + 1}{s^2 + 6s + 25}\right\}.$

We write

$$L^{-1}\left\{\frac{s + 1}{s^2 + 6s + 25}\right\} = L^{-1}\left\{\frac{s + 1}{(s + 3)^2 + 16}\right\}.$$

Then

$$L^{-1}\left\{\frac{s+1}{s^2+6s+25}\right\} = e^{-3t}L^{-1}\left\{\frac{s-2}{s^2+16}\right\}$$

$$= e^{-3t}\left[L^{-1}\left\{\frac{s}{s^2+16}\right\} - \tfrac{1}{2}L^{-1}\left\{\frac{4}{s^2+16}\right\}\right]$$

$$= e^{-3t}(\cos 4t - \tfrac{1}{2}\sin 4t).$$

## Exercises

In exercises 1 through 10 obtain $L^{-1}\{f(s)\}$ from the given $f(s)$.

**1.** $\dfrac{1}{s^2+2s+10}$.        ANS.   $\tfrac{1}{3}e^{-t}\sin 3t$.

**2.** $\dfrac{1}{s^2-4s+8}$.        ANS.   $\tfrac{1}{2}e^{2t}\sin 2t$.

**3.** $\dfrac{3s}{s^2+4s+13}$.        ANS.   $e^{-2t}(3\cos 3t - 2\sin 3t)$.

**4.** $\dfrac{s}{s^2+6s+13}$.        ANS.   $e^{-3t}(\cos 2t - \tfrac{3}{2}\sin 2t)$.

**5.** $\dfrac{1}{s^2+4s+4}$.        ANS.   $t e^{-2t}$.

**6.** $\dfrac{s}{s^2+4s+4}$.        ANS.   $e^{-2t}(1 - 2t)$.

**7.** $\dfrac{2s-3}{s^2-4s+8}$.        ANS.   $e^{2t}(2\cos 2t + \tfrac{1}{2}\sin 2t)$.

**8.** $\dfrac{3s+1}{s^2+6s+13}$.        ANS.   $e^{-3t}(3\cos 2t - 4\sin 2t)$.

**9.** $\dfrac{2s+3}{(s+4)^3}$.        ANS.   $e^{-4t}(2t - \tfrac{5}{2}t^2)$.

**10.** $\dfrac{s^2}{(s-1)^4}$.        ANS.   $e^t(t + t^2 + \tfrac{1}{6}t^3)$.

**11.** Show that for $n$ a nonnegative integer

$$L^{-1}\left\{\frac{1}{(s+a)^{n+1}}\right\} = \frac{t^n e^{-at}}{n!}.$$

**12.** Show that for $m > -1$,

$$L^{-1}\left\{\frac{1}{(s+a)^{m+1}}\right\} = \frac{t^m e^{-at}}{\Gamma(m+1)}.$$

**13.** Show that

$$L^{-1}\left\{\frac{1}{(s+a)^2+b^2}\right\} = \frac{1}{b}e^{-at}\sin bt.$$

**14.** Show that

$$L^{-1}\left\{\frac{s}{(s+a)^2+b^2}\right\} = \frac{1}{b}e^{-at}(b\cos bt - a\sin bt).$$

**15.** For $a > 0$, show that, from $L^{-1}\{f(s)\} = F(t)$, it follows that

$$L^{-1}\{f(as)\} = \frac{1}{a}F\left(\frac{t}{a}\right).$$

**16.** For $a > 0$, show that, from $L^{-1}\{f(s)\} = F(t)$, it follows that

$$L^{-1}\{f(as+b)\} = \frac{1}{a}\exp\left(-\frac{bt}{a}\right)F\left(\frac{t}{a}\right).$$

## 67. Partial fractions

In using the Laplace transform to solve differential equations, we often need to obtain the inverse transform of a rational fraction

$$\frac{N(s)}{D(s)}. \tag{1}$$

The numerator and denominator in (1) are polynomials in $s$ and the degree of $D(s)$ is larger than the degree of $N(s)$. The fraction (1) has the partial fractions expansion used in calculus. Because of the linearity of the inverse operator $L^{-1}$, the partial fractions expansion of (1) permits us to replace a complicated problem in obtaining an inverse transform with a set of simpler problems.

**EXAMPLE (a):**   Obtain $L^{-1}\left\{\dfrac{s^2-6}{s^3+4s^2+3s}\right\}.$

Since the denominator is a product of distinct linear factors, we know that constants $A, B, C$ exist such that

$$\frac{s^2-6}{s^3+4s^2+3s} = \frac{s^2-6}{s(s+1)(s+3)} = \frac{A}{s} + \frac{B}{s+1} + \frac{C}{s+3}.$$

Multiplying each term by the lowest common denominator, we obtain the identity

$$s^2 - 6 = A(s+1)(s+3) + Bs(s+3) + Cs(s+1), \tag{2}$$

from which we need to determine $A$, $B$, and $C$. Using the values $s = 0$, $-1$, $-3$ successively in (2), we get

$$s = 0: \qquad -6 = A(1)(3),$$
$$s = -1: \qquad -5 = B(-1)(2),$$
$$s = -3: \qquad 3 = C(-3)(-2),$$

from which $A = -2$, $B = \frac{5}{2}$, $C = \frac{1}{2}$. Therefore

$$\frac{s^2 - 6}{s^3 + 4s^2 + 3s} = \frac{-2}{s} + \frac{\frac{5}{2}}{s + 1} + \frac{\frac{1}{2}}{s + 3}.$$

Since $L^{-1}\left\{\dfrac{1}{s}\right\} = 1$ and $L^{-1}\left\{\dfrac{1}{s + a}\right\} = e^{-at}$, we get the desired result,

$$L^{-1}\left\{\frac{s^2 - 6}{s^3 + 4s^2 + 3s}\right\} = -2 + \tfrac{5}{2}e^{-t} + \tfrac{1}{2}e^{-3t}.$$

**EXAMPLE (b):**   Obtain $L^{-1}\left\{\dfrac{5s^3 - 6s - 3}{s^3(s + 1)^2}\right\}$.

Since the denominator contains repeated linear factors, we must assume partial fractions of the form shown:

$$\frac{5s^3 - 6s - 3}{s^3(s + 1)^2} = \frac{A_1}{s} + \frac{A_2}{s^2} + \frac{A_3}{s^3} + \frac{B_1}{s + 1} + \frac{B_2}{(s + 1)^2}. \qquad (3)$$

Corresponding to a denominator factor $(x - \gamma)^r$, we must in general assume $r$ partial fractions of the form

$$\frac{A_1}{x - \gamma} + \frac{A_2}{(x - \gamma)^2} + \cdots + \frac{A_r}{(x - \gamma)^r}.$$

From (3) we get

$$5s^3 - 6s - 3 = A_1 s^2(s + 1)^2 + A_2 s(s + 1)^2$$
$$+ A_3(s + 1)^2 + B_1 s^3(s + 1) + B_2 s^3, \qquad (4)$$

which must be an identity in $s$. To get the necessary five equations for the determination of $A_1, A_2, A_3, B_1, B_2$, two elementary methods are popular. Specific values of $s$ can be used in (4), or the coefficients of like powers of $s$ in the two members of (4) may be equated. We employ whatever combination of these methods yields simple equations to be solved for $A_1, A_2, \ldots, B_2$. From (4) we obtain

$$s = 0: \qquad -3 = A_3(1),$$
$$s = -1: \qquad -2 = B_2(-1),$$

$$\text{coeff. of } s^4: \qquad 0 = A_1 + B_1,$$

$$\text{coeff. of } s^3: \qquad 5 = 2A_1 + A_2 + B_1 + B_2,$$

$$\text{coeff. of } s: \qquad -6 = A_2 + 2A_3.$$

The above equations yield $A_1 = 3$, $A_2 = 0$, $A_3 = -3$, $B_1 = -3$, $B_2 = 2$. Therefore we find that

$$L^{-1}\left\{\frac{5s^3 - 6s - 3}{s^3(s+1)^2}\right\} = L^{-1}\left\{\frac{3}{s} - \frac{3}{s^3} - \frac{3}{s+1} + \frac{2}{(s+1)^2}\right\}$$

$$= 3 - \tfrac{3}{2}t^2 - 3e^{-t} + 2te^{-t}.$$

**EXAMPLE (c):** Obtain $L^{-1}\left\{\dfrac{16}{s(s^2+4)^2}\right\}$.

Since quadratic factors require the corresponding partial fractions to have linear numerators, we start with an expansion of the form

$$\frac{16}{s(s^2+4)^2} = \frac{A}{s} + \frac{B_1 s + C_1}{s^2 + 4} + \frac{B_2 s + C_2}{(s^2+4)^2}.$$

From the identity

$$16 = A(s^2 + 4)^2 + (B_1 s + C_1)s(s^2 + 4) + (B_2 s + C_2)s,$$

it is not difficult to find the values $A = 1$, $B_1 = -1$, $B_2 = -4$, $C_1 = 0$, $C_2 = 0$. We thus obtain

$$L^{-1}\left\{\frac{16}{s(s^2+4)^2}\right\} = L^{-1}\left\{\frac{1}{s} - \frac{s}{s^2+4} - \frac{4s}{(s^2+4)^2}\right\}$$

$$= 1 - \cos 2t - t \sin 2t.$$

### Exercises

In exercises 1 through 10, find an inverse transform of the given $f(s)$.

**1.** $\dfrac{1}{s^2 + as}$.        ANS. $\dfrac{1}{a}(1 - e^{-at})$.

**2.** $\dfrac{s+2}{s^2 - 6s + 8}$.        ANS. $3e^{4t} - 2e^{2t}$.

**3.** $\dfrac{2s^2 + 5s - 4}{s^3 + s^2 - 2s}$.        ANS. $2 + e^t - e^{-2t}$.

**4.** $\dfrac{2s^2 + 1}{s(s+1)^2}$.        ANS. $1 + e^{-t} - 3te^{-t}$.

**5.** $\dfrac{4s+4}{s^2(s-2)}$.        ANS. $3e^{2t} - 3 - 2t$.

**6.** $\dfrac{1}{s^3(s^2 + 1)}.$           ANS.   $\frac{1}{2}t^2 - 1 + \cos t.$

**7.** $\dfrac{5s - 2}{s^2(s + 2)(s - 1)}.$        ANS.   $t - 2 + e^t + e^{-2t}.$

**8.** $\dfrac{1}{(s^2 + a^2)(s^2 + b^2)}.$    $a^2 \neq b^2, ab \neq 0.$      ANS.   $\dfrac{b \sin at - a \sin bt}{ab(b^2 - a^2)}.$

**9.** $\dfrac{s}{(s^2 + a^2)(s^2 + b^2)},$    $a^2 \neq b^2, ab \neq 0.$      ANS.   $\dfrac{\cos at - \cos bt}{b^2 - a^2}.$

**10.** $\dfrac{s^2}{(s^2 + a^2)(s^2 + b^2)},$    $a^2 \neq b^2, ab \neq 0.$      ANS.   $\dfrac{a \sin at - b \sin bt}{a^2 - b^2}.$

## 68. Initial value problems

Because of Theorem 13, page 184, the Laplace operator will transform a linear differential equation with constant coefficients into an algebraic equation in the transformed function. If upon solving this algebraic equation for the transformed function we are able to obtain the inverse transform, we may have a solution of the original differential equation. Several examples will now be treated in detail so that we can get some feeling for the advantages and disadvantages of the transform method. One fact is apparent from the nature of the transform of derivatives: this method is most readily applied if the appropriate initial conditions are given along with the differential equation. If they are not, the algebra is more complicated.

EXAMPLE (a):   Solve the initial value problem

$$y''(t) + y(t) = 0; \qquad y(0) = 0, \; y'(0) = 1. \tag{1}$$

Applying the Laplace transform to both sides of the differential equation gives

$$L\{y'' + y\} = 0,$$

and because of the linearity of the transform

$$L\{y''\} + L\{y\} = 0.$$

An application of Theorem 13 now yields

$$s^2 L\{y(t)\} - 1 + L\{y(t)\} = 0,$$

an equation that may easily be solved for $L\{y(t)\}$. We have

$$L\{y(t)\} = \frac{1}{s^2 + 1}. \tag{2}$$

We know that $\sin t$ is a function that satisfies (2), and it is a simple matter to verify that $\sin t$ is the solution of (1).

EXAMPLE (b):   Solve the problem

$$y''(t) + \beta^2 y(t) = A \sin \omega t; \qquad y(0) = 1,\ y'(0) = 0. \tag{3}$$

Here $A$, $\beta$, $\omega$ are constants. Because $\beta = 0$ would make the problem one of elementary calculus and because a change in sign of $\beta$ or $\omega$ would not alter the character of the problem, we may assume that $\beta$ and $\omega$ are positive.
Let

$$L\{y(t)\} = u(s).$$

Then

$$L\{y'(t)\} = su(s) - 1,$$
$$L\{y''(t)\} = s^2 u(s) - s \cdot 1 - 0,$$

and application of the operator $L$ transforms the problem (3) into

$$s^2 u(s) - s + \beta^2 u(s) = \frac{A\omega}{s^2 + \omega^2},$$

from which

$$u(s) = \frac{s}{s^2 + \beta^2} + \frac{A\omega}{(s^2 + \beta^2)(s^2 + \omega^2)}. \tag{4}$$

We need the inverse transform of the right member of (4). The form of that inverse depends upon whether $\beta$ and $\omega$ are equal or unequal.
If $\omega \neq \beta$,

$$u(s) = \frac{s}{s^2 + \beta^2} + \frac{A\omega}{\beta^2 - \omega^2}\left(\frac{1}{s^2 + \omega^2} - \frac{1}{s^2 + \beta^2}\right)$$

$$= \frac{s}{s^2 + \beta^2} + \frac{A}{\beta(\beta^2 - \omega^2)}\left(\frac{\omega\beta}{s^2 + \omega^2} - \frac{\omega\beta}{s^2 + \beta^2}\right).$$

Now $y(t) = L^{-1}\{u(s)\}$ so, for $\omega \neq \beta$,

$$y(t) = \cos \beta t + \frac{A}{\beta(\beta^2 - \omega^2)}(\beta \sin \omega t - \omega \sin \beta t). \tag{5}$$

If $\omega = \beta$, the transform (4) becomes

$$u(s) = \frac{s}{s^2 + \beta^2} + \frac{A\beta}{(s^2 + \beta^2)^2}. \tag{6}$$

We know from equation (8) of Section 63 that

$$L^{-1}\left\{\frac{1}{(s^2 + \beta^2)^2}\right\} = \frac{1}{2\beta^3}(\sin \beta t - \beta t \cos \beta t).$$

Hence, for $\omega = \beta$,

$$y(t) = \cos \beta t + \frac{A}{2\beta^2}(\sin \beta t - \beta t \cos \beta t). \tag{7}$$

It is a simple matter to show that this function is indeed the solution of the given initial value problem.

Note that the initial conditions were satisfied automatically by this method when Theorem 13 was applied. We get not the general solution with arbitrary constants still to be determined but that particular solution which satisfies the desired initial conditions. The transform method also gives us some insight into the reason that the solution takes different forms according to whether $\omega$ and $\beta$ are equal or unequal.

EXAMPLE (c):   Solve the problem

$$x''(t) + 2x'(t) + x(t) = 3t\,e^{-t}; \qquad x(0) = 4, \quad x'(0) = 2. \tag{8}$$

Let $L\{x(t)\} = y(s)$. Then the operator $L$ converts (8) into

$$s^2 y(s) - 4s - 2 + 2[sy(s) - 4] + y(s) = \frac{3}{(s + 1)^2},$$

or

$$y(s) = \frac{4s + 10}{(s + 1)^2} + \frac{3}{(s + 1)^4}. \tag{9}$$

We may write

$$y(s) = \frac{4(s + 1) + 6}{(s + 1)^2} + \frac{3}{(s + 1)^4}$$

or

$$y(s) = \frac{4}{s + 1} + \frac{6}{(s + 1)^2} + \frac{3}{(s + 1)^4}.$$

Employing the inverse transform, we obtain

$$x(t) = (4 + 6t + \tfrac{1}{2}t^3)e^{-t}. \tag{10}$$

Again the knowledge of initial conditions contributed to the efficiency of our method. In obtaining and in using equation (9), those terms that came from the initial values $x(0)$ and $x'(0)$ were not combined with the term that came from the transform of the right member of the differential equation.

To combine such terms rarely simplifies and frequently complicates the task of obtaining the inverse transform.

From the solution (10) the student should obtain the derivatives

$$x'(t) = (2 - 6t + \tfrac{3}{2}t^2 - \tfrac{1}{2}t^3)e^{-t},$$

$$x''(t) = (-8 + 9t - 3t^2 + \tfrac{1}{2}t^3)e^{-t},$$

and thus verify that the $x$ of (10) satisfies both the differential equation and the initial conditions of the problem (8). Such verification not only checks our work but also removes any need to justify temporary assumptions about the right to use the Laplace transform theorems on the function $x(t)$ during the time that the function is still unknown.

EXAMPLE (d): Solve the problem

$$w''(x) + 2w'(x) + w(x) = x; \qquad w(0) = -3, w(1) = -1. \tag{11}$$

In this example the boundary conditions are not both of the initial condition type. Using $x$, rather than $t$, as independent variable, let

$$L\{w(x)\} = g(s). \tag{12}$$

We know $w(0) = -3$, but we also need $w'(0)$ in order to write the transform of $w''(x)$. Hence we put

$$w'(0) = B \tag{13}$$

and hope to determine $B$ later by using the condition that $w(1) = -1$.

The transformed problem is

$$s^2 g(s) - s(-3) - B + 2[sg(s) - (-3)] + g(s) = \frac{1}{s^2}$$

from which

$$g(s) = \frac{-3(s+1) + B - 3}{(s+1)^2} + \frac{1}{s^2(s+1)^2}. \tag{14}$$

But, by the usual partial fractions expansion,

$$\frac{1}{s^2(s+1)^2} = -\frac{2}{s} + \frac{1}{s^2} + \frac{2}{s+1} + \frac{1}{(s+1)^2},$$

so

$$g(s) = \frac{1}{s^2} - \frac{2}{s} - \frac{1}{s+1} + \frac{B-2}{(s+1)^2}, \tag{15}$$

from which we obtain

$$w(x) = x - 2 - e^{-x} + (B-2)x\,e^{-x}. \tag{16}$$

We have yet to impose the condition that $w(1) = -1$. From (16) with $x = 1$, we get

$$-1 = 1 - 2 - e^{-1} + (B - 2)e^{-1},$$

so $B = 3$.

Thus our final result is

$$w(x) = x - 2 - e^{-x} + xe^{-x}. \tag{17}$$

The problem in Example (d) may be solved efficiently by the methods of Chapter 7. See also exercises 23 through 44 below.

## *Exercises*

In exercises 1 through 22, solve the problem by the Laplace transform method. Verify that your solution satisfies the differential equation and the initial conditions.

**1.** $y' = e^t$; $y(0) = 2$.          ANS.   $y = e^t + 1$.

**2.** $y' = 2e^t$; $y(0) = -1$.          ANS.   $y = 2e^t - 3$.

**3.** $y' + y = e^{2t}$; $y(0) = 0$.        ANS.   $y = \frac{1}{3}e^{2t} - \frac{1}{3}e^{-t}$.

**4.** $y' - y = e^{-t}$; $y(0) = 1$.        ANS.   $y = \frac{3}{2}e^t - \frac{1}{2}e^{-t}$.

**5.** $y'' + a^2 y = 0$; $y(0) = 1$, $y'(0) = 0$.       ANS.   $y = \cos at$.

**6.** $y'' + a^2 y = 0$; $y(0) = 0$, $y'(0) = a$.       ANS.   $y = \sin at$.

**7.** $y'' - 3y' + 2y = e^{3t}$; $y(0) = y'(0) = 0$.    ANS.   $y = \frac{1}{2}e^t - e^{2t} + \frac{1}{2}e^{3t}$.

**8.** $y'' + y = e^{-t}$; $y(0) = y'(0) = 0$.    ANS.   $y = \frac{1}{2}(\sin t - \cos t + e^{-t})$.

**9.** $y'' - 2y' = -4$; $y(0) = 0$, $y'(0) = 4$.     ANS.   $y = e^{2t} + 2t - 1$.

**10.** $y'' + y' - 2y = -4$; $y(0) = 2$, $y'(0) = 3$.    ANS.   $y = 2 + e^t - e^{-2t}$.

**11.** $x''(t) - 4x'(t) + 4x(t) = 4e^{2t}$; $x(0) = -1$, $x'(0) = -4$.

                 ANS.   $x(t) = e^{2t}(2t^2 - 2t - 1)$.

**12.** $x''(t) + x(t) = 6 \sin 2t$; $x(0) = 3$, $x'(0) = 1$. ANS.   $x(t) = 5 \sin t + 3 \cos t - 2 \sin 2t$.

**13.** $y''(t) - y(t) = 4 \cos t$; $y(0) = 0$, $y'(0) = 1$.    ANS.   $y(t) = 2 \cosh t + \sinh t - 2 \cos t$.

**14.** $y''(t) - 6y'(t) + 9y(t) = 6t^2 e^{3t}$; $y(0) = y'(0) = 0$.      ANS.   $y(t) = \frac{1}{2}t^4 e^{3t}$.

**15.** $x''(t) + 4x(t) = t + 4$; $x(0) = 1$, $x'(0) = 0$.     ANS.   $x(t) = 1 + \frac{1}{4}t - \frac{1}{8}\sin 2t$.

**16.** $x''(t) - 2x'(t) = 6 - 4t$; $x(0) = 2$, $x'(0) = 0$.     ANS.   $x(t) = t^2 - 2t + 1 + e^{2t}$.

**17.** $x''(t) + x(t) = 4e^t$; $x(0) = 1$, $x'(0) = 3$.      ANS.   $x(t) = \sin t - \cos t + 2e^t$.

**18.** $x''(t) + x'(t) - 2x(t) = 6$; $x(0) = 1$, $x'(t) = 1$.      ANS.   $x(t) = e^{-2t} + 3e^t - 3$.

**19.** $y''(x) + 9y(x) = 40e^x$; $y(0) = 5$, $y'(0) = -2$. ANS.   $y(x) = 4e^x + \cos 3x - 2 \sin 3x$.

**20.** $y''(x) + y(x) = 4e^x$; $y(0) = 0$, $y'(0) = 0$.     ANS.   $y(x) = 2(e^x - \cos x - \sin x)$.

**21.** $x''(t) + 3x'(t) + 2x(t) = 4t^2$; $x(0) = 0$, $x'(0) = 0$.

             ANS.   $x(t) = 2t^2 - 6t + 7 - 8e^{-t} + e^{-2t}$.

**22.** $x''(t) - 4x'(t) + 4x(t) = 4 \cos 2t$; $x(0) = 2$, $x'(0) = 5$.

             ANS.   $x(t) = 2e^{2t}(1 + t) - \frac{1}{2}\sin 2t$.

In exercises 23 through 42, use the Laplace transform method with the realization that these exercises were not constructed with the Laplace transform technique in mind. Compare your work with that done in solving the same problems by the methods of Chapter 7. See pages 125–126.

**23.** Exercise 1.                      **24.** Exercise 2.

| | |
|---|---|
| **25.** Exercise 3. | **26.** Exercise 11. |
| **27.** Exercise 14. | **28.** Exercise 20. |
| **29.** Exercise 21. | **30.** Exercise 22. |
| **31.** Exercise 23. | **32.** Exercise 36. |
| **33.** Exercise 37. | **34.** Exercise 38. |
| **35.** Exercise 39. | **36.** Exercise 40. |
| **37.** Exercise 41. | **38.** Exercise 42. |
| **39.** Exercise 43. | **40.** Exercise 44. |
| **41.** Exercise 45. | **42.** Exercise 46. |

**43.** Solve the problem

$$x''(t) - 4x'(t) + 4x(t) = e^{2t}; \qquad x'(0) = 0, x(1) = 0.$$

ANS.    $x(t) = \frac{1}{2}(1 - t)^2\, e^{2t}.$

**44.** Solve the problem

$$x''(t) + 4x(t) = -8t^2; \qquad x(0) = 3, x(\tfrac{1}{4}\pi) = 0.$$

ANS.    $x(t) = 2\cos 2t + (\tfrac{1}{8}\pi^2 - 1)\sin 2t + 1 - 2t^2.$

## 69. A step function

Applications frequently deal with situations that change abruptly at specified times. We need a notation for a function that will suppress a given term up to a certain value of $t$ and insert the term for all larger $t$. The function we are about to introduce leads us to a powerful tool for constructing inverse transforms.

Let us define function $\alpha(t)$ by

$$\alpha(t) = 0, \qquad t < 0, \tag{1}$$
$$\phantom{\alpha(t)} = 1, \qquad t \geqq 0.$$

The graph of $\alpha(t)$ is shown in Figure 22.

FIGURE 22

The definition (1) says that $\alpha(t)$ is zero when the argument is negative and $\alpha(t)$ is unity when the argument is positive or is zero. It follows that

$$\alpha(t - c) = 0, \qquad t < c, \tag{2}$$
$$= 1, \qquad t \geq c.$$

The $\alpha$ function permits easy designation of the result of translating the graph of $F(t)$. If the graph of

$$y = F(t), \qquad t \geq 0, \tag{3}$$

is as shown in Figure 23, the graph of

$$y = \alpha(t - c)F(t - c), \qquad t \geq c, \tag{4}$$

is that shown in Figure 24. Furthermore, if $F(x)$ is defined for $-c \leq x < 0$, then $F(t - c)$ is defined for $0 \leq t < c$ and the $y$ of (4) is zero for $0 \leq t < c$ because of the negative argument in $\alpha(t - c)$. Notice that the values of $F(x)$ for negative $x$ have no bearing on this result because each value is multiplied by zero (from the $\alpha$); only the existence of $F$ for negative arguments is needed.

FIGURE 23

FIGURE 24

The Laplace transform of $\alpha(t - c)F(t - c)$ is related to that of $F(t)$. Consider

$$L\{\alpha(t - c)F(t - c)\} = \int_0^\infty e^{-st}\alpha(t - c)F(t - c)\,dt.$$

Since $\alpha(t - c) = 0$ for $0 \leq t < c$ and $\alpha(t - c) = 1$ for $t \geq c$, we get

$$L\{\alpha(t - c)F(t - c)\} = \int_c^\infty e^{-st}F(t - c)\,dt.$$

Now put $t - c = v$ in the integral to obtain

$$L\{\alpha(t - c)F(t - c)\} = \int_0^\infty e^{-s(c+v)}F(v)\,dv$$

$$= e^{-cs}\int_0^\infty e^{-sv}F(v)\,dv.$$

Since a definite integral is independent of the variable of integration,

$$\int_0^\infty e^{-sv} F(v)\, dv = \int_0^\infty e^{-st} F(t)\, dt = L\{F(t)\} = f(s).$$

Therefore we have shown that

$$L\{\alpha(t - c)F(t - c)\} = e^{-cs} L\{F(t)\} = e^{-cs} f(s). \tag{5}$$

**THEOREM 22:** *If $L^{-1}\{f(s)\} = F(t)$, if $c \geq 0$, and if $F(t)$ be assigned values (no matter what ones) for $-c \leq t < 0$,*

$$L^{-1}\{e^{-cs} f(s)\} = F(t - c)\alpha(t - c). \tag{6}$$

**EXAMPLE (a):** Find $L\{y(t)\}$ where (Figure 25)

$$y(t) = t^2, \qquad 0 < t < 2,$$
$$\quad\ = 6, \qquad\quad t > 2.$$

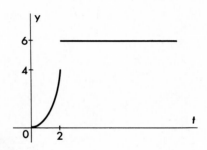

FIGURE 25

Here, direct use of the definition of a transform yields

$$L\{y(t)\} = \int_0^2 t^2 e^{-st}\, dt + \int_2^\infty 6 e^{-st}\, dt.$$

Although the above integrations are not difficult, we prefer to use the $\alpha$ function.

Since $\alpha(t - 2) = 0$ for $t < 2$ and $\alpha(t - 2) = 1$ for $t \geq 2$, we build the $y(t)$ in the following way. The crude trial

$$y_1 = t^2$$

works for $0 < t < 2$, but we wish to knock out the $t^2$ when $t > 2$. Hence we write

$$y_2 = t^2 - t^2 \alpha(t - 2).$$

This gives $t^2$ for $t < 2$ and zero for $t > 2$. Then we add the term $6\alpha(t - 2)$ and finally arrive at

$$y(t) = t^2 - t^2\alpha(t - 2) + 6\alpha(t - 2). \tag{7}$$

The $y$ of (7) is the $y$ of our example and, of course, it can be written at once after a little practice with the $\alpha$ function.

Unfortunately, the $y$ of (7) is not yet in the best form for our purpose. The theorem we wish to use gives us

$$L\{F(t - c)\alpha(t - c)\} = e^{-cs}f(s).$$

Therefore we must have the coefficient of $\alpha(t - 2)$ expressed as a function of $(t - 2)$. Since

$$-t^2 + 6 = -(t^2 - 4t + 4) - 4(t - 2) + 2,$$

$$y(t) = t^2 - (t - 2)^2\alpha(t - 2) - 4(t - 2)\alpha(t - 2) + 2\alpha(t - 2), \tag{8}$$

from which it follows at once that

$$L\{y(t)\} = \frac{2}{s^3} - \frac{2e^{-2s}}{s^3} - \frac{4e^{-2s}}{s^2} + \frac{2e^{-2s}}{s}.$$

EXAMPLE (b):  Find and sketch a function $g(t)$ for which

$$g(t) = L^{-1}\left\{\frac{3}{s} - \frac{4e^{-s}}{s^2} + \frac{4e^{-3s}}{s^2}\right\}.$$

We know that $L^{-1}\{4/s^2\} = 4t$. By Theorem 22 we then get

$$L^{-1}\left\{\frac{4e^{-s}}{s^2}\right\} = 4(t - 1)\alpha(t - 1)$$

and

$$L^{-1}\left\{\frac{4e^{-3s}}{s^2}\right\} = 4(t - 3)\alpha(t - 3).$$

We may therefore write

$$g(t) = 3 - 4(t - 1)\alpha(t - 1) + 4(t - 3)\alpha(t - 3). \tag{9}$$

To write $g(t)$ without the $\alpha$ function, consider first the interval

$$0 \le t < 1$$

in which $\alpha(t - 1) = 0$ and $\alpha(t - 3) = 0$. We find

$$g(t) = 3, \qquad 0 \le t < 1. \tag{10}$$

For $1 \le t < 3$, $\alpha(t - 1) = 1$, and $\alpha(t - 3) = 0$. Hence

$$g(t) = 3 - 4(t - 1) = 7 - 4t, \qquad 1 \le t < 3. \tag{11}$$

For $t \geq 3$, $\alpha(t - 1) = 1$ and $\alpha(t - 3) = 1$, so

$$g(t) = 3 - 4(t - 1) + 4(t - 3) = -5, \qquad t \geq 3. \tag{12}$$

Equations (10), (11), and (12) are equivalent to equation (9). The graph of $g(t)$ is shown in Figure 26.

FIGURE 26

EXAMPLE (c):   Solve the problem

$$x''(t) + 4x(t) = \psi(t); \qquad x(0) = 1, \, x'(0) = 0, \tag{13}$$

in which $\psi(t)$ is defined by

$$\psi(t) = 4t, \qquad 0 \leq t \leq 1, \tag{14}$$

$$= 4, \qquad t > 1.$$

We seek, of course, a solution valid in the range $t \geq 0$ in which the function $\psi(t)$ is defined.

In this problem another phase of the power of the Laplace transform method begins to emerge. The fact that the function $\psi(t)$ in the differential equation has discontinuous derivatives makes the use of the classical method of undetermined coefficients somewhat awkward, but such discontinuities do not interfere at all with the simplicity of the Laplace transform method.

In attacking this problem, let us put $L\{x(t)\} = h(s)$. We need to obtain $L\{\psi(t)\}$. In terms of the $\alpha$ function we may write, from (14),

$$\psi(t) = 4t - 4(t - 1)\alpha(t - 1), \qquad t \geq 0. \tag{15}$$

From (15) it follows that

$$L\{\psi(t)\} = \frac{4}{s^2} - \frac{4\,e^{-s}}{s^2}.$$

Therefore the application of the operator $L$ transforms problem (13) into

$$s^2 h(s) - s - 0 + 4h(s) = \frac{4}{s^2} - \frac{4e^{-s}}{s^2},$$

from which

$$h(s) = \frac{s}{s^2+4} + \frac{4}{s^2(s^2+4)} - \frac{4e^{-s}}{s^2(s^2+4)}. \qquad (16)$$

Now

$$\frac{4}{s^2(s^2+4)} = \frac{1}{s^2} - \frac{1}{s^2+4},$$

so (16) becomes

$$h(s) = \frac{s}{s^2+4} + \frac{1}{s^2} - \frac{1}{s^2+4} - \left(\frac{1}{s^2} - \frac{1}{s^2+4}\right)e^{-s}. \qquad (17)$$

Since $x(t) = L^{-1}\{h(s)\}$, we obtain the desired solution

$$x(t) = \cos 2t + t - \tfrac{1}{2}\sin 2t - [(t-1) - \tfrac{1}{2}\sin 2(t-1)]\alpha(t-1). \quad (18)$$

It is easy to verify our solution. From (18) it follows that

$$x'(t) = -2\sin 2t + 1 - \cos 2t - [1 - \cos 2(t-1)]\alpha(t-1), \qquad (19)$$

$$x''(t) = -4\cos 2t + 2\sin 2t - 2\sin 2(t-1)\alpha(t-1). \qquad (20)$$

Therefore $x(0) = 1$ and $x'(0) = 0$, as desired. Also, from (18) and (20), we get

$$x''(t) + 4x(t) = 4t - 4(t-1)\alpha(t-1) = \psi(t), \qquad t \geq 0.$$

## Exercises

In exercises 1 through 7 sketch the graph of the given function for $t \geq 0$.

**1.** $\alpha(t - c)$.
**2.** $\alpha(t - 1) + 2\alpha(t - 2) - 3\alpha(t - 4)$.
**3.** $(t - 3)\alpha(t - 3)$.
**4.** $\sin(t - \pi) \cdot \alpha(t - \pi)$.
**5.** $(t - 3)^2\alpha(t - 3)$.
**6.** $t^2 - (t - 1)^2\alpha(t - 1)$.
**7.** $t^2 - t^2\alpha(t - 2)$.

In exercises 8 through 15 express $F(t)$ in terms of the $\alpha$ function and find $L\{F(t)\}$.

**8.** $F(t) = 3, \qquad 0 < t < 1,$
$\qquad = t, \qquad\qquad t > 1.$

ANS. $\dfrac{3}{s} + e^{-s}\left(\dfrac{1}{s^2} - \dfrac{2}{s}\right).$

**9.** $F(t) = 4, \qquad 0 < t < 2,$
$\qquad = 2t - 1, \qquad t > 2.$

ANS. $\dfrac{4}{s} + e^{-2s}\left(\dfrac{2}{s^2} - \dfrac{1}{s}\right).$

10. $F(t) = t^2$,    $0 < t < 2$,
      $= 3$,        $t > 2$.
   ANS. $\dfrac{2}{s^3} - e^{-2s}\left(\dfrac{1}{s} + \dfrac{4}{s^2} + \dfrac{2}{s^3}\right)$.

11. $F(t) = t^2$,    $0 < t < 1$,
      $= 3$,      $1 < t < 2$,
      $= 0$,        $t > 2$.
   ANS. $\dfrac{2}{s^3} + e^{-s}\left(\dfrac{2}{s} - \dfrac{2}{s^2} - \dfrac{2}{s^3}\right) - \dfrac{3e^{-2s}}{s}$.

12. $F(t) = t^2$,    $0 < t < 2$,
      $= t - 1$,    $2 < t < 3$,
      $= 7$,        $t > 3$.
   ANS. $\dfrac{2}{s^3} - e^{-2s}\left(\dfrac{3}{s} + \dfrac{3}{s^2} + \dfrac{2}{s^3}\right) + e^{-3s}\left(\dfrac{5}{s} - \dfrac{1}{s^2}\right)$.

13. $F(t) = e^{-t}$,    $0 < t < 2$,
      $= 0$,        $t > 2$.
   ANS. $\dfrac{1 - \exp(-2s - 2)}{s + 1}$.

14. $F(t) = \sin 3t$,    $0 < t < \frac{1}{2}\pi$,
      $= 0$,         $t > \frac{1}{2}\pi$.
   ANS. $\dfrac{3 + s\exp(-\frac{1}{2}\pi s)}{s^2 + 9}$.

15. $F(t) = \sin 3t$,    $0 < t < \pi$,
      $= 0$,         $t > \pi$.
   ANS. $\dfrac{3(1 + e^{-\pi s})}{s^2 + 9}$.

16. Find and sketch an inverse Laplace transform of

$$\frac{5e^{-3s}}{s} - \frac{e^{-s}}{s}.$$

     ANS.   $F(t) = 5\alpha(t - 3) - \alpha(t - 1)$.

17. Evaluate $L^{-1}\left\{\dfrac{e^{-4s}}{(s + 2)^3}\right\}$.    ANS.   $\frac{1}{2}(t - 4)^2 \exp[-2(t - 4)]\alpha(t - 4)$.

18. If $F(t)$ is to be continuous for $t \geqq 0$ and

$$F(t) = L^{-1}\left\{\frac{e^{-3s}}{(s + 1)^3}\right\},$$

evaluate $F(2)$, $F(5)$, $F(7)$.      ANS.   $F(2) = 0$, $F(5) = 2e^{-2}$, $F(7) = 8e^{-4}$.

19. If $F(t)$ is to be continuous for $t \geqq 0$ and

$$F(t) = L^{-1}\left\{\frac{(1 - e^{-2s})(1 - 3e^{-2s})}{s^2}\right\},$$

evaluate $F(1)$, $F(3)$, $F(5)$.      ANS.   $F(1) = 1$, $F(3) = -1$, $F(5) = -4$.

20. Prove that $\psi(t, c) = \displaystyle\sum_{n=0}^{\infty} (-1)^n \alpha(t - nc)$ is the same function as was used in Example (a), Section 65. Note that for any specific $t$, the series is finite; no question of convergence is involved.

21. Obtain the transform of the half-wave rectification $F(t)$ of $\sin t$ by writing

$$F(t) = \sin t\,\psi(t, \pi)$$

in terms of the $\psi$ of exercise 20 above. Use the fact that

$$(-1)^n \sin t = \sin(t - n\pi).$$

Check your result with that of exercise 17, page 193.

In exercises 22 through 25, solve the problem using the Laplace transform. Verify that your solution satisfies the differential equation and the initial conditions.

**22.** $x''(t) + x(t) = F(t); x(0) = 0, x'(0) = 0$, in which

$$F(t) = 4, \qquad 0 \leq t \leq 2,$$

$$= t + 2, \qquad t > 2.$$

ANS.   $x(t) = 4 - 4\cos t + [(t - 2) - \sin(t - 2)]\alpha(t - 2)$.

**23.** $x''(t) + x(t) = H(t); x(0) = 1, x'(0) = 0$, in which

$$H(t) = 3, \qquad 0 \leq t \leq 4,$$

$$= 2t - 5, \qquad t > 4.$$

ANS.   $x(t) = 3 - 2\cos t + 2[t - 4 - \sin(t - 4)]\alpha(t - 4)$.

**24.** $x''(t) + x(t) = G(t); x(0) = 0, x'(0) = 1$, in which

$$G(t) = 1, \qquad 0 \leq t < \pi/2,$$

$$= 0, \qquad t \geq \pi/2.$$

ANS.   $x(t) = 1 - \cos t + \sin t - \alpha(t - \pi/2)(1 - \sin t)$.

**25.** $x''(t) + 4x(t) = M(t); x(0) = x'(0) = 0$, in which

$$M(t) = \sin t - \alpha(t - 2\pi)\sin(t - 2\pi).$$

ANS.   $x(t) = \frac{1}{6}[1 - \alpha(t - 2\pi)](2\sin t - \sin 2t)$.

**26.** Compute $y(\frac{1}{2}\pi)$ and $y(2 + \frac{1}{2}\pi)$ for the function $y(x)$ that satisfies the initial value problem

$$y''(x) + y(x) = (x - 2)\alpha(x - 2); \qquad y(0) = 0, y'(0) = 0.$$

ANS.   $y(\frac{1}{2}\pi) = 0, y(2 + \frac{1}{2}\pi) = \frac{1}{2}\pi - 1$.

**27.** Compute $x(1)$ and $x(4)$ for the function $x(t)$ that satisfies the initial value problem

$$x''(t) + 2x'(t) + x(t) = 2 + (t - 3)\alpha(t - 3); \qquad x(0) = 2, x'(0) = 1.$$

ANS.   $x(1) = 2 + e^{-1}, x(4) = 1 + 3e^{-1} + 4e^{-4}$.

## 70. A convolution theorem

We now seek a formula for the inverse transform of a product of transforms. Given

$$L^{-1}\{f(s)\} = F(t), \qquad L^{-1}\{g(s)\} = G(t), \tag{1}$$

in which $F(t)$ and $G(t)$ are assumed to be functions of class A, we shall obtain a formula for

$$L^{-1}\{f(s)g(s)\}. \tag{2}$$

Since $f(s)$ is the transform of $F(t)$, we may write

$$f(s) = \int_0^\infty e^{-st}F(t)\,dt. \tag{3}$$

Since $g(s)$ is the transform of $G(t)$,

$$g(s) = \int_0^\infty e^{-s\beta}G(\beta)\,d\beta, \tag{4}$$

in which, to avoid confusion, we have used $\beta$ (rather than $t$) as the variable of integration in the definite integral.

By equation (4), we have

$$f(s)g(s) = \int_0^\infty e^{-s\beta}f(s)G(\beta)\,d\beta. \tag{5}$$

On the right in (5) we encounter the product $e^{-s\beta}f(s)$. By Theorem 22, page 208, we know that from

$$L^{-1}\{f(s)\} = F(t) \tag{6}$$

it follows that

$$L^{-1}\{e^{-s\beta}f(s)\} = F(t - \beta)\alpha(t - \beta), \tag{7}$$

in which $\alpha$ is the step function discussed in Section 69. Equation (7) means that

$$e^{-s\beta}f(s) = \int_0^\infty e^{-st}F(t - \beta)\alpha(t - \beta)\,dt. \tag{8}$$

With the aid of (8) we may put equation (5) in the form

$$f(s)g(s) = \int_0^\infty \int_0^\infty e^{-st}G(\beta)F(t - \beta)\alpha(t - \beta)\,dt\,d\beta. \tag{9}$$

Since $\alpha(t - \beta) = 0$ for $0 < t < \beta$ and $\alpha(t - \beta) = 1$ for $t \geq \beta$, equation (9) may be rewritten as

$$f(s)g(s) = \int_0^\infty \int_\beta^\infty e^{-st}G(\beta)F(t - \beta)\,dt\,d\beta. \tag{10}$$

In (10), the integration in the $t\beta$-plane covers the shaded region shown in Figure 27. The elements are summed from $t = \beta$ to $t = \infty$ and then from $\beta = 0$ to $\beta = \infty$.

In advanced calculus it is shown that, because $F(t)$ and $G(t)$ are functions of class A, it is legitimate to interchange the order of integration on the right in equation (10). From Figure 27 we see that, in the new order of integration, the elements are to be summed from $\beta = 0$ to $\beta = t$ and then from $t = 0$

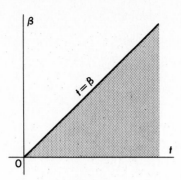

FIGURE 27

to $t = \infty$. We thus obtain

$$f(s)g(s) = \int_0^\infty \int_0^t e^{-st} G(\beta)F(t - \beta)\, d\beta\, dt,$$

or

$$f(s)g(s) = \int_0^\infty e^{-st} \left[ \int_0^t G(\beta)F(t - \beta)\, d\beta \right] dt. \tag{11}$$

Since the right member of (11) is precisely the Laplace transform of

$$\int_0^t G(\beta)F(t - \beta)\, d\beta,$$

we have arrived at the desired result, which is called the convolution theorem for the Laplace transform.

THEOREM 23:    *If $L^{-1}\{f(s)\} = F(t)$, if $L^{-1}\{g(s)\} = G(t)$, and if $F(t)$ and $G(t)$ are functions of class A (see page 181), then*

$$L^{-1}\{f(s)g(s)\} = \int_0^t G(\beta)F(t - \beta)\, d\beta. \tag{12}$$

It is easy to show that the right member of equation (12) is also a function of class A.

Of course $F$ and $G$ are interchangeable in (12) because $f$ and $g$ enter (12) symmetrically. We may replace (12) by

$$L^{-1}\{f(s)g(s)\} = \int_0^t F(\beta)G(t - \beta)\, d\beta. \tag{13}$$

a result which also follows from (12) by a change of variable of integration.

EXAMPLE (a):   Evaluate $L^{-1}\{f(s)/s\}$.

Let $L^{-1}\{f(s)\} = F(t)$. Since

$$L^{-1}\left\{\frac{1}{s}\right\} = 1,$$

we use Theorem 23 to conclude that

$$L^{-1}\left\{\frac{f(s)}{s}\right\} = \int_0^t F(\beta)\, d\beta.$$

EXAMPLE (b):   Solve the problem

$$x''(t) + k^2 x(t) = F(t); \qquad x(0) = A, x'(0) = B. \tag{14}$$

Here $k$, $A$, $B$ are constants and $F(t)$ is a function whose Laplace transform exists. Let

$$L\{x(t)\} = u(s), \qquad L\{F(t)\} = f(s).$$

Then the Laplace operator transforms problem (14) into

$$s^2 u(s) - As - B + k^2 u(s) = f(s),$$

$$u(s) = \frac{As + B}{s^2 + k^2} + \frac{f(s)}{s^2 + k^2}. \tag{15}$$

To get the inverse transform of the last term in (15), we use the convolution theorem. Thus we arrive at

$$x(t) = A \cos kt + \frac{B}{k} \sin kt + \frac{1}{k}\int_0^t F(t - \beta) \sin k\beta\, d\beta,$$

or

$$x(t) = A \cos kt + \frac{B}{k} \sin kt + \frac{1}{k}\int_0^t F(\beta) \sin k(t - \beta)\, d\beta. \tag{16}$$

Verification of the solution (16) is simple. Once that check has been performed, the need for the assumption that $F(t)$ has a Laplace transform is removed. It does not matter what method we use to get a solution (with certain exceptions naturally imposed during college examinations) if the validity of the result can be verified from the result itself.

## Exercises

In exercises 1 through 3 find the Laplace transform of the given convolution integral.

**1.** $\int_0^t (t - \beta) \sin 3\beta \, d\beta.$                     ANS.  $\dfrac{3}{s^2(s^2 + 9)}.$

**2.** $\int_0^t e^{-(t-\beta)} \sin \beta \, d\beta.$                     ANS.  $\dfrac{1}{(s + 1)(s^2 + 1)}.$

**3.** $\int_0^t (t - \beta)^3 e^\beta \, d\beta.$                     ANS.  $\dfrac{6}{s^4(s - 1)}.$

In exercises 4 through 7 find an inverse transform of the given $f(s)$ using the convolution theorem.

**4.** $\dfrac{1}{s(s^2 + k^2)}.$                     ANS.  $\dfrac{1}{k^2}(1 - \cos kt).$

**5.** $\dfrac{1}{s(s + 2)}.$                     ANS.  $\frac{1}{2}(1 - e^{-2t}).$

**6.** $\dfrac{4}{s^2(s - 2)}.$                     ANS.  $e^{2t} - 2t - 1.$

**7.** $\dfrac{1}{(s^2 + 1)^2}.$                     ANS.  $\frac{1}{2}(\sin t - t \cos t).$

**8.** Solve the problem

$$x''(t) + 2x'(t) + x(t) = F(t); \qquad x(0) = 0, \, x'(0) = 0.$$

ANS.  $x(t) = \int_0^t \beta \, e^{-\beta} F(t - \beta) \, d\beta.$

**9.** Solve the problem

$$y''(t) - k^2 y(t) = H(t); \qquad y(0) = 0, \, y'(0) = 0.$$

ANS.  $y(t) = \dfrac{1}{k}\int_0^t H(t - \beta) \sinh k\beta \, d\beta.$

**10.** Solve the problem

$$y''(t) + 4y'(t) + 13y(t) = F(t); \qquad y(0) = 0, \, y'(0) = 0.$$

**11.** Solve the problem

$$x''(t) + 6x'(t) + 9x(t) = F(t); \qquad x(0) = A, \, x'(0) = B.$$

ANS.  $x(t) = e^{-3t}[A + (B + 3A)t] + \int_0^t \beta \, e^{-3\beta} F(t - \beta) \, d\beta.$

## 71. Special integral equations

A differential equation may be loosely described as one that contains a derivative of a dependent variable; the equation contains a dependent variable under a derivative sign. An equation that contains a dependent variable under an integral sign is called an integral equation.

Because of the convolution theorem, the Laplace transform is an excellent tool for solving a very special class of integral equations. We know from Theorem 23 that if

$$L\{F(t)\} = f(s)$$

and

$$L\{G(t)\} = g(s),$$

then

$$L\left\{\int_0^t F(\beta)G(t - \beta)\,d\beta\right\} = f(s)g(s). \tag{1}$$

The relation (1) suggests the use of the Laplace transform in solving equations that contain convolution integrals.

EXAMPLE (a):   Find $F(t)$ from the integral equation

$$F(t) = 4t - 3\int_0^t F(\beta) \sin(t - \beta)\,d\beta. \tag{2}$$

The integral in (2) is in precisely the right form to permit the use of the convolution theorem. Let

$$L\{F(t)\} = f(s).$$

Then, because

$$L\{\sin t\} = \frac{1}{s^2 + 1},$$

application of Theorem 23 yields

$$L\left\{\int_0^t F(\beta) \sin(t - \beta)\,d\beta\right\} = \frac{f(s)}{s^2 + 1}.$$

Therefore, the Laplace operator converts equation (2) into

$$f(s) = \frac{4}{s^2} - \frac{3f(s)}{s^2 + 1}. \tag{3}$$

We need to obtain $f(s)$ from (3) and then $F(t)$ from $f(s)$. From (3) we get

$$\left(1 + \frac{3}{s^2 + 1}\right) f(s) = \frac{4}{s^2},$$

or

$$f(s) = \frac{4(s^2 + 1)}{s^2(s^2 + 4)} = \frac{1}{s^2} + \frac{3}{s^2 + 4}.$$

Therefore

$$F(t) = L^{-1}\left\{\frac{1}{s^2} + \frac{3}{s^2 + 4}\right\},$$

or

$$F(t) = t + \tfrac{3}{2} \sin 2t. \tag{4}$$

That the $F(t)$ of (4) is a solution of equation (2) may be verified directly. Such a check is frequently tedious. We shall show that for the $F$ of (4), the right-hand side of equation (2) reduces to the left-hand side of (2). Since

$$\text{RHS} = 4t - 3 \int_0^t (\beta + \tfrac{3}{2} \sin 2\beta) \sin (t - \beta) \, d\beta,$$

we integrate by parts with the choice shown in the table.

| $(\beta + \tfrac{3}{2} \sin 2\beta)$ | $\sin (t - \beta) \, d\beta$ |
|---|---|
| $(1 + 3 \cos 2\beta) \, d\beta$ | $\cos (t - \beta)$ |

It thus follows that

$$\text{RHS} = 4t - 3\left[ (\beta + \tfrac{3}{2} \sin 2\beta) \cos (t - \beta) \right]_0^t$$

$$+ 3 \int_0^t (1 + 3 \cos 2\beta) \cos (t - \beta) \, d\beta,$$

from which

$$\text{RHS} = 4t - 3(t + \tfrac{3}{2} \sin 2t) + 3 \int_0^t \cos (t - \beta) \, d\beta$$

$$+ 9 \int_0^t \cos 2\beta \cos (t - \beta) \, d\beta,$$

or

$$\text{RHS} = t - \tfrac{9}{2} \sin 2t - 3\left[ \sin (t - \beta) \right]_0^t + \tfrac{9}{2} \int_0^t [\cos (t + \beta) + \cos (t - 3\beta)] \, d\beta.$$

This leads us to the result

$$\text{RHS} = t - \tfrac{9}{2}\sin 2t + 3\sin t + \tfrac{9}{2}\left[\sin(t + \beta) - \tfrac{1}{3}\sin(t - 3\beta)\right]_0^t$$

$$= t - \tfrac{9}{2}\sin 2t + 3\sin t + \tfrac{9}{2}\sin 2t + \tfrac{3}{2}\sin 2t - \tfrac{9}{2}\sin t + \tfrac{3}{2}\sin t,$$

or

$$\text{RHS} = t + \tfrac{3}{2}\sin 2t = F(t) = \text{LHS},$$

as desired.

It is important to realize that the original equation

$$F(t) = 4t - 3\int_0^t F(\beta)\sin(t - \beta)\,d\beta \tag{2}$$

could equally well have been encountered in the equivalent form

$$F(t) = 4t - 3\int_0^t F(t - \beta)\sin\beta\,d\beta.$$

An essential ingredient for the success of the method being used is that the integral involved be in exactly the convolution integral form. We must have zero to the independent variable as limits of integration and an integrand that is the product of a function of the variable of integration by a function of the difference between the independent variable and the variable of integration. The fact that integrals of that form appear with significant frequency in physical problems keeps the topic of this section from being relegated to the role of a mathematical parlor game.

**EXAMPLE (b):** Solve the equation

$$g(x) = \tfrac{1}{2}x^2 - \int_0^x (x - y)g(y)\,dy. \tag{5}$$

Again the integral involved is one of the convolution type with $x$ playing the role of the independent variable. Let the Laplace transform of $g(x)$ be some as yet unknown function $h(z)$:

$$L\{g(x)\} = h(z). \tag{6}$$

Since $L\{\tfrac{1}{2}x^2\} = 1/z^3$ and $L\{x\} = 1/z^2$, we may apply the operator $L$ throughout (5) and obtain

$$h(z) = \frac{1}{z^3} - \frac{h(z)}{z^2},$$

from which

$$\left(1 + \frac{1}{z^2}\right)h(z) = \frac{1}{z^3},$$

or

$$h(z) = \frac{1}{z(z^2 + 1)} = \frac{z^2 + 1 - z^2}{z(z^2 + 1)} = \frac{1}{z} - \frac{z}{z^2 + 1}.$$

Then

$$g(x) = L^{-1}\left\{\frac{1}{z} - \frac{z}{z^2 + 1}\right\}$$

or

$$g(x) = 1 - \cos x. \tag{7}$$

Verification of (7) is simple. For the right member of (5) we get

$$\text{RHS} = \tfrac{1}{2}x^2 - \int_0^x (x - y)(1 - \cos y)\, dy$$

$$= \tfrac{1}{2}x^2 - \left[(x - y)(y - \sin y)\right]_0^x - \int_0^x (y - \sin y)\, dy$$

$$= \tfrac{1}{2}x^2 - 0 - \left[\tfrac{1}{2}y^2 + \cos y\right]_0^x$$

$$= \tfrac{1}{2}x^2 - \tfrac{1}{2}x^2 - \cos x + 1 = 1 - \cos x = \text{LHS}.$$

## Exercises

In exercises 1 through 4 solve the given equation and verify your solution.

**1.** $F(t) = 1 + 2 \int_0^t F(t - \beta)\, e^{-2\beta}\, d\beta.$      ANS.   $F(t) = 1 + 2t.$

**2.** $F(t) = 1 + \int_0^t F(\beta) \sin(t - \beta)\, d\beta.$      ANS.   $F(t) = 1 + \tfrac{1}{2}t^2.$

**3.** $F(t) = t + \int_0^t F(t - \beta)\, e^{-\beta}\, d\beta.$      ANS.   $F(t) = t + \tfrac{1}{2}t^2.$

**4.** $F(t) = 4t^2 - \int_0^t F(t - \beta)\, e^{-\beta}\, d\beta.$      ANS.   $F(t) = -1 + 2t + 2t^2 + e^{-2t}.$

In exercises 5 through 8 solve the given equation. If sufficient time is available, verify your solution.

**5.** $F(t) = t^3 + \displaystyle\int_0^t F(\beta) \sin(t - \beta) \, d\beta.$                       ANS.   $F(t) = t^3 + \frac{1}{20} t^5.$

**6.** $F(t) = 8t^2 - 3 \displaystyle\int_0^t F(\beta) \sin(t - \beta) \, d\beta.$

**7.** $F(t) = t^2 - 2 \displaystyle\int_0^t F(t - \beta) \sinh 2\beta \, d\beta.$                   ANS.   $F(t) = t^2 - \frac{1}{3} t^4.$

**8.** $F(t) = 1 + 2 \displaystyle\int_0^t F(t - \beta) \cos \beta \, d\beta.$                       ANS.   $F(t) = 1 + 2te^t.$


In exercises 9 through 12 solve the given equation.

**9.** $H(t) = 9 e^{2t} - 2 \displaystyle\int_0^t H(t - \beta) \cos \beta \, d\beta.$

**10.** $H(y) = y^2 + \displaystyle\int_0^y H(x) \sin(y - x) \, dx.$                       ANS.   $H(y) = y^2 + \frac{1}{12} y^4.$

**11.** $g(x) = e^{-x} - 2 \displaystyle\int_0^x g(\beta) \cos(x - \beta) \, d\beta.$                ANS.   $g(x) = e^{-x}(1 - x)^2.$

**12.** $y(t) = 6t + 4 \displaystyle\int_0^t (\beta - t)^2 y(\beta) \, d\beta.$   ANS.   $y(t) = e^{2t} - e^{-t}(\cos\sqrt{3}t - \sqrt{3} \sin \sqrt{3}t).$

**13.** Solve the following equation for $F(t)$ with the condition that $F(0) = 4$:

$$F'(t) = t + \int_0^t F(t - \beta) \cos \beta \, d\beta.$$                ANS.   $F(t) = 4 + \frac{5}{2} t^2 + \frac{1}{24} t^4.$

**14.** Solve the following equation for $F(t)$ with the condition that $F(0) = 0$:

$$F'(t) = \sin t + \int_0^t F(t - \beta) \cos \beta \, d\beta.$$                ANS.   $F(t) = \frac{1}{2} t^2.$

**15.** Show that the equation of exercise 3 above can be put in the form

$$e^t F(t) = t e^t + \int_0^t e^\beta F(\beta) \, d\beta. \tag{A}$$

Differentiate each member of (A) with respect to $t$ and thus replace the integral equation with a differential equation. Note that $F(0) = 0$. Find $F(t)$ by this method.

**16.** Solve the equation

$$\int_0^t F(t - \beta) e^{-\beta} \, d\beta = t$$

by two methods; use the convolution theorem and the basic idea introduced in exercise 15. Note that no differential equation need be solved in this instance.

## 72. Transform methods and the vibration of springs

All of the applications studied in Chapter 10 gave rise to linear differential equations with initial conditions. Those initial value problems were solved in that context by using the theory of linear differential equations developed in the earlier chapters. The same initial value problems may of course be solved by using Laplace transformations. We illustrate these techniques by reexamining some of the problems considered before.

EXAMPLE (a):   Solve the spring problem of Example (a), Section 52, with no damping but with $F(t) = A \sin \omega t$.
  As before, the problem to be solved is

$$x''(t) + \beta^2 x(t) = \sin \omega t, \tag{1}$$

with initial conditions

$$x(0) = x_0, \qquad x'(0) = v_0. \tag{2}$$

Let $L\{x(t)\} = u(s)$. Then (1) and (2) yield

$$s^2 u(s) - s x_0 - v_0 + \beta^2 u(s) = \frac{A\omega}{s^2 + \omega^2},$$

or

$$u(s) = \frac{s x_0 + v_0}{s^2 + \beta^2} + \frac{A\omega}{(s^2 + \beta^2)(s^2 + \omega^2)}. \tag{3}$$

The last term in (3) will lead to different inverse transforms according to whether $\omega = \beta$ or $\omega \neq \beta$. The case $\omega = \beta$ leads to resonance, which will be discussed in Example (d).
  If $\omega \neq \beta$, equation (3) yields

$$u(s) = \frac{s x_0 + v_0}{s^2 + \beta^2} + \frac{A\omega}{\omega^2 - \beta^2}\left(\frac{1}{s^2 + \beta^2} - \frac{1}{s^2 + \omega^2}\right). \tag{4}$$

From (4) it follows at once that

$$x(t) = x_0 \cos \beta t + v_0 \beta^{-1} \sin \beta t + \frac{A\omega}{\beta(\omega^2 - \beta^2)} \sin \beta t - \frac{A}{\omega^2 - \beta^2} \sin \omega t. \tag{5}$$

That the $x$ of (5) is a solution of the problem (1) and (2) is easily verified. A study of (5) is simple and leads at once to conclusions such as that $x(t)$ is bounded, and so on. The first two terms on the right in (5) yield the natural harmonic component of the motion, the last two terms form the forced component.

This is the same solution that was found for the same problem in Section 52 using a very different approach.

**EXAMPLE (b):** Solve the spring problem of Example (b), Section 52, using Laplace transformations.

The initial value problem is

$$x''(t) + 64x(t) = 0; \qquad x(0) = \tfrac{1}{3}, \qquad x'(0) = -2. \qquad (6)$$

We let $L\{x(t)\} = u(s)$ and conclude at once that

$$s^2u(s) - \tfrac{1}{3}s + 2 + 64u(s) = 0,$$

from which

$$u(s) = \frac{\tfrac{1}{3}s - 2}{s^2 + 64}.$$

Then

$$x(t) = \tfrac{1}{3}\cos 8t - \tfrac{1}{4}\sin 8t. \qquad (7)$$

**EXAMPLE (c):** A spring, with spring constant 0.75 lb/ft, lies on a long smooth (frictionless) table. A 6-lb weight is attached to the spring and is at rest (velocity zero) at the equilibrium position. A 1.5-lb force is applied to the support along the line of action of the spring for 4 sec and is then removed. Discuss the motion.

We must solve the problem

$$\tfrac{6}{32}x''(t) + \tfrac{3}{4}x(t) = H(t); \qquad x(0) = 0, \, x'(0) = 0, \qquad (8)$$

in which

$$H(t) = 1.5, \qquad 0 < t < 4,$$

$$= 0, \qquad\qquad t > 4.$$

Now $H(t) = 1.5[1 - \alpha(t - 4)]$ in terms of the $\alpha$ function of Section 69. Therefore we rewrite our problem (8) in the form.

$$x''(t) + 4x(t) = 8[1 - \alpha(t - 4)]; \qquad x(0) = 0, \, x'(0) = 0. \qquad (9)$$

Let $L\{x(t)\} = u(s)$. Then (9) yields

$$s^2u(s) + 4u(s) = \frac{8}{s}(1 - e^{-4s}),$$

or

$$u(s) = \frac{8(1 - e^{-4s})}{s(s^2 + 4)}$$

$$= 2\left(\frac{1}{s} - \frac{s}{s^2 + 4}\right)(1 - e^{-4s}).$$

The desired solution is

$$x(t) = 2(1 - \cos 2t) - 2[1 - \cos 2(t - 4)]\alpha(t - 4). \tag{10}$$

Of course, the solution (10) can be broken down into the two relations

for $0 \leq t \leq 4$,     $x(t) = 2(1 - \cos 2t)$,     (11)

for $t > 4$,     $x(t) = 2[\cos 2(t - 4) - \cos 2t]$,     (12)

if those forms seem simpler to use.

Verification of the solution (10), or (11) and (12), is direct. The student should show that

$$\lim_{t \to 4^-} x(t) = \lim_{t \to 4^+} x(t) = 2(1 - \cos 8) = 2.29$$

and

$$\lim_{t \to 4^-} x'(t) = \lim_{t \to 4^+} x'(t) = 4 \sin 8 = 3.96.$$

From (10) or (11) we see that in the range $0 < t < 4$, the maximum deviation of the weight from the starting point is $x = 4$ ft and occurs at $t = \frac{1}{2}\pi = 1.57$ sec. At $t = 4$, $x = 2.29$ ft, as shown above. For $t > 4$, equation (12) takes over and thereafter the motion is simple harmonic with a maximum $x$ of 3.03 ft. Indeed, for $t > 4$,

$$\max |x(t)| = 2\sqrt{(1 - \cos 8)^2 + \sin^2 8}$$

$$= 2\sqrt{2}\sqrt{1 - \cos 8}$$

$$= 2\sqrt{2.2910} = 3.03.$$

Example (c) is one type of problem for which the Laplace transform technique is particularly useful. Such problems can be solved by the older classical methods, but with much less simplicity and dispatch.

EXAMPLE (d):   Solve the problem of undamped vibration of a spring of Example (a) above in the case $\omega = \beta$.

Our problem is to solve

$$x''(t) + \beta^2 x(t) = A \sin \beta t; \qquad x(0) = x_0, x'(0) = v_0, \tag{13}$$

with the aid of

$$u(s) = \frac{sx_0 + v_0}{s^2 + \beta^2} + \frac{A\beta}{(s^2 + \beta^2)^2}. \tag{14}$$

We already know, from page 187, that

$$L^{-1}\left\{\frac{1}{(s^2 + \beta^2)^2}\right\} = \frac{1}{2\beta^3}(\sin \beta t - \beta t \cos \beta t).$$

Therefore (14) leads us to the solution

$$x(t) = x_0 \cos \beta t + \frac{v_0}{\beta} \sin \beta t + \frac{A}{2\beta^2}(\sin \beta t - \beta t \cos \beta t). \qquad (15)$$

Again this solution is the same as the solution obtained in equation (5), Section 53, and we have resonance occurring.

EXAMPLE (e):   Solve the problem of the example of Section 54,

$$\tfrac{12}{32}x''(t) + 0.6x'(t) + 24x(t) = 0; \qquad x(0) = \tfrac{1}{3}, x'(0) = -2. \qquad (16)$$

Put $L\{x(t)\} = u(s)$. Then (16) yields

$$(s^2 + 1.6s + 64)u(s) = \tfrac{1}{3}(s - 4.4),$$

from which we obtain

$$x(t) = \tfrac{1}{3}L^{-1}\left\{\frac{s - 4.4}{(s + 0.8)^2 + 63.36}\right\}$$

$$= \tfrac{1}{3}\exp(-0.8t)L^{-1}\left\{\frac{s - 5.2}{s^2 + 63.36}\right\}.$$

Therefore the desired solution is

$$x(t) = \exp(-0.8t)(0.33 \cos 8.0t - 0.22 \sin 8.0t), \qquad (17)$$

a portion of its graph being shown in Figure 16.

### Exercises

Each of the exercises of Chapter 10 is an appropriate exercise here. It would be instructive to solve a problem both with and without the Laplace transform and to compare the two methods.

## 73. The deflection of beams

As a further example of an application in which transform methods are useful, we consider a beam of length $2c$, as shown in Figure 28. Denote distance from one end of the beam by $x$, the deflection of the beam by $y$. If the beam is subjected to a vertical load $W(x)$, the deflection $y$ must satisfy the equation

$$EI\frac{d^4y}{dx^4} = W(x), \qquad \text{for } 0 < x < 2c, \qquad (1)$$

in which $E$, the modulus of elasticity, and $I$, a moment of inertia, are known constants associated with the particular beam.

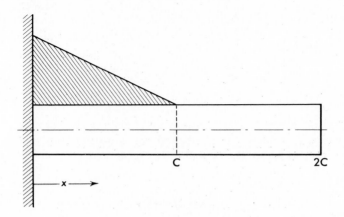

FIGURE 28

The slope of the curve of deflection is $y'(x)$, the bending moment is $EIy''(x)$, and the shearing force is $EIy'''(x)$. Common boundary conditions are of the following types:

(a) Beam imbedded in a support: $y = 0$ and $y' = 0$ at the point.
(b) Beam simply supported: $y = 0$ and $y'' = 0$ at the point.
(c) Beam free: $y'' = 0$ and $y''' = 0$ at the point.

Problems in the transverse displacement of a beam take the form of the differential equation (1) with boundary conditions at each end of the beam. Such problems can be solved by integration with the use of a little algebra. There are, however, two reasons for employing our transform method in such problems. Frequently the load function, or its derivative, is discontinuous. Beam problems also give us a chance to examine a useful device in which a problem over a finite range is solved with the aid of an associated problem over an infinite range.

EXAMPLE: Find the displacement $y$ throughout the beam of Figure 28, in which the load is assumed to decrease uniformly from $w_0$ at $x = 0$ to zero at $x = c$ and to remain zero from $x = c$ to $x = 2c$. The weight of the beam is to be negligible. The beam is imbedded at $x = 0$ and free at $x = 2c$. We are to solve the problem

$$EI\frac{d^4 y}{dx^4} = \frac{w_0}{c}[c - x + (x - c)\alpha(x - c)], \qquad \text{for } 0 < x < 2c; \qquad (2)$$

$$y(0) = 0, \qquad y'(0) = 0; \tag{3}$$

$$y''(2c) = 0, \qquad y'''(2c) = 0. \tag{4}$$

The student should verify that the right member of (2) is the stipulated load function

$$W(x) = \frac{w_0}{c}(c - x), \qquad \text{for } 0 \leq x \leq c, \tag{5}$$

$$= 0, \qquad \text{for } c < x \leq 2c.$$

To apply the transform technique, with $x$ playing the role for which we usually employ $t$, we need first to extend the range of $x$ so it will run from 0 to $\infty$. That is, instead of the problem (2), (3), (4), we shall solve the problem consisting of

$$EI\frac{d^4y}{dx^4} = H(x), \qquad \text{for } 0 < x < \infty, \tag{6}$$

and the conditions (3) and (4). In (6) the function $H(x)$ is to be chosen by us except that $H(x)$ must agree with $W(x)$ over the range $0 < x < 2c$. The solution of the problem (6), (3), (4) will then be used only in the range $0 \leq x \leq 2c$. Of the various choices for $H(x)$, it seems simplest to use

$$H(x) = \frac{w_0}{c}[c - x + (x - c)\alpha(x - c)], \qquad \text{for } 0 < x < \infty. \tag{7}$$

That is, in practice we ordinarily retain the equation (2) and merely extend the range from $0 < x < 2c$ to $0 < x < \infty$. The student must, however, keep in mind that we cannot apply the Laplace operator to the function $W(x)$ of (5), since that function is not defined over the entire range $0 < x < \infty$. We shall solve (6) and conclude that the solution is valid for (2) on the range $0 \leq x \leq 2c$, over which (2) and (6) are identical.

Let

$$L\{EI\ y(x)\} = u(s);$$

$$u(s) = EI \int_0^\infty e^{-sx}y(x)\,dx. \tag{8}$$

To transform $EIy^{(4)}(x)$ we need to use the values of $EIy(x)$ and its first three derivatives at $x = 0$. From (3) we know that

$$EIy(0) = 0, \qquad EIy'(0) = 0.$$

Put

$$EIy''(0) = A, \qquad EIy'''(0) = B. \tag{9}$$

The constants $A$ and $B$ must be determined by using the conditions (4).

By our usual methods we obtain, for the $H(x)$ of (7),

$$L\{H(x)\} = \frac{w_0}{c} L\{c - x + (x - c)\alpha(x - c)\}$$

$$= \frac{w_0}{c}\left(\frac{c}{s} - \frac{1}{s^2} + \frac{e^{-cs}}{s^2}\right).$$

Thus the differential equation (6) is transformed into

$$s^4 u(s) - s^3 \cdot 0 - s^2 \cdot 0 - s \cdot A - B = \frac{w_0}{c}\left(\frac{c}{s} - \frac{1}{s^2} + \frac{e^{-cs}}{s^2}\right),$$

from which we get

$$u(s) = \frac{A}{s^3} + \frac{B}{s^4} + \frac{w_0}{c}\left(\frac{c}{s^5} - \frac{1}{s^6} + \frac{e^{-cs}}{s^6}\right). \tag{10}$$

Now $L^{-1}\{u(s)\} = EIy(x)$. Hence

$$EIy(x) = \tfrac{1}{2}Ax^2 + \tfrac{1}{6}Bx^3 + \frac{w_0}{120c}[5cx^4 - x^5 + (x - c)^5\alpha(x - c)]. \tag{11}$$

From (11) we obtain

$$EIy'(x) = Ax + \tfrac{1}{2}Bx^2 + \frac{w_0}{24c}[4cx^3 - x^4 + (x - 4)^4\alpha(x - c)], \tag{12}$$

$$EIy''(x) = A + Bx + \frac{w_0}{6c}[3cx^2 - x^3 + (x - c)^3\alpha(x - c)], \tag{13}$$

$$EIy'''(x) = B + \frac{w_0}{2c}[2cx - x^2 + (x - c)^2\alpha(x - c)]. \tag{14}$$

By differentiating both members of equations (14), we can see that the $y$ of (11) is a solution of (6) over the infinite range and, more important, a solution of (2) over the range $0 < x < 2c$.

With the aid of equations (11) through (14) we can now determine $A$ and $B$ to make the $y$ satisfy appropriate conditions at $x = 2c$, whether the beam be free, imbedded, or pin-supported there. In our example the beam is to be free at $x = 2c$; the solution is to satisfy the conditions

$$y''(2c) = 0, \qquad y'''(2c) = 0. \tag{4}$$

Using (13) and (14), and a little work, we find that (4) requires $A = \tfrac{1}{6}w_0c^2$, $B = -\tfrac{1}{2}w_0c$. We are thus led to the solution

$$EIy(x) = \tfrac{1}{12}w_0c^2x^2 - \tfrac{1}{12}w_0cx^3 + \frac{w_0}{120c}[5cx^4 - x^5 + (x - c)^5\alpha(x - c)], \tag{15}$$

for $0 \leqq x \leqq 2c$.

The student should verify by differentiations and appropriate substitutions that the $y$ of (15) satisfies the original differential equation (2) and boundary conditions (3) and (4).

From (15) we can obtain whatever information we wish. For example, at $x = \frac{1}{2}c$ the bending moment is

$$EIy''(\tfrac{1}{2}c) = w_0c^2[\tfrac{1}{6} - \tfrac{1}{4} + \tfrac{1}{6}(-\tfrac{1}{8} + \tfrac{3}{4} + 0)] = \tfrac{1}{48}w_0c^2.$$

## Exercises

In exercises 1 through 4, find the $y$ that satisfies equation (1), page 226, with the given load function $W(x)$ and the given conditions at the ends of the beam. [See (a), (b), (c), page 227.] Verify your solutions.

1. $W(x)$ as in the example introduced on page 227, beam imbedded at both $x = 0$ and $x = 2c$.

   ANS.    $EIy(x) = \tfrac{23}{480}w_0c^2x^2 - \tfrac{3}{40}w_0cx^3$

   $$+ \frac{w_0}{120c}[5cx^4 - x^5 + (x - c)^5\alpha(x - c)].$$

2. $W(x) = 0$,        for $0 < x < \tfrac{1}{2}c$,

   $= w_0$,       for $\tfrac{1}{2}c < x < \tfrac{3}{2}c$,

   $= 0$,        for $\tfrac{3}{2}c < x < 2c$;

   beam imbedded at $x = 0$, free at $x = 2c$.
   ANS.    $EIy(x) = \tfrac{1}{2}w_0c^2x^2 - \tfrac{1}{6}w_0cx^3$

   $$+\tfrac{1}{24}w_0[(x - \tfrac{1}{2}c)^4\alpha(x - \tfrac{1}{2}c) - (x - \tfrac{3}{2}c)^4\alpha(x - \tfrac{3}{2}c)].$$

3. $W(x) = w_0[1 - \alpha(x - c)]$ (describe the load); beam to be imbedded at $x = 0$ and pin-supported (simply supported) at $x = 2c$.
   ANS.    $EIy(x) = \tfrac{9}{64}w_0c^2x^2 - \tfrac{19}{128}w_0cx^3 + \tfrac{1}{24}w_0[x^4 - (x - c)^4\alpha(x - c)].$

4. $W(x) = \dfrac{w_0}{c}(2c - x)$,        for $0 < x < c$,

   $= w_0$,         for $c < x < 2c$;

   beam to be imbedded at $x = 0$ and free at $x = 2c$.
   ANS.    $EIy(x) = \tfrac{13}{12}w_0c^2x^2 - \tfrac{5}{12}w_0cx^3 + \tfrac{1}{24}w_0x^4$

   $$+ \frac{w_0}{120c}[5cx^4 - x^5 + (x - c)^5\alpha(x - c)].$$

## TABLE OF TRANSFORMS

Whenever $n$ is used, it denotes a nonnegative integer. The range of validity may be determined from the appropriate text material. Many other transforms will be found in the examples and exercises.

| $f(s) = L\{F(t)\}$ | $F(t)$ |
|---|---|
| $f(s - a)$ | $e^{at} F(t)$ |
| $f(as + b)$ | $\dfrac{1}{a}\exp\left(-\dfrac{bt}{a}\right)F\left(\dfrac{t}{a}\right)$ |
| $\dfrac{1}{s}e^{-cs}, \quad c > 0$ | $\alpha(t - c) = 0, \quad 0 \le t < c,$ <br> $= 1, \quad t \ge c$ |
| $e^{-cs}f(s), \quad c > 0$ | $F(t - c)\alpha(t - c)$ |
| $f_1(s)f_2(s)$ | $\displaystyle\int_0^t F_1(\beta)F_2(t - \beta)\,d\beta$ |
| $\dfrac{1}{s}$ | $1$ |
| $\dfrac{1}{s^{n+1}}$ | $\dfrac{t^n}{n!}$ |
| $\dfrac{1}{s^{x+1}}, \quad x > -1$ | $\dfrac{t^x}{\Gamma(x + 1)}$ |
| $s^{-1/2}$ | $(\pi t)^{-1/2}$ |
| $\dfrac{1}{s + a}$ | $e^{-at}$ |
| $\dfrac{1}{(s + a)^{n+1}}$ | $\dfrac{t^n e^{-at}}{n!}$ |
| $\dfrac{k}{s^2 + k^2}$ | $\sin kt$ |
| $\dfrac{s}{s^2 + k^2}$ | $\cos kt$ |
| $\dfrac{k}{s^2 - k^2}$ | $\sinh kt$ |
| $\dfrac{s}{s^2 - k^2}$ | $\cosh kt$ |

| $f(s) = L\{F(t)\}$ | $F(t)$ |
|:---:|:---:|
| $\dfrac{2k^3}{(s^2 + k^2)^2}$ | $\sin kt - kt \cos kt$ |
| $\dfrac{2ks}{(s^2 + k^2)^2}$ | $t \sin kt$ |
| $\ln\left(1 + \dfrac{1}{s}\right)$ | $\dfrac{1 - e^{-t}}{t}$ |
| $\ln\dfrac{s + k}{s - k}$ | $\dfrac{2\sinh kt}{t}$ |
| $\ln\left(1 - \dfrac{k^2}{s^2}\right)$ | $\dfrac{2}{t}(1 - \cosh kt)$ |
| $\ln\left(1 + \dfrac{k^2}{s^2}\right)$ | $\dfrac{2}{t}(1 - \cos kt)$ |
| $\arctan\dfrac{k}{s}$ | $\dfrac{\sin kt}{t}$ |

# Linear Systems of Equations

## 74. Introduction

We shall see in the next chapter that certain problems in the studies of electrical circuits and arms races lead naturally to systems of linear differential equations with constant coefficients. Although the subject of systems of equations can be studied in a wider context involving coefficients that are not constant, we shall not do so in this book.

## 75. Elementary elimination calculus

Let us consider the problem of finding solutions for the system of equations

$$y'' - y + 5v' = x, \tag{1}$$
$$2y' - v'' + 4v = 2,$$

with independent variable $x$ and dependent variables $y$ and $v$. It is reasonable to attack the system (1) by eliminating one dependent variable to obtain a single equation in the other dependent variable. Using $D = d/dx$, write the system (1) in the form

$$(D^2 - 1)y + 5Dv = x, \tag{2}$$

$$2Dy - (D^2 - 4)v = 2.$$

Then, elimination of one dependent variable is straightforward. We may, for instance, operate upon the first equation with $2D$ and upon the second equation with $(D^2 - 1)$, and then subtract one from the other, obtaining

$$[10D^2 + (D^2 - 1)(D^2 - 4)]v = 2Dx - (D^2 - 1)2,$$

or

$$(D^4 + 5D^2 + 4)v = 4. \tag{3}$$

In a similar manner $v$ may be eliminated; the resultant equation for $y$ is

$$[(D^2 - 1)(D^2 - 4) + 10D^2]y = (D^2 - 4)x + 5D(2),$$

or

$$(D^4 + 5D^2 + 4)y = -4x. \tag{4}$$

From equations (3) and (4) it follows at once that

$$v = 1 + a_1 \cos x + a_2 \sin x + a_3 \cos 2x + a_4 \sin 2x \tag{5}$$

and

$$y = -x + b_1 \cos x + b_2 \sin x + b_3 \cos 2x + b_4 \sin 2x. \tag{6}$$

The $a$'s and $b$'s have yet to be chosen to make (5) and (6) satisfy the original equations, rather than just the equations (3) and (4), which resulted from the original ones after certain eliminations were performed.

Combining (by substitution) the $v$ of (5) and the $y$ of (6) with the first equation of the system (2) leads to the identity

$$x - 2b_1 \cos x - 2b_2 \sin x - 5b_3 \cos 2x - 5b_4 \sin 2x \tag{7}$$

$$- 5a_1 \sin x + 5a_2 \cos x - 10a_3 \sin 2x + 10a_4 \cos 2x \equiv x.$$

That (7) be an identity in $x$ demands that

$$-2b_1 + 5a_2 = 0, \tag{8}$$

$$-2b_2 - 5a_1 = 0,$$

$$-5b_3 + 10a_4 = 0,$$

$$-5b_4 - 10a_3 = 0.$$

Relations between the $a$'s and $b$'s equivalent to the relations (8) follow from substitution of the $v$ of (5) and the $y$ of (6) into the second equation of the system (2).

We conclude that a set of solutions of the system (2) is

$$v = 1 + a_1 \cos x + a_2 \sin x + a_3 \cos 2x + a_4 \sin 2x, \tag{9}$$

$$y = -x + \tfrac{5}{2}a_2 \cos x - \tfrac{5}{2}a_1 \sin x + 2a_4 \cos 2x - 2a_3 \sin 2x,$$

in which $a_1, a_2, a_3, a_4$ are arbitrary constants.

The equations (3) and (4) for $v$ and $y$ can be written with the aid of determinants. From the system (2) above we may write at once

$$\begin{vmatrix} (D^2 - 1) & 5D \\ 2D & -(D^2 - 4) \end{vmatrix} v = \begin{vmatrix} (D^2 - 1) & x \\ 2D & 2 \end{vmatrix},$$

which reduces to equation (3) above, if care is used in the interpretation of the right-hand member. The determinant on the right is to be interpreted as

$$(D^2 - 1)(2) - 2D(x),$$

not as the differential operator $2(D^2 - 1) - x(2D)$.

Determinants are extremely useful in any treatment of the theory of systems of linear equations. For many simple systems that arise in practice, no such powerful tool is needed.

There are other techniques for treating a system such as (2). For example, we may first obtain (3) and from it $v$ as in equation (5). Next we wish to find an equation giving $y$ in terms of $v$; that is, we seek to eliminate from the system (2) those terms that involve derivatives of $y$. From (2) we obtain the two equations

$$(2D^2 - 2)y + 10Dv = 2x, \tag{10}$$

$$2D^2y - (D^3 - 4D)v = 0, \tag{11}$$

the latter by operating with $D$ on each member of the second equation of the system (2). From (10) and (11) it follows at once that

$$2y - D^3v - 6Dv = -2x,$$

or

$$y = -x + \tfrac{1}{2}D^3v + 3Dv. \tag{12}$$

The $v$ of equation (5) may now be used in (12) to compute the $y$, which was given in the solution (9). In this method of solution there is no need to obtain (4), (6), (7), or (8).

## Exercises

Use elementary elimination calculus to solve the following systems of equations.

**1.** $u' = 4u - v$,
$\quad v' = -4u + 4v$.

ANS. $u = c_1 e^{2x} + c_2 e^{6x}$,
$\qquad v = 2c_1 e^{2x} - 2c_2 e^{6x}$.

**2.** $w' = w - y - z$,
$\quad y' = y + 3z$,
$\quad z' = 3y + z$.

ANS. $w = c_1 e^x + 2c_2 e^{4x}$,
$\qquad y = -3c_2 e^{4x} + c_3 e^{-2x}$,
$\qquad z = -3c_2 e^{4x} - c_3 e^{-2x}$.

**3.** $y' = 2y + z$,
$\quad z' = -4y + 2z$.

ANS. $y = c_1 e^{2t} \cos 2t + b_2 e^{2t} \sin 2t$,
$\qquad z = -2c_1 e^{2t} \sin 2t + 2b_2 e^{2t} \cos 2t$.

**4.** $dx/dt = y$,
$\quad dy/dt = -4x + 4y$.

ANS. $x = c_1 e^{2t} + c_2 t e^{2t}$,
$\qquad y = 2c_1 e^{2t} + c_2(2t e^{2t} + e^{2t})$.

**5.** $v' - 2v + 2w' = 2 - 4 e^{2x}$,
$\quad 2v' - 3v + 3w' - w = 0$.

ANS. $v = a_1 e^x + a_2 e^{-2x} + 5 e^{2x} - 1$,
$\qquad w = \frac{1}{2}a_1 e^x - a_2 e^{-2x} - e^{2x} + 3$.

**6.** $(3D + 2)v + (D - 6)w = 5 e^x$,
$\quad (4D + 2)v + (D - 8)w = 5 e^x + 2x - 3$.

ANS. $v = a_1 \cos 2x + a_2 \sin 2x + 2 e^x - 3x + 5$,
$\qquad w = a_2 \cos 2x - a_1 \sin 2x + e^x - x$.

**7.** $(D^2 + 6)y + Dv = 0$,
$\quad (D + 2)y + (D - 2)v = 2$.

ANS. $y = c_1 e^{3x} + c_2 \cos 2x + c_3 \sin 2x$,
$\qquad v = -1 - 5c_1 e^{3x} + c_3 \cos 2x - c_2 \sin 2x$.

**8.** $D^2 y - (2D - 1)v = 1$,
$\quad (2D + 1)y + (D^2 - 4)v = 0$.

ANS. $v = 1 + a_1 e^x + a_2 e^{-x} + a_3 \cos x + a_4 \sin x$,
$\qquad y = 4 + a_1 e^x - 3a_2 e^{-x} + (a_3 - 2a_4) \cos x + (2a_3 + a_4) \sin x$.

**9.** $(D^2 - 3D)y - (D - 2)z = 14x + 7$,
$\quad (D - 3)y + Dz = 1$.

ANS. $y = 2 + a_1 e^x + a_2 e^{3x} + a_3 e^{-2x}$,
$\qquad z = 7 + 7x + 2a_1 e^x - \frac{5}{2}a_3 e^{-2x}$.

**10.** $(D^3 + D^2 - 1)u + (D^3 + 2D^2 + 3D + 1)v = 3 - x$,
$\quad (D - 1)u + (D + 1)v = 3 - x$.

ANS. $u = 2x + a_1 + a_2 \cos x + a_3 \sin x$,
$\qquad v = x + a_1 - a_3 \cos x + a_2 \sin x$.

**11.** $(D^2 + 1)y + 4(D - 1)v = 4 e^x$,
$\quad (D - 1)y + (D + 9)v = 0$; when $x = 0$, $y = 5$, $y' = 0$, $v = \frac{1}{2}$.

ANS. $y = 2 e^x + 2 e^{-x} + e^{-2x}(\cos x + 2 \sin x)$,
$\qquad v = \frac{1}{2}e^{-x} + e^{-2x} \sin x$.

**12.** $\dfrac{d^2 x}{dt^2} + x - 2\dfrac{dy}{dt} = 2t$,

$\quad 2\dfrac{dx}{dt} - x + \dfrac{dy}{dt} - 2y = 7$.

ANS. $x = 2t - 2 + b_1 e^t + b_2 e^{-t} + b_3 e^{-2t}$,
$\qquad y = -t - 1 + b_1 e^t - b_2 e^{-t} - \frac{5}{4}b_3 e^{-2t}$.

**13.** $2(D + 1)y + (D - 1)w = x + 1,$
   $(D + 3)y + (D + 1)w = 4x + 14.$

ANS.   $y = x + 2 + c_1 e^{-x} \cos 2x + c_2 e^{-x} \sin 2x,$
   $w = x + 6 + (c_2 - c_1) e^{-x} \cos 2x - (c_1 + c_2) e^{-x} \sin 2x.$

**14.** $(D + 1)y + (D - 4)v = 6 \cos x,$
   $(D - 1)y + (D^2 + 4)v = -6 \sin x.$

ANS.   $y = 3 \cos x + 3 \sin x + a_1 + a_2 \cos 3x + a_3 \sin 3x,$
   $v = \tfrac{1}{4}a_1 - \tfrac{1}{5}(a_2 - 3a_3) \cos 3x - \tfrac{1}{5}(3a_2 + a_3) \sin 3x.$

**15.** $2Du + (D - 1)v + (D + 2)w = 0,$
   $(D + 2)u + (2D - 3)v - (D - 6)w = 0,$
   $2Du - (D + 3)v - Dw = 0.$

ANS.   $u = 3c_1 e^{2x} + 3c_2 e^{-x} + 3c_3 e^{x/2},$
   $v = 4c_1 e^{2x} - 5c_2 e^{-x} + c_3 e^{x/2},$
   $w = -4c_1 e^{2x} - 4c_2 e^{-x} - c_3 e^{x/2}.$

**16.** $D^2y + (D - 1)v = 0,$
   $(2D - 1)y + (D - 1)w = 0,$
   $(D + 3)y + (D - 4)v + 3w = 0.$

ANS.   $y = a_1 + a_2 e^x + 3a_3 e^{4x},$
   $v = b_2 e^x - a_2 x e^x - 16a_3 e^{4x},$
   $w = -a_1 + (b_2 - a_2) e^x - a_2 x e^x - 7a_3 e^{4x}.$

## 76. First order systems with constant coefficients

In the following sections we shall show how matrix algebra can be used to reduce the problem of solving systems of differential equations to an algebraic routine. Before we do that it is important to realize that a system of linear equations of order higher than the first can be written in terms of a first-order system.

Consider for example the single equation

$$y'' + 2y' - y = e^x. \tag{1}$$

If we let $u = y'$ then equation (1) becomes

$$u' = y - 2u + e^x.$$

In other words the single second-order equation (1) has been replaced by the first-order system

$$y' = u, \tag{2}$$

$$u' = y - 2u + e^x.$$

In a similar manner the third-order equation

$$y''' + 2y'' - y' + 3y = x \tag{3}$$

can be written as a system of first order equations by choosing new variables

$$u = y', \quad \text{and} \quad v = u' = y''.$$

Then equation (3) becomes

$$v' = -2v + u - 3y + x,$$

and we can consider the first-order system

$$y' = u,$$

$$u' = v, \tag{4}$$

$$v' = u - 2v - 3y + x,$$

as being equivalent to (3).

The system of second-order equations (1) in Section 75 can be replaced by a first-order system if we let $u = v'$ and $w = y'$ so that

$$u' = 4v + 2w - 2,$$

$$v' = u,$$

$$w' = -5u + y + x, \tag{5}$$

$$y' = w.$$

## *Exercises*

In exercises 1 through 5, replace the given equation by a system of first order equations.

**1.** $y'' - 6y' + 8y = x + 2.$       ANS.   $y' = u,$
$$u' = 6u - 8y + x + 2.$$

**2.** $y'' + 4y' + 4y = e^x.$       ANS.   $y' = u,$
$$u' = -4u - 4y + e^x.$$

**3.** $y'' + py' + qy = f(x).$       ANS.   $y' = u,$
$$u' = -pu - qy + f(x).$$

**4.** $y''' + py'' + qy' + ry = f(x).$       ANS.   $y' = u,$
$$u' = v,$$
$$v' = -pv - qu - ry + f(x).$$

**5.** $y^{(4)} - y = 0.$       ANS.   $y' = u,$
$$u' = v,$$
$$v' = w,$$
$$w' = y.$$

In exercises 6 through 9, replace the given system by an equivalent system of first order equations.

**6.** Exercise 5, Section 75.
**7.** Exercise 6, Section 75.
**8.** Exercise 7, Section 75.
**9.** Exercise 8, Section 75.

## 77. Solution of a first order system

Consider the first order system

$$\frac{dx}{dt} = y, \tag{1}$$

$$\frac{dy}{dt} = -2x + 3y.$$

We can rewrite this system in the form

$$Dx - y = 0, \tag{2}$$

$$2x + (D - 3)y = 0.$$

Operating on the first equation with the operator $D - 3$ and adding the two equations eliminates the variable $y$ to give

$$(D^2 - 3D + 2)x = 0. \tag{3}$$

In a similar manner we can eliminate $x$ from system (3) to obtain

$$(D^2 - 3D + 2)y = 0. \tag{4}$$

Thus we realize that the solutions of equations (1) are of the form

$$x = c_1 e^{2t} + c_2 e^t,$$

$$y = c_3 e^{2t} + c_4 e^t,$$

where there are some relations among the four constants $c_1, c_2, c_3, c_4$ that we can determine by substitution back into equations (1).

An alternative way of viewing the nature of the solutions of system (1) is to expect from the beginning the existence of solutions of the form

$$x = c_1 e^{mt}, \tag{5}$$

$$y = c_2 e^{mt},$$

where the constants $c_1, c_2$, and $m$ must be determined by substitution into (1). If we do this we find

$$c_1 m e^{mt} = c_2 e^{mt},$$

and

$$c_2 m e^{mt} = -2c_1 e^{mt} + 3c_2 e^{mt},$$

or

$$-mc_1 + c_2 = 0, \tag{6}$$

$$-2c_1 + (3 - m)c_2 = 0.$$

The system (6) can have nontrivial solutions for $c_1$ and $c_2$ only if the determinant

$$\begin{vmatrix} -m & 1 \\ -2 & 3-m \end{vmatrix} \tag{7}$$

is zero. That is,

$$m^2 - 3m + 2 = (m-1)(m-2) = 0.$$

Moreover, for the choice $m = 1$, the system (6) yields the condition $c_2 = c_1$, and for $m = 2$ we would be forced to take $c_2 = 2c_1$. Thus there would be two distinct solutions of the form of equations (5), namely

$$\begin{aligned} x &= c_1 e^t, \\ y &= c_1 e^t, \end{aligned} \quad \text{and} \quad \begin{aligned} x &= c_1 e^{2t}, \\ y &= 2c_1 e^{2t}. \end{aligned} \tag{8}$$

A careful perusal of what we have done here should lead one to suspect that the elementary algebra problem of finding nontrivial solutions to the system (6) holds the entire key to the problem of solving the system of differential equations (1). The formalization of this procedure is best accomplished with the assistance of some vector and matrix notation. The next section summarizes the minimum amount of matrix algebra required for our purposes.

## 78. Some matrix algebra

We shall presume at this point that the student is familiar with the elementary calculus of vector functions. The basic ideas involved are deduced from the definition

$$\frac{d}{dt}\bigl(f_1(t), f_2(t), \dots, f_n(t)\bigr) = \left(\frac{df_1}{dt}, \frac{df_2}{dt}, \dots, \frac{df_n}{dt}\right) \tag{1}$$

and from the properties of vector addition, multiplication of vectors by numbers, and the scalar product of vectors. Most elementary calculus texts and courses provide all of the required algebra.

It is not always the case that students who have completed a course in elementary calculus are familiar with matrix algebra. We therefore include here a brief introduction suitable for our purposes.

A *matrix* is a rectangular array of numbers. For example, the following arrays are matrices:

$$A = \begin{pmatrix} 2 & 1 \\ 1 & 3 \end{pmatrix}, \qquad B = \begin{pmatrix} 3 & 2 \\ 1 & 1 \\ 2 & -1 \end{pmatrix}, \qquad C = \begin{pmatrix} 1 & 2 & 1 \\ 3 & 1 & 2 \end{pmatrix}.$$

Each of the numbers in a matrix is called an *element* of that matrix.

A matrix is said to be of *dimension* $n \times m$ ($n$ by $m$) if it has $n$ rows and $m$ columns. Thus the general $n \times m$ matrix can be written

$$\begin{pmatrix} a_{11} & a_{12} \ldots a_{1m} \\ a_{21} & a_{22} \ldots a_{2m} \\ & \vdots \\ a_{n1} & a_{n2} \ldots a_{nm} \end{pmatrix},$$

where $a_{ij}$ indicates the number in the $i$th row and $j$th column.

Two matrices of the same dimension are said to be *equal* if the corresponding elements are equal.

*Addition* is defined only for two matrices of the same dimension, and is defined elementwise. For example, the sum of two $2 \times 3$ matrices is accomplished as follows:

$$\begin{pmatrix} a & b & c \\ d & e & f \end{pmatrix} + \begin{pmatrix} g & h & i \\ j & k & l \end{pmatrix} = \begin{pmatrix} a+g & b+h & c+i \\ d+j & e+k & f+l \end{pmatrix}. \tag{2}$$

Any matrix may be multiplied by a number by multiplying each of its elements by that number. For example,

$$k\begin{pmatrix} a & b \\ c & d \\ e & f \end{pmatrix} = \begin{pmatrix} ka & kb \\ kc & kd \\ ke & kf \end{pmatrix}. \tag{3}$$

The student should recognize that the algebra of matrices, which is dictated by the definitions in the preceding paragraphs, is essentially the same kind of algebra as the algebra of vectors. This comes from the fact that the operations of addition and multiplication by numbers are done elementwise both for matrix algebra and vector algebra. Indeed, one can regard the vector $(a, b, c)$ as a $1 \times 3$ matrix and then interpret the usual vector algebra as a special case of the matrix algebra described above. The only caution to be observed is that the matrices

$$(a \quad b \quad c) \text{ and } \begin{pmatrix} a \\ b \\ c \end{pmatrix}$$

are very different from the matrix point of view, although we may wish to identify them as the same vector in some physical or geometrical context. We shall call the first vector a row vector and the second a column vector.

In what follows we shall often form the product of a row vector, a matrix of dimension $1 \times n$, with a column vector, a matrix of dimension $n \times 1$. The product is the familiar scalar product of elementary calculus

$$(a_1 \quad a_2 \quad \ldots \quad a_n) \cdot \begin{pmatrix} b_1 \\ b_2 \\ \vdots \\ b_n \end{pmatrix} = a_1 b_1 + a_2 b_2 + \cdots + a_n b_n, \tag{4}$$

with the insistence that the row vector always be written on the left and the column vector on the right.

The *product* of two matrices can now be defined in terms of scalar products of row and column vectors. An $n \times m$ matrix and a $p \times q$ matrix can be multiplied only if $m = p$; that is, the number of columns of the first matrix must equal the number of rows in the second matrix. The resulting product matrix has dimension $n \times q$. The definition is most easily described as follows:

$$A \cdot B = \begin{pmatrix} A_1 \cdot B^1 & A_1 \cdot B^2 & \ldots & A_1 \cdot B^q \\ A_2 \cdot B^1 & A_2 \cdot B^2 & \ldots & A_2 \cdot B^q \\ & & \vdots & \\ A_n \cdot B^1 & A_n \cdot B^2 & \ldots & A_n \cdot B^q \end{pmatrix}, \tag{5}$$

where $A_i$ is the $i$th row vector of the matrix $A$ and $B^j$ is the $j$th column vector of the matrix $B$. Thus the element in the $i$th row and $j$th column of the product is the ordinary scalar product of the vectors $A_i$ and $B^j$.

A few examples will help fix these definitions in our minds.

EXAMPLE (a):

$$\begin{pmatrix} 2 & 1 \\ -1 & 3 \end{pmatrix} + 2 \begin{pmatrix} -1 & 2 \\ 1 & 1 \end{pmatrix} = \begin{pmatrix} 2 & 1 \\ -1 & 3 \end{pmatrix} + \begin{pmatrix} -2 & 4 \\ 2 & 2 \end{pmatrix} = \begin{pmatrix} 0 & 5 \\ 1 & 5 \end{pmatrix}.$$

EXAMPLE (b):

$$\begin{pmatrix} t^2 + t \\ t^2 - t \end{pmatrix} = \begin{pmatrix} t^2 \\ t^2 \end{pmatrix} + \begin{pmatrix} t \\ -t \end{pmatrix} = \begin{pmatrix} 1 \\ 1 \end{pmatrix} t^2 + \begin{pmatrix} 1 \\ -1 \end{pmatrix} t.$$

EXAMPLE (c):

$$\begin{pmatrix} 2 & 1 \\ -1 & 3 \end{pmatrix}\begin{pmatrix} x \\ y \end{pmatrix} = \begin{pmatrix} 2x + y \\ -x + 3y \end{pmatrix}.$$

EXAMPLE (d):

$$\begin{pmatrix} a & b \\ c & d \end{pmatrix} \cdot \begin{pmatrix} p & q \\ r & s \end{pmatrix} = \begin{pmatrix} ap + br & aq + bs \\ cp + dr & cq + ds \end{pmatrix}.$$

EXAMPLE (e):

$$\begin{pmatrix} 2 & 1 \\ 1 & -1 \end{pmatrix} \cdot \begin{pmatrix} 3 & 1 \\ 4 & 1 \end{pmatrix} = \begin{pmatrix} 2\cdot 3 + 1\cdot 4 & 2\cdot 1 + 1\cdot 1 \\ 1\cdot 3 + (-1)\cdot 4 & 1\cdot 1 + (-1)\cdot 1 \end{pmatrix} = \begin{pmatrix} 10 & 3 \\ -1 & 0 \end{pmatrix}.$$

EXAMPLE (f):   The product

$$\begin{pmatrix} 1 & 2 \\ 1 & 3 \end{pmatrix} \cdot \begin{pmatrix} 1 & 1 \\ 2 & 1 \\ 3 & 1 \end{pmatrix}$$

is not defined because the number of columns in the first matrix is two and the number of rows in the second matrix is not two. On the other hand

$$\begin{pmatrix} 1 & 1 \\ 2 & 1 \\ 3 & 1 \end{pmatrix} \cdot \begin{pmatrix} 1 & 2 \\ 1 & 3 \end{pmatrix} = \begin{pmatrix} 2 & 5 \\ 3 & 7 \\ 4 & 9 \end{pmatrix}$$

is defined, a fact that demonstrates that matrix multiplication is not commutative.

In elementary vector algebra we often find it convenient to designate a vector by a single symbol instead of expressing the vector in terms of its components. We shall frequently use a single symbol to refer to a matrix. Particular caution must be observed when using such abbreviations to be sure that the dimensions of the objects involved are appropriate in terms of the definitions in our algebra. For example, the equation

$$\frac{dX}{dt} = X' = AX,$$

can be interpreted as a system of first-order linear equations with constant coefficients, if we interpret $X$ as an $n$-dimensional column vector function of $t$

and $A$ as an $n \times n$ matrix of real numbers. Thus

$$\frac{dx}{dt} = 2x + y,$$

$$\frac{dy}{dt} = x - y,$$

can be written

$$X' = AX,$$

where

$$A = \begin{pmatrix} 2 & 1 \\ 1 & -1 \end{pmatrix}, \qquad X = \begin{pmatrix} x \\ y \end{pmatrix}, \qquad X' = \begin{pmatrix} \dfrac{dx}{dt} \\ \dfrac{dy}{dt} \end{pmatrix}.$$

We shall use capital letters here for matrices and lower case letters for numbers. In particular, we shall use a large zero for a matrix, all of whose elements are zero wherever the dimension of that zero matrix is clear from the context.

EXAMPLE (g):   If

$$A = \begin{pmatrix} 2 & 1 & 1 \\ 3 & 2 & 1 \\ 1 & 4 & 1 \end{pmatrix}$$

and $X' - AX = O$, then $X$ must be a three-dimensional column vector function and $O$ must be the three-dimensional column vector

$$\begin{pmatrix} 0 \\ 0 \\ 0 \end{pmatrix}.$$

We shall single out one further kind of matrix for special attention. If we multiply any $2 \times 2$ matrix $A$ by the matrix

$$I = \begin{pmatrix} 1 & 0 \\ 0 & 1 \end{pmatrix},$$

we clearly obtain the results

$$AI = A \qquad \text{and} \qquad IA = A.$$

A similar observation can be made for any $n \times n$ matrix providing $I$ is interpreted as the $n \times n$ matrix modelled after the two-dimensional case, namely

$$I = (a_{ij}), \qquad \text{where} \quad a_{ij} = 0, \text{ if } i \neq j,$$
$$\text{and} \qquad a_{ij} = 1, \text{ if } i = j.$$

Again, we shall use the symbol $I$ wherever the dimension is clear from the context.

The algebraic structure that one obtains as a consequence of the above definitions is called the algebra of matrices. Some of the basic theorems of this algebra are listed below. Most are easily proved in the cases where the dimensions are low and rather tedious for large matrices. We shall ask the student to prove some of the theorems in the exercises that follow. In each theorem it is presumed that the matrices involved have dimensions for which the indicated operations are defined.

$$(A + B) + C = A + (B + C). \tag{6}$$

$$A + B = B + A. \tag{7}$$

$$A + O = A. \tag{8}$$

$$A + (-1)A = O. \tag{9}$$

$$(AB)C = A(BC). \tag{10}$$

$$A(B + C) = AB + AC. \tag{11}$$

$$IA = AI = A. \tag{12}$$

$$k(A + B) = kA + kB. \tag{13}$$

$$kO = O. \tag{14}$$

$$k(AB) = (kA)B = A(kB). \tag{15}$$

As a final theorem for this section we state, without proof, a basic theorem from elementary algebra that we have often used in previous chapters.

THEOREM 24: *Let $A$ be an $n \times n$ matrix of constant real numbers and let $X$ be an n-dimensional column vector. The system of equations*

$$AX = O$$

*has nontrivial solutions, that is $X \neq O$, if and only if the determinant of $A$ is zero.*

## Exercises

In exercises 1 through 8 find the requested matrix given the following matrices:

$$A = \begin{pmatrix} 1 & 2 \\ 3 & 1 \end{pmatrix}, \qquad B = \begin{pmatrix} 2 & 0 \\ 1 & -1 \end{pmatrix}, \qquad C = \begin{pmatrix} 1 & -1 \\ 1 & 2 \end{pmatrix}.$$

**1.** $A + 2B$.

ANS. $\begin{pmatrix} 5 & 2 \\ 5 & -1 \end{pmatrix}$.

**2.** $2A + C$.

**3.** $AB + 2I$.

ANS. $\begin{pmatrix} 6 & -2 \\ 7 & 1 \end{pmatrix}$.

**4.** $AC + BI$.

**5.** $C - 2I$.

ANS. $\begin{pmatrix} -1 & -1 \\ 1 & 0 \end{pmatrix}$.

**6.** $AB + C$.

**7.** $AC - B$.

ANS. $\begin{pmatrix} 1 & 3 \\ 3 & 0 \end{pmatrix}$.

**8.** $AB$ and $BA$.

**9.** Prove $D + E = E + D$ for any matrices $D$ and $E$ for which the sum is defined.

**10.** Prove the distributive law of equation (11) above for $2 \times 2$ matrices.

**11.** Prove the theorem of equation (8).

**12.** Prove the theorem of equation (9).

**13.** Prove the theorem of equation (13).

**14.** Prove the theorem of equation (15) for $2 \times 2$ matrices.

**15.** Show that the system of equations

$$\begin{pmatrix} 1 & 2 \\ 1 & -1 \end{pmatrix} \begin{pmatrix} x \\ y \end{pmatrix} = O$$

has no nontrivial solutions.

Find all of the solutions of the systems in exercises 16 through 21.

**16.** $\begin{pmatrix} 1 & 2 \\ 2 & 4 \end{pmatrix} \begin{pmatrix} x \\ y \end{pmatrix} = O.$

ANS. $\begin{pmatrix} x \\ y \end{pmatrix} = c \begin{pmatrix} -2 \\ 1 \end{pmatrix}.$

**17.** $\begin{pmatrix} 2 & 1 \\ -4 & -2 \end{pmatrix} \begin{pmatrix} x \\ y \end{pmatrix} = O.$

ANS. $\begin{pmatrix} x \\ y \end{pmatrix} = c \begin{pmatrix} 1 \\ -2 \end{pmatrix}.$

**18.** $\begin{pmatrix} 1 & 4 \\ -1 & 1 \end{pmatrix} \begin{pmatrix} x \\ y \end{pmatrix} = O.$

ANS. $\begin{pmatrix} x \\ y \end{pmatrix} = O.$

**19.** $\begin{pmatrix} 1 & 2 & 1 \\ 1 & -1 & 1 \\ 2 & 1 & 2 \end{pmatrix} \begin{pmatrix} x \\ y \\ z \end{pmatrix} = O.$

ANS. $\begin{pmatrix} x \\ y \\ z \end{pmatrix} = c \begin{pmatrix} 1 \\ 0 \\ -1 \end{pmatrix}.$

**20.** $\begin{pmatrix} 1 & -1 & 2 \\ 2 & 1 & 3 \\ 0 & -1 & 1 \end{pmatrix} \begin{pmatrix} x \\ y \\ z \end{pmatrix} = O.$ ANS. $\begin{pmatrix} x \\ y \\ z \end{pmatrix} = O.$

**21.** $\begin{pmatrix} 2 & 1 & 3 \\ -1 & 1 & 2 \\ 5 & 1 & 4 \end{pmatrix} \begin{pmatrix} x \\ y \\ z \end{pmatrix} = O.$ ANS. $\begin{pmatrix} x \\ y \\ z \end{pmatrix} = c \begin{pmatrix} -1 \\ -7 \\ 3 \end{pmatrix}.$

In exercises 22 through 25, write the given system of differential equations as a matrix equation.

**22.** $\dfrac{dx}{dt} = 2x + 3y,$

$\dfrac{dy}{dt} = x - y.$

ANS. $X' = AX$, where

$A = \begin{pmatrix} 2 & 3 \\ 1 & -1 \end{pmatrix}.$

**23.** $\dfrac{dx}{dt} = x - y + z + t,$

$\dfrac{dy}{dt} = x + 2y - z + 1,$

$\dfrac{dz}{dt} = 2x - y + z + e^t.$

ANS. $X' = AX + B$, where

$A = \begin{pmatrix} 1 & -1 & 1 \\ 1 & 2 & -1 \\ 2 & -1 & 1 \end{pmatrix}$ and $B = \begin{pmatrix} t \\ 1 \\ e^t \end{pmatrix}.$

**24.** $\dfrac{dx}{dt} = 2x - y + e^t,$

$\dfrac{dy}{dt} = x + y + t.$

**25.** $\dfrac{dx}{dt} = tx + y + \sin t,$

$\dfrac{dy}{dt} = t^2 x + ty + 1.$

## 79. First-order systems revisited

We return now to the consideration of first-order linear systems of equations with constant coefficients. Let $X$ be an $n$-dimensional column vector function of $t$ and let $A$ be an $n \times n$ matrix of real numbers. Further suppose $B(t)$ is a column vector whose components are known functions of $t$. Then the vector equation

$$X' = AX + B \qquad (1)$$

represents a system of $n$ equations in the $n$ unknown component functions of $X$. If $B = O$, we say that the system is homogeneous. For the present we will restrict our attention to homogeneous systems and return to the non-homogeneous case in a later section.

From our past experience we have reason to believe that the homogeneous system

$$X' = AX \tag{2}$$

may have solutions of the form

$$X = C\, e^{mt}, \tag{3}$$

where $C$ is a constant column vector and $m$ is some number that we wish to determine. Substitution of $X$ into system (2) yields

$$Cm\, e^{mt} = AC\, e^{mt},$$

which can be written

$$(AC - mC)\, e^{mt} = O.$$

We can rewrite this equation again, remembering that $C = IC$, in the form

$$(A - mI)C\, e^{mt} = O. \tag{4}$$

Equation (4) is to be satisfied for all real values of $t$, a condition that can be satisfied only if

$$(A - mI)C = O. \tag{5}$$

The theorem at the end of the previous section states that the algebraic system (5) has nontrivial solutions only if the determinant of $A - mI$ is zero; that is,

$$|A - mI| = 0. \tag{6}$$

Equation (6) is a polynomial equation of degree $n$ in the unknown number $m$. We observe that the polynomial $|A - mI|$ depends only on the matrix $A$.

The polynomial $|A - mI|$ is called the *characteristic polynomial* of the matrix $A$ and equation (6) is called the *characteristic equation* of the matrix $A$.

The roots of the characteristic equation of $A$ are called *eigenvalues* of the matrix $A$.

A nonzero vector $C_1$, which is a solution of equation (5) for a particular eigenvalue $m_1$, is called an *eigenvector* of the matrix $A$ corresponding to the eigenvalue $m_1$.

Thus we see that the first step in solving the homogeneous system (2) is to find the eigenvalues and the corresponding eigenvectors of the matrix $A$.

We may also suspect from our experience with the roots of the auxiliary equation in Chapter 6, that the nature of the solutions of equation (3) will depend on whether the eigenvalues are real and distinct, complex, or repeated. We shall deal with these cases separately.

EXAMPLE (a): Let us now reconsider the system of equations of Section 77. In matrix notation we have

$$X' = AX, \qquad \text{where } A = \begin{pmatrix} 0 & 1 \\ -2 & 3 \end{pmatrix}. \tag{7}$$

The characteristic equation of $A$ is

$$\left| \begin{pmatrix} 0 & 1 \\ -2 & 3 \end{pmatrix} - m \begin{pmatrix} 1 & 0 \\ 0 & 1 \end{pmatrix} \right| = \begin{vmatrix} -m & 1 \\ -2 & 3-m \end{vmatrix} = m^2 - 3m + 2 = 0.$$

The eigenvalues are the distinct real numbers $m_1 = 1$ and $m_2 = 2$.

For $m_1 = 1$, equation (5) becomes

$$\begin{pmatrix} -1 & 1 \\ -2 & 2 \end{pmatrix} \begin{pmatrix} c_1 \\ c_2 \end{pmatrix} = O,$$

so that $-c_1 + c_2 = 0$ and we have

$$C = \begin{pmatrix} c_1 \\ c_2 \end{pmatrix} = \begin{pmatrix} c_1 \\ c_1 \end{pmatrix} = c_1 \begin{pmatrix} 1 \\ 1 \end{pmatrix}.$$

Thus corresponding to the eigenvalue $m_1 = 1$, there is a set of eigenvectors whose elements are scalar multiples of the column vector $\begin{pmatrix} 1 \\ 1 \end{pmatrix}$.

Similarly for $m_2 = 2$ equation (5) is

$$\begin{pmatrix} -2 & 1 \\ -2 & 1 \end{pmatrix} \begin{pmatrix} c_1 \\ c_2 \end{pmatrix} = O,$$

so that $-2c_1 + c_2 = 0$ and

$$C = \begin{pmatrix} c_1 \\ c_2 \end{pmatrix} = \begin{pmatrix} c_1 \\ 2c_1 \end{pmatrix} = c_1 \begin{pmatrix} 1 \\ 2 \end{pmatrix}.$$

That is, the eigenvectors are multiples of the vector $\begin{pmatrix} 1 \\ 2 \end{pmatrix}$.

Thus we have obtained two distinct sets of solutions for system (7),

$$X_1 = b_1 \begin{pmatrix} 1 \\ 1 \end{pmatrix} e^t \qquad \text{and} \qquad X_2 = b_2 \begin{pmatrix} 1 \\ 2 \end{pmatrix} e^{2t}, \tag{8}$$

where $b_1$ and $b_2$ are arbitrary constants. It is a simple matter to verify that these vector functions are solutions of system (7). It is quite another matter to make the claim that every solution of system (7) is some combination of the two solutions we have found. The fact that this is true requires the support of several important definitions and theorems. The student should observe that these definitions and theorems closely parallel the theoretical development in Chapter 5. We shall state the appropriate theorems without proof.

A set of $m$ constant vectors of dimension $n$

$$\{X_1, X_2, \ldots, X_m\}$$

is *linearly independent* if

$$c_1 X_1 + c_2 X_2 + \cdots + c_m X_m = O$$

implies that $c_1 = c_2 = \cdots = c_m = 0$.

A set of vector functions of $t$

$$\{X_1(t), X_2(t), \ldots, X_m(t)\}$$

is linearly independent on an interval $a < t < b$ if

$$c_1 X_1(t) + c_2 X_2(t) + \cdots + c_m X_m(t) = O$$

for all $t$ on the interval implies that

$$c_1 = c_2 = \cdots = c_m = 0.$$

**THEOREM 25:**   *If $X_1(t), \ldots, X_m(t)$ are each solutions of a homogeneous linear system $X' = AX$ then $c_1 X_1(t) + c_2 X_2(t) + \cdots + c_m X_m(t)$ is a solution of the same system for arbitrary constants $c_1, \ldots, c_m$.*

**THEOREM 26:**   *If $A$ is an $n \times n$ matrix of real numbers and $\{X_1, \ldots, X_n\}$ is a linearly independent set of solutions of the system $X' = AX$ on the interval $a < t < b$, then any solution of the system is a unique linear combination of the set $\{X_1, \ldots, X_n\}$.*

If the set of $n$-dimensional column vectors $\{X_1(t), \ldots, X_n(t)\}$ is considered as an $n \times n$ matrix

$$(X_1(t) \qquad X_2(t) \ldots X_n(t))$$

then the determinant

$$|X_1(t) \qquad X_2(t) \ldots X_n(t)|$$

is called the *Wronskian* of the set of vectors.

**THEOREM 27:**   *The set of $n$-dimensional column vectors $\{X_1(t), \ldots, X_n(t)\}$ is linearly independent at $t = t_0$ if, and only if, the Wronskian of the set is not zero at $t = t_0$; that is,*

$$W\{X_1(t_0), X_2(t_0), \ldots, X_n(t_0)\} = |X_1(t_0) \qquad X_2(t_0) \ldots X_n(t_0)| \neq 0.$$

**THEOREM 28:**   *If the vector functions $X_1(t), \ldots, X_n(t)$ are solutions of the system $X' = AX$ for all $t$ on the interval $a < t < b$ where $A$ is an $n \times n$ matrix,* *then the set* $\{X_1(t), \ldots, X_n(t)\}$ *is linearly independent on* $a < t < b$ *if, and only if,* $W\{(X_1(t_0), \ldots, X_n(t_0)\} \neq 0$ *for some* $t_0$ *on the interval* $a < t < b$.

THEOREM 29: *If* $X_1(t), \ldots, X_n(t)$ *are linearly independent solutions of the n-dimensional homogeneous system* $X' = AX$ *on the interval* $a < t < b$ *and if* $X_p(t)$ *is any solution of the nonhomogeneous system* $X' = AX + B(t)$ *on the interval* $a < t < b$, *then any solution of the nonhomogeneous system can be written*

$$X(t) = c_1 X_1(t) + \cdots + c_n X_n(t) + X_p(t)$$

*for a unique choice of the constants* $c_1, \ldots, c_n$.

We return to Example (a). In equation (8) we presented two sets of solutions of the system (7). If we pick $b_1 = b_2 = 1$ and consider the Wronskian of the resulting set, we have

$$W\{X_1, X_2\} = \begin{vmatrix} e^t & e^{2t} \\ e^t & 2\,e^{2t} \end{vmatrix} = \begin{vmatrix} 1 & 1 \\ 1 & 2 \end{vmatrix} e^{3t} = e^{3t}.$$

Since the Wronskian does not vanish for any value of $t$, we conclude that the solutions $X_1$ and $X_2$ are linearly independent on any interval and it follows that the general solution of the system (7) is

$$X(t) = c_1 \begin{pmatrix} 1 \\ 1 \end{pmatrix} e^t + c_2 \begin{pmatrix} 1 \\ 2 \end{pmatrix} e^{2t}.$$

EXAMPLE (b):    Consider the system

$$X' = AX \qquad \text{for } A = \begin{pmatrix} 4 & -1 \\ -4 & 4 \end{pmatrix}. \tag{9}$$

The characteristic equation of $A$ is

$$\begin{vmatrix} 4 - m & -1 \\ -4 & 4 - m \end{vmatrix} = m^2 - 8m + 12 = 0.$$

Therefore the eigenvalues of $A$ are $m_1 = 2$ and $m_2 = 6$.

For $m_1 = 2$, we compute nontrivial solutions of

$$\begin{pmatrix} 2 & -1 \\ -4 & 2 \end{pmatrix} \begin{pmatrix} c_1 \\ c_2 \end{pmatrix} = O.$$

Thus $2c_1 - c_2 = 0$. One such solution is obtained by choosing $c_1 = 1$ to give the eigenvector $\begin{pmatrix} 1 \\ 2 \end{pmatrix}$. It follows that $X_1 = \begin{pmatrix} 1 \\ 2 \end{pmatrix} e^{2t}$ is a solution of the system (9).

For $m_2 = 6$, the system

$$\begin{pmatrix} -2 & -1 \\ -4 & -2 \end{pmatrix} \begin{pmatrix} c_1 \\ c_2 \end{pmatrix} = O,$$

upon choosing $c_1 = 1$, leads to an eigenvector $\begin{pmatrix} 1 \\ -2 \end{pmatrix}$ for the matrix $A$ and a second solution

$$X_2 = \begin{pmatrix} 1 \\ -2 \end{pmatrix} e^{6t}.$$

The Wronskian of $X_1$ and $X_2$ is

$$W\{X_1, X_2\} = \begin{vmatrix} e^{2t} & e^{6t} \\ 2\,e^{2t} & -2\,e^{6t} \end{vmatrix} = -4\,e^{8t}.$$

Because $W\{X_1, X_2\}$ is never zero, it follows from Theorem 28 that $X_1$ and $X_2$ are linearly independent. By Theorem 26 we see that the general solution of the system $X' = AX$ is

$$X = c_1 \begin{pmatrix} 1 \\ 2 \end{pmatrix} e^{2t} + c_2 \begin{pmatrix} 1 \\ -2 \end{pmatrix} e^{6t}.$$

EXAMPLE (c):   Solve the system

$$X' = AX \qquad \text{for } A = \begin{pmatrix} 1 & -1 & -1 \\ 0 & 1 & 3 \\ 0 & 3 & 1 \end{pmatrix}. \tag{10}$$

The characteristic equation of $A$ is

$$\begin{vmatrix} 1 - m & -1 & -1 \\ 0 & 1 - m & 3 \\ 0 & 3 & 1 - m \end{vmatrix} = (1 - m)(m - 4)(m + 2) = 0. \tag{11}$$

Choosing the eigenvalue $m_1 = 1$ leads to

$$\begin{pmatrix} 0 & -1 & -1 \\ 0 & 0 & 3 \\ 0 & 3 & 0 \end{pmatrix} \begin{pmatrix} c_1 \\ c_2 \\ c_3 \end{pmatrix} = O,$$

which requires that $c_2 = c_3 = 0$ but leaves $c_1$ arbitrary. Thus

$$X_1 = \begin{pmatrix} 1 \\ 0 \\ 0 \end{pmatrix} e^t.$$

is one solution of (10).

The eigenvalue $m_2 = 4$ requires

$$\begin{pmatrix} -3 & -1 & -1 \\ 0 & -3 & 3 \\ 0 & 3 & -3 \end{pmatrix} \begin{pmatrix} c_1 \\ c_2 \\ c_3 \end{pmatrix} = 0,$$

or

$$3c_1 + c_2 + c_3 = 0,$$
$$-c_2 + c_3 = 0.$$

One solution of this system is the eigenvector $\begin{pmatrix} 2 \\ -3 \\ -3 \end{pmatrix}$ from which we obtain

$$X_2 = \begin{pmatrix} 2 \\ -3 \\ -3 \end{pmatrix} e^{4t}$$ as a second solution to the system (10).

Finally, choosing $m_3 = -2$ from equation (11) we have

$$\begin{pmatrix} 3 & -1 & -1 \\ 0 & 3 & 3 \\ 0 & 3 & 3 \end{pmatrix} \begin{pmatrix} c_1 \\ c_2 \\ c_3 \end{pmatrix} = 0,$$

which yields the eigenvector $\begin{pmatrix} 0 \\ 1 \\ -1 \end{pmatrix}$ and the solution

$$X_3 = \begin{pmatrix} 0 \\ 1 \\ -1 \end{pmatrix} e^{-2t}.$$

To establish the linear independence of the three solutions $X_1, X_2, X_3$ we compute the Wronskian at $t = 0$; that is,

$$W\{X_1(0), X_2(0), X_3(0)\} = \begin{vmatrix} 1 & 2 & 0 \\ 0 & -3 & 1 \\ 0 & -3 & -1 \end{vmatrix} = 6.$$

Because $W \neq 0$, it follows from Theorem 28 that the solutions are linearly independent on any interval. Thus the general solution of system (10) is

$$X(t) = c_1 \begin{pmatrix} 1 \\ 0 \\ 0 \end{pmatrix} e^t + c_2 \begin{pmatrix} 2 \\ -3 \\ -3 \end{pmatrix} e^{4t} + c_3 \begin{pmatrix} 0 \\ 1 \\ -1 \end{pmatrix} e^{-2t}.$$

EXAMPLE (d): We conclude this section with an example to illustrate why the use of the word *Wronskian*, in the context of a set of solutions of a system of first-order linear differential equations, is consistent with the usage of the same word, made in Section 27 in the context of a set of solutions of a single $n$th-order linear differential equation.

Consider the second-order linear equation

$$[D^2 - (a + b)D + ab]x = 0, \qquad a \neq b, \tag{12}$$

in which $D = d/dt$. The operator factors into $(D - a)(D - b)$ and the functions $e^{at}$ and $e^{bt}$ are therefore solutions of equation (12). The Wronskian of these solutions, as defined in Section 27, is the determinant

$$W\{e^{at}, e^{bt}\} = \begin{vmatrix} e^{at} & e^{bt} \\ ae^{at} & be^{bt} \end{vmatrix}, \tag{13}$$

where the functions in the second row are the derivatives of the functions in the first row.

Equation (12) may be converted to a system of first-order equations by setting $Dx = y$ so that (12) becomes

$$D^2 x = Dy = (a + b)y - abx.$$

Thus (12) is equivalent to the system

$$x' = y$$

$$y' = -abx + (a + b)y,$$

or

$$\begin{pmatrix} x \\ y \end{pmatrix}' = \begin{pmatrix} 0 & 1 \\ -ab & a+b \end{pmatrix} \begin{pmatrix} x \\ y \end{pmatrix}. \tag{14}$$

If we apply the technique of the current section, we write the characteristic equation of the matrix of system (14)

$$\begin{vmatrix} -m & 1 \\ -ab & a+b-m \end{vmatrix} = 0,$$

which reduces to

$$m^2 - (a + b)m + ab = 0. \tag{15}$$

It is important to observe that the characteristic polynomial of (15) and the operator polynomial of (12) have the same form, hence the same roots.

From equation (15) we obtain the eigenvalues $m_1 = a$ and $m_2 = b$. Choosing the eigenvalue $m_1 = a$ leads to

$$\begin{pmatrix} -a & 1 \\ -ab & b \end{pmatrix} \begin{pmatrix} c_1 \\ c_2 \end{pmatrix} = 0,$$

so that $c_2 = ac_1$ and

$$\begin{pmatrix} c_1 \\ c_2 \end{pmatrix} = \begin{pmatrix} c_1 \\ ac_1 \end{pmatrix} = c_1 \begin{pmatrix} 1 \\ a \end{pmatrix}.$$

Thus $\begin{pmatrix} 1 \\ a \end{pmatrix} e^{at}$ is one solution of (14).

Choosing the eigenvalue $m_2 = b$ leads to

$$\begin{pmatrix} -b & 1 \\ -ab & a \end{pmatrix} \begin{pmatrix} c_1 \\ c_2 \end{pmatrix} = 0,$$

so that $c_2 = bc_1$ and

$$\begin{pmatrix} c_1 \\ c_2 \end{pmatrix} = \begin{pmatrix} c_1 \\ bc_1 \end{pmatrix} = c_1 \begin{pmatrix} 1 \\ b \end{pmatrix}.$$

Thus $\begin{pmatrix} 1 \\ b \end{pmatrix} e^{bt}$ is a second solution of (14).

In the context of the current section, the Wronskian of these two solutions is

$$W\{X_1(t), X_2(t)\} = \begin{vmatrix} e^{at} & e^{bt} \\ ae^{at} & be^{bt} \end{vmatrix}. \tag{16}$$

Thus we see that the expressions given in (13) and (16), while coming from entirely different contexts are the same, and the word *Wronskian* is used in both contexts for that expression.

The examples that we have considered have all involved matrices whose eigenvalues are distinct real numbers. In each case the eigenvectors corresponding to distinct eigenvalues turned out to be linearly independent. That was no accident. It is possible to prove a theorem to that effect.

THEOREM 30:  *If $m_1, m_2, \ldots, m_s$ are distinct eigenvalues of an $n \times n$ matrix A and if $X_1, X_2, \ldots, X_s$ are corresponding eigenvectors, then the set $\{X_1, \ldots, X_s\}$ is linearly independent.*

The definitions and theorems of this section have been stated without providing the proof required to understand them clearly. It is hoped that this will serve as a motivation for a study of linear algebra, where the definitions and theorems become more easily understood.

### Exercises

In exercises 1 through 7, find the general solution of the system $X' = AX$ for the given matrix $A$. In each case check on the linear independence of solutions by examining the Wronskian.

1. $A = \begin{pmatrix} 8 & -3 \\ 16 & -8 \end{pmatrix}$. 

ANS. $X = c_1 \begin{pmatrix} 3 \\ 4 \end{pmatrix} e^{4t} + c_2 \begin{pmatrix} 1 \\ 4 \end{pmatrix} e^{-4t}$.

2. $A = \begin{pmatrix} 1 & 0 \\ -2 & 2 \end{pmatrix}$.

ANS. $X = c_1 \begin{pmatrix} 1 \\ 2 \end{pmatrix} e^{t} + c_2 \begin{pmatrix} 0 \\ 1 \end{pmatrix} e^{2t}$.

3. $A = \begin{pmatrix} 4 & 3 \\ -4 & -4 \end{pmatrix}$.

ANS. $X = c_1 \begin{pmatrix} 3 \\ -2 \end{pmatrix} e^{2t} + c_2 \begin{pmatrix} 1 \\ -2 \end{pmatrix} e^{-2t}$.

4. $A = \begin{pmatrix} 3 & 3 \\ -1 & -1 \end{pmatrix}$.

ANS. $X = c_1 \begin{pmatrix} 1 \\ -1 \end{pmatrix} + c_2 \begin{pmatrix} 3 \\ -1 \end{pmatrix} e^{2t}$.

5. $A = \begin{pmatrix} 2 & 3 \\ 1 & -2 \end{pmatrix}$.

ANS. $X = c_1 \begin{pmatrix} 3 \\ -2 + \sqrt{7} \end{pmatrix} e^{\sqrt{7}t} + c_2 \begin{pmatrix} -3 \\ 2 + \sqrt{7} \end{pmatrix} e^{-\sqrt{7}t}$.

6. $A = \begin{pmatrix} 12 & -15 \\ 4 & -4 \end{pmatrix}$.

ANS. $X = c_1 \begin{pmatrix} 3 \\ 2 \end{pmatrix} e^{2t} + c_2 \begin{pmatrix} 5 \\ 2 \end{pmatrix} e^{6t}$.

7. $A = \begin{pmatrix} 1 & 2 & -1 \\ 0 & -1 & 3 \\ 0 & 0 & 2 \end{pmatrix}$.

ANS. $X = c_1 \begin{pmatrix} 1 \\ 0 \\ 0 \end{pmatrix} e^{t} + c_2 \begin{pmatrix} 1 \\ -1 \\ 0 \end{pmatrix} e^{-t} + c_3 \begin{pmatrix} 1 \\ 1 \\ 1 \end{pmatrix} e^{2t}$.

8. $A = \begin{pmatrix} 1 & 2 & -1 \\ 2 & 1 & 1 \\ -1 & 1 & 0 \end{pmatrix}$.

ANS. $X = c_1 \begin{pmatrix} 1 \\ -1 \\ -2 \end{pmatrix} e^{t} + c_2 \begin{pmatrix} 1 \\ 1 \\ 0 \end{pmatrix} e^{3t} + c_3 \begin{pmatrix} 1 \\ -1 \\ 1 \end{pmatrix} e^{-2t}$.

## 80. Complex eigenvalues

In Section 79 we carefully avoided systems for which the eigenvalues were complex numbers. We now consider some examples in which complex numbers occur.

EXAMPLE (a): Solve the system

$$\begin{pmatrix} x \\ y \end{pmatrix}' = \begin{pmatrix} 2 & -5 \\ 2 & -4 \end{pmatrix} \begin{pmatrix} x \\ y \end{pmatrix}. \tag{1}$$

The characteristic equation of the matrix in system (1) is

$$\begin{vmatrix} 2 - m & -5 \\ 2 & -4 - m \end{vmatrix} = m^2 + 2m + 2 = 0, \tag{2}$$

with eigenvalues $m_1 = -1 + i$ and $m_2 = -1 - i$.
For $m_1 = -1 + i$ we must satisfy the system

$$\begin{pmatrix} 3 - i & -5 \\ 2 & -3 - i \end{pmatrix} \begin{pmatrix} c_1 \\ c_2 \end{pmatrix} = O,$$

which requires that

$$c_2 = \frac{3 - i}{5}c_1.$$

One solution is obtained by choosing $c_1 = 5$. Thus an eigenvector corresponding to the eigenvalue $m_1$ is $\begin{pmatrix} 5 \\ 3 - i \end{pmatrix}$ with the complex vector function

$$X_1 = \begin{pmatrix} 5 \\ 3 - i \end{pmatrix} e^{(-1 + i)t}, \tag{3}$$

at least formally a solution of system (1).

The second eigenvalue $m_2 = -1 - i$ leads in a similar way to a second solution,

$$X_2 = \begin{pmatrix} 5 \\ 3 + i \end{pmatrix} e^{(-1 - i)t}. \tag{4}$$

The two solutions can be combined to give

$$X = c_1 \begin{pmatrix} 5 \\ 3 - i \end{pmatrix} e^{(-1 + i)t} + c_2 \begin{pmatrix} 5 \\ 3 + i \end{pmatrix} e^{(-1 - i)t}. \tag{5}$$

The presentation of a solution in this form should be reminiscent of the situation in Chapter 6, where we were solving single linear equations with constant coefficients. We will proceed in much the same way as we did there by making use of Euler's formula

$$e^{(a + bi)t} = e^{at}(\cos bt + i \sin bt). \tag{6}$$

Formally changing the form of equation (5) gives us

$$X = c_1 \begin{pmatrix} 5 \\ 3 - i \end{pmatrix} e^{-t}(\cos t + i \sin t) + c_2 \begin{pmatrix} 5 \\ 3 + i \end{pmatrix} e^{-t}(\cos t - i \sin t),$$

and after combining real and imaginary parts

$$X = e^{-t}\left[(c_1 + c_2)\begin{pmatrix} 5 \cos t \\ 3 \cos t + \sin t \end{pmatrix} + i(c_1 - c_2)\begin{pmatrix} 5 \sin t \\ -\cos t + 3 \sin t \end{pmatrix}\right]. \tag{7}$$

If we let $b_1 = c_1 + c_2$ and $b_2 = i(c_1 - c_2)$ equation (7) can be written finally as

$$X = e^{-t}\left[b_1\left\{\begin{pmatrix} 5 \\ 3 \end{pmatrix} \cos t - \begin{pmatrix} 0 \\ -1 \end{pmatrix} \sin t\right\} + b_2\left\{\begin{pmatrix} 0 \\ -1 \end{pmatrix} \cos t + \begin{pmatrix} 5 \\ 3 \end{pmatrix} \sin t\right\}\right]. \tag{8}$$

The linear independence of the two solutions in (8) can be established by computing the Wronskian at $t = 0$. The student should show that $W(0) = -5$.

We make the following observations from the above example:

(a) Since the matrix of system (1) is real, the eigenvalues occur in conjugate pairs.

(b) The eigenvectors corresponding to conjugate eigenvalues are also conjugates of one another.

(c) The first eigenvector,

$$B = \begin{pmatrix} 5 + 0i \\ 3 - 1i \end{pmatrix} = \begin{pmatrix} 5 \\ 3 \end{pmatrix} + \begin{pmatrix} 0 \\ -1 \end{pmatrix} i = \text{Re } B + i \text{ Im } B,$$

appears in the solution (8) in the form

$$X = e^{-t}[b_1\{\text{Re } B \cos t - \text{Im } B \sin t\} + b_2\{\text{Im } B \cos t + \text{Re } B \sin t\}]. \quad (9)$$

(d) The Wronskian of the two solutions in (9) at $t = 0$ is given by the determinant $W = |\text{Re } B \text{ Im } B|$.

In exercises 12 to 14 below the student is asked to show the applicability of observations (a) to (d) to the general system of two equations in two unknowns.

EXAMPLE (b): Solve the system

$$\begin{pmatrix} x \\ y \end{pmatrix}' = \begin{pmatrix} 2 & 1 \\ -4 & 2 \end{pmatrix} \begin{pmatrix} x \\ y \end{pmatrix}, \quad (10)$$

making use of the observations made in Example (a).

The characteristic equation

$$\begin{vmatrix} 2 - m & 1 \\ -4 & 2 - m \end{vmatrix} = m^2 - 4m + 8 = 0,$$

has conjugate roots $m_1 = 2 + 2i$ and $m_2 = 2 - 2i$. An eigenvector corresponding to $m_1$ is

$$\begin{pmatrix} 1 \\ 2i \end{pmatrix} = \begin{pmatrix} 1 \\ 0 \end{pmatrix} + \begin{pmatrix} 0 \\ 2 \end{pmatrix} i.$$

If we accept the results of the observations in Example (a), observing that

$$W(0) = \begin{vmatrix} 1 & 0 \\ 0 & 2 \end{vmatrix} = 2 \neq 0,$$

we conclude that the general solution of system (10) is

$$X = e^{2t} \left[ b_1 \left\{ \begin{pmatrix} 1 \\ 0 \end{pmatrix} \cos 2t - \begin{pmatrix} 0 \\ 2 \end{pmatrix} \sin 2t \right\} + b_2 \left\{ \begin{pmatrix} 0 \\ 2 \end{pmatrix} \cos 2t + \begin{pmatrix} 1 \\ 0 \end{pmatrix} \sin 2t \right\} \right].$$

**EXAMPLE (c):**  We now consider a system of three equations in three unknowns given by

$$X' = AX, \quad \text{where } A = \begin{pmatrix} 1 & 2 & -1 \\ 0 & 1 & 1 \\ 0 & -1 & 1 \end{pmatrix}. \tag{11}$$

The characteristic equation

$$(1 - m)(m^2 - 2m + 2) = 0$$

has roots $m_1 = 1$, $m_2 = 1 + i$, and $m_3 = 1 - i$.

The eigenvalue $m_1 = 1$ has the vector $\begin{pmatrix} 1 \\ 0 \\ 0 \end{pmatrix}$ as an eigenvector, giving one solution of (11) as

$$X_1 = \begin{pmatrix} 1 \\ 0 \\ 0 \end{pmatrix} e^t.$$

The eigenvalue $m_2 = 1 + i$ yields an eigenvector

$$\begin{pmatrix} 2 - i \\ 0 + i \\ -1 + 0i \end{pmatrix} = \begin{pmatrix} 2 \\ 0 \\ -1 \end{pmatrix} + i \begin{pmatrix} -1 \\ 1 \\ 0 \end{pmatrix}.$$

The general solution can be written

$$X = c_1 \begin{pmatrix} 1 \\ 0 \\ 0 \end{pmatrix} e^t + e^t \left[ c_2 \left\{ \begin{pmatrix} 2 \\ 0 \\ -1 \end{pmatrix} \cos t - \begin{pmatrix} -1 \\ 1 \\ 0 \end{pmatrix} \sin t \right\} \right.$$

$$\left. + c_3 \left\{ \begin{pmatrix} -1 \\ 1 \\ 0 \end{pmatrix} \cos t + \begin{pmatrix} 2 \\ 0 \\ -1 \end{pmatrix} \sin t \right\} \right].$$

The linear independence of the solutions at $t = 0$ is guaranteed by evaluating the Wronskian at $t = 0$. Its value is

$$W(0) = \begin{vmatrix} 1 & 2 & -1 \\ 0 & 0 & 1 \\ 0 & -1 & 0 \end{vmatrix} = 1 \neq 0.$$

## Exercises

In exercises 1 through 7 find the general solution of the system $X' = AX$ for the given matrix $A$.

**1.** $A = \begin{pmatrix} 4 & 5 \\ -4 & -4 \end{pmatrix}.$

ANS. $X = c_1 \left[ \begin{pmatrix} 5 \\ -4 \end{pmatrix} \cos 2t - \begin{pmatrix} 0 \\ 2 \end{pmatrix} \sin 2t \right] + c_2 \left[ \begin{pmatrix} 0 \\ 2 \end{pmatrix} \cos 2t + \begin{pmatrix} 5 \\ -4 \end{pmatrix} \sin 2t \right].$

**2.** $A = \begin{pmatrix} 4 & 1 \\ -8 & 8 \end{pmatrix}.$

ANS. $X = c_1 e^{6t} \left[ \begin{pmatrix} 1 \\ 2 \end{pmatrix} \cos 2t + \begin{pmatrix} 0 \\ -2 \end{pmatrix} \sin 2t \right] + c_2 e^{6t} \left[ \begin{pmatrix} 0 \\ 2 \end{pmatrix} \cos 2t + \begin{pmatrix} 1 \\ 2 \end{pmatrix} \sin 2t \right].$

**3.** $A = \begin{pmatrix} 4 & -13 \\ 2 & -6 \end{pmatrix}.$

ANS. $X = c_1 e^{-t} \left[ \begin{pmatrix} 13 \\ 5 \end{pmatrix} \cos t - \begin{pmatrix} 0 \\ -1 \end{pmatrix} \sin t \right] + c_2 e^{-t} \left[ \begin{pmatrix} 0 \\ -1 \end{pmatrix} \cos t + \begin{pmatrix} 13 \\ 5 \end{pmatrix} \sin t \right].$

**4.** $A = \begin{pmatrix} 3 & 5 \\ -1 & -1 \end{pmatrix}.$

ANS. $X = c_1 e^{t} \left[ \begin{pmatrix} 5 \\ -2 \end{pmatrix} \cos t - \begin{pmatrix} 0 \\ 1 \end{pmatrix} \sin t \right] + c_2 e^{t} \left[ \begin{pmatrix} 0 \\ 1 \end{pmatrix} \cos t + \begin{pmatrix} 5 \\ -2 \end{pmatrix} \sin t \right].$

**5.** $A = \begin{pmatrix} 12 & -17 \\ 4 & -4 \end{pmatrix}.$

ANS. $X = c_1 e^{4t} \left[ \begin{pmatrix} 17 \\ 8 \end{pmatrix} \cos 2t - \begin{pmatrix} 0 \\ -2 \end{pmatrix} \sin 2t \right] + c_2 e^{4t} \left[ \begin{pmatrix} 0 \\ -2 \end{pmatrix} \cos 2t + \begin{pmatrix} 17 \\ 8 \end{pmatrix} \sin 2t \right].$

**6.** $A = \begin{pmatrix} 8 & -5 \\ 16 & -8 \end{pmatrix}.$

ANS. $X = c_1 \left[ \begin{pmatrix} 5 \\ 8 \end{pmatrix} \cos 4t - \begin{pmatrix} 0 \\ -4 \end{pmatrix} \sin 4t \right] + c_2 \left[ \begin{pmatrix} 0 \\ -4 \end{pmatrix} \cos 4t + \begin{pmatrix} 5 \\ 8 \end{pmatrix} \sin 4t \right].$

**7.** $A = \begin{pmatrix} 1 & 0 & 0 \\ 2 & 1 & -2 \\ 3 & 2 & 1 \end{pmatrix}.$

$$\text{ANS.} \quad X = c_1 \begin{pmatrix} 2 \\ -3 \\ 2 \end{pmatrix} e^t + c_2\, e^t \left[ \begin{pmatrix} 0 \\ 1 \\ 0 \end{pmatrix} \cos 2t - \begin{pmatrix} 0 \\ 0 \\ -1 \end{pmatrix} \sin 2t \right]$$

$$+ c_3\, e^t \left[ \begin{pmatrix} 0 \\ 0 \\ -1 \end{pmatrix} \cos 2t + \begin{pmatrix} 0 \\ 1 \\ 0 \end{pmatrix} \sin 2t \right].$$

Following the example of Section 76, replace each of the following equations by a system of first-order equations. Solve that system using matrix techniques and check your answers by solving the original equation directly.

**8.** $y^{(4)} - y = 0.$
**9.** $y'' + 2y' + 2y = 0.$
**10.** $y''' - 3y'' + 4y' - 2y = 0.$
**11.** $y'' + 4y = 0.$

In exercises 12 through 15 we consider the general homogeneous system with real coefficients

$$\begin{pmatrix} x \\ y \end{pmatrix}' = \begin{pmatrix} a & b \\ c & d \end{pmatrix} \begin{pmatrix} x \\ y \end{pmatrix}. \tag{A}$$

**12.** Find the eigenvalues for the matrix of (A) and show that complex eigenvalues occur only if $(a - d)^2 + 4bc < 0$. In particular, note that complex eigenvalues occur as conjugate pairs and that they occur only if $b$ and $c$ are not zero.
**13.** For the system (A) suppose the eigenvalues are complex numbers $p + qi$ and $p - qi$, where $q \neq 0$. Show that the corresponding eigenvectors are conjugate pairs.
**14.** Show that the observation made in equation (9) above is true in the complex case.
**15.** Find the value of the Wronskian in the complex case for $t = 0$ and show that it is not zero.

## 81. Repeated eigenvalues

We now consider an example in which the characteristic equation has repeated roots.

EXAMPLE (a): Solve the system

$$X' = AX, \quad \text{for } A = \begin{pmatrix} 0 & 1 \\ -4 & 4 \end{pmatrix}. \tag{1}$$

The characteristic equation of $A$ is

$$\begin{vmatrix} -m & 1 \\ -4 & 4 - m \end{vmatrix} = (m - 2)^2 = 0.$$

For the eigenvalue $m_1 = 2$ we obtain the solution

$$X_1 = \binom{1}{2} e^{2t}. \tag{2}$$

A second solution $X_2$, independent of $X_1$, is not immediately available because the eigenvalue $m_1$ is a double root of the characteristic equation. From our experience with repeated roots in Chapter 6, we may be tempted to guess that a second solution has the form

$$X_2 = \binom{c_1}{c_2} t\, e^{2t}. \tag{3}$$

However, a substitution back into equation (1) quickly shows that the only solution of this form is the trivial solution with $c_1 = c_2 = 0$.

Another suggestion might be made from our previous experience; that is, to try to find a second solution of the form

$$X_2 = \binom{c_1(t)}{c_2(t)} e^{2t}, \tag{4}$$

where we are essentially using a variation of parameters technique. Direct substitution of (4) into (1) gives

$$\binom{c_1(t)}{c_2(t)} 2\, e^{2t} + \binom{c_1'(t)}{c_2'(t)} e^{2t} = \begin{pmatrix} 0 & 1 \\ -4 & 4 \end{pmatrix} \binom{c_1(t)}{c_2(t)} e^{2t}.$$

We may rewrite this system of equations in the form

$$\binom{c_1'(t)}{c_2'(t)} = \begin{pmatrix} -2 & 1 \\ -4 & 2 \end{pmatrix} \binom{c_1(t)}{c_2(t)}. \tag{5}$$

System (5) can be rewritten

$$
\begin{aligned}
c_1'(t) &= -2c_1(t) + c_2(t), \\
c_2'(t) &= -4c_1(t) + 2c_2(t),
\end{aligned}
\tag{6}
$$

from which we conclude that $c_2'(t) = 2c_1'(t)$. Integrating, we obtain $c_2(t) = 2c_1(t) + a$, for arbitrary constant $a$. Substituting back into the first equation of (6) gives

$$c_1'(t) = a, \qquad \text{or} \qquad c_1(t) = at + b.$$

Thus we obtain a set of solutions of equations (5)

$$c_1(t) = at + b, \qquad \text{and} \qquad c_2(t) = 2at + 2b + a,$$

with arbitrary constant values for $a$ and $b$. Equation (4) now becomes

$$X_2 = \begin{pmatrix} 1 \\ 2 \end{pmatrix} at\, e^{2t} + \begin{pmatrix} 1 \\ 2 \end{pmatrix} b\, e^{2t} + \begin{pmatrix} 0 \\ 1 \end{pmatrix} a\, e^{2t}.$$

If we were to choose $a = 0$ and $b = 1$, this solution would be the same as $X_1$. Instead, we will choose $a = 1$ and $b = 0$ to give

$$X_2 = \begin{pmatrix} 1 \\ 2 \end{pmatrix} t\, e^{2t} + \begin{pmatrix} 0 \\ 1 \end{pmatrix} e^{2t}.$$

At $t = 0$ the Wronskian

$$W\{X_1(0), X_2(0)\} = \begin{vmatrix} 1 & 0 \\ 2 & 1 \end{vmatrix} = 1 \neq 0,$$

so the two solutions are linearly independent.

The general solution of system (1) is therefore

$$X = c_1 \begin{pmatrix} 1 \\ 2 \end{pmatrix} e^{2t} + c_2 \left[ \begin{pmatrix} 1 \\ 2 \end{pmatrix} t\, e^{2t} + \begin{pmatrix} 0 \\ 1 \end{pmatrix} e^{2t} \right].$$

In retrospect we note that the guess we made earlier in equation (3), although incorrect, was nevertheless not very far from the truth. We are now in a position to make a more reasonable assumption about the nature of a second solution in the case of repeated roots.

**EXAMPLE (b):**   Solve the system

$$\begin{pmatrix} x \\ y \end{pmatrix}' = \begin{pmatrix} 8 & -1 \\ 4 & 12 \end{pmatrix} \begin{pmatrix} x \\ y \end{pmatrix}. \tag{7}$$

The eigenvalues are the roots of the equation

$$\begin{vmatrix} 8 - m & -1 \\ 4 & 12 - m \end{vmatrix} = m^2 - 20m + 100 = (m - 10)^2 = 0.$$

Therefore, one solution is given by the eigenvalue $m_1 = 10$. This solution is

$$X_1 = \begin{pmatrix} 1 \\ -2 \end{pmatrix} e^{10t}. \tag{8}$$

Guided by our experience in Example (a) we now seek a second solution of the form

$$X_2 = \begin{pmatrix} 1 \\ -2 \end{pmatrix} t\, e^{10t} + \begin{pmatrix} c_3 \\ c_4 \end{pmatrix} e^{10t}. \tag{9}$$

Substitution of (9) into system (7) yields

$$\binom{1}{-2}10t\,e^{10t} + \binom{1}{-2}e^{10t} + \binom{c_3}{c_4}10\,e^{10t}$$

$$= \begin{pmatrix} 8 & -1 \\ 4 & 12 \end{pmatrix}\binom{1}{-2}t\,e^{10t} + \begin{pmatrix} 8 & -1 \\ 4 & 12 \end{pmatrix}\binom{c_3}{c_4}e^{10t}.$$

We note that the terms involving $t\,e^{10t}$ cancel each other leaving us with

$$\begin{pmatrix} -2 & -1 \\ 4 & 2 \end{pmatrix}\binom{c_3}{c_4}e^{10t} = \binom{1}{-2}e^{10t},$$

or

$$\begin{pmatrix} -2 & -1 \\ 4 & 2 \end{pmatrix}\binom{c_3}{c_4} = \binom{1}{-2}. \tag{10}$$

One solution of system (10) is $c_3 = 0$ and $c_4 = -1$. Therefore,

$$X_2 = \binom{1}{-2}t\,e^{10t} + \binom{0}{-1}e^{10t}$$

is a second solution of system (7). The general solution of system (7) is

$$X = c_1\binom{1}{-2}e^{10t} + c_2\,e^{10t}\left[\binom{1}{-2}t + \binom{0}{-1}\right].$$

**EXAMPLE (c):** Solve the system

$$D^2 y + (D-1)v = 0,$$

$$(2D-1)y + (D-1)w = 0, \tag{11}$$

$$(D+3)y + (D-4)v + 3w = 0.$$

In order to reduce (11) to a system of first-order equations, we let $Dy = u$. Then system (11) can be written

$$Du = u - 3v + 3w + 3y,$$

$$Dv = -u + 4v - 3w - 3y,$$

$$Dw = -2u \qquad + w + y, \tag{12}$$

$$Dy = u.$$

The matrix of system (12) has characteristic equation $m(m-4)(m-1)^2 = 0$. The eigenvalues $m_1 = 0$, $m_2 = 4$, and $m_3 = 1$ give rise to solutions

$$X_1 = \begin{pmatrix} 0 \\ 0 \\ -1 \\ 1 \end{pmatrix}, \qquad X_2 = \begin{pmatrix} 12 \\ -16 \\ -7 \\ 3 \end{pmatrix}e^{4t}, \qquad X_3 = \begin{pmatrix} 0 \\ 1 \\ 1 \\ 0 \end{pmatrix}e^{t}.$$

If we assume that the repeated root $m_3 = 1$ will yield a solution of the form

$$X_4 = \begin{pmatrix} 0 \\ 1 \\ 1 \\ 0 \end{pmatrix} t e^t + \begin{pmatrix} c_5 \\ c_6 \\ c_7 \\ c_8 \end{pmatrix} e^t,$$

a direct substitution into (12) will yield one set of values for the constants $c_5$ to $c_8$,

$$\begin{pmatrix} c_5 \\ c_6 \\ c_7 \\ c_8 \end{pmatrix} = \begin{pmatrix} -1 \\ 0 \\ 1 \\ -1 \end{pmatrix}.$$

Thus the desired solution is

$$X_4 = \begin{pmatrix} 0 \\ 1 \\ 1 \\ 0 \end{pmatrix} t e^t + \begin{pmatrix} -1 \\ 0 \\ 1 \\ -1 \end{pmatrix} e^t.$$

Finally, the solution of system (11) is

$$\begin{pmatrix} v \\ w \\ y \end{pmatrix} = c_1 \begin{pmatrix} 1 \\ 1 \\ 0 \end{pmatrix} e^t + c_2 \left[ \begin{pmatrix} 1 \\ 1 \\ 0 \end{pmatrix} t e^t + \begin{pmatrix} 0 \\ 1 \\ -1 \end{pmatrix} e^t \right] + c_3 \begin{pmatrix} 0 \\ -1 \\ 1 \end{pmatrix} + c_4 \begin{pmatrix} -16 \\ -7 \\ 3 \end{pmatrix} e^{4t}.$$

The student is asked to fill in the details of this example in exercise 10 below.

EXAMPLE (d): We now consider an example in which a repeated root of the characteristic equation of a matrix gives rise to two linearly independent eigenvectors of that matrix and thus avoids the complications encountered in Examples (a), (b), and (c) above.

The eigenvalues of the matrix of the system

$$\begin{pmatrix} x \\ y \\ z \end{pmatrix}' = \begin{pmatrix} 0 & 1 & 1 \\ 1 & 0 & 1 \\ 1 & 1 & 0 \end{pmatrix} \begin{pmatrix} x \\ y \\ z \end{pmatrix} \tag{13}$$

are the roots of the equation

$$\begin{vmatrix} -m & 1 & 1 \\ 1 & -m & 1 \\ 1 & 1 & -m \end{vmatrix} = -(m + 1)^2(m - 2) = 0.$$

For the repeated root $m = -1$ we seek nontrivial solutions of the system

$$\begin{pmatrix} 1 & 1 & 1 \\ 1 & 1 & 1 \\ 1 & 1 & 1 \end{pmatrix} \begin{pmatrix} c_1 \\ c_2 \\ c_3 \end{pmatrix} = O,$$

that is, solutions of $c_1 + c_2 + c_3 = 0$. Thus

$$C = \begin{pmatrix} c_1 \\ c_2 \\ c_3 \end{pmatrix} = \begin{pmatrix} c_1 \\ c_2 \\ -c_1 - c_2 \end{pmatrix} = c_1 \begin{pmatrix} 1 \\ 0 \\ -1 \end{pmatrix} + c_2 \begin{pmatrix} 0 \\ 1 \\ -1 \end{pmatrix}.$$

It follows that both

$$\begin{pmatrix} 1 \\ 0 \\ -1 \end{pmatrix} e^{-t} \quad \text{and} \quad \begin{pmatrix} 0 \\ 1 \\ -1 \end{pmatrix} e^{-t}$$

are solutions of system (13).

The second eigenvalue, $m = 2$, requires us to find nontrivial solutions of the system of equations

$$\begin{pmatrix} -2 & 1 & 1 \\ 1 & -2 & 1 \\ 1 & 1 & -2 \end{pmatrix} \begin{pmatrix} c_1 \\ c_2 \\ c_3 \end{pmatrix} = O.$$

That is,

$$\begin{aligned} -2c_1 + c_2 + c_3 &= 0 \\ c_1 - 2c_2 + c_3 &= 0 \\ c_1 + c_2 - 2c_3 &= 0. \end{aligned}$$

Elementary elimination of $c_2$ from the first and last equations and $c_1$ from the second and third equations leave us with

$$c_1 = c_3 \quad \text{and} \quad c_2 = c_3$$

so that

$$C = \begin{pmatrix} c_1 \\ c_2 \\ c_3 \end{pmatrix} = c_1 \begin{pmatrix} 1 \\ 1 \\ 1 \end{pmatrix}.$$

Hence a third solution of system (13) is

$$\begin{pmatrix} 1 \\ 1 \\ 1 \end{pmatrix} e^{2t}.$$

The linear independence of the three solutions is established by examining their Wronskian at $t = 0$,

$$W\{X_1(0), X_2(0), X_3(0)\} = \begin{vmatrix} 1 & 0 & 1 \\ 0 & 1 & 1 \\ -1 & -1 & 1 \end{vmatrix} = 3 \neq 0.$$

The general solution is therefore

$$\begin{pmatrix} x \\ y \\ z \end{pmatrix} = \left[ c_1 \begin{pmatrix} 1 \\ 0 \\ -1 \end{pmatrix} + c_2 \begin{pmatrix} 0 \\ 1 \\ -1 \end{pmatrix} \right] e^{-t} + c_3 \begin{pmatrix} 1 \\ 1 \\ 1 \end{pmatrix} e^{2t}.$$

**EXAMPLE (e):**   Solve the system

$$\begin{pmatrix} x \\ y \\ z \end{pmatrix}' = \begin{pmatrix} 2 & 1 & 2 \\ 1 & 2 & 2 \\ 1 & 1 & 3 \end{pmatrix} \begin{pmatrix} x \\ y \\ z \end{pmatrix}. \tag{14}$$

The characteristic equation

$$\begin{vmatrix} 2-m & 1 & 2 \\ 1 & 2-m & 2 \\ 1 & 1 & 3-m \end{vmatrix} = -(m-1)^2(m-5) = 0$$

has roots 1 and 5. The repeated eigenvalue 1 gives rise to the equation $c_1 = -c_2 - 2c_3$. Consequently

$$C = \begin{pmatrix} c_1 \\ c_2 \\ c_3 \end{pmatrix} = \begin{pmatrix} -c_2 - 2c_3 \\ c_2 \\ c_3 \end{pmatrix} = c_2 \begin{pmatrix} -1 \\ 1 \\ 0 \end{pmatrix} + c_3 \begin{pmatrix} -2 \\ 0 \\ 1 \end{pmatrix},$$

and two solutions of system (14) are

$$\begin{pmatrix} -1 \\ 1 \\ 0 \end{pmatrix} e^t \quad \text{and} \quad \begin{pmatrix} -2 \\ -0 \\ 1 \end{pmatrix} e^t.$$

The eigenvalue 5 forces us to solve the system

$$\begin{aligned} -3c_1 + c_2 + 2c_3 &= 0 \\ c_1 - 3c_2 + 2c_3 &= 0 \\ c_1 + c_2 - 2c_3 &= 0, \end{aligned}$$

which by elementary elimination yields $c_1 = c_2 = c_3$ and the eigenvector

$\begin{pmatrix} 1 \\ 1 \\ 1 \end{pmatrix}$. The Wronskian of the three solutions

$$\begin{pmatrix} -1 \\ 1 \\ 0 \end{pmatrix} e^t, \qquad \begin{pmatrix} -2 \\ 0 \\ 1 \end{pmatrix} e^t, \qquad \begin{pmatrix} 1 \\ 1 \\ 1 \end{pmatrix} e^{5t}$$

at $t = 0$ has value 4. Thus the general solution of system (14) is

$$\begin{pmatrix} x \\ y \\ z \end{pmatrix} = c_1 \begin{pmatrix} -1 \\ 1 \\ 0 \end{pmatrix} e^t + c_2 \begin{pmatrix} -2 \\ 0 \\ 1 \end{pmatrix} e^t + c_3 \begin{pmatrix} 1 \\ 1 \\ 1 \end{pmatrix} e^{5t}.$$

## Exercises

In exercises 1 through 9, solve the system $X' = AX$.

1. $A = \begin{pmatrix} 4 & 1 \\ -4 & 8 \end{pmatrix}$.      ANS.   $X = c_1 \begin{pmatrix} 1 \\ 2 \end{pmatrix} e^{6t} + c_2 \left[ \begin{pmatrix} 1 \\ 2 \end{pmatrix} t + \begin{pmatrix} 0 \\ 1 \end{pmatrix} \right] e^{6t}$.

2. $A = \begin{pmatrix} 4 & -9 \\ 4 & -8 \end{pmatrix}$.      ANS.   $X = c_1 \begin{pmatrix} 3 \\ 2 \end{pmatrix} e^{-2t} + c_2 \left[ \begin{pmatrix} 6 \\ 4 \end{pmatrix} t + \begin{pmatrix} 1 \\ 0 \end{pmatrix} \right] e^{-2t}$.

3. $A = \begin{pmatrix} 2 & -1 \\ 4 & 6 \end{pmatrix}$.      ANS.   $X = c_1 \begin{pmatrix} 1 \\ -2 \end{pmatrix} e^{4t} + c_2 \left[ \begin{pmatrix} 1 \\ -2 \end{pmatrix} t + \begin{pmatrix} 0 \\ -1 \end{pmatrix} \right] e^{4t}$.

4. $A = \begin{pmatrix} 1 & -2 \\ 2 & -3 \end{pmatrix}$.      ANS.   $X = c_1 \begin{pmatrix} 1 \\ 1 \end{pmatrix} e^{-t} + c_2 \left[ \begin{pmatrix} 2 \\ 2 \end{pmatrix} t + \begin{pmatrix} 2 \\ 1 \end{pmatrix} \right] e^{-t}$.

5. $A = \begin{pmatrix} 1 & 2 & -1 \\ 0 & 1 & 1 \\ 0 & 0 & 2 \end{pmatrix}$.    ANS.   $X = c_1 \begin{pmatrix} 1 \\ 1 \\ 1 \end{pmatrix} e^{2t} + c_2 \begin{pmatrix} 1 \\ 0 \\ 0 \end{pmatrix} e^t + c_3 \left[ \begin{pmatrix} 2 \\ 0 \\ 0 \end{pmatrix} t + \begin{pmatrix} 0 \\ 1 \\ 0 \end{pmatrix} \right] e^t$.

6. $A = \begin{pmatrix} 2 & 1 & -1 \\ 0 & -1 & 2 \\ 0 & 0 & -1 \end{pmatrix}$.

      ANS.   $X = c_1 \begin{pmatrix} 1 \\ 0 \\ 0 \end{pmatrix} e^{2t} + c_2 \begin{pmatrix} 1 \\ -3 \\ 0 \end{pmatrix} e^{-t} + c_3 \left[ \begin{pmatrix} 6 \\ -18 \\ 0 \end{pmatrix} t + \begin{pmatrix} -1 \\ 0 \\ -9 \end{pmatrix} \right] e^{-t}$.

**7.** $A = \begin{pmatrix} 0 & 3 & 1 \\ 1 & 2 & 1 \\ 1 & 3 & 0 \end{pmatrix}$.  ANS. $X = \left[ c_1 \begin{pmatrix} 1 \\ 0 \\ -1 \end{pmatrix} + c_2 \begin{pmatrix} 0 \\ 1 \\ -3 \end{pmatrix} \right] e^{-t} + \left[ c_3 \begin{pmatrix} 1 \\ 1 \\ 1 \end{pmatrix} e^{4t} \right]$.

**8.** $A = \begin{pmatrix} 12 & 2 & -2 \\ 5 & 3 & -1 \\ 5 & 1 & 1 \end{pmatrix}$.  ANS. $X = c_1 \begin{pmatrix} 2 \\ 1 \\ 1 \end{pmatrix} e^{12t} + \left[ c_2 \begin{pmatrix} 1 \\ 0 \\ 5 \end{pmatrix} + c_3 \begin{pmatrix} 1 \\ -5 \\ 0 \end{pmatrix} \right] e^{2t}$.

**9.** $A = \begin{pmatrix} 0 & -1 & 3 \\ 2 & -3 & 3 \\ 2 & -1 & 1 \end{pmatrix}$.  ANS. $X = c_1 \begin{pmatrix} 1 \\ 1 \\ 1 \end{pmatrix} e^{2t} + \left[ c_2 \begin{pmatrix} 3 \\ 0 \\ -2 \end{pmatrix} + c_3 \begin{pmatrix} 1 \\ 2 \\ 0 \end{pmatrix} \right] e^{-2t}$.

**10.** Complete the details in Example (c) of this section.

**11.** Discuss in complete detail the possible solutions of the system $X' = AX$ if $A$ is the diagonal matrix

$$A = \begin{pmatrix} a & 0 & 0 \\ 0 & b & 0 \\ 0 & 0 & c \end{pmatrix}.$$

**12.** Consider the system $X' = AX$ for

$$A = \begin{pmatrix} a & b \\ c & d \end{pmatrix}.$$

(a) Show that the characteristic equation of $A$ has a repeated root only if $(a - d)^2 + 4bc = 0$.

(b) Show that if $a \neq d$ and if $(a - d)^2 + 4bc = 0$ then the complete solution of the system is

$$X = c_1 \begin{pmatrix} 2b \\ d - a \end{pmatrix} e^{\frac{1}{2}(a+d)t} + c_2 \left[ \begin{pmatrix} 2b \\ d - a \end{pmatrix} t + \begin{pmatrix} 0 \\ 2 \end{pmatrix} \right] e^{\frac{1}{2}(a+d)t}.$$

(c) Completely discuss the solution in case

$$(a - d)^2 + 4bc = 0 \text{ and } a = d.$$

## 82. Nonhomogeneous systems

Now that we have some understanding of homogeneous systems with constant coefficients we turn our attention to systems that are nonhomogeneous. Consider the system

$$X' = AX + B, \tag{1}$$

where $A$ is a constant $n \times n$ matrix and $B$ is a vector function of $t$. Theorem 29 on page 251 indicates that we need to find a particular solution $X_p$ of system (1) and add it to the general solution of the associated homogeneous system. We will use a variation of parameters technique to find the particular solution $X_p$.

EXAMPLE (a): Consider the system

$$\begin{pmatrix} x \\ y \end{pmatrix}' = \begin{pmatrix} 0 & 1 \\ -2 & 3 \end{pmatrix} \begin{pmatrix} x \\ y \end{pmatrix} + \begin{pmatrix} f(t) \\ g(t) \end{pmatrix}. \tag{2}$$

In Example (a) of Section 79 we found the general solution of the homogeneous system

$$\begin{pmatrix} x \\ y \end{pmatrix}' = \begin{pmatrix} 0 & 1 \\ -2 & 3 \end{pmatrix} \begin{pmatrix} x \\ y \end{pmatrix} \tag{3}$$

to be

$$\begin{pmatrix} x \\ y \end{pmatrix}_c = a_1 \begin{pmatrix} 1 \\ 1 \end{pmatrix} e^t + a_2 \begin{pmatrix} 1 \\ 2 \end{pmatrix} e^{2t}, \tag{4}$$

where $a_1$ and $a_2$ are arbitrary constants.

We now seek a solution of system (2) of the form

$$\begin{pmatrix} x \\ y \end{pmatrix}_p = a_1(t) \begin{pmatrix} 1 \\ 1 \end{pmatrix} e^t + a_2(t) \begin{pmatrix} 1 \\ 2 \end{pmatrix} e^{2t}. \tag{5}$$

Direct substitution into (2) gives

$$a_1(t) \begin{pmatrix} 1 \\ 1 \end{pmatrix} e^t + 2a_2(t) \begin{pmatrix} 1 \\ 2 \end{pmatrix} e^{2t} + a_1'(t) \begin{pmatrix} 1 \\ 1 \end{pmatrix} e^t + a_2'(t) \begin{pmatrix} 1 \\ 2 \end{pmatrix} e^{2t}$$

$$= \begin{pmatrix} 0 & 1 \\ -2 & 3 \end{pmatrix} \begin{pmatrix} 1 \\ 1 \end{pmatrix} a_1(t) e^t + \begin{pmatrix} 0 & 1 \\ -2 & 3 \end{pmatrix} \begin{pmatrix} 1 \\ 2 \end{pmatrix} a_2(t) e^{2t} + \begin{pmatrix} f(t) \\ g(t) \end{pmatrix}, \tag{6}$$

or more simply

$$a_1'(t) \begin{pmatrix} 1 \\ 1 \end{pmatrix} e^t + a_2'(t) \begin{pmatrix} 1 \\ 2 \end{pmatrix} e^{2t} = \begin{pmatrix} f(t) \\ g(t) \end{pmatrix}. \tag{7}$$

The other terms in (6) cancel each other precisely because $\begin{pmatrix} x \\ y \end{pmatrix}_c$ is a solution of the homogeneous system (3). Equation (7) can now be written

$$\begin{pmatrix} 1 & 1 \\ 1 & 2 \end{pmatrix} \begin{pmatrix} a_1'(t) e^t \\ a_2'(t) e^{2t} \end{pmatrix} = \begin{pmatrix} f(t) \\ g(t) \end{pmatrix}.$$

Using Cramer's rule, we find

$$a'_1(t) e^t = \frac{\begin{vmatrix} f(t) & 1 \\ g(t) & 2 \end{vmatrix}}{\begin{vmatrix} 1 & 1 \\ 1 & 2 \end{vmatrix}} = 2f(t) - g(t),$$

$$a'_2(t) e^{2t} = \frac{\begin{vmatrix} 1 & f(t) \\ 1 & g(t) \end{vmatrix}}{\begin{vmatrix} 1 & 1 \\ 1 & 2 \end{vmatrix}} = g(t) - f(t).$$

Thus

$$a'_1(t) = [2f(t) - g(t)] e^{-t},$$
$$a'_2(t) = [g(t) - f(t)] e^{-2t}.$$

If, for example, $f(t) = e^t$ and $g(t) = 1$, we have

$$a'_1(t) = (2 e^t - 1) e^{-t} = 2 - e^{-t},$$
$$a'_2(t) = (1 - e^t) e^{-2t} = e^{-2t} - e^{-t},$$

so that

$$a_1(t) = 2t + e^{-t},$$
$$a_2(t) = -\tfrac{1}{2} e^{-2t} + e^{-t}.$$

The particular solution (5) is

$$\begin{pmatrix} x \\ y \end{pmatrix}_p = (2t + e^{-t}) \begin{pmatrix} 1 \\ 1 \end{pmatrix} e^t + (-\tfrac{1}{2} e^{-2t} + e^{-t}) \begin{pmatrix} 1 \\ 2 \end{pmatrix} e^{2t},$$

or

$$X_p = \begin{pmatrix} x \\ y \end{pmatrix}_p = (2t e^t + 1) \begin{pmatrix} 1 \\ 1 \end{pmatrix} + (e^t - \tfrac{1}{2}) \begin{pmatrix} 1 \\ 2 \end{pmatrix}.$$

The general solution of system (2) for $f(t) = e^t$ and $g(t) = 1$ is therefore

$$X = a_1 \begin{pmatrix} 1 \\ 1 \end{pmatrix} e^t + a_2 \begin{pmatrix} 1 \\ 2 \end{pmatrix} e^{2t} + (2t e^t + 1) \begin{pmatrix} 1 \\ 1 \end{pmatrix} + (e^t - \tfrac{1}{2}) \begin{pmatrix} 1 \\ 2 \end{pmatrix}.$$

EXAMPLE (b): Solve the system

$$X' = AX + B \qquad \text{for } A = \begin{pmatrix} 2 & 1 \\ -4 & 2 \end{pmatrix} \text{ and } B = \begin{pmatrix} 3 e^{2t} \\ t e^{2t} \end{pmatrix}. \tag{8}$$

The associated homogeneous problem is the same as Example (b) of Section 80. The general solution of the homogeneous system is

$$X_c = e^{2t}\left[ b_1\left\{ \begin{pmatrix} 1 \\ 0 \end{pmatrix} \cos 2t - \begin{pmatrix} 0 \\ 2 \end{pmatrix} \sin 2t \right\} + b_2\left\{ \begin{pmatrix} 0 \\ 2 \end{pmatrix} \cos 2t + \begin{pmatrix} 1 \\ 0 \end{pmatrix} \sin 2t \right\} \right]. \quad (9)$$

We seek a particular solution of the nonhomogeneous system of the form

$$X_p = e^{2t}\left[ b_1(t)\left\{ \begin{pmatrix} 1 \\ 0 \end{pmatrix} \cos 2t - \begin{pmatrix} 0 \\ 2 \end{pmatrix} \sin 2t \right\} + b_2(t)\left\{ \begin{pmatrix} 0 \\ 2 \end{pmatrix} \cos 2t + \begin{pmatrix} 1 \\ 0 \end{pmatrix} \sin 2t \right\} \right].$$

Omitting the terms that cancel one another when $X_p$ is substituted into (8), we have

$$e^{2t}\left[ b_1'(t)\left\{ \begin{pmatrix} 1 \\ 0 \end{pmatrix} \cos 2t - \begin{pmatrix} 0 \\ 2 \end{pmatrix} \sin 2t \right\} + b_2'(t)\left\{ \begin{pmatrix} 0 \\ 2 \end{pmatrix} \cos 2t + \begin{pmatrix} 1 \\ 0 \end{pmatrix} \sin 2t \right\} \right]$$

$$= \begin{pmatrix} 3\,e^{2t} \\ t\,e^{2t} \end{pmatrix}.$$

We may rewrite this system in the form

$$\begin{pmatrix} \cos 2t & \sin 2t \\ -2\sin 2t & 2\cos 2t \end{pmatrix} \begin{pmatrix} b_1'(t) \\ b_2'(t) \end{pmatrix} = \begin{pmatrix} 3 \\ t \end{pmatrix}.$$

Solving for $b_1'(t)$ and $b_2'(t)$ yields

$$b_1'(t) = \frac{1}{2}\begin{vmatrix} 3 & \sin 2t \\ t & 2\cos 2t \end{vmatrix} = \frac{1}{2}(6\cos 2t - t\sin 2t),$$

$$b_2'(t) = \frac{1}{2}\begin{vmatrix} \cos 2t & 3 \\ -2\sin 2t & t \end{vmatrix} = \frac{1}{2}(t\cos 2t + 6\sin 2t).$$

Integration of these functions gives

$$b_1(t) = \tfrac{1}{8}(2t\cos 2t + 11\sin 2t),$$

$$b_2(t) = \tfrac{1}{8}(2t\sin 2t - 11\cos 2t).$$

One particular solution of system (8) is therefore

$$X_p = e^{2t}\begin{pmatrix} b_1(t)\cos 2t + b_2(t)\sin 2t \\ -2b_1(t)\sin 2t + 2b_2(t)\cos 2t \end{pmatrix} = e^{2t}\begin{pmatrix} \tfrac{1}{4}t \\ -\tfrac{11}{4} \end{pmatrix},$$

or more simply

$$X_p = \tfrac{1}{4}e^{2t}\begin{pmatrix} t \\ -11 \end{pmatrix}.$$

The general solution of system (8) is

$$X = X_c + \tfrac{1}{4}e^{2t}\begin{pmatrix} t \\ -11 \end{pmatrix},$$

where $X_c$ is given in equation (9).

## Exercises

Find the general solution of each of the following systems.

1. $\begin{pmatrix} x \\ y \end{pmatrix}' = \begin{pmatrix} 0 & 1 \\ -2 & 3 \end{pmatrix} \begin{pmatrix} x \\ y \end{pmatrix} + \begin{pmatrix} e^t \\ 2 \end{pmatrix}.$   See Example (a), Section 82.

$$\text{ANS.} \quad X_p = t\,e^t \begin{pmatrix} 2 \\ 2 \end{pmatrix} + e^t \begin{pmatrix} 1 \\ 2 \end{pmatrix} + \begin{pmatrix} 1 \\ 0 \end{pmatrix}.$$

2. $\begin{pmatrix} x \\ y \end{pmatrix}' = \begin{pmatrix} 2 & 1 \\ -4 & 2 \end{pmatrix} \begin{pmatrix} x \\ y \end{pmatrix} + \begin{pmatrix} t\,e^{2t} \\ -e^{2t} \end{pmatrix}.$   See Example (b), Section 82.

$$\text{ANS.} \quad X_p = t\,e^{2t} \begin{pmatrix} 0 \\ -1 \end{pmatrix}.$$

3. $\begin{pmatrix} x \\ y \end{pmatrix}' = \begin{pmatrix} 3 & 3 \\ -1 & -1 \end{pmatrix} \begin{pmatrix} x \\ y \end{pmatrix} + \begin{pmatrix} t \\ 1 \end{pmatrix}.$   $\text{ANS.} \quad X_p = \frac{1}{8} \begin{pmatrix} -2t^2 - 18t - 6 \\ 2t^2 + 14t \end{pmatrix}.$

4. $\begin{pmatrix} x \\ y \end{pmatrix}' = \begin{pmatrix} 3 & 3 \\ -1 & -1 \end{pmatrix} \begin{pmatrix} x \\ y \end{pmatrix} + \begin{pmatrix} e^{-t} \\ e^{2t} \end{pmatrix}.$   See exercise 3.

$$\text{ANS.} \quad X_p = -\frac{1}{3} e^{-t} \begin{pmatrix} 0 \\ 1 \end{pmatrix} - \frac{1}{4} e^{2t} \begin{pmatrix} 3 - 6t \\ -3 + 2t \end{pmatrix}.$$

5. $\begin{pmatrix} x \\ y \end{pmatrix}' = \begin{pmatrix} 0 & 1 \\ -2 & 3 \end{pmatrix} \begin{pmatrix} x \\ y \end{pmatrix} + \begin{pmatrix} 0 \\ 3 \end{pmatrix} e^t.$   $\text{ANS.} \quad X_p = -3t\,e^t \begin{pmatrix} 1 \\ 1 \end{pmatrix} - 3\,e^t \begin{pmatrix} 1 \\ 2 \end{pmatrix}.$

6. $\begin{pmatrix} x \\ y \end{pmatrix}' = \begin{pmatrix} 4 & 1 \\ -4 & 8 \end{pmatrix} \begin{pmatrix} x \\ y \end{pmatrix} + \begin{pmatrix} 1 \\ 6t \end{pmatrix} e^{6t}.$   $\text{ANS.} \quad X_p = \begin{pmatrix} t^3 - t^2 + t \\ 2t^3 + t^2 \end{pmatrix} e^{6t}.$

## 83. Arms races

An interesting application that leads to a system of linear differential equations is the study of arms races. The presentation made here is often called the Richardson model since it was first proposed by the English meteorologist L. F. Richardson (1881–1953).*

Consider the problem of two countries with expenditures for armaments $x$ and $y$ measured in billions of dollars. We presume that $x$ and $y$ are functions of time measured in years. The Richardson model then makes the following assumptions:

---

* See, for instance, T. L. Saaty, *Mathematical Models of Arms Control and Disarmament* (New York: John Wiley & Sons, Inc., 1968).

(a) The expenditure for armaments of each country will increase at a rate that is proportional to the other country's expenditure.
(b) The expenditure for armaments of each country will decrease at a rate that is proportional to its own expenditure.
(c) The rate of change of arms expenditure for a country has a constant component that measures the level of antagonism of that country toward the other.
(d) The effects of the three previous assumptions are additive.

These assumptions lead to the system

$$\frac{dx}{dt} = ay - px + r,$$
$$\frac{dy}{dt} = bx - qy + s. \tag{1}$$

The constants $a$, $b$, $p$, and $q$ are positive, but the numbers $r$ and $s$ may have any value, positive values arising if the countries have internal attitudes of distrust for each other.

In matrix notation the problem may be written

$$X' = AX + B, \qquad X(0) = \begin{pmatrix} x_0 \\ y_0 \end{pmatrix}, \tag{2}$$

where

$$X(t) = \begin{pmatrix} x(t) \\ y(t) \end{pmatrix}, \qquad A = \begin{pmatrix} -p & a \\ b & -q \end{pmatrix}, \qquad B = \begin{pmatrix} r \\ s \end{pmatrix}.$$

As we have seen, the nature of the solutions of the system will depend on the eigenvalues of the matrix $A$, that is, on the roots of the characteristic equation

$$\begin{vmatrix} -p - m & a \\ b & -q - m \end{vmatrix} = m^2 + (p + q)m + (pq - ab) = 0.$$

These roots are

$$\frac{-(p + q) \pm \sqrt{(p + q)^2 - 4(pq - ab)}}{2} = \frac{-(p + q) \pm \sqrt{(p - q)^2 + 4ab}}{2},$$

and, since $a$ and $b$ are positive, the eigenvalues are real and distinct. Because $p > 0$ and $q > 0$, it follows that if $pq - ab > 0$, then the two eigenvalues are both negative, but if $pq - ab < 0$, then the eigenvalues will have opposite signs. The presence of a positive eigenvalue is disturbing since it will lead to an exponential function that becomes unbounded as time increases, a situation that may result in a runaway arms race.

We now examine several different examples to illustrate the possible consequences of Richardson's model.

EXAMPLE (a): Consider a situation in which the parameters in equations (2) are $a = 4$, $b = 2$, $p = 3$, $q = 1$, $r = 2$, $s = 2$, $x_0 = 4$, and $y_0 = 1$, that is

$$X' = \begin{pmatrix} -3 & 4 \\ 2 & -1 \end{pmatrix} X + \begin{pmatrix} 2 \\ 2 \end{pmatrix} \quad \text{and} \quad X(0) = \begin{pmatrix} 4 \\ 1 \end{pmatrix}.$$

The characteristic equation of the matrix is

$$\begin{vmatrix} -3 - m & 4 \\ 2 & -1 - m \end{vmatrix} = m^2 + 4m - 5 = 0,$$

so that the eigenvalues are $m_1 = 1$ and $m_2 = -5$.

For $m_1 = 1$, we compute the nontrivial solutions of the system

$$\begin{pmatrix} -4 & 4 \\ 2 & -2 \end{pmatrix} \begin{pmatrix} c_1 \\ c_2 \end{pmatrix} = O.$$

Thus $c_1 = c_2$. One such solution is obtained by taking $c_1 = 1$ to give the eigenvector $\begin{pmatrix} 1 \\ 1 \end{pmatrix}$.

For $m_2 = -5$, the system

$$\begin{pmatrix} 2 & 4 \\ 2 & 4 \end{pmatrix} \begin{pmatrix} c_1 \\ c_2 \end{pmatrix} = O$$

requires $c_1 + 2c_2 = 0$. Taking $c_2 = -1$ yields the eigenvector $\begin{pmatrix} 2 \\ -1 \end{pmatrix}$.

The general solution of the homogeneous system $X' = AX$ is therefore

$$X(t) = c_1 \begin{pmatrix} 1 \\ 1 \end{pmatrix} e^t + c_2 \begin{pmatrix} 2 \\ -1 \end{pmatrix} e^{-5t}.$$

The nonhomogeneous system $X' = AX + B$ has a constant solution of the form $\begin{pmatrix} e \\ f \end{pmatrix}$. Substitution into the system gives

$$\begin{pmatrix} -3 & 4 \\ 2 & -1 \end{pmatrix} \begin{pmatrix} e \\ f \end{pmatrix} + \begin{pmatrix} 2 \\ 2 \end{pmatrix} = O,$$

a system with solution $\begin{pmatrix} -2 \\ -2 \end{pmatrix}$. Thus the general solution of the nonhomogeneous system is

$$X(t) = c_1 \begin{pmatrix} 1 \\ 1 \end{pmatrix} e^t + c_2 \begin{pmatrix} 2 \\ -1 \end{pmatrix} e^{-5t} + \begin{pmatrix} -2 \\ -2 \end{pmatrix}.$$

The initial condition $X(0) = \begin{pmatrix} 4 \\ 1 \end{pmatrix}$ now requires that

$$\begin{pmatrix} 4 \\ 1 \end{pmatrix} = c_1 \begin{pmatrix} 1 \\ 1 \end{pmatrix} + c_2 \begin{pmatrix} 2 \\ -1 \end{pmatrix} + \begin{pmatrix} -2 \\ -2 \end{pmatrix},$$

so that $c_1 = 4$ and $c_2 = 1$. The final solution is therefore

$$X(t) = 4 \begin{pmatrix} 1 \\ 1 \end{pmatrix} e^t + \begin{pmatrix} 2 \\ -1 \end{pmatrix} e^{-5t} + \begin{pmatrix} -2 \\ -2 \end{pmatrix},$$

or

$$x(t) = 4e^t + 2e^{-5t} - 2,$$
$$y(t) = 4e^t - e^{-5t} - 2.$$

We have a runaway arms race.

**EXAMPLE (b):** As a second example of an arms race, we take the following values for the parameters in equation (2): $a = 4, b = 2, p = 3, q = 1, r = -2, s = -2, x_0 = 2, y_0 = \frac{1}{2}$. The system of differential equations has the same solution as in Example (a) except for the sign of the particular solution. Thus the general solution is

$$X(t) = c_1 \begin{pmatrix} 1 \\ 1 \end{pmatrix} e^t + c_2 \begin{pmatrix} 2 \\ -1 \end{pmatrix} e^{-5t} + \begin{pmatrix} 2 \\ 2 \end{pmatrix}.$$

The initial conditions now require that

$$\begin{pmatrix} 2 \\ \frac{1}{2} \end{pmatrix} = c_1 \begin{pmatrix} 1 \\ 1 \end{pmatrix} + c_2 \begin{pmatrix} 2 \\ -1 \end{pmatrix} + \begin{pmatrix} 2 \\ 2 \end{pmatrix},$$

from which we obtain $c_1 = -1$ and $c_2 = \frac{1}{2}$. The solution is

$$x(t) = -e^t + e^{-5t} + 2,$$
$$y(t) = -e^t - \tfrac{1}{2}e^{-5t} + 2,$$

and each party will eventually decrease its arms expenditure to zero, a condition of disarmament.

**EXAMPLE (c):** Let us now change the values of the parameters in system (2) to $a = 3, b = 1, p = 4, q = 2, r = 6, s = 1$ with initial conditions $x_0 = 0$ and $y_0 = 0$. The system to be solved becomes

$$\begin{pmatrix} x \\ y \end{pmatrix}' = \begin{pmatrix} -4 & 3 \\ 1 & -2 \end{pmatrix} \begin{pmatrix} x \\ y \end{pmatrix} + \begin{pmatrix} 6 \\ 1 \end{pmatrix}.$$

A particular solution of this system is the vector $\begin{pmatrix} 3 \\ 2 \end{pmatrix}$ and the eigenvalues of

the matrix are $-1$ and $-5$ with corresponding eigenvectors $\begin{pmatrix} 1 \\ 1 \end{pmatrix}$ and $\begin{pmatrix} 3 \\ -1 \end{pmatrix}$.

The general solution of the system is therefore

$$\begin{pmatrix} x \\ y \end{pmatrix} = c_1 \begin{pmatrix} 1 \\ 1 \end{pmatrix} e^{-t} + c_2 \begin{pmatrix} 3 \\ -1 \end{pmatrix} e^{-5t} + \begin{pmatrix} 3 \\ 2 \end{pmatrix}.$$

The initial conditions $x_0 = y_0 = 0$ require

$$\begin{pmatrix} 0 \\ 0 \end{pmatrix} = c_1 \begin{pmatrix} 1 \\ 1 \end{pmatrix} + c_2 \begin{pmatrix} 3 \\ -1 \end{pmatrix} + \begin{pmatrix} 3 \\ 2 \end{pmatrix},$$

an equation that is satisfied only if $c_1 = -\frac{9}{4}$ and $c_2 = -\frac{1}{4}$. Thus the solution of the initial value problem is

$$\begin{pmatrix} x \\ y \end{pmatrix} = \frac{-9}{4}\begin{pmatrix} 1 \\ 1 \end{pmatrix} e^{-t} - \frac{1}{4}\begin{pmatrix} 3 \\ -1 \end{pmatrix} e^{-5t} + \begin{pmatrix} 3 \\ 2 \end{pmatrix}. \tag{3}$$

It is also true that

$$\begin{pmatrix} x \\ y \end{pmatrix}' = \frac{9}{4}\begin{pmatrix} 1 \\ 1 \end{pmatrix} e^{-t} + \frac{5}{4}\begin{pmatrix} 3 \\ -1 \end{pmatrix} e^{-5t}. \tag{4}$$

We may now interpret equations (3) and (4) as an arms race with each party starting with zero expenditure but with $dx/dt = 6$ and $dy/dt = 1$, both positive quantities. Because of the negative exponents, the rates at which the expenditures are changing will tend toward zero and the arms expenditures will approach $x = 3$ and $y = 2$. There will be a stabilized arms race.

## Exercises

For Richardson's model as described by equations (2), solve the following special cases, noting in each exercise whether there will be a stable arms race, a runaway arms race, or disarmament.

1. $a = 2, b = 4, p = 5, q = 3, r = 1, s = 2, x_0 = 8, y_0 = 7$.
   ANS.  $x(t) = 4e^{-t} + 3e^{-7t} + 1$, $y(t) = 8e^{-t} - 3e^{-7t} + 2$. Stable arms race.
2. What effect does the changing of the initial values $x_0$ and $y_0$ have on the stability of the solution in exercise 1?
3. $a = 4, b = 4, p = 2, q = 2, r = 8, s = 2, x_0 = 5, y_0 = 2$.
   ANS.  $x(t) = e^{-6t} + 6e^{2t} - 2$, $y(t) = -e^{-6t} + 6e^{2t} - 3$. Runaway arms race.
4. Show that the solution in exercise 3 will remain unstable if the initial values are changed to any other nonnegative values.
5. For $a = 4, b = 4, p = 2, q = 2, r = -2, s = -2$, show that there will be disarmament if $x_0 + y_0 < 2$ and a runaway arms race if $x_0 + y_0 > 2$.

6. For $a = 4$, $b = 4$, $p = 2$, $q = 2$, $r > 0$, $s > 0$, show that there will be a runaway arms race for any nonnegative $x_0$ and $y_0$.

7. For $a = 4$, $b = 4$, $p = 2$, $q = 2$, $r < 0$, $s < 0$, show that there will be a runaway arms race if $x_0 + y_0 > \dfrac{-r - s}{2}$.

8. Show that if $pq - ab > 0$, $r > 0$, and $s > 0$, there will be a stable solution to the arms race.

9. Show that if $pq - ab < 0$, $r > 0$, and $s > 0$, there will be a runaway arms race.

10. Show that if $pq - ab > 0$, $r < 0$, and $s < 0$, there will be disarmament.

## 84. The Laplace transform

The Laplace operator can be used to transform a system of linear differential equations with constant coefficients into a system of algebraic equations.

EXAMPLE (a):   Solve the system of equations

$$x''(t) - x(t) + 5y'(t) = t, \tag{1}$$

$$y''(t) - 4y(t) - 2x'(t) = -2, \tag{2}$$

with the initial conditions

$$x(0) = 0, \qquad x'(0) = 0, \qquad y(0) = 0, \qquad y'(0) = 0. \tag{3}$$

Let $L\{x(t)\} = u(s)$ and $L\{y(t)\} = v(s)$. Then, application of the Laplace operator transforms the problem into that of solving a pair of simultaneous algebraic equations:

$$(s^2 - 1)u(s) + 5sv(s) = \frac{1}{s^2}, \tag{4}$$

$$-2su(s) + (s^2 - 4)v(s) = -\frac{2}{s}. \tag{5}$$

We solve equations (4) and (5) to obtain

$$u(s) = \frac{11s^2 - 4}{s^2(s^2 + 1)(s^2 + 4)}, \tag{6}$$

$$v(s) = \frac{-2s^2 + 4}{s(s^2 + 1)(s^2 + 4)}. \tag{7}$$

Seeking the inverse transforms of $u$ and $v$, we first expand the right members of (6) and (7) into partial fractions:

$$u(s) = -\frac{1}{s^2} + \frac{5}{s^2 + 1} - \frac{4}{s^2 + 4}, \tag{8}$$

$$v(s) = \frac{1}{s} - \frac{2s}{s^2 + 1} + \frac{s}{s^2 + 4}. \tag{9}$$

Since $x(t) = L^{-1}\{u(s)\}$ and $y(t) = L^{-1}\{v(s)\}$, we get the desired result

$$x(t) = -t + 5 \sin t - 2 \sin 2t, \tag{10}$$

$$y(t) = 1 - 2 \cos t + \cos 2t, \tag{11}$$

which is easily verified by direct substitution into (1), (2), and (3).

The foregoing procedure is simple in concept, but of course its practical use will depend on our ability to find the inverse transforms of $u(s)$ and $v(s)$. On the other hand, the use of the transform theory can help us gain insight into the general theory of systems of linear differential equations. We illustrate this idea in the following example.

EXAMPLE (b):   Solve the system

$$\frac{dx}{dt} = y + F(t), \tag{12}$$

$$\frac{dy}{dt} = x + G(t). \tag{13}$$

Here we presume that the transforms of the functions $x(t)$, $y(t)$, $F(t)$, and $G(t)$ all exist and are given by $u(s)$, $v(s)$, $f(s)$, and $g(s)$, respectively. We then have

$$su(s) - c_1 = v(s) + f(s), \tag{14}$$

$$sv(s) - c_2 = u(s) + g(s), \tag{15}$$

where $c_1$ and $c_2$ represent the initial values of $x(t)$ and $y(t)$. We can rewrite equations (14) and (15) in the form

$$su(s) - v(s) = c_1 + f(s), \tag{16}$$

$$-u(s) + sv(s) = c_2 + g(s). \tag{17}$$

Using Cramer's rule, the solution of the algebraic system (16) and (17) may be written

$$u(s) = \frac{\begin{vmatrix} c_1 + f(s) & -1 \\ c_2 + g(s) & s \end{vmatrix}}{s^2 - 1} = \frac{\begin{vmatrix} c_1 & -1 \\ c_2 & s \end{vmatrix}}{s^2 - 1} + \frac{\begin{vmatrix} f(s) & -1 \\ g(s) & s \end{vmatrix}}{s^2 - 1}, \tag{18}$$

$$v(s) = \frac{\begin{vmatrix} s & c_1 + f(s) \\ -1 & c_2 + g(s) \end{vmatrix}}{s^2 - 1} = \frac{\begin{vmatrix} s & c_1 \\ -1 & c_2 \end{vmatrix}}{s^2 - 1} + \frac{\begin{vmatrix} s & f(s) \\ -1 & g(s) \end{vmatrix}}{s^2 - 1}. \tag{19}$$

A careful examination of equations (18) and (19) reveals several important properties of the pair of functions, $x(t)$ and $y(t)$, we are seeking. First, each of these functions can be considered as the sum of two functions; that is,

$$x(t) = x_c(t) + x_p(t),$$

$$y(t) = y_c(t) + y_p(t).$$

The notation used is intended to remind us of a similar situation we met when we were dealing with a single linear differential equation with one dependent variable. We note that the pair of functions, $x_c(t)$ and $y_c(t)$, is a solution of the system (12) and (13) in the case where $F(t) = G(t) = 0$ (and of course $f(s) = g(s) = 0$). Moreover, each of the functions $x_c(t)$ and $y_c(t)$ involves the constants $c_1$ and $c_2$ that are the initial values of $x(t)$ and $y(t)$.

On the other hand, the functions $x_p(t)$ and $y_p(t)$, although independent of the initial conditions, are intimately related to the functions $F(t)$ and $G(t)$.

Although it is possible to obtain the functions $x(t)$ and $y(t)$ from (18) and (19) by using convolution integrals (see exercise 9 below), let us simplify the problem by considering a particular case.

**EXAMPLE (c):** Solve the system of equations (12) and (13) if $F(t) = 1$ and $G(t) = t$.

Equations (18) and (19) now become

$$u(s) = \frac{\begin{vmatrix} c_1 & -1 \\ c_2 & s \end{vmatrix}}{s^2 - 1} + \frac{\begin{vmatrix} \dfrac{1}{s} & -1 \\ \dfrac{1}{s^2} & s \end{vmatrix}}{s^2 - 1}, \tag{20}$$

$$v(s) = \frac{\begin{vmatrix} s & c_1 \\ -1 & c_2 \end{vmatrix}}{s^2 - 1} + \frac{\begin{vmatrix} s & \dfrac{1}{s} \\ -1 & \dfrac{1}{s^2} \end{vmatrix}}{s^2 - 1}. \tag{21}$$

These equations may be written

$$u(s) = \frac{c_1 + c_2}{2(s - 1)} + \frac{c_1 - c_2}{2(s + 1)} - \frac{1}{s^2} + \frac{1}{s - 1} - \frac{1}{s + 1}, \tag{22}$$

$$v(s) = \frac{c_1 + c_2}{2(s-1)} - \frac{c_1 - c_2}{2(s+1)} - \frac{2}{s} + \frac{1}{s-1} + \frac{1}{s+1}, \tag{23}$$

where we have taken care to keep the two parts of each of these functions separate. The inversion of equations (22) and (23) yields

$$x(t) = \frac{c_1 + c_2}{2} e^t + \frac{c_1 - c_2}{2} e^{-t} - t + e^t - e^{-t}, \tag{24}$$

$$y(t) = \frac{c_1 + c_2}{2} e^t - \frac{c_1 - c_2}{2} e^{-t} - 2 + e^t + e^{-t}. \tag{25}$$

The reader should now verify directly that the pair of functions, $x_c(t)$ and $y_c(t)$, which involve the initial values $c_1$ and $c_2$, is a solution of the system

$$\frac{dx}{dt} = y \quad \text{and} \quad \frac{dy}{dt} = x.$$

This system results from placing $F(t) = G(t) = 0$ in equations (12) and (13). The reader should also verify that the remaining part of the solution, $x_p(t)$ and $y_p(t)$, is a particular solution of the system (12) and (13) with $F(t) = 1$ and $G(t) = t$. Finally, it is easy to show that $x(0) = c_1$ and $y(0) = c_2$.

## *Exercises*

In exercises 1 through 8, use the Laplace transform method to solve the given system.

1. $x''(t) - 3x'(t) - y'(t) + 2y(t) = 14t + 3$,
   $x'(t) - 3x(t) + y'(t) = 1 ; x(0) = 0, x'(0) = 0, y(0) = 6.5$.
         ANS.    $x(t) = 2 - \frac{1}{2} e^t - \frac{1}{2} e^{3t} - e^{-2t}$,
                     $y(t) = 7t + 5 - e^t + \frac{5}{2} e^{-2t}$.

2. $2x'(t) + 2x(t) + y'(t) - y(t) = 3t$,
   $x'(t) + x(t) + y'(t) + y(t) = 1 ; x(0) = 1, y(0) = 3$.
         ANS.    $x(t) = t + 3 e^{-t} - 2 e^{-3t}$,
                     $y(t) = 1 - t + 2 e^{-3t}$.

3. $x'(t) - 2x(t) - y'(t) - y(t) = 6 e^{3t}$,
   $2x'(t) - 3x(t) + y'(t) - 3y(t) = 6 e^{3t} ; x(0) = 3, y(0) = 0$.
         ANS.    $x(t) = (1 + 2t) e^t + 2 e^{3t}$,
                     $y(t) = (1 - t) e^t - e^{3t}$.

4. $x''(t) + 2x(t) - y'(t) = 2t + 5$,
   $x'(t) - x(t) + y'(t) + y(t) = -2t - 1 ; x(0) = 3, x'(0) = 0, y(0) = -3$.
         ANS.    $x(t) = t + 2 + e^{-2t} + \sin t$,
                     $y(t) = 1 - t - 3 e^{-2t} - \cos t$.

5. The equations of Example (a) of Section 84 with initial conditions $x(0) = 0$, $x'(0) = 0, y(0) = 1, y'(0) = 0$.
         ANS.    $x(t) = -t - \frac{5}{3} \sin t + \frac{4}{3} \sin 2t$,
                     $y(t) = 1 + \frac{2}{3} \cos t - \frac{2}{3} \cos 2t$.

6. The equations of Example (a) of Section 84 with initial conditions $x(0) = 9$, $x'(0) = 2, y(0) = 1, y'(0) = 0$.

ANS. $x(t) = -t + 15 \cos t - 5 \sin t - 6 \cos 2t + 4 \sin 2t$,
$y(t) = 1 + 2 \cos t + 6 \sin t - 2 \cos 2t - 3 \sin 2t$.

7. $x''(t) + y'(t) - y(t) = 0$,
$2x'(t) - x(t) + z'(t) - z(t) = 0$,
$x'(t) + 3x(t) + y'(t) - 4y(t) + 3z(t) = 0$;
$x(0) = 0, x'(0) = 1, y(0) = 0, z(0) = 0$.

8. $x''(t) - x(t) + 5y'(t) = \beta(t)$,
$y''(t) - 4y(t) - 2x'(t) = 0$,
in which $\beta(t) = 6t$, $\quad 0 \le t \le 2$,
$\qquad\quad = 12$, $\qquad t > 2$;
$x(0) = 0, x'(0) = 0, y(0) = 0, y'(0) = 0$.

ANS. $x(t) = -2(3t - 5 \sin t + \sin 2t)$
$+ 2[3(t - 2) - 5 \sin (t - 2) + \sin 2(t - 2)]\alpha(t - 2)$,
$y(t) = 3 - 4 \cos t + \cos 2t$
$-[3 - 4 \cos (t - 2) + \cos 2(t - 2)]\alpha(t - 2)$.

9. Write the solution of the system of Example (b) in Section 84 in terms of convolution integrals.

10. Use the results of exercise 9 to obtain the solution of Example (c) in Section 84.

In exercises 11 and 12, write the solution in terms of convolution integrals.

11. $x'(t) - 2y(t) = F(t)$,
$y'(t) + x(t) = G(t); x(0) = 1, y(0) = 0$.

12. $2x'(t) + 3y(t) = F(t)$,
$y'(t) + 2x(t) = G(t); x(0) = 2, y(0) = 1$.

13. Consider the initial value problem
$x'(t) = ax + by + f(t)$,
$y'(t) = cx + dy + g(t); x(0) = c_1, y(0) = c_2$,
where $a, b, c, d$, and $c_1, c_2$ are constants. Use an argument similar to that of Example (b) of Section 84 to show that the solution, if it exists, should have the form

$$x(t) = x_c(t) + x_p(t),$$
$$y(t) = y_c(t) + y_p(t),$$

where $x_c(t)$ and $y_c(t)$ depend on $c_1$ and $c_2$ whereas $x_p(t)$ and $y_p(t)$ depend on $f(t)$ and $g(t)$.

14. For the initial value problem of exercise 13, state a reasonable definition for a first-order homogeneous linear system with constant coefficients. Now state a theorem that expresses the results of exercise 13 for the solution of a nonhomogeneous system.

15. Consider the initial value problem

$$x'(t) + 2y(t) = 0,$$
$$x''(t) + 2y'(t) + 2y(t) = 2 e^t;$$
$$x(0) = 1, x'(0) = 0, y(0) = 0.$$

(a) Show that the Laplace transform method produces

$$x = 3 - 2 e^t, \qquad y = e^t.$$

(b) Verify that these functions satisfy the differential equations but do not satisfy the initial conditions.

(c) By elementary elimination, show that a solution of the system of differential equations has the form

$$x = c_1 - 2 e^t, \qquad y = e^t,$$

and thus the initial conditions given are not compatible with the system of equations.

# Electric Circuits and Networks

## 85. Circuits

The basic laws governing the flow of electric current in a circuit or a network will be given here without derivation. The notation used is common to most texts in electrical engineering and is:

$t$ (sec) = time
$Q$ (coulombs) = quantity of electricity; for example, charge on a capacitor
$I$ (amperes) = current, time rate of flow of electricity
$E$ (volts) = electromotive force or voltage
$R$ (ohms) = resistance
$L$ (henrys) = inductance
$C$ (farads) = capacitance.

By the definition of $Q$ and $I$ it follows that

$$I(t) = Q'(t).$$

The current at each point in a network may be determined by solving the equations that result from applying Kirchhoff's laws:

(a) *The sum of the currents into (or away from) any point is zero.*
(b) *Around any closed path the sum of the instantaneous voltage drops in a specified direction is zero.*

A circuit is treated as a network containing only one closed path. Figure 29 exhibits an "*RLC* circuit" with some of the customary conventions for indicating various elements.

FIGURE 29

For a circuit, Kirchhoff's current law (a) indicates merely that the current is the same throughout. That law plays a larger role in networks, as we shall see later.

To apply Kirchhoff's voltage law (b), it is necessary to know the contributions of each of the idealized elements in Figure 29. The voltage drop across the resistance is $RI$, that across the inductance is $LI'(t)$, and that across the capacitor is $C^{-1}Q(t)$. The impressed electromotive force $E(t)$ is contributing a voltage rise.

Assume that at time $t = 0$ the switch shown in Figure 29 is to be closed. At $t = 0$ there is no current flowing, $I(0) = 0$ and, if the capacitor is initially without charge, $Q(0) = 0$. From Kirchhoff's law (b), we get the differential equation

$$LI'(t) + RI(t) + C^{-1}Q(t) = E(t), \tag{1}$$

in which

$$I(t) = Q'(t). \tag{2}$$

Equations (1) and (2), with the initial conditions

$$I(0) = 0, \qquad Q(0) = 0, \tag{3}$$

constitute the problem to be solved.

The function $I(t)$ may be eliminated from (1), (2), (3) to obtain the initial value problem

$$LQ''(t) + RQ'(t) + C^{-1}Q(t) = E(t); \qquad Q(0) = 0, Q'(0) = 0. \tag{4}$$

It follows that the circuit problem is equivalent to a problem in damped vibrations of a spring (Section 54). The resistance term $RQ'(t)$ corresponds to the damping term in vibration problems. The analogies between electrical and mechanical systems are useful in practice.

Initial value problems of the type given in equation (4) may be solved either by the general theory of linear equations with constant coefficients or by the use of the Laplace transform. We present here an example using each technique.

FIGURE 30

EXAMPLE (a):  In the $RL$ circuit shown in Figure 30, find the current $I(t)$ if the current at $t = 0$ is zero and $E$ is a constant.

From equations (1) and (3) we have

$$LI'(t) + RI(t) = E; \qquad I(0) = 0. \tag{5}$$

This first-order linear equation may be written

$$\left(D + \frac{R}{L}\right)I = \frac{E}{L}$$

for which the general solution is

$$I(t) = \frac{E}{R} + c_1 \exp\left(-\frac{R}{L}t\right).$$

The initial condition $I(0) = 0$ requires that

$$0 = \frac{E}{R} + c_1,$$

so that finally

$$I(t) = \frac{E}{R}\left[1 - \exp\left(\frac{-R}{L}t\right)\right].$$

EXAMPLE (b):  In the $RL$ circuit with the schematic diagram shown in Figure 30, let the switch be closed at $t = 0$. At some later time, $t = t_0$, the direct current element, the constant $E$, is to be removed from the circuit, which remains closed. Find the current for all $t > 0$.

The initial value problem to be solved is

$$LI'(t) + RI(t) = E(t); \qquad I(0) = 0, \tag{6}$$

$$E(t) = E[1 - \alpha(t - t_0)]. \tag{7}$$

Let the Laplace transform of $I(t)$ be $i(s)$. We know the transform of $E(t)$. Therefore we obtain the transformed problem

$$sLi(s) + Ri(s) = \frac{E}{s}[1 - \exp(-t_0 s)], \tag{8}$$

from which

$$i(s) = \frac{E[1 - \exp(-t_0 s)]}{s(sL + R)}. \tag{9}$$

Now

$$\frac{1}{s(sL + R)} = \frac{1}{R}\left(\frac{1}{s} - \frac{1}{s + RL^{-1}}\right),$$

so

$$i(s) = \frac{E}{R}\left(\frac{1}{s} - \frac{1}{s + RL^{-1}}\right)[1 - \exp(-t_0 s)].$$

Therefore

$$I(t) = \frac{E}{R}\left[1 - \alpha(t - t_0) - \exp\left(-\frac{R}{L}t\right) + \exp\left\{-\frac{R}{L}(t - t_0)\right\}\alpha(t - t_0)\right]. \tag{10}$$

The student should verify (10) and show that it can be written

$$\text{For } 0 \leq t \leq t_0, \qquad I(t) = \frac{E}{R}\left[1 - \exp\left(-\frac{Rt}{L}\right)\right]; \tag{11}$$

$$\text{For } t > t_0, \qquad I(t) = I(t_0)\exp\left[-\frac{R}{L}(t - t_0)\right]. \tag{12}$$

## 86. Simple networks

Systems of equations occur naturally in the application of Kirchhoff's laws to electric networks. We consider in this section two extremely simple networks to indicate how the techniques of Chapter 13 can be applied. In each example we will first use matrix techniques to determine the solution and then solve the problem again using the Laplace transformation.

**EXAMPLE (a):**   Determine the character of the currents $I_1(t)$, $I_2(t)$, and $I_3(t)$ in the network having the schematic diagram shown in Figure 31, under the assumption that when the switch is closed the currents are each zero.

FIGURE 31

In a network, we apply Kirchhoff's laws, page 285, to obtain a system of equations to determine the currents. Since there are three dependent variables, $I_1, I_2, I_3$, we need three equations.

From the current law it follows that

$$I_1 = I_2 + I_3. \tag{1}$$

Application of the voltage law to the circuit on the left in Figure 31 yields

$$R_1 I_1 + L_2 I_2' = E. \tag{2}$$

Using the voltage law on the outside circuit, we get

$$R_1 I_1 + R_3 I_3 + L_3 I_3' = E. \tag{3}$$

Still another equation can be obtained from the circuit on the right in Figure 31:

$$R_3 I_3 + L_3 I_3' - L_2 I_2' = 0. \tag{4}$$

Equation (4) also follows at once from equations (2) and (3); it may be used instead of either (2) or (3).

We wish to obtain the currents from the initial value problem consisting of equations (1), (2), and (3) and the conditions $I_1(0) = 0$, $I_2(0) = 0$, and $I_3(0) = 0$. One of the three initial conditions is redundant because of equation (1).

If we eliminate $I_1$ from equations (1), (2), (3) we can write

$$I_2' = -\frac{R_1}{L_2} I_2 - \frac{R_1}{L_2} I_3 + \frac{E}{L_2},$$

$$I_3' = -\frac{R_1}{L_3}I_2 - \frac{R_1 + R_3}{L_3}I_3 + \frac{E}{L_3},$$

or in matrix notation,

$$\begin{pmatrix} I_2 \\ I_3 \end{pmatrix}' = \begin{pmatrix} -\dfrac{R_1}{L_2} & -\dfrac{R_1}{L_2} \\ -\dfrac{R_1}{L_3} & -\dfrac{R_1 + R_3}{L_3} \end{pmatrix} \begin{pmatrix} I_2 \\ I_3 \end{pmatrix} + \begin{pmatrix} \dfrac{E}{L_2} \\ \dfrac{E}{L_3} \end{pmatrix} \qquad (5)$$

The characteristic equation of the matrix of system (5) is therefore

$$\begin{vmatrix} -\dfrac{R_1}{L_2} - m & -\dfrac{R_1}{L_2} \\ -\dfrac{R_1}{L_3} & -\dfrac{R_1 + R_3}{L_3} - m \end{vmatrix} = 0,$$

or

$$\Delta = L_2 L_3 m^2 + (R_1 L_2 + R_3 L_2 + R_1 L_3)m + R_1 R_3 = 0. \qquad (6)$$

We are interested in the factors of the characteristic polynomial $\Delta$. Equation (6) has no positive roots. The discriminant of $\Delta$ is

$$(R_1 L_2 + R_3 L_2 + R_1 L_3)^2 - 4L_2 L_3 R_1 R_3$$

and may be written

$$(R_1 L_2)^2 + 2R_1 L_2(R_3 L_2 + R_1 L_3) + (R_3 L_2 + R_1 L_3)^2 - 4L_2 L_3 R_1 R_3,$$

which equals

$$(R_1 L_2)^2 + 2R_1 L_2(R_3 L_2 + R_1 L_3) + (R_3 L_2 - R_1 L_3)^2$$

and is therefore positive. Thus we see that equation (6) has two distinct negative roots. Call them $(-a_1)$ and $(-a_2)$. It follows that

$$\Delta = L_2 L_3 (m + a_1)(m + a_2)$$

and that the eigenvalues of the matrix of system (5) are $(-a_1)$ and $(-a_2)$. Corresponding to these eigenvalues, we obtain eigenvectors

$$\begin{pmatrix} R_1 \\ a_1 L_2 - R_1 \end{pmatrix} \quad \text{and} \quad \begin{pmatrix} R_1 \\ a_2 L_2 - R_1 \end{pmatrix}. \qquad (7)$$

It follows that the general solution of the homogeneous system associated with (5) is

$$\begin{pmatrix} I_2 \\ I_3 \end{pmatrix}_c = c_1 \begin{pmatrix} R_1 \\ a_1 L_2 - R_1 \end{pmatrix} e^{-a_1 t} + c_2 \begin{pmatrix} R_1 \\ a_2 L_2 - R_1 \end{pmatrix} e^{-a_2 t}.$$

It should be clear in system (5) that there exists a particular solution of the form

$$\begin{pmatrix} I_2 \\ I_3 \end{pmatrix}_p = \begin{pmatrix} B_1 \\ B_2 \end{pmatrix},$$

where $B_1$ and $B_2$ are constants. Direct substitution into (5) yields

$$\begin{pmatrix} I_2 \\ I_3 \end{pmatrix}_p = \frac{E}{R_1}\begin{pmatrix} 1 \\ 0 \end{pmatrix}.$$

We have therefore found the general solution of system (5) to be

$$\begin{pmatrix} I_2 \\ I_3 \end{pmatrix} = c_1\begin{pmatrix} R_1 \\ a_1 L_2 - R_1 \end{pmatrix}e^{-a_1 t} + c_2\begin{pmatrix} R_1 \\ a_2 L_2 - R_1 \end{pmatrix}e^{-a_2 t} + \frac{E}{R_1}\begin{pmatrix} 1 \\ 0 \end{pmatrix}. \qquad (8)$$

The initial conditions $I_1(0) = I_2(0) = 0$ now require

$$c_1\begin{pmatrix} R_1 \\ a_1 L_2 - R_1 \end{pmatrix} + c_2\begin{pmatrix} R_1 \\ a_2 L_2 - R_1 \end{pmatrix} = -\frac{E}{R_1}\begin{pmatrix} 1 \\ 0 \end{pmatrix} \qquad (9)$$

as the system which must be satisfied by $c_1$ and $c_2$. The solution of equation (9) is

$$c_1 = -\frac{E(a_2 L_2 - R_1)}{R_1^2 L_2(a_2 - a_1)} \quad \text{and} \quad c_2 = \frac{E(a_1 L_2 - R_1)}{R_1^2 L_2(a_2 - a_1)}. \qquad (10)$$

The solution of the initial value problem is given by the insertion of these constants into equation (8). Finally, the current $I_1$ is easily obtained as the sum of $I_2$ and $I_3$

$$I_1 = -\frac{Ea_1(a_2 L_2 - R_1)}{R_1^2(a_2 - a_1)}e^{-a_1 t} + \frac{Ea_2(a_1 L_2 - R_1)}{R_1^2(a_2 - a_1)}e^{-a_2 t} + \frac{E}{R_1}. \qquad (11)$$

**EXAMPLE (b):** Determine the current $I_1(t)$ of Example (a) by using Laplace transform techniques.

To retain the conventional symbol $L$ for the number of henrys inductance of the circuit, we shall in this section denote by $L_t$ the Laplace operator for which $L$ is used in all other sections of the book.

Let $L_t\{I_k(t)\} = i_k(s)$ for each $k = 1, 2, 3$. Then the operator $L_t$ transforms the problem of solving equations (1), (2), and (3) into the algebraic problem of solving the equations

$$i_1 - i_2 - i_3 = 0, \qquad (12)$$

$$R_1 i_1 + s L_2 i_2 = \frac{E}{s}, \tag{13}$$

$$R_1 i_1 + (R_3 + s L_3) i_3 = \frac{E}{s}. \tag{14}$$

Since we desire only $i_1(s)$, let us use Cramer's rule to write the solution

$$i_1(s) = \frac{\begin{vmatrix} 0 & -1 & -1 \\ \dfrac{E}{s} & s L_2 & 0 \\ \dfrac{E}{s} & 0 & R_3 + s L_3 \end{vmatrix}}{\Delta} = \frac{E}{s} \cdot \frac{R_3 + s(L_2 + L_3)}{\Delta}, \tag{15}$$

in which

$$\Delta = \begin{vmatrix} 1 & -1 & -1 \\ R_1 & s L_2 & 0 \\ R_1 & 0 & R_3 + s L_3 \end{vmatrix},$$

or

$$\Delta = L_2 L_3 s^2 + (R_1 L_2 + R_3 L_2 + R_1 L_3)s + R_1 R_3. \tag{16}$$

It is important to recognize that this polynomial in $s$ is the same as the characteristic polynomial obtained in Example (a) in equation (6). Therefore the remarks made there concerning the roots of $\Delta$ hold here also. That is,

$$\Delta = L_2 L_3 (s + a_1)(s + a_2),$$

where $a_1$ and $a_2$ are distinct positive real numbers. We therefore have, from (15),

$$i_1(s) = \frac{E}{s} \cdot \frac{R_3 + s(L_2 + L_3)}{L_2 L_3 (s + a_1)(s + a_2)}. \tag{17}$$

The right member of equation (17) has a partial fractions expansion

$$i_1(s) = \frac{A_0}{s} + \frac{A_1}{s + a_1} + \frac{A_2}{s + a_2},$$

so that

$$I_1(t) = A_0 + A_1 e^{-a_1 t} + A_2 e^{-a_2 t}. \tag{18}$$

Some rather tedious algebra will determine the constants $A_0, A_1, A_2$ and show that equation (18) is identical to equation (11).

FIGURE 32

EXAMPLE (c):   For the network shown in Figure 32, set up the equations for the determination of the currents $I_1, I_2, I_3$, and the charge $Q_3$. Assume that when the switch is closed all currents and charges are zero. Find the characteristic polynomial for the matrix of the resultant system.

Using Kirchhoff's laws we write the equations

$$I_1 = I_2 + I_3, \tag{19}$$

$$R_1 I_1 + L_2 \frac{dI_2}{dt} = E \sin \omega t, \tag{20}$$

$$R_1 I_1 + R_3 I_3 + \frac{1}{C_3} Q_3 = E \sin \omega t; \tag{21}$$

and the definition of current as time rate of change of charge yields

$$I_3 = \frac{dQ_3}{dt}. \tag{22}$$

Our problem consists of the four equations (19) through (22) with the initial conditions that

$$I_2(0) = 0, \qquad I_3(0) = 0, \qquad Q_3(0) = 0. \tag{23}$$

If we use equations (19) and (22) to eliminate $I_3$ and $Q_3$ from the system, we obtain

$$\frac{dI_1}{dt} = -\frac{C_3 R_1 R_3 + L_2}{C_3 L_2 (R_1 + R_3)} I_1 + \frac{1}{C_3 (R_1 + R_3)} I_2$$

$$+ \frac{E\omega}{R_1 + R_3} \cos \omega t + \frac{ER_3}{L_2(R_1 + R_3)} \sin \omega t,$$

$$\frac{dI_2}{dt} = -\frac{R_1}{L_2} I_1 + \frac{E}{L_2} \sin \omega t.$$

The matrix of the associated homogeneous system is

$$\begin{pmatrix} -\dfrac{C_3R_1R_3 + L_2}{C_3L_2(R_1 + R_3)} & \dfrac{1}{C_3(R_1 + R_3)} \\ -\dfrac{R_1}{L_2} & 0 \end{pmatrix}.$$

Thus the characteristic polynomial is

$$\begin{vmatrix} -\dfrac{C_3R_1R_3 + L_2}{C_3L_2(R_1 + R_3)} - m & \dfrac{1}{C_3(R_1 + R_3)} \\ -\dfrac{R_1}{L_2} & -m \end{vmatrix}$$

$$= m^2 + \frac{C_3R_1R_3 + L_2}{C_3L_2(R_1 + R_3)}m + \frac{R_1}{C_3L_2(R_1 + R_3)}. \tag{24}$$

EXAMPLE (d): In Example (c) obtain the Laplace transform of the initial value problem.

Let $L_t\{I_k(t)\} = i_k(s), k = 1, 2, 3,$ and $L_t\{Q_3(t)\} = q_3(s)$. Then the transformed problem becomes

$$i_1 - i_2 - i_3 = 0, \tag{25}$$

$$R_1i_1 + sL_2i_2 = \frac{E\omega}{s^2 + \omega^2}, \tag{26}$$

$$R_1i_1 + R_3i_3 + \frac{1}{C_3}q_3 = \frac{E\omega}{s^2 + \omega^2}, \tag{27}$$

$$i_3 = sq_3. \tag{28}$$

If we eliminate $q_3$ by substitution from equation (28) into equation (27), we obtain a system of three equations in the three unknowns $i_1, i_2, i_3$. The nature of the solutions of this system is governed by the determinant

$$\Delta = \begin{vmatrix} 1 & -1 & -1 \\ R_1 & sL_2 & 0 \\ R_1 & 0 & R_3 + \dfrac{1}{C_3s} \end{vmatrix}.$$

Expanding this determinant gives

$$\Delta = \frac{L_2(R_1 + R_3)}{s}\left[s^2 + \frac{C_3R_1R_3 + L_2}{C_3L_2(R_1 + R_3)}s + \frac{R_1}{C_3L_2(R_1 + R_3)}\right]. \tag{29}$$

The student should compare the quadratic polynomial in equation (29) with the characteristic polynomial of equation (24) in Example (c) and realize that the nature of the solutions of the initial value problem is dictated in each example by the same polynomial.

## Exercises

1. For the $RL$ circuit of Figure 30, page 286, find the current $I$ if the direct current element $E$ is not removed from the circuit.

   ANS.   $I = ER^{-1}[1 - \exp(-RtL^{-1})]$.

2. Solve exercise 1 if the direct-current element is replaced by an alternating-current element $E \cos \omega t$. For convenience, use the notation

$$Z^2 = R^2 + \omega^2 L^2,$$

   in which $Z$ is called the steady-state impedance of this circuit.

   ANS.   $I = EZ^{-2}[\omega L \sin \omega t + R \cos \omega t - R \exp(-RtL^{-1})]$.

3. Solve exercise 2, replacing $E \cos \omega t$ with $E \sin \omega t$.

FIGURE 33

4. Figure 33 shows an $RC$ circuit with an alternating-current element inserted. Assume that the switch is closed at $t = 0$, at which time $Q = 0$ and $I = 0$. Use the notation

$$Z^2 = R^2 + (\omega C)^{-2},$$

   where $Z$ is the steady-state impedance of this circuit. Find $I$ for $t > 0$.

   ANS.   $I = EZ^{-2}[R \sin \omega t + (\omega C)^{-1} \cos \omega t - (\omega C)^{-1} \exp(-tR^{-1}C^{-1})]$.

5. In Figure 33, replace the alternating-current element with a direct-current element $E = 50$ volts and use $R = 10$ ohms, $C = 4(10)^{-4}$ farad. Assume that when the switch is closed (at $t = 0$) the charge on the capacitor is 0.015 coulomb. Find the initial current in the circuit and the current for $t > 0$.

   ANS.   $I(0) = 1.25$(amp), $I(t) = 1.25 \exp(-250t)$(amp).

6. In Figure 29, page 285, find $I(t)$ if $E(t) = 60$ volts, $R = 40$ ohms, $C = 5(10)^{-5}$ farad, $L = 0.02$ henry. Assume that $I(0) = 0$, $Q(0) = 0$.

   ANS.   $I = 3000t \exp(-1000t)$(amp).

7. In exercise 6, find the maximum current.   ANS.   $I_{max} = 3 e^{-1}$(amp).

In exercises 8 through 11, use Figure 29, page 285, with $E(t) = E \sin \omega t$ and with the following notations used to simplify the appearance of the formulas:

$$a = \frac{R}{2L}, \qquad b^2 = a^2 - \frac{1}{LC}, \qquad \beta^2 = \frac{1}{LC} - a^2,$$

$$\gamma = \omega L - \frac{1}{\omega C}, \qquad Z^2 = R^2 + \gamma^2.$$

The quantity $Z$ is the steady-state impedance for an $RLC$ circuit. In each of exercises 8 to 11, find $I(t)$ assuming that $I(0) = 0$ and $Q(0) = 0$.

**8.** Assume that $4L < R^2 C$.

> ANS.    $I = EZ^{-2}(R \sin \omega t - \gamma \cos \omega t) + \frac{1}{2}Eb^{-1}Z^{-2}[\{\gamma(a + b)$
>                 $- \omega R\} \exp\{-(a - b)t\} + \{\omega R - \gamma(a - b)\} \exp\{-(a + b)t\}].$

**9.** Assume that $R^2 C < 4L$.

> ANS.    $I = EZ^{-2}(R \sin \omega t - \gamma \cos \omega t)$
>                 $+ E\beta^{-1}Z^{-2} e^{-at}[\beta\gamma \cos \beta t - a(\gamma + 2\omega^{-1}C^{-1}) \sin \beta t].$

**10.** Assume that $R^2 C = 4L$.

> ANS.    $I = EZ^{-2}(R \sin \omega t - \gamma \cos \omega t) + E\omega^{-1}Z^{-2} e^{-at}[\gamma\omega - a(\gamma\omega + aR)t].$

**11.** Show that the answer to exercise 10 can be put in the form

$$I = EZ^{-2}(R \sin \omega t - \gamma \cos \omega t) + EZ^{-2} e^{-at}[\gamma + (a\gamma - R\omega)t].$$

**12.** In exercise 4, replace the alternating current element $E \sin \omega t$ with

$$E[\alpha(t - t_0) - \alpha(t - t_1)], \qquad t_1 > t_0 > 0.$$

Graph the new electromotive force (emf). Determine the current in the circuit.

> ANS.    $I(t) = \dfrac{E}{R}\left[\exp\left(-\dfrac{t - t_0}{RC}\right)\alpha(t - t_0) - \exp\left(-\dfrac{t - t_1}{RC}\right)\alpha(t - t_1)\right].$

FIGURE 34

**13.** In Figure 34, let $E = 60$ volts, $R_1 = 10$ ohms, $R_3 = 20$ ohms, and $C_2 = 5(10)^{-4}$ farad. Determine the currents if, when the switch is closed, the capacitor carries a charge of 0.03 coulomb.

> ANS.    $I_1 = 2(1 - e^{-300t})$, $I_2 = -3 e^{-300t}$, $I_3 = 2 + e^{-300t}$.

**14.** In exercise 13, let the initial charge on the capacitor be 0.01 coulomb, but leave the rest of the problem unchanged.

> ANS.    $I_1 = 2(1 + e^{-300t})$, $I_2 = 3 e^{-300t}$, $I_3 = 2 - e^{-300t}$.

FIGURE 35

**15.** For the network in Figure 35, set up the equations for the determination of the charge $Q_3$ and the currents $I_1, I_2, I_3$. Assume all four of those quantities to be zero at time zero. Use either the Laplace transform or matrix algebra to show that the nature of the solutions depends on the nature of the roots of the polynomial

$$C_3 L_2 (R_1 + R_3) m^2 + [C_3 (R_1 R_2 + R_2 R_3 + R_3 R_1) + L_2] m + R_1 + R_3.$$

FIGURE 36

**16.** For the network in Figure 36 set up the equations for the determination of the currents. Assume all currents to be zero at time zero. Use either the Laplace transform or matrix algebra to discuss the character of $I_1(t)$ without explicitly finding the function.

# The Existence and
# Uniqueness of Solutions

## 87. Preliminary remarks

The methods of Chapter 2 are strictly dependent on certain special properties (variables separable, exactness, and so on), that may or may not be
possessed by an individual equation. It is intuitively plausible that no collection of methods can be found that would permit the explicit solution, in the
sense of Chapter 2, of all first-order differential equations. We may seek
solutions in other forms, employing infinite series or other limiting processes;
we may resort to numerical approximations.

Confronted with this situation, a mathematician reacts by searching for
what is known as an existence theorem. He seeks to determine conditions
sufficient to insure the existence of a solution that has certain properties.
In Chapter 2 we stated such a theorem, and we now wish to examine it more
closely.

## 88. An existence and uniqueness theorem

Consider the equation of order one

$$\frac{dy}{dx} = f(x, y). \tag{1}$$

Let $T$ denote the rectangular region defined by

$$|x - x_0| \leq a \qquad \text{and} \qquad |y - y_0| \leq b,$$

a region with the point $(x_0, y_0)$ at its center. Let the function $f$ in equation (1) and the function $\partial f / \partial y$ be continuous at each point in $T$. Then there exists an interval, $|x - x_0| \leq h$, and a function $\phi(x)$ that have the following properties:

(a) $y = \phi(x)$ is a solution of equation (1) on the interval $|x - x_0| \leq h$.
(b) On the interval $|x - x_0| \leq h$, $\phi(x)$ satisfies the inequality

$$|\phi(x) - y_0| \leq b.$$

(c) $\phi(x_0) = y_0$.
(d) $\phi(x)$ is unique on the interval $|x - x_0| \leq h$ in the sense that it is the only function that has all of the properties (a), (b), and (c).

The interval $|x - x_0| \leq h$ may or may not need to be smaller than the interval $|x - x_0| \leq a$ over which conditions were imposed on $f$ and $\partial f / \partial y$.

In rough language the theorem states that, if $f(x, y)$ is sufficiently well behaved near the point $(x_0, y_0)$, the differential equation (1) has a solution that passes through the point $(x_0, y_0)$ and that solution is unique near $(x_0, y_0)$.

A proof of this fundamental theorem is presented in the next three sections. In essence the proof involves showing that a certain sequence of functions has a limit and that the limiting function is the desired solution. The sequence considered will be defined as follows:

$$y_0(x) = y_0,$$

$$y_1(x) = y_0 + \int_{x_0}^{x} f(t, y_0(t))\, dt,$$

$$y_2(x) = y_0 + \int_{x_0}^{x} f(t, y_1(t))\, dt, \tag{2}$$

$$\vdots$$

$$y_n(x) = y_0 + \int_{x_0}^{x} f(t, y_{n-1}(t))\, dt.$$

So that the proof may appear more reasonable, we first consider some examples of the proof for special differential equations.

EXAMPLE (a):   Show that the sequence of functions defined in equations (2) converges to a solution for the initial value problem

$$\frac{dy}{dx} = y; \qquad x_0 = 0, y_0 = 1.$$ (3)

We find that

$$y_0(x) = 1,$$

$$y_1(x) = 1 + \int_0^x dt = 1 + x,$$

$$y_2(x) = 1 + \int_0^x (1 + t)\, dt = 1 + x + \frac{x^2}{2},$$

$$y_3(x) = 1 + \int_0^x \left(1 + t + \frac{t^2}{2}\right) dt = 1 + x + \frac{x^2}{2} + \frac{x^3}{3!}.$$

From the pattern that is developing, it is easy to conjecture that

$$y_n(x) = \sum_{k=0}^{n} \frac{x^k}{k!}.$$

Indeed, this is easy to prove by induction. Moreover, the limit of this sequence exists for every real number $x$ because the limit is nothing more than the Maclaurin series expansion for $e^x$, which converges for every $x$. That is,

$$\phi(x) = \lim_{n \to \infty} y_n(x) = \sum_{k=0}^{\infty} \frac{x^k}{k!} = e^x.$$

It is a simple matter to verify that $e^x$ is a solution to the initial value problem (3).

EXAMPLE (b):   Find a solution of the initial value problem

$$\frac{dy}{dx} = x^2; \qquad x_0 = 2, y_0 = 1.$$ (4)

The sequence defined in (2) above now becomes

$$y_0(x) = 1$$

$$y_1(x) = 1 + \int_2^x t^2\, dt = \frac{x^3}{3} - \frac{5}{3},$$

$$y_2(x) = 1 + \int_2^x t^2 \, dt = \frac{x^3}{3} - \frac{5}{3},$$

$$\vdots$$

$$y_n(x) = 1 + \int_2^x t^2 \, dt = \frac{x^3}{3} - \frac{5}{3}.$$

Clearly the limit of this sequence is $x^3/3 - \frac{5}{3}$, and this function is a solution of (4).

## Exercises

In each of the following exercises, determine the limit of the sequence defined in (2) above. Verify that the function you obtain is a solution of the initial value problem.

1. $y' = x; x_0 = 2, y_0 = 1.$
2. $y' = y; x_0 = 0, y_0 = 2.$
3. $y' = 2y; x_0 = 0, y_0 = 1.$
4. $y' = x + y; x_0 = 0, y_0 = 1.$

## 89. A Lipschitz condition

We have assumed in the hypothesis of the foregoing existence theorem that the function $f$ and its derivative $\partial f/\partial y$ are continuous in the rectangle $T$. Thus if $(x, y_1)$ and $(x, y_2)$ are points in $T$, the mean value theorem applies to $f$ as a function of $y$. Hence, there exists a number $y^*$ between $y_1$ and $y_2$ such that

$$f(x, y_1) - f(x, y_2) = \frac{\partial f}{\partial y}(x, y^*)(y_1 - y_2).$$

The assumption that $\partial f/\partial y$ is continuous in $T$ allows us to assert that $\partial f/\partial y$ is bounded there. That is, there exists a number $K > 0$ such that

$$\left| \frac{\partial f}{\partial y} \right| \leq K,$$

for every point in $T$. Since $(x, y^*)$ is in $T$, it follows that

$$|f(x, y_1) - f(x, y_2)| = \left| \frac{\partial f}{\partial y}(x, y^*) \right| \cdot |y_1 - y_2|,$$

$$|f(x, y_1) - f(x, y_2)| \leq K|y_1 - y_2|, \tag{1}$$

for every pair of points $(x, y_1)$ and $(x, y_2)$ in $T$.

The inequality (1) is called a "Lipschitz condition" for the function $f$. We have shown that under the hypotheses of our existence theorem, the Lipschitz condition (1) holds for every pair of points $(x, y_1)$ and $(x, y_2)$ in $T$.

In the proof in Section 90 we shall actually use the Lipschitz condition rather than the hypothesized continuity of $\partial f/\partial y$. Thus, we could restate the existence theorem in terms of condition (1) instead of assuming $\partial f/\partial y$ is continuous in $T$.

## 90. A proof of the existence theorem

One hypothesis of the existence theorem of Section 88 is that $f$ is continuous in the rectangle $T$. It follows that $f$ must be bounded in $T$. Let $M > 0$ be a number such that $|f(x, y)| \leq M$ for every point in $T$. We now take $h$ to be the smaller of the two numbers $a$ and $b/M$, and define the rectangle $R$ to be the set of points $(x, y)$ for which

$$|x - x_0| \leq h \quad \text{and} \quad |y - y_0| \leq b.$$

Clearly $R$ is a subset of $T$.

As indicated in Section 88, we now consider the sequence of functions

$$y_n(x) = y_0 + \int_{x_0}^{x} f(t, y_{n-1}(t)) \, dt \tag{1}$$

and prove the following lemma:

LEMMA 1: *If $|x - x_0| \leq h$ then*

$$|y_n(x) - y_0| \leq b,$$

*for $n = 1, 2, 3, \ldots$.*

The proof of this lemma will be accomplished by induction. First of all, if $|x - x_0| \leq h$ we have

$$|y_1(x) - y_0| = \left| \int_{x_0}^{x} f(t, y_0) \, dt \right|$$

$$\leq M \left| \int_{x_0}^{x} dt \right|$$

$$\leq M|x - x_0|$$

$$\leq Mh$$

$$\leq b.$$

If we now assume that for $|x - x_0| \leqq h$, $|y_k(x) - y_0| \leqq b$, it follows that the point $[x, y_k(x)]$ is in $R$ so that $|f(x, y_k(x))| \leqq M$. Thus

$$|y_{k+1}(x) - y_0| \leqq \left| \int_{x_0}^x f(t, y_k(t))\, dt \right|$$

$$\leqq M \left| \int_{x_0}^x dt \right|$$

$$\leqq Mh$$

$$\leqq b.$$

By induction we can now assert the validity of the lemma.

Lemma 1 may be stated in a slightly different way: if $|x - x_0| \leqq h$, then the points $[x, y_n(x)]$, $n = 0, 1, 2, \ldots$, are in $R$. The Lipschitz condition of Section 89 may now be used to deduce the following lemma.

LEMMA 2:   *If* $|x - x_0| \leqq h$, *then*

$$|f(x, y_n(x)) - f(x, y_{n-1}(x))| \leqq K|y_n(x) - y_{n-1}(x)|,$$

*for* $n = 1, 2, 3, \ldots$.

We are now in a position to give an inductive proof of still another lemma.

LEMMA 3:   *If* $|x - x_0| \leqq h$, *then*

$$|y_n(x) - y_{n-1}(x)| \leqq \frac{MK^{n-1}|x - x_0|^n}{n!} \leqq \frac{MK^{n-1}h^n}{n!},$$

*for* $n = 1, 2, 3, \ldots$.

For the case $n = 1$, we have from the proof of Lemma 1,

$$|y_1(x) - y_0| \leqq M|x - x_0|.$$

Assuming that

$$|y_{n-1}(x) - y_{n-2}(x)| \leqq \frac{MK^{n-2}|x - x_0|^{n-1}}{(n-1)!}, \tag{2}$$

we now must show that

$$|y_n(x) - y_{n-1}(x)| \leqq \frac{MK^{n-1}|x - x_0|^n}{n!}.$$

We will prove this for the case $x_0 \leq x \leq x_0 + h$. From Lemma 2 we have

$$|y_n(x) - y_{n-1}(x)| = \left| \int_{x_0}^{x} [f(t, y_{n-1}(t)) - f(t, y_{n-2}(t))] \, dt \right|$$

$$\leq \int_{x_0}^{x} |f(t, y_{n-1}(t)) - f(t, y_{n-2}(t))| \, dt$$

$$\leq K \int_{x_0}^{x} |y_{n-1}(t) - y_{n-2}(t)| \, dt.$$

Using the hypothesis (2) we conclude that

$$|y_n(x) - y_{n-1}(x)| \leq \frac{MK^{n-1}}{(n-1)!} \int_{x_0}^{x} (t - x_0)^{n-1} \, dt,$$

or

$$|y_n(x) - y_{n-1}(x)| \leq \frac{MK^{n-1}}{n!} |x - x_0|^n. \tag{3}$$

For the case $x_0 - h \leq x \leq x_0$, the same type of argument will yield the same result. The proof of Lemma 3 is thus complete.

To utilize the results of Lemma 3 we now compare the two infinite series

$$\sum_{n=1}^{\infty} [y_n(x) - y_{n-1}(x)] \qquad \text{and} \qquad \sum_{n=1}^{\infty} \frac{MK^{n-1}h^n}{n!}.$$

The second of these series is an absolutely convergent series. Moreover, by Lemma 3, the second series dominates the first series. Hence, by the Weierstrass $M$ test the series

$$\sum_{n=1}^{\infty} [y_n(x) - y_{n-1}(x)] \tag{4}$$

converges absolutely and uniformly on the interval $|x - x_0| \leq h$. If we consider the $k$th partial sum of the series (4)

$$\sum_{n=1}^{k} [y_n(x) - y_{n-1}(x)] = [y_1(x) - y_0(x)] + [y_2(x) - y_1(x)] + \cdots$$

$$+ [y_k(x) - y_{k-1}(x)],$$

we see that

$$\sum_{n=1}^{k} [y_n(x) - y_{n-1}(x)] = y_k(x).$$

That is, the statement that the series (4) converges absolutely and uniformly is equivalent to the statement that the sequence $y_n(x)$ converges uniformly on the interval

$$|x - x_0| \leqq h.$$

If we now define

$$\phi(x) = \lim_{n \to \infty} y_n(x)$$

and recall from the definition of the sequence $y_n(x)$ that each $y_n(x)$ is continuous on $|x - x_0| \leqq h$, it follows (since the convergence is uniform) that $\phi(x)$ is also continuous and

$$\phi(x) = \lim_{n \to \infty} y_n(x) = y_0 + \lim_{n \to \infty} \int_{x_0}^x f(t, y_{n-1}(t)) \, dt.$$

Because of the continuity of $f$ and the uniform convergence of the sequence $y_n(x)$, we may interchange the order of the two limiting processes to show that $\phi(x)$ is a solution of the integral equation

$$\phi(x) = y_0 + \int_{x_0}^x f(t, \phi(t)) \, dt. \tag{5}$$

It follows immediately upon differentiation of equation (5) that $\phi(x)$ is a solution of the differential equation $dy/dx = f(x, y)$ on the interval $|x - x_0| \leqq h$. Furthermore, it is clear from equation (5) that $\phi(x_0) = y_0$.

Finally, since we have shown in Lemma 1 that $|y_n(x) - y_0| \leqq b$ for each $n$ and for $|x - x_0| \leqq h$, it follows that the same inequality must hold for $\phi(x) = \lim_{n \to \infty} y_n(x)$. That is, if $|x - x_0| \leqq h$, then $|\phi(x) - y_0| \leqq b$.

Thus we have completed the proof of parts (a), (b), and (c) of the existence theorem of Section 88.

## 91. A proof of the uniqueness theorem

We must now show that the function $\phi(x)$ obtained in Section 90 is unique. Suppose there is another function $Y(x)$, such that $dY/dx = f[x, Y(x)]$, $Y(x_0) = y_0$, and $|Y(x) - y_0| \leqq b$ for $|x - x_0| \leqq h$. Then we may write

$$Y(x) = y_0 + \int_{x_0}^x f(t, Y(t)) \, dt.$$

If we compare $Y(x)$ to the functions of the sequence $y_n(x)$ of Section 90 we see that

$$|Y(x) - y_n(x)| \leqq \left| \int_{x_0}^x [f(t, Y(t)) - f(t, y_{n-1}(t))] \, dt \right|. \tag{1}$$

We shall now show that as $n \to \infty$ the integral on the right side of (1) approaches zero for $|x - x_0| \leq h$. It will then follow that $Y(x) = \lim\limits_{n \to \infty} y_n(x)$, so that finally $Y(x) \equiv \phi(x)$ on the interval $|x - x_0| \leq h$.

For any $x$ on the interval $|x - x_0| \leq h$ it is true that $[x, Y(x)]$ and $[x, y_{n-1}(x)]$ are in the rectangle $R$, hence the Lipschitz condition of Section 89 will allow us to change (1) into

$$|Y(x) - y_n(x)| \leq K \int_{x_0}^{x} |Y(t) - y_{n-1}(t)| \, dt. \tag{2}$$

We now proceed by an inductive proof and limit our attention to values of $x$ greater than $x_0$. (A similar argument obtains the same result for $x_0 - h \leq x \leq x_0$.) For $n = 1$, we have

$$|Y(x) - y_1(x)| \leq K \int_{x_0}^{x} |Y(t) - y_0| \, dt$$

$$\leq Kb(x - x_0).$$

We wish also to show that the assumption

$$|Y(x) - y_{n-1}(x)| \leq \frac{K^{n-1} b(x - x_0)^{n-1}}{(n-1)!}$$

leads to the conclusion

$$|Y(x) - y_n(x)| \leq \frac{K^n b(x - x_0)^n}{n!}. \tag{3}$$

This will complete an inductive argument for the relation (3). We have for $x_0 \leq x \leq x_0 + h$,

$$|Y(x) - y_n(x)| \leq \int_{x_0}^{x} |f(t, Y(t)) - f(t, y_{n-1}(t))| \, dt$$

$$\leq K \int_{x_0}^{x} |Y(t) - y_{n-1}(t)| \, dt$$

$$\leq \frac{K^n b}{(n-1)!} \int_{x_0}^{x} (t - x_0)^{n-1} \, dt$$

$$\leq \frac{K^n b}{n!} (x - x_0)^n,$$

thus completing the proof of relation (3).

For $|x - x_0| \leq h$ we have from the inequality (3),

$$|Y(x) - y_n(x)| \leq \frac{K^n b h^n}{n!}. \tag{4}$$

As $n \to \infty$ the expression on the right side of relation (4) approaches zero. Hence it follows that for $|x - x_0| \leqq h$, $y_n(x) \to Y(x)$. Thus $Y(x)$ must be the same function $\phi(x)$ we obtained in Section 90. That is, the solution $\phi(x)$ is unique.

## 92. Other existence theorems

The existence theorem we have proved in the preceding sections for a first-order equation can be extended to equations of higher order. The simplest such extension is to equations of second order that can be written in the form

$$y'' = f(x, y, y'). \tag{1}$$

It is natural to expect the theorem to involve continuity requirements on the function $f$ and its partial derivatives. The theorem may be stated as follows:

THEOREM 31: *If the function $f$ of equation (1) and its partial derivatives with respect to $y$ and $y'$ are continuous functions in a region $T$ defined by*

$$|x - x_0| \leqq a, \qquad |y - y_0| \leqq b, \qquad |y' - y_0'| \leqq c,$$

*then there exists an interval $|x - x_0| \leqq h$ and a unique function $\phi(x)$ such that $\phi(x)$ is a solution of (1) for all $x$ in the interval $|x - x_0| \leqq h$, $\phi(x_0) = y_0$, and $\phi'(x_0) = y_0'$.*

A proof of this theorem that is quite similar to the proof given in Sections 90 and 91 can be found in Ince.* The generalization of the theorem to equations of higher order is direct.

* E. L. Ince, *Ordinary Differential Equations* (London: Longmans, Green & Co., 1927), Chapter 3.

# 16

# Nonlinear Equations

## 93. Preliminary remarks

The existence and uniqueness theorem of Chapter 15 made no distinction between linear and nonlinear differential equations. We know from our study in the earlier chapters of this book, however, that the methods we have found for actually determining solutions of a given equation often depend on the equation being a linear one. For example, in Chapter 2 we found that certain particular kinds of first-order nonlinear equations can be solved; that is, if the equation is exact, separable, homogeneous, and so on. On the other hand, if a first-order equation is linear, we have a method which can produce all possible solutions of the differential equation.

The fact is, there is no general method for solving first-order nonlinear differential equations, even if the existence of such solutions can be shown by the theorems of Chapter 15. Indeed, the determination of such solutions is often difficult if not impossible.

In this chapter we shall briefly discuss a few of the special difficulties that arise with nonlinear equations and a few techniques that will find solutions for certain particular types of equations.

## 94. Factoring the left member

To give illustration to the kind of complexity that may arise in nonlinear situations, we consider first a relatively simple complication. For an equation of the form

$$f(x, y, y') = 0, \tag{1}$$

it may be possible to factor the left member. The problem of solving (1) is then replaced by two or more problems of simpler type. The latter may be capable of solution by the methods of Chapters 2 and 4.

Since $y'$ will be raised to powers in the example and exercises, let us simplify the printing and writing by a common device, using $p$ for $y'$:

$$p = \frac{dy}{dx}.$$

EXAMPLE:   Solve the differential equation

$$xyp^2 + (x + y)p + 1 = 0. \tag{2}$$

The left member of equation (2) is readily factored. Thus (2) leads to

$$(xp + 1)(yp + 1) = 0,$$

from which it follows that either

$$yp + 1 = 0 \tag{3}$$

or

$$xp + 1 = 0. \tag{4}$$

From equation (3) in the form

$$y \, dy + dx = 0$$

it follows that

$$y^2 = -2(x - c_1). \tag{5}$$

Equation (4) may be written

$$x \, dy + dx = 0$$

from which, for $x \neq 0$,

$$dy + \frac{dx}{x} = 0,$$

so

$$y = -\ln |c_2 x|. \tag{6}$$

We say, and it is very rough language, that the solutions of (2) are (5) and (6). Particular solutions may be made up from these solutions; they may be drawn from (5) alone, from (6) alone, or conceivably pieced together by using (5) in some intervals and (6) in others. At a point where a solution from (5) is to be joined with a solution from (6), the slope must remain continuous (see exercise 21 below), so the piecing together must take place along the line $y = x$. Note (see exercise 24) that the second derivative, which does not enter the differential equation, need not be continuous.

The existence of these three sets of particular solutions of (2), that is, solutions from (5), from (6), or from (5) and (6), leads to an interesting phenomenon in initial value problems. Consider the problem of finding a solution of (2) such that the solution passes through the point $(-\frac{1}{2}, 2)$. If the result is to be valid for the interval $-1 < x < -\frac{1}{4}$, there are two answers, which will be found in exercise 25, following. If the result is to be valid for $-1 < x < \frac{1}{2}$, there is only one answer (exercise 26), one of the two answers to exercise 25. If the result is to be valid in $-1 < x < 2$, there is only one answer (exercise 27).

## Exercises

In exercises 1 through 18, find the solutions in the sense of (5) and (6) above.

**1.** $x^2p^2 - y^2 = 0.$  ANS.  $y = c_1x, xy = c_2.$

**2.** $xp^2 - (2x + 3y)p + 6y = 0.$  ANS.  $y = c_1x^3, y = 2x + c_2.$

**3.** $x^2p^2 - 5xyp + 6y^2 = 0.$  ANS.  $y = c_1x^2, y = c_2x^3.$

**4.** $x^2p^2 + xp - y^2 - y = 0.$  ANS.  $y = c_1x, x(y + 1) = c_2.$

**5.** $xp^2 + (1 - x^2y)p - xy = 0.$  ANS.  $y = c_1 \exp(\frac{1}{2}x^2), y = -\ln|c_2x|.$

**6.** $p^2 - (x^2y + 3)p + 3x^2y = 0.$  ANS.  $y = 3x + c_1, x^3 = 3\ln|c_2y|.$

**7.** $xp^2 - (1 + xy)p + y = 0.$  ANS.  $y = \ln|c_1x|, x = \ln|c_2y|.$

**8.** $p^2 - x^2y^2 = 0.$  ANS.  $x^2 = 2\ln|c_1y|, x^2 = -2\ln|c_2y|.$

**9.** $(x + y)^2p^2 = y^2.$  ANS.  $x = y\ln|c_1y|, y(2x + y) = c_2.$

**10.** $yp^2 + (x - y^2)p - xy = 0.$  ANS.  $x^2 + y^2 = c_1^2, y = c_2 e^x.$

**11.** $p^2 - xy(x + y)p + x^3y^3 = 0.$  ANS.  $y(x^2 + c_1) = -2, x^3 = 3\ln|c_2y|.$

**12.** $(4x - y)p^2 + 6(x - y)p + 2x - 5y = 0.$

  ANS.  $x + y = c_1, (2x + y)^2 = c_2(y - x).$

**13.** $(x - y)^2p^2 = y^2.$  ANS.  $x = -y\ln|c_1y|, y(2x - y) = c_2.$

**14.** $xyp^2 + (xy^2 - 1)p - y = 0.$  ANS.  $y^2 = 2\ln|c_1x|, x = -\ln|c_2y|.$

**15.** $(x^2 + y^2)^2p^2 = 4x^2y^2.$  ANS.  $y^2 - x^2 = c_1y, y(3x^2 + y^2) = c_2.$

**16.** $(y + x)^2p^2 + (2y^2 + xy - x^2)p + y(y - x) = 0.$

  ANS.  $y^2 + 2xy = c_1, y^2 + 2xy - x^2 = c_2.$

**17.** $xy(x^2 + y^2)(p^2 - 1) = p(x^4 + x^2y^2 + y^4).$

  ANS.  $y^2(y^2 + 2x^2) = c_1, y^2 = 2x^2 \ln|c_2x|.$

**18.** $xp^3 - (x^2 + x + y)p^2 + (x^2 + xy + y)p - xy = 0.$

  ANS.  $y = c_1x, y = x + c_2, x^2 = 2(y - c_3).$

Exercises 19 through 27 refer to the example of this section. There the differential equation

$$xyp^2 + (x + y)p + 1 = 0 \tag{2}$$

was shown to have the solutions

$$y^2 = -2(x - c_1) \tag{5}$$

and

$$y = -\ln |c_2 x|. \tag{6}$$

**19.** Show that of the family (5) above, the only curve that passes through the point $(1, 1)$ is $y = (3 - 2x)^{1/2}$ and that this solution is valid for $x < \frac{3}{2}$.

**20.** Show that of the family (6) above, the only curve that passes through the point $(1, 1)$ is $y = 1 - \ln x$ and that this solution is valid for $0 < x$.

**21.** Show that the function defined by

$$y = (2 - 2x)^{1/2} \qquad \text{for } x \leq 1,$$

$$y = -\ln x \qquad \text{for } x \geq 1,$$

is a solution of equation (2) for all $x \neq 1$, but fails to have a derivative at $x = 1$ and is therefore not a solution there.

**22.** Show that if a solution of (2) is to be pieced together from (5) and (6), then the slopes of the curves must be equal where the pieces join. Show that the pieces must therefore be joined at a point on the line $y = x$.

**23.** Show that the function determined by

$$y = (3 - 2x)^{1/2} \qquad \text{for } x \leq 1,$$

$$y = 1 - \ln x \qquad \text{for } 1 \leq x$$

is a solution of equation (2) and is valid for all $x$. The interesting portion of this curve is shown in Figure 37.

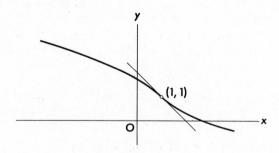

FIGURE 37

**24.** Show for the solution given in exercise 23 that $y''$ is not continuous at $x = 1$. Show that as $x \to 1^-$, $y'' \to -1$, and as $x \to 1^+$, $y'' \to +1$.

**25.** Find those solutions of (2) which are valid in $-1 < x < -\frac{1}{4}$ and each of which has its graph passing through the point $(-\frac{1}{2}, 2)$.

ANS.    $y = (3 - 2x)^{1/2}$; and $y = 2 - \ln 2 - \ln(-x)$.

**26.** Find that solution of (2) which is valid for $-1 < x < \frac{1}{2}$ and has its graph passing through the point $(-\frac{1}{2}, 2)$.        ANS.    $y = (3 - 2x)^{1/2}$.

**27.** Find that solution of (2) which is valid for $-1 < x < 2$ and has its graph passing through the point $(-\frac{1}{2}, 2)$.        ANS.    $y = (3 - 2x)^{1/2}$ for $x \leqq 1$.

$$y = 1 - \ln x \text{ for } 1 \leqq x.$$

## 95. Singular solutions

Let us solve the differential equation

$$y^2 p^2 - a^2 + y^2 = 0. \tag{1}$$

Here

$$yp = \pm\sqrt{a^2 - y^2}$$

so we may write

$$\frac{y\,dy}{\sqrt{a^2 - y^2}} = dx, \tag{2}$$

or

$$-\frac{y\,dy}{\sqrt{a^2 - y^2}} = dx, \tag{3}$$

or (if the division by $\sqrt{a^2 - y^2}$ cannot be effected)

$$a^2 - y^2 = 0. \tag{4}$$

From (2) it follows that

$$x = c_1 - \sqrt{a^2 - y^2}, \tag{5}$$

while from (3) that

$$x = c_2 + \sqrt{a^2 - y^2}, \tag{6}$$

and from (4) that

$$y = a \quad \text{or} \quad y = -a. \tag{7}$$

Graphically, the solutions (5) are left-hand semicircles with radius $a$ and centered on the $x$-axis; the solutions (6) are right-hand semicircles of radius

$a$, centered on the $x$-axis. We may combine (5) and (6) into

$$(x - c)^2 + y^2 = a^2, \tag{8}$$

which we might be tempted to call the "general" solution of (1). However, from either of equations (7) we get $p = 0$, so that $y = a$ and $y = -a$ are both solutions of equation (1), but neither of these functions is a special case of (8).

We therefore see that the use of the term "general" solution for the functions defined implicitly by (8) is not consistent with the usage of the term as applied to linear differential equations. For linear equations any solution was a particular case of the general solution. It is perhaps unfortunate that the words "general solution" are used for the one-parameter family of solutions defined by (8). The particular solutions $y = a$ and $y = -a$ are called singular solutions. (It should be clear that linear equations cannot have singular solutions.)

A *singular solution* of a nonlinear first-order differential equation is any solution that

(a) is not a special case of the general solution, and
(b) is, at each of its points, tangent to some element of the one-parameter family that is the general solution.

Figure 38 shows several elements of the family of circles given by equation (8) and also shows the two lines representing $y = a$ and $y = -a$. At each point of either line, the line is tangent to an element of the family of circles. A curve which at each of its points is tangent to an element of a one-parameter family of curves is called an *envelope* of that family.

FIGURE 38

## 96. The $c$-discriminant equation

Consider the differential equation of first order,

$$f(x, y, p) = 0; \qquad p = \frac{dy}{dx}, \tag{1}$$

in which the left member is a polynomial in $x$, $y$, and $p$. It may not be possible to factor the left member into factors that are themselves polynomials in $x$, $y$, and $p$. Then the equation is said to be irreducible.

The general solution of (1) will be a one-parameter family,

$$\phi(x, y, c) = 0. \tag{2}$$

A singular solution, if it exists, for equation (1) must be an envelope of the family (2). Each point on the envelope is a point of tangency of the envelope with some element of the family (2) and is determined by the value of $c$ that identifies that element of the family. Then the envelope has parametric equations, $x = x(c)$ and $y = y(c)$, with the $c$ of equation (2) as the parameter. The functions $x(c)$ and $y(c)$ are as yet unknown to us. But the $x$ and $y$ of the point of contact must also satisfy equation (2), from which we get, by differentiation with respect to $c$, the equation

$$\frac{\partial \phi}{\partial x}\frac{dx}{dc} + \frac{\partial \phi}{\partial y}\frac{dy}{dc} + \frac{\partial \phi}{\partial c} = 0. \tag{3}$$

The slope of the envelope and the slope of the family element concerned must be equal at the point of contact. That slope can be determined by differentiating equation (2) with respect to $x$, keeping $c$ constant. Thus it follows that

$$\frac{\partial \phi}{\partial x} + \frac{\partial \phi}{\partial y}\frac{dy}{dx} = 0. \tag{4}$$

Equations (3) and (4) both hold at the point of contact and from them it follows that

$$\frac{\partial \phi}{\partial c} = 0. \tag{5}$$

We now have two equations, $\phi = 0$ and $\partial \phi / \partial c = 0$, which must be satisfied by $x$, $y$, and $c$. These two equations may be taken as the desired parametric equations. They contain any envelope which may exist for the original family of curves, $\phi = 0$. Fortunately, there is no need for us to put these equations into the form $x = x(c)$, $y = y(c)$.

The equation that results from the elimination of $c$ from the equations $\phi = 0$ and $\partial \phi / \partial c = 0$ is called the *c-discriminant equation*\* of the family $\phi = 0$. It is a necessary and sufficient condition that the equation

$$\phi(x, y, c) = 0, \tag{2}$$

considered as an equation in $c$, have at least two of its roots equal.

---

\* The $c$-discriminant equation may contain a locus of cusps of the elements of the general solution and a locus of nodes of those elements, as well as the envelope which aroused our interest in it.

There is nothing in our work to guarantee that the $c$-discriminant equation, or any part of it, will yield a solution of the differential equation. To get the $c$-discriminant equation we need the general solution. During the process of obtaining the general solution, we find also the singular solution, if there is one.

## 97. The $p$-discriminant equation

Suppose that in the irreducible differential equation

$$f(x, y, p) = 0 \tag{1}$$

the polynomial $f$ is of degree $n$ in $p$. There will be $n$ roots of equation (1), each yielding a result of the form

$$p = g(x, y). \tag{2}$$

If at a point $(x_0, y_0)$ the equation (1) has, as an equation in $p$, all its roots distinct, then near $(x_0, y_0)$ there will be $n$ distinct equations of the type of equation (2). Near $(x_0, y_0)$ the right members of these $n$ equations will be single-valued and may satisfy the conditions of the existence theorem described in Chapter 15. But if at $(x_0, y_0)$ equation (1) has at least two of its roots equal, then at least two of the $n$ equations like (2) will have right members assuming the same value at $(x_0, y_0)$. For such equations there is no region, no matter how small, surrounding $(x_0, y_0)$ in which the right member is single-valued. Hence the existence theorem of Chapter 15 cannot be applied when equation (1) has two or more equal roots as an equation in $p$. Therefore we must give separate consideration to the locus of points $(x, y)$ for which (1) has at least two of its roots equal.

The condition that equation (1) have at least two equal roots as an equation in $p$ is that both $f = 0$ and $\partial f/\partial p = 0$. These two equations in the three variables $x$, $y$, and $p$ are parametric equations of a curve in the $xy$-plane with $p$ playing the role of parameter. The equation that results when $p$ is eliminated from the parametric equations $f = 0$ and $\partial f/\partial p = 0$ is called the *p-discriminant equation*.

If an envelope of the general solution of $f = 0$ exists, it will be contained in the $p$-discriminant equation. No proof is included here.* For us the $p$-discriminant equation is useful in two ways. When a singular solution is obtained in the natural course of solving an equation, the $p$-discriminant equation furnishes us with a check. If none of our methods of attack leads to a general solution, then the $p$-discriminant equation offers functions that

---

* For more detail on singular solutions and the discriminants see E. L. Ince, *Ordinary Differential Equations* (London: Longmans, Green & Co., 1927), pp. 82–92.

may be particular (including singular) solutions of the differential equation. Then the *p*-discriminant equation should be tested for possible solutions of the differential equation. Such particular solutions make, of course, no contribution toward finding the general solution.

The *p*-discriminant equation may contain singular solutions, solutions that are not singular, and functions that are not solutions at all.

## Exercises

1. For the quadratic equation

$$f = Ap^2 + Bp + C = 0,$$

with $A, B, C$ functions of $x$ and $y$, show that the *p*-discriminant obtained by eliminating $p$ from $f = 0$ and $\partial f / \partial p = 0$ is the familiar equation

$$B^2 - 4AC = 0.$$

2. For the cubic $p^3 + Ap + B = 0$, show that the *p*-discriminant equation is $4A^3 + 27B^2 = 0$.
3. For the cubic $p^3 + Ap^2 + B = 0$, show that the *p*-discriminant equation is $B(4A^3 + 27B) = 0$.
4. Set up the condition that the equation $x^3 p^2 + x^2 yp + 4 = 0$ have equal roots as a quadratic in $p$. Compare with the singular solution $xy^2 = 16$.
5. Show that the condition that the equation $xyp^2 + (x + y)p + 1 = 0$ of the example of Section 94 have equal roots in $p$ is $(x - y)^2 = 0$ and that the latter equation does not yield a solution of the differential equation. Was there a singular solution?
6. For the equation $y^2 p^2 - a^2 + y^2 = 0$ of Section 95, find the condition for equal roots in $p$ and compare with the singular solution.
7. For the differential equation of exercise 6 show that the function defined by

$$y = [a^2 - (x + 2a)^2]^{1/2} \qquad \text{for } -3a < x \leqq -2a,$$

$$y = a \qquad \text{for } -2a \leqq x \leqq 2a,$$

$$y = [a^2 - (x - 2a)^2]^{1/2} \qquad \text{for } 2a \leqq x < 3a,$$

is a solution. Sketch the graph and show how it was pieced together from the general solution and the singular solution given in equations (5), (6), and (7) of Section 95.

In exercises 8 through 16, obtain (a) the *p*-discriminant equation and (b) those solutions of the differential equation that are contained in the *p*-discriminant.

8. $xp^2 - 2yp + 4x = 0.$      ANS. (a) $y^2 - 4x^2 = 0$; (b) $y = 2x, y = -2x$.
9. $3x^4 p^2 - xp - y = 0.$      ANS. (a) $x^2(1 + 12x^2 y) = 0$; (b) $x = 0, 12x^2 y = -1$.
10. $p^2 - xp - y = 0.$      ANS. (a) $x^2 + 4y = 0$; (b) None.
11. $p^2 - xp + y = 0.$      ANS. (a) $x^2 - 4y = 0$; (b) $x^2 - 4y = 0$.
12. $p^2 + 4x^5 p - 12x^4 y = 0.$      ANS. (a) $x^4(x^6 + 3y) = 0$; (b) $3y = -x^6$.
13. $4y^3 p^2 - 4xp + y = 0.$      ANS. (a) $(y^2 - x)(y^2 + x) = 0$; (b) Same as (a).

**14.** $4y^3p^2 + 4xp + y = 0$. See also exercise 13.        ANS.    (a) $(y^2 - x)(y^2 + x) = 0$;
                                                                (b) None.

**15.** $p^3 + xp^2 - y = 0$.                ANS.    (a) $y(4x^3 - 27y) = 0$; (b) $y = 0$.

**16.** $y^4p^3 - 6xp + 2y = 0$.            ANS.    (a) $y^8(y^2 - 2x)(y^4 + 2xy^2 + 4x^2) = 0$;
                                                            (b) $y = 0, y^2 = 2x$.

**17.** For the differential equation of exercise 4 above, the general solution will be found
to be $cxy + 4x + c^2 = 0$. Find the condition that this quadratic equation in $c$
have equal roots. Compare that condition with the singular solution.

## 98. Eliminating the dependent variable

Suppose the equation

$$f(x, y, p) = 0; \qquad p = \frac{dy}{dx}, \tag{1}$$

is of a form such that we can readily solve it for the dependent variable $y$
and write

$$y = g(x, p). \tag{2}$$

We can differentiate equation (2) with respect to $x$ and, since $dy/dx = p$,
get an equation

$$h\left(x, p, \frac{dp}{dx}\right) = 0 \tag{3}$$

involving only $x$ and $p$. If we can solve equation (3), we will have two equations
relating $x$, $y$, and $p$, namely, equation (2) and the solution of (3). These
together form parametric equations of the solution of (1) with $p$ now considered a parameter. Or, if $p$ be eliminated between (2) and the solution
of (3), then a solution in the nonparametric form is obtained.

EXAMPLE:   Solve the differential equation

$$xp^2 - 3yp + 9x^2 = 0, \qquad \text{for } x > 0. \tag{4}$$

Rewrite (4) as

$$3y = xp + 9x^2/p. \tag{5}$$

Then differentiate both members of (5) with respect to $x$, using the fact that
$dy/dx = p$, thus getting

$$3p = p + \frac{18x}{p} + \left(x - \frac{9x^2}{p^2}\right)\frac{dp}{dx},$$

or

$$2p\left(1 - \frac{9x}{p^2}\right) = x\left(1 - \frac{9x}{p^2}\right)\frac{dp}{dx}. \qquad (6)$$

From (6) it follows that either

$$1 - \frac{9x}{p^2} = 0 \qquad (7)$$

or

$$2p = x\frac{dp}{dx}. \qquad (8)$$

First consider (8), which leads to

$$2\frac{dx}{x} = \frac{dp}{p}$$

so that

$$p = cx^2. \qquad (9)$$

Therefore equations (4) and (9), with $p$ as a parameter, constitute a solution of (4) looked upon as a differential equation with $p = dy/dx$.

In this example it is easy to eliminate $p$ from equations (4) and (9), so we perform that elimination. The result is

$$x \cdot c^2 x^4 - 3y \cdot cx^2 + 9x^2 = 0.$$

Since $x > 0$, we have

$$c^2 x^3 - 3cy + 9 = 0,$$

or

$$3cy = c^2 x^3 + 9.$$

Now put $c = 3k$ to get

$$ky = k^2 x^3 + 1. \qquad (10)$$

Equation (10) with $k$ as an arbitrary constant is called the general solution of the differential equation (4).

We have yet to deal with equation (7). Note that (7) is an algebraic relation between $x$ and $p$ in contrast to the differential relation (8), which we have already used. We reason that the elimination of $p$ from (7) and (4) may lead to a solution of the differential equation (4) and that the solution will not involve an arbitrary constant. From (7) it is seen that $p = 3x^{1/2}$ or $p = -3x^{1/2}$. Either of these expressions for $p$ may be substituted into (4) and will lead to

$$y^2 = 4x^3. \qquad (11)$$

It is not difficult to show that equation (11) defines two solutions of the differential equation. These solutions are not special cases of the general solution (10). They are singular solutions. Equation (11) has for its graph the envelope of the family of curves given by equation (10). The solutions defined by (11) are also easily obtained from the $p$-discriminant equation.

## 99. Clairaut's equation

Any differential equation of the form

$$y = px + f(p), \tag{1}$$

where $f(p)$ contains neither $x$ nor $y$ explicitly, can be solved at once by the method of Section 98. Equation (1) is called *Clairaut's equation*.

Let us differentiate both members of (1) with respect to $x$, thus getting

$$p = p + [x + f'(p)]p',$$

$$[x + f'(p)]\frac{dp}{dx} = 0. \tag{2}$$

Then either

$$\frac{dp}{dx} = 0 \tag{3}$$

or

$$x + f'(p) = 0. \tag{4}$$

The solution of the differential equation (3) is, of course, $p = c$, where $c$ is an arbitrary constant. Returning to the differential equation (1), we can now write its general solution as

$$y = cx + f(c), \tag{5}$$

a result easily verified by direct substitution into the differential equation (1). Note that (5) is the equation of a family of straight lines.

Now consider equation (4). Since $f(p)$ and $f'(p)$ are known functions of $p$, equations (4) and (1) together constitute a set of parametric equations giving $x$ and $y$ in terms of the parameter $p$. Indeed, from equation (4) it follows that

$$x = -f'(p), \tag{6}$$

which, combined with equation (1), yields

$$y = f(p) - pf'(p). \tag{7}$$

If $f(p)$ is not a linear function of $p$ and not a constant, it can be shown (exercises 1 and 2 below) that (6) and (7) are parametric equations of a non-linear solution of the differential equation (1). Since the general solution (5) represents a straight line for each value of $c$, the solution (6) and (7) cannot be a special case of (5); it is a singular solution.

EXAMPLE (a):   Solve the differential equation

$$y = px + p^3. \tag{8}$$

Since (8) is a Clairaut equation, we can write its general solution

$$y = cx + c^3$$

at once.

Then using (6) and (7) we obtain the parametric equations

$$x = -3p^2, \qquad y = -2p^3, \tag{9}$$

of the singular solutions. The parameter $p$ may be eliminated from equations (9), yielding the form

$$27y^2 = -4x^3 \tag{10}$$

for the singular solutions. See Figure 39.

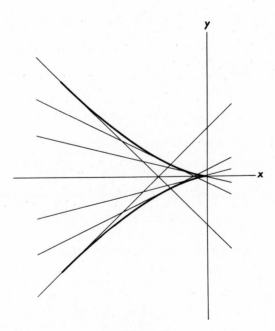

FIGURE 39

EXAMPLE (b):   Solve the differential equation

$$(x^2 - 1)p^2 - 2xyp + y^2 - 1 = 0. \tag{11}$$

Rewrite (11) as

$$x^2p^2 - 2xyp + y^2 - 1 - p^2 = 0.$$

Then it is clear that the equation is of the form

$$(y - xp)^2 - 1 - p^2 = 0 \tag{12}$$

and so could be broken up into two equations, each of Clairaut's form. Then the general solution of (11) is obtained by replacing $p$ everywhere in it by an arbitrary constant $c$. That is,

$$(x^2 - 1)c^2 - 2xyc + y^2 - 1 = 0 \tag{13}$$

is the general solution of (11). The solution (13) is composed of two families of straight lines,

$$y = c_1x + \sqrt{1 + c_1^2} \tag{14}$$

and

$$y = c_2x - \sqrt{1 + c_2^2}. \tag{15}$$

From the $p$-discriminant equation for (11) we obtain at once the singular solutions defined by

$$x^2 + y^2 = 1. \tag{16}$$

## Exercises

1. Let $\alpha$ be a parameter and prove that if $f''(\alpha)$ exists, then

$$x = -f'(\alpha), \qquad y = f(\alpha) - \alpha f'(\alpha) \tag{A}$$

is a solution of the differential equation $y = px + f(p)$. *Hint:* use $dx$ and $dy$ to get $p$ in terms of $\alpha$ and then show that $y - px - f(p)$ vanishes identically.
2. Prove that if $f''(\alpha) \neq 0$, then (A) above is not a special case of the general solution $y = cx + f(c)$. *Hint:* show that the slope of the graph of one solution depends upon $x$ whereas the slope of the graph of the other does not depend upon $x$.

In exercises 3 through 30, find the general solution and also the singular solution, if it exists.

3. $p^2 + x^3p - 2x^2y = 0$.           ANS.   $c^2 + cx^2 = 2y$; sing. sol., $8y = -x^4$.
4. $p^2 + 4x^5p - 12x^4y = 0$.         ANS.   $12y = c(c + 4x^3)$; sing. sol., $3y = -x^6$.
5. $2xp^3 - 6yp^2 + x^4 = 0$.          ANS.   $2c^3x^3 = 1 - 6c^2y$; sing. sol., $2y = x^2$.
6. $p^2 - xp + y = 0$.                 ANS.   $y = cx - c^2$; sing. sol., $x^2 = 4y$.

**7.** $y = px + kp^2$.     ANS.   $y = cx + kc^2$; sing. sol., $x^2 = -4ky$.

**8.** $x^8p^2 + 3xp + 9y = 0$.     ANS.   $x^3(y + c^2) + c = 0$; sing. sol., $4x^6y = 1$.

**9.** $x^4p^2 + 2x^3yp - 4 = 0$.     ANS.   $x^2(1 + cy) = c^2$.

**10.** $xp^2 - 2yp + 4x = 0$.     ANS.   $x^2 = c(y - c)$; sing. sol., $y = 2x$ and $y = -2x$.

**11.** $3x^4p^2 - xp - y = 0$.     ANS.   $xy = c(3cx - 1)$; sing. sol., $12x^2y = -1$.

**12.** $xp^2 + (x - y)p + 1 - y = 0$.
    ANS.   $xc^2 + (x - y)c + 1 - y = 0$; sing. sol., $(x + y)^2 = 4x$.

**13.** $p(xp - y + k) + a = 0$.
    ANS.   $c(xc - y + k) + a = 0$; sing. sol., $(y - k)^2 = 4ax$.

**14.** $x^6p^3 - 3xp - 3y = 0$.     ANS.   $3xy = c(xc^2 - 3)$; sing. sol., $9x^3y^2 = 4$.

**15.** $y = x^6p^3 - xp$.     ANS.   $xy = c(c^2x - 1)$; sing. sol., $27x^3y^2 = 4$.

**16.** $xp^4 - 2yp^3 + 12x^3 = 0$.     ANS.   $2c^3y = c^4x^2 + 12$; sing. sol., $3y^2 = \pm 8x^3$.

**17.** $xp^3 - yp^2 + 1 = 0$.     ANS.   $xc^3 - yc^2 + 1 = 0$; sing. sol., $4y^3 = 27x^2$.

**18.** $y = px + p^n$; for $n \neq 0, n \neq 1$.
    ANS.   $y = cx + c^n$; sing. sol., $\left(\dfrac{x}{n}\right)^n = -\left(\dfrac{y}{n-1}\right)^{n-1}$.

**19.** $p^2 - xp - y = 0$.     ANS.   $3x = 2p + cp^{-1/2}$ and $3y = p^2 - cp^{1/2}$.

**20.** $2p^3 + xp - 2y = 0$.     ANS.   $x = 2p(3p + c)$ and $y = p^2(4p + c)$.

**21.** $2p^2 + xp - 2y = 0$.     ANS.   $x = 4p \ln|pc|$ and $y = p^2[1 + 2 \ln|pc|]$.

**22.** $p^3 + 2xp - y = 0$.     ANS.   $4x = -3p^2 + cp^{-2}$ and $2y = -p^3 + cp^{-1}$.

**23.** $4xp^2 - 3yp + 3 = 0$.     ANS.   $2x = 3p^{-2} + cp^{-4}$ and $3y = 9p^{-1} + 2cp^{-3}$.

**24.** $p^3 - xp + 2y = 0$.     ANS.   $x = p(c - 3p)$ and $2y = p^2(c - 4p)$.

**25.** $5p^2 + 6xp - 2y = 0$.     ANS.   $p^3(x + p)^2 = c$ and $2y = 6xp + 5p^2$.

**26.** $2xp^2 + (2x - y)p + 1 - y = 0$.     ANS.   $p^2x = (1 + p)^{-1} + \ln|c(1 + p)|$ and
    $py = 1 + (1 + p)^{-1} + 2 \ln|c(1 + p)|$.

**27.** $5p^2 + 3xp - y = 0$.     ANS.   $p^3(x + 2p)^2 = c$ and $y = 3xp + 5p^2$.

**28.** $p^2 + 3xp - y = 0$.     ANS.   $p^3(5x + 2p)^2 = c$ and $y = 3xp + p^2$.

**29.** $y = xp + x^3p^2$.     ANS.   $x^2 = cp^{-4/3} - 2p^{-1}$ and $y = xp + x^3p^2$.

**30.** $8y = 3x^2 + p^2$.     ANS.   $(p - 3x)^3 = c(p - x)$ and $8y = 3x^2 + p^2$.

## 100. Dependent variable missing

Consider a second-order equation,

$$f(x, y', y'') = 0, \tag{1}$$

which does not contain the dependent variable $y$ explicitly. Let us put

$$y' = p.$$

Then

$$y'' = \frac{dp}{dx}$$

and equation (1) may be replaced by

$$f\left(x, p, \frac{dp}{dx}\right) = 0,$$   (2)

an equation of order one in $p$. If we can find $p$ from equation (2), then $y$ can be obtained from $y' = p$ by an integration.

EXAMPLE:   Solve the equation

$$xy'' - (y')^3 - y' = 0$$   (3)

of Example (b), page 12.

Because $y$ does not appear explicitly in the differential equation (3), put $y' = p$. Then

$$y'' = \frac{dp}{dx},$$

so equation (3) becomes

$$x\frac{dp}{dx} - p^3 - p = 0.$$

Separation of variables leads to

$$\frac{dp}{p(p^2 + 1)} = \frac{dx}{x}$$

or

$$\frac{dp}{p} - \frac{p\,dp}{p^2 + 1} = \frac{dx}{x},$$

from which

$$\ln|p| - \tfrac{1}{2}\ln(p^2 + 1) + \ln|c_1| = \ln|x|$$   (4)

follows.

Equation (4) yields

$$c_1 p(p^2 + 1)^{-1/2} = x,$$   (5)

which we wish to solve for $p$. From (5) we conclude that

$$c_1^2 p^2 = x^2(1 + p^2),$$

$$p^2 = \frac{x^2}{c_1^2 - x^2}.$$

But $p = y'$, so we have

$$dy = \pm\frac{x\,dx}{\sqrt{c_1^2 - x^2}}.$$   (6)

The solutions of (6) are

$$y - c_2 = \mp(c_1^2 - x^2)^{1/2},$$

or

$$x^2 + (y - c_2)^2 = c_1^2. \tag{7}$$

Equation (7) is the desired general solution of the differential equation (3). Note that in dividing by $p$ early in the work we might have discarded the solutions $y = k(p = 0)$, where $k$ is constant. But (7) can be put in the form

$$c_3(x^2 + y^2) + c_4y + 1 = 0, \tag{8}$$

with new arbitrary constants $c_3$ and $c_4$. Then the choice $c_3 = 0, c_4 = -1/k$ yields the solution $y = k$.

## 101. Independent variable missing

A second-order equation

$$f(y, y', y'') = 0 \tag{1}$$

in which the independent variable $x$ does not appear explicitly can be reduced to a first-order equation in $y$ and $y'$. Put

$$y' = p,$$

then

$$y'' = \frac{dp}{dx} = \frac{dy}{dx}\frac{dp}{dy} = p\frac{dp}{dy},$$

so equation (1) becomes

$$f\left(y, p, p\frac{dp}{dy}\right) = 0. \tag{2}$$

We try to determine $p$ in terms of $y$ from equation (2) and then substitute the result into $y' = p$.

EXAMPLE:   Solve the equation

$$yy'' + (y')^2 + 1 = 0 \tag{3}$$

of exercise 8, page 14.

Since the independent variable does not appear explicitly in equation (3), we put $y' = p$ and obtain

$$y'' = p\frac{dp}{dy}$$

as before. Then equation (3) becomes

$$yp\frac{dp}{dy} + p^2 + 1 = 0, \tag{4}$$

in which the variables $p$ and $y$ are easily separated.

From (4) it follows that

$$\frac{p\,dp}{p^2 + 1} + \frac{dy}{y} = 0,$$

from which

$$\tfrac{1}{2}\ln(p^2 + 1) + \ln|y| = \ln|c_1|,$$

so

$$p^2 + 1 = c_1^2 y^{-2}. \tag{5}$$

We solve equation (5) for $p$ and find that

$$p = \pm\frac{(c_1^2 - y^2)^{1/2}}{y}.$$

Therefore

$$\frac{dy}{dx} = \pm\frac{(c_1^2 - y^2)^{1/2}}{y}$$

or

$$\pm y(c_1^2 - y^2)^{-1/2}\,dy = dx.$$

Then

$$\mp(c_1^2 - y^2)^{1/2} = x - c_2$$

from which we obtain the final result

$$(x - c_2)^2 + y^2 = c_1^2.$$

## Exercises

1. $y'' = x(y')^3$.    ANS.    $x = c_1\sin(y + c_2)$.
2. $x^2 y'' + (y')^2 - 2xy' = 0$; when $x = 2$, $y = 5$, $y' = -4$.
    ANS.    $y = \tfrac{1}{2}x^2 + 3x - 3 + 9\ln(3 - x)$.
3. $x^2 y'' + (y')^2 - 2xy' = 0$; when $x = 2$, $y = 5$, $y' = 2$.    ANS.    $x^2 = 2(y - 3)$.
4. $yy'' + (y')^2 = 0$.    ANS.    See exercise 20, page 15.
5. $y^2 y'' + (y')^3 = 0$.    ANS.    $x = c_1 y - \ln|c_2 y|$.
6. $(y + 1)y'' = (y')^2$.    ANS.    $y + 1 = c_2 e^{c_1 x}$.
7. $2ay'' + (y')^3 = 0$.    ANS.    $(y - c_2)^2 = 4a(x - c_1)$.
8. Do exercise 7 by another method.

**9.** $xy'' = y' + x^5$; when $x = 1$, $y = \frac{1}{2}$, $y' = 1$.      ANS.   $24y = x^6 + 9x^2 + 2$.

**10.** $xy'' + y' + x = 0$; when $x = 2$, $y = -1$, $y' = -\frac{1}{2}$.    ANS.   $y = -\frac{1}{4}x^2 + \ln(\frac{1}{2}x)$.

**11.** $y'' = 2y(y')^3$.                              ANS.   $y^3 = 3(c_2 - x - c_1 y)$.

**12.** $yy'' + (y')^3 - (y')^2 = 0$.             ANS.   $x = y - c_1 \ln |c_2 y|$.

**13.** $y'' + \beta^2 y = 0$. Check your result by solving the equation in two ways.

**14.** $yy'' + (y')^3 = 0$.                    ANS.   $x = c_1 + y \ln |c_2 y|$.

**15.** $y'' \cos x = y'$.                 ANS.   $y = c_2 + c_1 \ln(1 - \sin x)$.

**16.** $y'' = x(y')^2$; when $x = 2$, $y = \frac{1}{4}\pi$, $y' = -\frac{1}{4}$.       ANS.   $x = 2 \cot y$.

**17.** $y'' = x(y')^2$; when $x = 0$, $y = 1$, $y' = \frac{1}{2}$.      ANS.   $y = 1 + \frac{1}{2} \ln \dfrac{2 + x}{2 - x}$.

**18.** $y'' = -e^{-2y}$; when $x = 3$, $y = 0$, $y' = 1$.      ANS.   $y = \ln(x - 2)$.

**19.** $y'' = -e^{-2y}$; when $x = 3$, $y = 0$, $y' = -1$.      ANS.   $y = \ln(4 - x)$.

**20.** $2y'' = \sin 2y$; when $x = 0$, $y = \pi/2$, $y' = 1$.    ANS.   $x = -\ln(\csc y + \cot y)$.

**21.** $2y'' = \sin 2y$; when $x = 0$, $y = -\pi/2$, $y' = 1$.    ANS.   $x = \ln(-\csc y - \cot y)$.

**22.** Show that if you can perform the integrations encountered, then you can solve any equation of the form $y'' = f(y)$.

**23.** $x^3 y'' - x^2 y' = 3 - x^2$.         ANS.   $y = x^{-1} + x + c_1 x^2 + c_2$.

**24.** $y'' = (y')^2$.      ANS.   $y = -\ln |c_2(c_1 - x)|$; or $x = c_1 + c_3 e^{-y}$.

**25.** $y'' = e^x (y')^2$.           ANS.   $c_1 y + c_2 = -\ln |c_1 e^{-x} - 1|$.

**26.** $2y'' = (y')^3 \sin 2x$; when $x = 0$, $y = 1$, $y' = 1$.    ANS.   $y = 1 + \ln(\sec x + \tan x)$.

**27.** $x^2 y'' + (y')^2 = 0$.         ANS.   $c_1^2 y = c_1 x + \ln |c_2(c_1 x - 1)|$.

**28.** $y'' = 1 + (y')^2$.           ANS.   $e^y \cos(x + c_1) = c_2$.

**29.** Do exercise 28 by another method.

**30.** $y'' = [1 + (y')^2]^{3/2}$. Solve in three ways, by considering the geometric significance of the equation, and by the methods of this chapter.

**31.** $yy'' = (y')^2[1 - y' \sin y - yy' \cos y]$.      ANS.   $x = c_1 \ln |c_2 y| - \cos y$.

**32.** $(1 + y^2)y'' + (y')^3 + y' = 0$.     ANS.   $x = c_2 + c_1 y - (1 + c_1^2) \ln |y + c_1|$.

**33.** $[yy'' + 1 + (y')^2]^2 = [1 + (y')^2]^3$.     ANS.   $(y - c_1)^2 + (x - c_2)^2 = c_1^2$.

**34.** $x^2 y'' = y'(2x - y')$; when $x = -1$, $y = 5$, $y' = 1$.

           ANS.   $2y - 1 = (x - 2)^2 + 8 \ln(x + 2)$.

**35.** $x^2 y'' = y'(3x - 2y')$.        ANS.   $2y = x^2 + c_2 - c_1 \ln |x^2 + c_1|$.

**36.** $xy'' = y'(2 - 3xy')$.           ANS.   $3y = c_2 + \ln |x^3 + c_1|$.

**37.** $x^4 y'' = y'(y' + x^3)$; when $x = 1$, $y = 2$, $y' = 1$.

           ANS.   $y = 1 + x^2 - \ln \left( \dfrac{1 + x^2}{2} \right)$.

**38.** $y'' = 2x + (x^2 - y')^2$.         ANS.   $3y = x^3 + c_2 - 3 \ln |x + c_1|$.

**39.** $(y'')^2 - 2y'' + (y')^2 - 2xy' + x^2 = 0$; when $x = 0$, $y = \frac{1}{2}$ and $y' = 1$.

           ANS.   $2y = 1 + x^2 + 2 \sin x$.

**40.** $(y'')^2 - xy'' + y' = 0$.      ANS.   General solution: $2y = c_1 x^2 - 2c_1^2 x + c_2$; family of singular solutions: $12y = x^3 + k$.

**41.** $(y'')^3 = 12y'(xy'' - 2y')$.      ANS.   General solution: $y = c_1(x - c_1)^3 + c_2$; family of singular solutions: $9y = x^4 + k$.

**42.** $3yy'y'' = (y')^3 - 1$.         ANS.   $27c_1(y + c_1)^2 = 8(x + c_2)^3$.

**43.** $4y(y')^2 y'' = (y')^4 + 3$.       ANS.   $256c_1(y - c_1)^3 = 243(x - c_2)^4$.

FIGURE 40

## 102. The catenary

Let a cable of uniformly distributed weight $w$ (lb/ft) be suspended between two supports at points $A$ and $B$ as indicated in Figure 40. The cable will sag and there will be a lowest point $V$ as indicated in the figure. We wish to determine the curve formed by the suspended cable. That curve is called the *catenary*.

Choose coordinate axes as shown in Figure 41, the $y$-axis vertical through the point $V$ and the $x$-axis horizontal and passing at a distance $y_0$ (to be chosen later) below $V$. Let $s$ represent length (ft) of the cable measured from $V$ to the variable point $P$ with coordinates $(x, y)$. Then the portion of the cable from $V$ to $P$ is subject to the three forces shown in Figure 41. Those forces are: (a) the gravitational force $ws$ (lb) acting downward through the center of gravity of the portion of the cable from $V$ to $P$, (b) the tension $T_1$ (lb) acting tangentially at $P$, and (c) the tension $T_2$ (lb) acting horizontally (again tangentially) at $V$. The tension $T_1$ is a variable; the tension $T_2$ is constant.

FIGURE 41

Since equilibrium is assumed, the algebraic sum of the vertical components of these forces is zero and the algebraic sum of the horizontal components of these forces is also zero. Therefore, if $\theta$ is the angle of inclination, from the horizontal, of the tangent to the curve at the point $(x, y)$, we have

$$T_1 \sin \theta - ws = 0 \tag{1}$$

and

$$T_1 \cos \theta - T_2 = 0. \tag{2}$$

But $\tan \theta$ is the slope of the curve of the cable, so

$$\tan \theta = \frac{dy}{dx}. \tag{3}$$

We may eliminate the variable tension $T_1$ from equations (1) and (2) and obtain

$$\tan \theta = \frac{ws}{T_2}. \tag{4}$$

The constant $T_2/w$ has the dimension of a length. Put $T_2/w = a$ (ft). Then equation (4) becomes

$$\tan \theta = \frac{s}{a}. \tag{5}$$

From equations (3) and (5) we see that

$$\frac{s}{a} = \frac{dy}{dx}. \tag{6}$$

Now we know from calculus that since $s$ is the length of arc of the curve, then

$$\frac{ds}{dx} = \sqrt{1 + \left(\frac{dy}{dx}\right)^2}. \tag{7}$$

From (6) we get

$$\frac{1}{a}\frac{ds}{dx} = \frac{d^2y}{dx^2},$$

so the elimination of $s$ yields the differential equation

$$\frac{d^2y}{dx^2} = \frac{1}{a}\sqrt{1 + \left(\frac{dy}{dx}\right)^2}. \tag{8}$$

The desired equation of the curve assumed by the suspended cable is that solution of the differential equation (8) which also satisfies the initial conditions

$$\text{when } x = 0, \qquad y = y_0 \text{ and } \frac{dy}{dx} = 0. \tag{9}$$

Equation (8) fits into either of the types studied in this chapter. It is left as an exercise for the student to solve the differential equation (8) with the conditions (9) and arrive at the result

$$y = a \cosh \frac{x}{a} + y_0 - a. \tag{10}$$

Then, of course, the sensible choice $y_0 = a$ is made, so the equation of the desired curve (the catenary) is

$$y = a \cosh \frac{x}{a}.$$

## Miscellaneous Exercises

1. $x^3 p^2 + x^2 yp + 4 = 0.$  ANS.  $cxy + 4x + c^2 = 0$; sing. sol., $xy^2 = 16.$
2. $6xp^2 - (3x + 2y)p + y = 0.$  ANS.  $y^3 = c_1 x, 2y = x + c_2.$
3. $9p^2 + 3xy^4 p + y^5 = 0.$  ANS.  $cy^3(x - c) = 1$; sing. sol., $x^2 y^3 = 4.$
4. $4y^3 p^2 - 4xp + y = 0.$  ANS.  $y^4 = 4c(x - c)$; sing. sol., $y^2 = x.$
5. $x^6 p^2 - 2xp - 4y = 0.$  ANS.  $x^2(y - c^2) = c$; sing. sol., $4x^4 y = -1.$
6. $5p^2 + 6xp - 2y = 0.$  ANS.  $x = cp^{-3/2} - p$ and $2y = 6cp^{-1/2} - p^2.$
7. Do exercise 6 by another method.
8. $y^2 p^2 - y(x + 1)p + x = 0.$  ANS.  $x^2 - y^2 = c_1, y^2 = 2(x - c_2).$
9. $4x^5 p^2 + 12x^4 yp + 9 = 0.$  ANS.  $x^3(2cy - 1) = c^2$; sing. sol., $x^3 y^2 = 1.$
10. $4y^2 p^3 - 2xp + y = 0.$  ANS.  $y^2 = 2c(x - 2c^2)$; sing. sol., $8x^3 = 27y^4.$
11. $p^4 + xp - 3y = 0.$  ANS.  $5x = 4p^3 + cp^{1/2}$ and $15y = 9p^4 + cp^{3/2}.$
12. Do exercise 11 by another method.
13. $xp^2 + (k - x - y)p + y = 0$, the equation of exercise 5, page 14.

   ANS.  $xc^2 + (k - x - y)c + y = 0$; sing. sol., $(x - y)^2 - 2k(x + y) + k^2 = 0.$

14. $x^2 p^3 - 2xyp^2 + y^2 p + 1 = 0.$

   ANS.  $x^2 c^3 - 2xyc^2 + y^2 c + 1 = 0$; sing. sol., $27x = -4y^3.$

15. $16xp^2 + 8yp + y^6 = 0.$  ANS.  $y^2(c^2 x + 1) = 2c$; sing. sol., $xy^4 = 1.$
16. $xp^2 - (x^2 + 1)p + x = 0.$  ANS.  $x^2 = 2(y - c_1), y = \ln |c_2 x|.$
17. $p^3 - 2xp - y = 0.$  ANS.  $8x = 3p^2 + cp^{-2/3}$ and $4y = p^3 - cp^{1/3}.$
18. Do exercise 17 by another method.
19. $9xy^4 p^2 - 3y^5 p - 1 = 0.$  ANS.  $cy^3 = c^2 x - 1$; sing. sol., $y^6 = -4x.$
20. $x^2 p^2 - (2xy + 1)p + y^2 + 1 = 0.$

   ANS.  $x^2 c^2 - (2xy + 1)c + y^2 + 1 = 0$; sing. sol., $4x^2 - 4xy - 1 = 0.$

21. $x^6 p^2 = 8(2y + xp).$  ANS.  $c^2 x^2 = 8(2x^2 y - c)$; sing. sol., $x^4 y = -1.$

**22.** $x^2p^2 = (x - y)^2$.      ANS.    $x(x - 2y) = c_1, y = -x \ln |c_2 x|$.

**23.** $xp^3 - 2yp^2 + 4x^2 = 0$. See exercise 10 above.

           ANS.    $x^2 = 4c(y - 8c^2)$; sing., sol., $8y^3 = 27x^4$.

**24.** $(p + 1)^2(y - px) = 1$.

           ANS.    $(c + 1)^2(y - cx) = 1$; sing. sol., $4(x + y)^3 = 27x^2$.

**25.** $p^3 - p^2 + xp - y = 0$.

           ANS.    $y = cx + c^3 - c^2$; sing. sol. with parametric equations,

$$x = 2\alpha - 3\alpha^2 \text{ and } y = \alpha^2 - 2\alpha^3.$$

**26.** $xp^2 + y(1 - x)p - y^2 = 0$.      ANS.    $xy = c_1, x = \ln |c_2 y|$.

**27.** $yp^2 - (x + y)p + y = 0$.

           ANS.    $py = c \exp(p^{-1})$ and $px = y(p^2 - p + 1)$; sing. sol., $y = x$.

# Power Series Solutions

## 103. Linear equations and power series

The solution of linear equations with constant coefficients can be accomplished by the methods developed earlier in the book. The general linear equation of the first order yields to an integrating factor as was seen in Chapter 2. For linear ordinary differential equations with variable coefficients and of order greater than one, probably the most generally effective method of attack is that based upon the use of power series.

To simplify the work and the statement of theorems, the equations treated here will be restricted to those with polynomial coefficients. The difficulties to be encountered, the methods of attack, and the results accomplished all remain essentially unchanged when the coefficients are permitted to be functions that have power series expansions valid about some point. (Such functions are called *analytic functions*.)

Consider the homogeneous linear equation of the second order,

$$b_0(x)y'' + b_1(x)y' + b_2(x)y = 0, \tag{1}$$

with polynomial coefficients. If $b_0(x)$ does not vanish at $x = 0$, then in some interval about $x = 0$, staying away from the nearest point where $b_0(x)$ does vanish, it is safe to divide throughout by $b_0(x)$. Thus we replace equation (1) by

$$y'' + p(x)y' + q(x)y = 0, \tag{2}$$

in which the coefficients $p(x)$, $q(x)$ are rational functions of $x$ with denominators that do not vanish at $x = 0$.

We shall now show that it is reasonable to expect* a solution of (2) that is a power series in $x$ and that contains two arbitrary constants. Let $y = y(x)$ be a solution of equation (2). We assign arbitrarily the values of $y$ and $y'$ at $x = 0$; $y(0) = A$, $y'(0) = B$.

Equation (2) yields

$$y''(x) = -p(x)y'(x) - q(x)y(x), \tag{3}$$

so $y''(0)$ may be computed directly, because $p(x)$ and $q(x)$ are well behaved at $x = 0$. From equation (3) we get

$$y'''(x) = -p(x)y''(x) - p'(x)y'(x) - q(x)y'(x) - q'(x)y(x), \tag{4}$$

so $y'''(0)$ can be computed once $y''(0)$ is known.

The above process can be continued as long as we wish; therefore, we can determine successively $y^{(n)}(0)$ for as many integral values of $n$ as may be desired. Now, by Maclaurin's formula in calculus,

$$y(x) = y(0) + \sum_{n=1}^{\infty} y^{(n)}(0)\frac{x^n}{n!}; \tag{5}$$

that is, the right member of (5) will converge to the value $y(x)$ throughout some interval about $x = 0$ if $y(x)$ is sufficiently well behaved at and near $x = 0$. Thus we can determine the function $y(x)$ and are led to a solution in power series form.

For actually obtaining the solutions for specific equations, we shall study another method, to be illustrated in examples, a technique far superior to the brute-force method used above. What we have gained from the present discussion is the knowledge that it is reasonable to seek a power series solution. Once we know that, it remains only to develop good methods for finding the solution and theorems regarding the validity of the results so found.

---

* This is no proof. For proof see, for instance, E. D. Rainville, *Intermediate Differential Equations*, 2nd ed. (New York: Macmillan Publishing Co., Inc., 1964), pp. 67–71.

## 104. Convergence of power series

From calculus we know that the power series

$$\sum_{n=0}^{\infty} a_n x^n$$

converges either at $x = 0$ only, or for all finite $x$, or the series converges in an interval $-R < x < R$ and diverges outside that interval. Unless the series converges at only one point, it represents, where it does converge, a function $f(x)$ in the sense that the series has at each value $x$ the sum $f(x)$.

If

$$f(x) = \sum_{n=0}^{\infty} a_n x^n, \qquad -R < x < R,$$

then also

$$f'(x) = \sum_{n=0}^{\infty} n a_n x^{n-1}, \qquad -R < x < R,$$

and

$$\int_0^x f(y)\,dy = \sum_{n=0}^{\infty} \frac{a_n x^{n+1}}{n+1}, \qquad -R < x < R.$$

That is, the series is termwise differentiable and integrable in the sense that the series of the derivatives of the separate terms converges to the derivative of the sum of the original series and similarly for integration. It is important that the interval of convergence remains unchanged. We are not concerned here with convergence behavior at the endpoints of that interval.

Let us look more closely into the reason that a series has a particular interval of convergence rather than some other one. An elementary example from calculus is

$$\frac{1}{1-x} = \sum_{n=0}^{\infty} x^n, \qquad -1 < x < 1. \tag{1}$$

It is reasonable to suspect that the misbehavior of the function $1/(1-x)$ at $x = 1$ is what lies behind the fact that the interval of convergence terminates at $x = 1$. That it must extend that far is not so evident. The point $x = -1$ has no bearing in this instance; the interval is terminated at that end by the requirement that it be symmetric about $x = 0$.

Now let us replace $x$ in (1) by $(-x^2)$ to get

$$\frac{1}{1+x^2} = \sum_{n=0}^{\infty} (-1)^n x^{2n}, \qquad -1 < x < 1. \tag{2}$$

Again the interval of convergence is $-1 < x < 1$, but the function $f(x) = 1/(1 + x^2)$ is well behaved for all real $x$. What stopped the interval of convergence at $x = 1$ is not so clear. The fact is that we need to consider $x$ as a complex variable to understand what is going on here.

We use the ordinary Argand diagram for complex numbers. Let $x = a + ib$, $a$ and $b$ real, with $i = \sqrt{-1}$, and associate with the point $(a, b)$ in the plane the number $a + ib$. Now mark on the diagram those points (values of $x$) for which the function $1/(1 + x^2)$ does not exist. The points are $x = i$, $x = -i$, where the denominator $(1 + x^2)$ vanishes. In books on functions of a complex variable it is shown that the series in (2) converges for all values of $x$ inside the circle shown in Figure 42. The interval of convergence given in (2) is merely a cross section of the region of convergence in the complex plane.

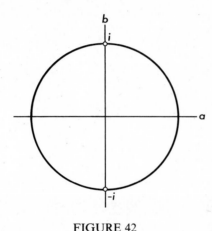

FIGURE 42

A power series in a complex variable $x$ always has as its region of convergence the interior of a circle, if we are willing to admit the extreme cases in which the circle degenerates into a single point or expands over the whole complex plane.

Points at which the denominator of a rational function vanishes are the most elementary examples of singularities of an analytic function. The circle of convergence of a power series cannot have inside it a singularity of the function represented by the series. The circle of convergence has its center at the origin and passes through the singularity nearest the origin.

The function

$$\frac{x - 2}{(x - 3)(x - 4)(x^2 + 25)}$$

has a denominator which vanishes at $x = 3, 4, 5i, -5i$. The function then has a power series expansion valid inside a circle (Figure 43) with center at $x = 0$ and extending as far as the nearest ($x = 3$) of the points where the function misbehaves. For real $x$, the interval of convergence is $-3 < x < 3$.

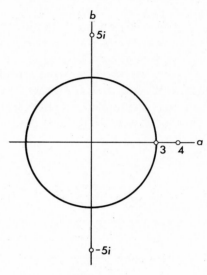

FIGURE 43

## 105. Ordinary points and singular points

For a linear differential equation

$$b_0(x)y^{(n)} + b_1(x)y^{(n-1)} + \cdots + b_n(x)y = R(x) \tag{1}$$

with polynomial coefficients, the point $x = x_0$ is called an *ordinary point* of the equation if $b_0(x_0) \neq 0$. A *singular point* of the linear equation (1) is any point $x = x_1$ for which $b_0(x_1) = 0$. In this chapter we shall obtain power series solutions valid near an ordinary point of a linear equation. In the next chapter we shall get power series solutions valid near a certain kind of singular point of the equation.

A knowledge of the location of the singular points of a differential equation will be useful to us later. Any point that is not a singular point is an ordinary point, so we list only the former.

We have left undiscussed the matter of a "point at infinity" in the complex plane, a concept of great utility. It is not necessary for this elementary discussion. As a result of this omission, however, it is necessary for us to attach the words "in the finite plane" to any statement purporting to list all singular points of a differential equation. The concept of a point at infinity will be introduced in Section 117.

The differential equation

$$(1 - x^2)y'' - 6xy' - 4y = 0 \tag{2}$$

has $x = 1$ and $x = -1$ as its only singular points in the finite complex plane. The equation

$$y'' + 2xy' + y = 0$$

has no singular points in the finite plane. The equation

$$xy'' + y' + xy = 0$$

has the origin, $x = 0$, as the only singular point in the finite plane.

### Exercises

For each equation, list all of the singular points in the finite plane.

1. $(x^2 + 4)y'' - 6xy' + 3y = 0$.                    ANS.    $x = 2i, -2i$.
2. $x(3 - x)y'' - (3 - x)y' + 4xy = 0$.               ANS.    $x = 0, 3$.
3. $4y'' + 3xy' + 2y = 0$.                            ANS.    None.
4. $x(x - 1)^2 y'' + 3xy' + (x - 1)y = 0$.            ANS.    $x = 0, 1$.
5. $x^2 y'' + xy' + (1 - x^2)y = 0$.
6. $x^4 y'' + y = 0$.
7. $(1 + x^2)y'' - 2xy' + 6y = 0$.
8. $(x^2 - 4x + 3)y'' + x^2 y' - 4y = 0$.
9. $x^2(1 - x)^3 y'' + (1 + 2x)y = 0$.
10. $6xy'' + (1 - x^2)y' + 2y = 0$.
11. $4xy'' + y = 0$.
12. $4y'' + y = 0$.
13. $x^2(x^2 - 9)y'' + 3xy' - y = 0$.
14. $x^2(1 + 4x^2)y'' - 4xy' + y = 0$.
15. $(2x + 1)(x - 3)y'' - y' + (2x + 1)y = 0$.
16. $x^3(x^2 - 4)^2 y'' + 2(x^2 - 4)y' - xy = 0$.
17. $x(x^2 + 1)^2 y'' - y = 0$.
18. $(x^2 + 6x + 8)y'' + 3y = 0$.
19. $(4x + 1)y'' + 3xy' + y = 0$.

## 106. Validity of the solutions near an ordinary point

Suppose $x = 0$ is an ordinary point of the linear equation

$$b_0(x)y'' + b_1(x)y' + b_2(x)y = 0. \tag{1}$$

It is proved in more advanced books that there is a solution

$$y = \sum_{n=0}^{\infty} a_n x^n \tag{2}$$

that contains two arbitrary constants, namely, $a_0$ and $a_1$, and converges inside a circle with center at $x = 0$ and extending out to the singular point (or points) nearest $x = 0$. If the differential equation has no singular points in the finite plane, then the solution (2) is valid for all finite $x$. There remains, of course, the job of finding $a_n$ for $n \geq 2$. That is a major part of our work in solving a particular equation.

It is necessary to realize that the theorem quoted states that the series involved converges inside a certain circle. It does not state that the series diverges outside that circle. In a particular instance it may be that the circle of convergence happens to extend farther than the minimum given by the theorem. In any case, the circle of convergence passes through a singular point of the equation—it may not be the nearest singular point.

## 107. Solutions near an ordinary point

In solving numerical equations, the technique employed in the following examples will be found useful.

EXAMPLE (a): Solve the equation

$$y'' + 4y = 0 \tag{1}$$

near the ordinary point $x = 0$.

The equation we are considering has no singular points in the finite plane. Hence we may expect to find a solution

$$y = \sum_{n=0}^{\infty} a_n x^n \tag{2}$$

valid for all real $x$ and with $a_0$ and $a_1$ arbitrary. Substituting the series into (1) gives us

$$\sum_{n=0}^{\infty} n(n-1)a_n x^{n-2} + 4 \sum_{n=0}^{\infty} a_n x^n = 0. \tag{3}$$

We now change the indexing of the terms in the second sum in (3), so that the series will involve $x^{n-2}$ in its general term. Thus (3) becomes

$$\sum_{n=0}^{\infty} n(n-1)a_n x^{n-2} + 4 \sum_{n=2}^{\infty} a_{n-2} x^{n-2} = 0. \tag{4}$$

We can now add the two series to obtain

$$\sum_{n=2}^{\infty} [n(n-1)a_n + 4a_{n-2}]x^{n-2} = 0, \tag{5}$$

because the first two terms of the first sum in (4) are zero.

We now use the fact that, for a power series to vanish identically over any interval, each coefficient in the series must be zero. Thus, for (5) to be valid in some interval, it must be true that

$$n(n-1)a_n + 4a_{n-2} = 0, \qquad \text{for } n \geq 2.$$

This relation may be written (since $n \geq 2$)

$$a_n = \frac{-4a_{n-2}}{n(n-1)}. \tag{6}$$

Relation (6) may be used to obtain $a_n$ for $n \geq 2$ in terms of $a_0$ and $a_1$ which are left arbitrary. We have

$$a_2 = \frac{-4a_0}{2 \cdot 1} \qquad\qquad a_3 = \frac{-4a_1}{3 \cdot 2}$$

$$a_4 = \frac{-4a_2}{4 \cdot 3} \qquad\qquad a_5 = \frac{-4a_3}{5 \cdot 4}$$

$$\vdots \qquad\qquad\qquad \vdots$$

$$a_{2k} = \frac{-4a_{2k-2}}{2k(2k-1)} \qquad a_{2k+1} = \frac{-4a_{2k-1}}{(2k+1)(2k)}.$$

In writing out these particular cases of equation (6) we have taken pains to keep the $a$'s with even or odd subscripts in separate columns. If we now multiply the corresponding members of the equations of the first column, we obtain

$$a_2 a_4 \cdots a_{2k} = \frac{(-1)^k 4^k}{(2k)!} a_0 a_2 \cdots a_{2k-2},$$

which simplifies to

$$a_{2k} = \frac{(-1)^k 4^k}{(2k)!} a_0, \qquad \text{for } k \geq 1. \tag{7}$$

A similar argument applied to the right column in the foregoing array gives us

$$a_{2k+1} = \frac{(-1)^k 4^k a_1}{(2k+1)!}, \qquad \text{for } k \geq 1. \tag{8}$$

We now wish to substitute the expressions for the $a$'s back into the assumed series for $y$,

$$y = \sum_{n=0}^{\infty} a_n x^n. \tag{2}$$

Since we have different forms for $a_{2k}$ and $a_{2k+1}$ we first rewrite (2) in the form

$$y = a_0 + \sum_{k=1}^{\infty} a_{2k} x^{2k} + a_1 x + \sum_{k=1}^{\infty} a_{2k+1} x^{2k+1},$$

and then we use (7) and (8) to obtain

$$y = a_0 \left[ 1 + \sum_{k=1}^{\infty} \frac{(-1)^k 4^k x^{2k}}{(2k)!} \right] + a_1 \left[ x + \sum_{k=1}^{\infty} \frac{(-1)^k 4^k x^{2k+1}}{(2k+1)!} \right]. \tag{9}$$

It is possible to rewrite equation (9) in the form

$$y = a_0 \left[ 1 + \sum_{k=1}^{\infty} \frac{(-1)^k (2x)^{2k}}{(2k)!} \right] + \tfrac{1}{2} a_1 \left[ 2x + \sum_{k=1}^{\infty} \frac{(-1)^k (2x)^{2k+1}}{(2k+1)!} \right]. \tag{10}$$

The two series in (10) are the Maclaurin series for the functions $\cos 2x$ and $\sin 2x$, so that finally we may write

$$y = a_0 \cos 2x + \tfrac{1}{2} a_1 \sin 2x.$$

Thus we have shown that the solution of equation (1) is a linear combination of $\cos 2x$ and $\sin 2x$, a fact that could have been obtained immediately by the methods of Chapter 6.

**EXAMPLE (b):**   Solve the equation

$$(1 - x^2) y'' - 6xy' - 4y = 0 \tag{11}$$

near the ordinary point $x = 0$.

The only singular points that this equation has in the finite plane are $x = 1$ and $x = -1$. Hence we know in advance that there is a solution

$$y = \sum_{n=0}^{\infty} a_n x^n \tag{12}$$

valid in $|x| < 1$ and with $a_0$ and $a_1$ arbitrary.

To determine the $a_n$, $n > 1$, we substitute the $y$ of equation (12) into the left member of (11). We get

$$\sum_{n=0}^{\infty} n(n-1) a_n x^{n-2} - \sum_{n=0}^{\infty} n(n-1) a_n x^n - \sum_{n=0}^{\infty} 6n a_n x^n - \sum_{n=0}^{\infty} 4a_n x^n = 0,$$

or

$$\sum_{n=0}^{\infty} n(n-1)a_n x^{n-2} - \sum_{n=0}^{\infty} (n^2 + 5n + 4)a_n x^n = 0, \tag{13}$$

in which we have combined series that contained the same powers of $x$.

Next let us factor the coefficient in the second series in equation (13), writing

$$\sum_{n=0}^{\infty} n(n-1)a_n x^{n-2} - \sum_{n=0}^{\infty} (n+1)(n+4)a_n x^n = 0. \tag{14}$$

Relations for the determination of the $a_n$ will be obtained by using the fact that, for a power series to vanish identically over any interval, each coefficient in the series must be zero. Therefore we wish next to write the two series in equation (14) in a form in which the exponents on $x$ will be the same so we can easily pick off the coefficient of each power of $x$.

Let us shift the index in the second series, replacing $n$ everywhere by $(n - 2)$. Then the summation which started with the old $n = 0$ will now start with $n - 2 = 0$, or the new $n = 2$. Thus we obtain

$$\sum_{n=0}^{\infty} n(n-1)a_n x^{n-2} - \sum_{n=2}^{\infty} (n-1)(n+2)a_{n-2} x^{n-2} = 0. \tag{15}$$

In equation (15) the coefficient of each separate power of $x$ must be zero. For $n = 0$ and $n = 1$, the second series has not yet started, so we get contributions from the first series only. In detail, we have

$$n = 0: \qquad 0 \cdot a_0 = 0,$$

$$n = 1: \qquad 0 \cdot a_1 = 0,$$

$$n \geq 2: \qquad n(n-1)a_n - (n-1)(n+2)a_{n-2} = 0.$$

As was expected, $a_0$ and $a_1$ are arbitrary. The relation for $n \geq 2$ can be used to determine the other $a$'s in terms of $a_0$ and $a_1$. Since

$$n(n-1) \neq 0$$

for $n \geq 2$, we can write

$$n \geq 2: \qquad a_n = \frac{n+2}{n} a_{n-2}. \tag{16}$$

Equation (16) is called a *recurrence relation*. It gives $a_n$ in terms of preceding $a$'s. In this particular case, each $a$ is determined by the $a$ with subscript two lower than its own and consequently, eventually, by either $a_0$ or $a_1$, according to whether the original $a$ had an even or an odd subscript.

A recurrence relation is a special kind of *difference equation.* In difference equations the arguments of the unknown function (the subscripts in our relations) need not differ by integers. There are books and courses on difference equations and the calculus of finite differences paralleling the books and courses on differential equations and calculus.

It is convenient to arrange the iterated instances of the relation (16) in two vertical columns (two columns because the subscript in (16) differ by two), thus using successively $n = 2, 4, 6, \ldots$, and $n = 3, 5, 7, \ldots$, to obtain

$$a_2 = \frac{4}{2} a_0 \qquad\qquad a_3 = \frac{5}{3} a_1$$

$$a_4 = \frac{6}{4} a_2 \qquad\qquad a_5 = \frac{7}{5} a_3$$

$$a_6 = \frac{8}{6} a_4 \qquad\qquad a_7 = \frac{9}{7} a_5$$

$$\vdots \qquad\qquad\qquad \vdots$$

$$a_{2k} = \frac{2k + 2}{2k} a_{2k-2} \qquad a_{2k+1} = \frac{2k + 3}{2k + 1} a_{2k-1}.$$

Next, obtain the product of corresponding members of the equations in the first column. The result,

$$k \geqq 1: \qquad a_2 a_4 a_6 \cdots a_{2k} = \frac{4 \cdot 6 \cdot 8 \cdots (2k + 2)}{2 \cdot 4 \cdot 6 \cdots (2k)} a_0 a_2 a_4 \cdots a_{2k-2},$$

simplifies at once to

$$k \geqq 1: \qquad a_{2k} = (k + 1) a_0,$$

thus giving us each $a$ with an even subscript in terms of $a_0$.

Similarly, from the right column in the array above we get

$$k \geqq 1: \qquad a_{2k+1} = \frac{5 \cdot 7 \cdot 9 \cdots (2k + 3)}{3 \cdot 5 \cdot 7 \cdots (2k + 1)} a_1,$$

or

$$k \geqq 1: \qquad a_{2k+1} = \frac{2k + 3}{3} a_1,$$

thus giving us each $a$ with an odd subscript in terms of $a_1$.

We need next to substitute the expressions we have obtained for the $a$'s into the assumed series for $y$,

$$y = \sum_{n=0}^{\infty} a_n x^n. \tag{12}$$

The nature of our expressions for the $a$'s, depending on whether the subscript is odd or even, dictates that we should first split the series in (12) into two series, one containing all the terms with even subscripts and the other containing all the terms with odd subscripts. We write

$$y = \left[ a_0 + \sum_{k=1}^{\infty} a_{2k}x^{2k} \right] + \left[ a_1x + \sum_{k=1}^{\infty} a_{2k+1}x^{2k+1} \right],$$

and then use our known results for $a_{2k}$ and $a_{2k+1}$ to obtain the general solution in the form

$$y = a_0 \left[ 1 + \sum_{k=1}^{\infty} (k+1)x^{2k} \right] + a_1 \left[ x + \sum_{k=1}^{\infty} \frac{2k+3}{3}x^{2k+1} \right]. \qquad (17)$$

These series converge at least for $|x| < 1$, as we know from the theory. That they converge there and only there can be verified by applying elementary convergence tests.

It happens in this example that the solution (17) may be written more simply as

$$y = a_0 \sum_{k=0}^{\infty} (k+1)x^{2k} + a_1 \sum_{k=0}^{\infty} \frac{2k+3}{3}x^{2k+1}. \qquad (18)$$

Indeed, the series can be summed in terms of elementary functions,

$$y = \frac{a_0}{(1-x^2)^2} + \frac{a_1(3x - x^3)}{3(1-x^2)^2}.$$

Such simplifications may be important when they can be accomplished in a particular problem, but it must be realized that our goal was to obtain equation (17) and to know where it is a valid solution. Additional steps taken after that goal is reached are frequently irrelevant to the essential desire to find a computable solution of the differential equation.

The length of the work in solving this particular equation is largely due to detailed steps, many of which will be taken mentally as we acquire more experience.

EXAMPLE (c):   Solve the equation

$$y'' + (x-1)^2 y' - 4(x-1)y = 0 \qquad (19)$$

about the ordinary point $x = 1$.

To solve an equation "about the point $x = x_0$" means to obtain solutions valid in a region surrounding the point, solutions expressed in powers of $(x - x_0)$.

We first translate the axes, putting $x - 1 = v$. Then equation (19) becomes

$$\frac{d^2y}{dv^2} + v^2\frac{dy}{dv} - 4vy = 0. \tag{20}$$

Always in a pure translation, $x - x_0 = v$, we have $dy/dx = dy/dv$, and so on.
As usual we put

$$y = \sum_{n=0}^{\infty} a_n v^n \tag{21}$$

and from (20) obtain

$$\sum_{n=0}^{\infty} n(n-1)a_n v^{n-2} + \sum_{n=0}^{\infty} na_n v^{n+1} - \sum_{n=0}^{\infty} 4a_n v^{n+1} = 0. \tag{22}$$

Collecting like terms in (22) yields

$$\sum_{n=0}^{\infty} n(n-1)a_n v^{n-2} + \sum_{n=0}^{\infty} (n-4)a_n v^{n+1} = 0,$$

which, with a shift of index from $n$ to $(n-3)$ in the second series, gives

$$\sum_{n=0}^{\infty} n(n-1)a_n v^{n-2} + \sum_{n=3}^{\infty} (n-7)a_{n-3} v^{n-2} = 0. \tag{23}$$

Therefore $a_0$ and $a_1$ are arbitrary and for the remainder we have

$$n = 2: \qquad 2a_2 = 0,$$

$$n \geq 3: \qquad n(n-1)a_n + (n-7)a_{n-3} = 0,$$

$$a_n = -\frac{n-7}{n(n-1)}a_{n-3}.$$

This time the $a$'s fall into three groups, those that come from $a_0$, from $a_1$, and from $a_2$. We use three columns:

| $a_0$ arbitrary | $a_1$ arbitrary | $a_2 = 0$ |
|---|---|---|
| $a_3 = -\dfrac{-4}{3\cdot 2}a_0$ | $a_4 = -\dfrac{-3}{4\cdot 3}a_1$ | $a_5 = -\dfrac{-2}{5\cdot 4}a_2 = 0$ |
| $a_6 = -\dfrac{-1}{6\cdot 5}a_3$ | $a_7 = -\dfrac{0}{7\cdot 6}a_4 = 0$ | $a_8 = (\ )a_5 = 0$ |
| $a_9 = -\dfrac{2}{9\cdot 8}a_6$ | $a_{10} = -\dfrac{3}{10\cdot 9}a_7 = 0$ | $a_{11} = 0$ |
| $\vdots$ | $\vdots$ | $\vdots$ |
| $a_{3k} = -\dfrac{3k-7}{3k(3k-1)}a_{3k-3}$ | $a_{3k+1} = 0, k \geq 2$ | $a_{3k+2} = 0, k \geq 1.$ |

With the usual multiplication scheme, the first column yields

$$k \geq 1: \qquad a_{3k} = \frac{(-1)^k[(-4)(-1) \cdot 2 \cdots (3k-7)]a_0}{[3 \cdot 6 \cdot 9 \cdots (3k)][2 \cdot 5 \cdot 8 \cdots (3k-1)]}.$$

For the $a$'s which are determined by $a_1$, we see that $a_4 = \frac{1}{4}a_1$ but that each of the others is zero. Since $a_2 = 0$, all the $a$'s proportional to it, $a_5$, $a_8$, and so on, are also zero.

For $y$ we now have

$$y = a_0 \left[ 1 + \sum_{k=1}^{\infty} \frac{(-1)^k[(-4)(-1) \cdot 2 \cdots (3k-7)]v^{3k}}{[3 \cdot 6 \cdot 9 \cdots (3k)][2 \cdot 5 \cdot 8 \cdots (3k-1)]} \right] + a_1(v + \tfrac{1}{4}v^4).$$

Since $v = x - 1$, the solution appears as

$$y = a_0 \left[ 1 + \sum_{k=1}^{\infty} \frac{(-1)^k[(-4)(-1) \cdot 2 \cdots (3k-7)](x-1)^{3k}}{[3 \cdot 6 \cdot 9 \cdots (3k)][2 \cdot 5 \cdot 8 \cdots (3k-1)]} \right]$$
$$+ a_1[(x-1) + \tfrac{1}{4}(x-1)^4]. \tag{24}$$

The original differential equation has no singular point in the finite plane, so the series in (24) is convergent for all finite $x$. In computations, of course, it is most useful in the neighborhood of the point $x = 1$.

The coefficient of $(x-1)^{3k}$ is sufficiently complicated to warrant attempts to simplify it. In the product $3 \cdot 6 \cdot 9 \cdots (3k)$, there are $k$ factors, each a multiple of 3. Thus we arrive at

$$3 \cdot 6 \cdot 9 \cdots (3k) = 3^k(1 \cdot 2 \cdot 3 \cdots k) = 3^k k!.$$

Furthermore, all but the first two factors inside the square bracket in the numerator also appear in the denominator. With a little more argument, testing the terms $k = 0, 1, 2$ because the factors to be cancelled do not appear until $k > 2$, it can be shown that

$$y = a_0 \sum_{k=0}^{\infty} \frac{4(-1)^k(x-1)^{3k}}{3^k(3k-1)(3k-4)k!} + a_1[(x-1) + \tfrac{1}{4}(x-1)^4]. \tag{25}$$

In the exercises below, the equations are mostly homogeneous and of second order. Raising the order of the equation introduces nothing except additional labor, as can be seen by doing exercise 18. A nonhomogeneous equation with right member having a power series expansion is theoretically no worse to handle than a homogeneous one; it is merely a matter of equating coefficients in two power series.

The treatment of equations leading to recurrence relations involving more than two different $a$'s is left for Chapter 18.

## Exercises

Unless it is otherwise requested, find the general solution valid near the origin. Always state the region of validity of the solution.

1. Solve the equation $y'' + y = 0$ both by series and by elementary methods and compare your answers.
2. Solve the equation $y'' - 4y = 0$ by series and by elementary methods.
3. $y'' + 3xy' + 3y = 0$.

$$\text{ANS.} \quad y = a_0\left[1 + \sum_{k=1}^{\infty} \frac{(-3)^k x^{2k}}{2^k k!}\right] + a_1\left[x + \sum_{k=1}^{\infty} \frac{(-3)^k x^{2k+1}}{3\cdot5\cdot7\cdots(2k+1)}\right];$$

valid for all finite $x$.

4. $(1 + 4x^2)y'' - 8y = 0$.

$$\text{ANS.} \quad y = a_0(1 + 4x^2) + a_1\sum_{k=0}^{\infty} \frac{(-1)^{k+1}2^{2k}x^{2k+1}}{4k^2 - 1}; \text{ valid for } |x| < \tfrac{1}{2}.$$

5. $(1 + x^2)y'' - 4xy' + 6y = 0$.

$$\text{ANS.} \quad y = a_0(1 - 3x^2) + a_1(x - \tfrac{1}{3}x^3); \text{ valid for all finite } x.$$

6. $(1 + x^2)y'' + 10xy' + 20y = 0$.

$$\text{ANS.} \quad y = \frac{a_0}{3}\sum_{k=0}^{\infty}(-1)^k(k+1)(2k+1)(2k+3)x^{2k}$$

$$+ \frac{a_1}{6}\sum_{k=0}^{\infty}(-1)^k(k+1)(k+2)(2k+3)x^{2k+1};$$

valid for $|x| < 1$.

7. $(x^2 + 4)y'' + 2xy' - 12y = 0$. $\quad$ ANS. $\quad y = a_0\left[1 + \sum_{k=1}^{\infty} \frac{3(-1)^k(k+1)x^{2k}}{2^{2k}(2k-1)(2k-3)}\right]$

$$+ a_1(x + \tfrac{5}{12}x^3); \text{ valid for } |x| < 2.$$

8. $(x^2 - 9)y'' + 3xy' - 3y = 0$.

$$\text{ANS.} \quad y = a_0\left[1 - \sum_{k=1}^{\infty} \frac{[3\cdot5\cdot7\cdots(2k+1)]x^{2k}}{(18)^k(2k-1)k!}\right] + a_1 x; \text{ valid for } |x| < 3.$$

9. $y'' + 2xy' + 5y = 0$. $\quad$ ANS. $\quad y = a_0\left[1 + \sum_{k=1}^{\infty} \frac{(-1)^k[5\cdot9\cdot13\cdots(4k+1)]x^{2k}}{(2k)!}\right]$

$$+ a_1\left[x + \sum_{k=1}^{\infty} \frac{(-1)^k[7\cdot11\cdot15\cdots(4k+3)]x^{2k+1}}{(2k+1)!}\right];$$

valid for all finite $x$.

10. $(x^2 + 4)y'' + 6xy' + 4y = 0$.

$$\text{ANS.} \quad y = a_0\left[1 + \sum_{k=1}^{\infty} \frac{(-1)^k(k+1)x^{2k}}{2^{2k}}\right] + a_1\left[x + \sum_{k=1}^{\infty} \frac{(-1)^k(2k+3)x^{2k+1}}{3\cdot2^{2k}}\right];$$

valid for $|x| < 2$.

**11.** $2y'' + xy' - 4y = 0.$

ANS.    $y = a_0(1 + x^2 + \tfrac{1}{12}x^4) + a_1 \sum_{k=0}^{\infty} \dfrac{3(-1)^k x^{2k+1}}{2^{2k} k! (2k-3)(2k-1)(2k+1)};$

valid for all finite $x$.

**12.** $(1 + 2x^2)y'' - 5xy' + 3y = 0.$

ANS.    $y = a_0 \left[ 1 + \sum_{k=1}^{\infty} \dfrac{3(-1)^k[(-1)\cdot 3 \cdot 7 \cdots (4k-5)]x^{2k}}{2^k k! (2k-3)(2k-1)} \right] + a_1(x + \tfrac{1}{3}x^3);$

valid for $|x| < 1/\sqrt{2}$.

**13.** $y'' + x^2 y = 0.$          ANS.    $y = a_0 \left[ 1 + \sum_{k=1}^{\infty} \dfrac{(-1)^k x^{4k}}{2^{2k} k! \cdot 3 \cdot 7 \cdot 11 \cdots (4k-1)} \right]$

$+ \, a_1 \left[ x + \sum_{k=1}^{\infty} \dfrac{(-1)^k x^{4k+1}}{2^{2k} k! \cdot 5 \cdot 9 \cdot 13 \cdots (4k+1)} \right];$

valid for all finite $x$.

**14.** $y'' - 2(x + 3)y' - 3y = 0.$ Solve about $x = -3$.

ANS.    $y = a_0 \left[ 1 + \sum_{k=1}^{\infty} \dfrac{3 \cdot 7 \cdot 11 \cdots (4k-1)(x+3)^{2k}}{(2k)!} \right]$

$+ \, a_1 \left[ (x+3) + \sum_{k=1}^{\infty} \dfrac{5 \cdot 9 \cdot 13 \cdots (4k+1)(x+3)^{2k+1}}{(2k+1)!} \right];$

valid for all finite $x$.

**15.** $y'' + (x - 2)y = 0.$ Solve about $x = 2$.

ANS.    $y = a_0 \left[ 1 + \sum_{k=1}^{\infty} \dfrac{(-1)^k (x-2)^{3k}}{3^k k! [2 \cdot 5 \cdot 8 \cdots (3k-1)]} \right]$

$+ \, a_1 \left[ (x-2) + \sum_{k=1}^{\infty} \dfrac{(-1)^k (x-2)^{3k+1}}{3^k k! [4 \cdot 7 \cdot 10 \cdots (3k+1)]} \right];$

valid for all finite $x$.

**16.** $(1 - 4x^2)y'' + 6xy' - 4y = 0.$

ANS.    $y = a_0(1 + 2x^2) + a_1 \left[ x - \sum_{k=1}^{\infty} \dfrac{1 \cdot 5 \cdot 9 \cdots (4k-3)x^{2k+1}}{k!(4k^2-1)} \right];$

valid for $|x| < \tfrac{1}{2}$.

**17.** $(1 + 2x^2)y'' + 3xy' - 3y = 0.$

ANS.    $y = a_1 x + a_0 \left[ 1 + \sum_{k=1}^{\infty} \dfrac{(-1)^{k+1} 3 \cdot 7 \cdot 11 \cdots (4k-1)x^{2k}}{2^k(2k-1)k!} \right];$

valid for $|x| < 1/\sqrt{2}$.

**18.** $y''' + x^2 y'' + 5xy' + 3y = 0.$

ANS. $y = a_0 \left[ 1 + \sum_{k=1}^{\infty} \frac{(-1)^k x^{3k}}{2 \cdot 5 \cdot 8 \cdots (3k-1)} \right] + a_1 \left[ x + \sum_{k=1}^{\infty} \frac{(-1)^k x^{3k+1}}{3^k k!} \right]$

$+ a_2 \left[ x^2 + \sum_{k=1}^{\infty} \frac{(-1)^k x^{3k+2}}{4 \cdot 7 \cdot 10 \cdots (3k+1)} \right]$; valid for all finite $x$.

**19.** $y'' + xy' + 3y = x^2.$ ANS. $y = -\frac{2}{15} + \frac{1}{5}x^2 + a_0 \sum_{k=0}^{\infty} \frac{(-1)^k (2k+1) x^{2k}}{2^k k!}$

$+ a_1 \sum_{k=0}^{\infty} \frac{(-1)^k (k+1) x^{2k+1}}{1 \cdot 3 \cdot 5 \cdots (2k+1)};$

valid for all finite $x$.

**20.** $y'' + 2xy' + 2y = 0.$
**21.** $y'' + 3xy' + 7y = 0.$
**22.** $2y'' + 9xy' - 36y = 0.$
**23.** $(x^2 + 4)y'' + xy' - 9y = 0.$
**24.** $(x^2 + 4)y'' + 3xy' - 8y = 0.$
**25.** $(1 + 9x^2)y'' - 18y = 0.$
**26.** $(x^2 - 2x + 2)y'' - 4(x-1)y' + 6y = 0$; solve about $x = 1.$
**27.** $(1 + 3x^2)y'' + 13xy' + 7y = 0.$
**28.** $(1 + 2x^2)y'' + 11xy' + 9y = 0.$

# Solutions Near Regular Singular Points

### 108. Regular singular points

Suppose that the point $x = x_0$ is a singular point of the equation

$$b_0(x)y'' + b_1(x)y' + b_2(x)y = 0 \tag{1}$$

with polynomial coefficients. Then $b_0(x_0) = 0$, so $b_0(x)$ has a factor $(x - x_0)$ to some power.

Let us put equation (1) into the form

$$y'' + p(x)y' + q(x)y = 0. \tag{2}$$

Because $x = x_0$ is a singular point and because $p(x)$ and $q(x)$ are rational functions of $x$, at least one (maybe both) of $p(x)$ and $q(x)$ has a denominator that contains the factor $(x - x_0)$.

In what follows we assume that both $p(x)$ and $q(x)$ have been reduced so that in each case the numerator and denominator contain no common factors.

If $x = x_0$ is a singular point of equation (2), if the denominator of $p(x)$ does not contain the factor $(x - x_0)$ to a power higher than one, and if the denominator of $q(x)$ does not contain the factor $(x - x_0)$ to a power higher than two, then $x = x_0$ is called a *regular singular point* (R.S.P.) of equation (2).

If $x = x_0$ is a singular point but is not a regular singular point, then it is called an *irregular singular point* (I.S.P.).

EXAMPLE (a):  Classify the singular points, in the finite plane, of the equation

$$x(x - 1)^2(x + 2)y'' + x^2y' - (x^3 + 2x - 1)y = 0. \tag{3}$$

For this equation

$$p(x) = \frac{x}{(x - 1)^2(x + 2)}$$

and

$$q(x) = \frac{-(x^3 + 2x - 1)}{x(x - 1)^2(x + 2)}.$$

The singular points in the finite plane are $x = 0, 1, -2$. Consider $x = 0$. The factor $x$ is absent from the denominator of $p(x)$ and it appears to the first power in the denominator of $q(x)$. Hence $x = 0$ is a regular singular point of equation (3).

Now consider $x = 1$. The factor $(x - 1)$ appears to the second power in the denominator of $p(x)$. That is a higher power than is permitted in the defintion of a regular singular point. Hence, it does not matter how $(x - 1)$ appears in $q(x)$; the point $x = 1$ is an irregular singular point.

The factor $(x + 2)$ appears to the first power in the denominator of $p(x)$, just as high as is permitted, and to the first power also in the denominator of $q(x)$; therefore $x = -2$ is a regular singular point.

In summary, equation (3) has in the finite plane the following singular points: regular singular points at $x = 0, x = -2$; irregular singular point at $x = 1$. The methods of Section 117 will show that (3) has also an irregular singular point "at infinity."

EXAMPLE (b):  Classify the singular points in the finite plane for the equation

$$x^4(x^2 + 1)(x - 1)^2y'' + 4x^3(x - 1)y' + (x + 1)y = 0.$$

Here

$$p(x) = \frac{4}{x(x^2 + 1)(x - 1)} = \frac{4}{x(x - i)(x + i)(x - 1)}$$

and

$$q(x) = \frac{x + 1}{x^4(x + i)(x - i)(x - 1)^2}.$$

Therefore the desired classification is:

$$\text{R.S.P. at } x = i, -i, 1; \quad \text{I.S.P. at } x = 0.$$

Singular points of a linear equation of higher order are classified in much the same way. For instance, the singular point $x = x_0$ of the equation

$$y''' + p_1(x)y'' + p_2(x)y' + p_3(x)y = 0$$

is called regular if the factor $(x - x_0)$ does not appear in the denominator of $p_1(x)$ to a power higher than one, of $p_2(x)$ to a power higher than two, of $p_3(x)$ to a power higher than three. If it is not regular, a singular point is irregular.

This chapter is devoted to the solution of linear equations near regular singular points. Solutions near irregular singular points present a great deal more difficulty and are not studied in this book.

## *Exercises*

For each equation, locate and classify all its singular points in the finite plane. (See Section 117 for the concept of a singular point "at infinity.")

**1.** $x^3(x - 1)y'' + (x - 1)y' + 4xy = 0.$      ANS.   R.S.P. at $x = 1$; I.S.P. at $x = 0.$
**2.** $x^2(x^2 - 4)y'' + 2x^3y' + 3y = 0.$      ANS.   R.S.P. at $x = 0, 2, -2$; no I.S.P.
**3.** $y'' + xy = 0.$      ANS.   No S.P. (in the finite plane).
**4.** $x^2y'' + y = 0.$      ANS.   R.S.P. at $x = 0$; no I.S.P.
**5.** $x^4y'' + y = 0.$      ANS.   No R.S.P.; I.S.P. at $x = 0.$
**6.** $(x^2 + 1)(x - 4)^3y'' + (x - 4)^2y' + y = 0.$
     ANS.   R.S.P. at $x = i, -i$; I.S.P. at $x = 4.$
**7.** $x^2(x - 2)y'' + 3(x - 2)y' + y = 0.$      ANS.   R.S.P. at $x = 2$; I.S.P. at $x = 0.$
**8.** $x^2(x - 4)^2y'' + 3xy' - (x - 4)y = 0.$      ANS.   R.S.P. at $x = 0$; I.S.P. at $x = 4.$
**9.** $x^2(x + 2)y'' + (x + 2)y' + 4y = 0.$
**10.** $x(x + 3)y'' + y' - y = 0.$
**11.** $x^3y'' + 4y = 0.$
**12.** $(x - 1)(x + 2)y'' + 5(x + 2)y' + x^2y = 0.$
**13.** $(1 + 4x^2)y'' + 6xy' - 9y = 0.$
**14.** $(1 + 4x^2)^2y'' + 6x(1 + 4x^2)y' - 9y = 0.$
**15.** $(1 + 4x^2)^2y'' + 6xy' - 9y = 0.$
**16.** $(x - 1)^2(x + 4)^2y'' + (x + 4)y' + 7y = 0.$
**17.** $(2x + 1)^4y'' + (2x + 1)y' - 8y = 0.$
**18.** $x^4y'' + 2x^3y' + 4y = 0.$
**19.** Exercise 1, p. 335.            **20.** Exercise 2, p. 335.
**21.** Exercise 3, p. 335.            **22.** Exercise 4, p. 335.

## 109. The indicial equation

As in Chapter 17, whenever we wish to obtain solutions about a point other than $x = 0$, we first translate the origin to that point and then proceed with the usual technique. Hence we concentrate our attention on solutions valid about $x = 0$. We shall restrict our study to the interval $x > 0$ and, if we then wish to find solutions of the same differential equation valid for $x < 0$, we can do so most simply by substituting $x = -u$ and studying the resulting equation on the interval $u > 0$.

Let $x = 0$ be a regular singular point of the equation

$$y'' + p(x)y' + q(x)y = 0 \tag{1}$$

where $p$ and $q$ are rational functions of $x$. Then $p(x)$ cannot have in its denominator the factor $x$ to a power higher than one. Therefore

$$p(x) = \frac{r(x)}{x}$$

where $r(x)$ is a rational function of $x$ and $r(x)$ exists at $x = 0$. We know that such a rational function, this $r(x)$, has a power series expansion about $x = 0$. Then there exists the expansion

$$p(x) = \frac{p_0}{x} + p_1 + p_2 x + p_3 x^2 + \cdots, \tag{2}$$

valid in some interval $0 < x < a$.

By a similar argument we find that there exists an expansion

$$q(x) = \frac{q_0}{x^2} + \frac{q_1}{x} + q_2 + q_3 x + q_4 x^2 + \cdots \tag{3}$$

valid in some interval $0 < x < b$.

We shall see in a formal manner that it is reasonable to expect equation (1) to have a solution of the form

$$y = \sum_{n=0}^{\infty} a_n x^{n+c} = a_0 x^c + a_1 x^{1+c} + a_2 x^{2+c} + \cdots \tag{4}$$

valid in an interval $0 < x < h$, where $h$ is less than both $a$ and $b$.

If we put the series for $y$, $p(x)$, and $q(x)$ into equation (1) and consider only the first few terms, we get

$$c(c - 1)a_0 x^{c-2} + (1 + c)ca_1 x^{c-1} + (2 + c)(1 + c)a_2 x^c + \cdots$$

$$+ \left[ \frac{p_0}{x} + p_1 + p_2 x + \cdots \right] [ca_0 x^{c-1} + (1 + c)a_1 x^c + (2 + c)a_2 x^{1+c} + \cdots]$$

$$+ \left[ \frac{q_0}{x^2} + \frac{q_1}{x} + q_2 + \cdots \right] [a_0 x^c + a_1 x^{1+c} + a_2 x^{2+c} + \cdots] = 0.$$

Performing the indicated multiplications, we find that we have

$$c(c - 1)a_0 x^{c-2} + (1 + c)ca_1 x^{c-1} + (2 + c)(1 + c)a_2 x^c + \cdots$$

$$+ p_0 c a_0 x^{c-2} + [p_0(1 + c)a_1 + p_1 c a_0]x^{c-1} + \cdots + q_0 a_0 x^{c-2}$$

$$+ [q_0 a_1 + q_1 a_0]x^{c-1} + \cdots = 0.$$

From the fact that the coefficient of $x^{c-2}$ must vanish we obtain

$$[c(c - 1) + p_0 c + q_0]a_0 = 0. \tag{5}$$

We may insist that $a_0 \neq 0$ because $a_0$ is the coefficient of the lowest power of $x$ appearing in the solution (4), no matter what that lowest power is. So from (5) it follows that

$$c^2 + (p_0 - 1)c + q_0 = 0, \tag{6}$$

which is called the *indicial equation* (at $x = 0$). The $p_0$ and $q_0$ are known constants; equation (6) is a quadratic equation giving us two roots, $c = c_1$ and $c = c_2$.

To distinguish between the roots of the indicial equation, we shall denote by $c_1$ the root whose real part is not smaller than the real part of the other root. Thus, if the roots are real, $c_1 \geqq c_2$; if the roots are imaginary, $\mathrm{Re}\,(c_1) \geqq \mathrm{Re}\,(c_2)$. For brevity we call $c_1$ the "larger" root.

Superficially, it appears that there should be two solutions of the form (4), one from each of these values of $c$. In each solution the $a_0$ should be arbitrary and the succeeding $a$'s should be determined by equating to zero the co-efficients of the higher powers of $x(x^{c-1}, x^c, x^{1+c}$, and so on) in the identity just above equation (5).

This superficial conclusion is correct if the difference of the roots $c_1$ and $c_2$ is not integral. If that difference is integral, however, a logarithmic term may enter the solution. The reasons for this strange behavior will be made clear when we develop a method for obtaining the solutions.

## 110. Form and validity of the solutions near a regular singular point

Let $x = 0$ be a regular singular point of the equation

$$y'' + p(x)y' + q(x)y = 0. \tag{1}$$

Then the functions $xp(x)$ and $x^2q(x)$ have Maclaurin series expansions that are valid in some common interval $0 < x < b$. It can be proved that equation (1) always has a general solution either of the form

$$y = A \sum_{n=0}^{\infty} a_n x^{n+c_1} + B \sum_{n=0}^{\infty} b_n x^{n+c_2} \tag{2}$$

or of the form

$$y = (A + B \ln x) \sum_{n=0}^{\infty} a_n x^{n+c_1} + B \sum_{n=0}^{\infty} b_n x^{n+c_2}, \tag{3}$$

in which $A$ and $B$ are arbitrary constants. Furthermore, the infinite series that occur in the above solutions converge in the interval $0 < x < b$.

## 111. Indicial equation with difference of roots nonintegral

The equation

$$2xy'' + (1 + x)y' - 2y = 0 \tag{1}$$

has a regular singular point at $x = 0$ and no other singular points for finite $x$. Let us assume that there is a solution of the form

$$y = \sum_{n=0}^{\infty} a_n x^{n+c}, \qquad x > 0. \tag{2}$$

Direct substitution of this $y$ into (1) yields

$$\sum_{n=0}^{\infty} 2(n + c)(n + c - 1)a_n x^{n+c-1} + \sum_{n=0}^{\infty} (n + c)a_n x^{n+c-1}$$
$$+ \sum_{n=0}^{\infty} (n + c)a_n x^{n+c} - 2 \sum_{n=0}^{\infty} a_n x^{n+c} = 0,$$

or

$$\sum_{n=0}^{\infty} (n + c)(2n + 2c - 1)a_n x^{n+c-1} + \sum_{n=0}^{\infty} (n + c - 2)a_n x^{n+c} = 0. \tag{3}$$

Having collected like terms, we next shift index to bring all the exponents of $x$ down to the smallest one present. This choice is used to get a recurrence relation for $a_n$ rather than one for $a_{n+1}$ or some other $a$. In equation (3), we replace the index $n$ in the second summation by $(n - 1)$, thus getting

$$\sum_{n=0}^{\infty} (n + c)(2n + 2c - 1)a_n x^{n+c-1} + \sum_{n=1}^{\infty} (n + c - 3)a_{n-1} x^{n+c-1} = 0. \tag{4}$$

Once more we reason that the total coefficient of each power of $x$ in the left member of (4) must vanish. The second summation does not start its contribution until $n = 1$. Hence the equations for the determination of $c$ and the $a$'s are

$$n = 0: \qquad c(2c - 1)a_0 = 0,$$

$$n \geqq 1: \qquad (n + c)(2n + 2c - 1)a_n + (n + c - 3)a_{n-1} = 0.$$

Since we may without loss of generality assume $a_0 \neq 0$, the indicial equation, that which determines $c$, is

$$c(2c - 1) = 0. \tag{5}$$

*The indicial equation always comes from the $n = 0$ term when the technique being presented in this book is employed.*

From (5) we see that $c_1 = \frac{1}{2}$ and $c_2 = 0$. The difference of the roots is $s = c_1 - c_2 = \frac{1}{2}$, which is nonintegral. When $s$ is not an integer, the method we are using always gives two linearly independent solutions of the form (2), one with each choice of $c$.

Let us return to the recurrence relation using the value $c = c_1 = \frac{1}{2}$. We have

$$n \geqq 1: \qquad (n + \tfrac{1}{2})(2n + 1 - 1)a_n + (n + \tfrac{1}{2} - 3)a_{n-1} = 0,$$

$$n \geqq 1: \qquad a_n = -\frac{(2n - 5)a_{n-1}}{2n(2n + 1)}.$$

As usual we use a vertical array and then form the product to get a formula for $a_n$. We have

$$\begin{array}{c}
\underline{\qquad a_0 \text{ arb} \qquad} \\[4pt]
a_1 = -\dfrac{(-3)a_0}{2 \cdot 3} \\[10pt]
a_2 = -\dfrac{(-1)a_1}{4 \cdot 5} \\[10pt]
a_3 = -\dfrac{(1)a_2}{6 \cdot 7} \\[6pt]
\vdots \\[6pt]
a_n = -\dfrac{(2n - 5)a_{n-1}}{2n(2n + 1)},
\end{array}$$

so the product yields, for $n \geq 1$,

$$a_n = \frac{(-1)^n[(-3)(-1)(1) \cdots (2n-5)]a_0}{[2 \cdot 4 \cdot 6 \cdots (2n)][3 \cdot 5 \cdot 7 \cdots (2n+1)]}. \tag{6}$$

The formula (6) may be simplified to the form

$$a_n = \frac{(-1)^n \cdot 3a_0}{2^n n!(2n-3)(2n-1)(2n+1)}. \tag{7}$$

Using $a_0 = 1$, the $a_n$ from (7), and the pertinent value of $c$, $c_1 = \frac{1}{2}$, we may now write a particular solution. It is

$$y_1 = x^{1/2} + \sum_{n=1}^{\infty} \frac{(-1)^n 3 x^{n+1/2}}{2^n n!(2n-3)(2n-1)(2n+1)}. \tag{8}$$

The notation $y_1$ is to emphasize that this particular solution corresponds to the root $c_1$ of the indicial equation. Our next task will be to get a particular solution $y_2$ corresponding to the smaller root $c_2$. Then the general solution, if it is desired, may be written at once as

$$y = Ay_1 + By_2$$

with $A$ and $B$ arbitrary constants.

In returning to the recurrence relation just above the indicial equation (5) with the intention of using $c = c_2 = 0$, it is evident that the $a$'s will be different from those with $c = c_1$. Hence it is wise to change notation. Let us use $b$'s instead of $a$'s. With $c = 0$, the recurrence relation becomes

$$n \geq 1: \qquad n(2n-1)b_n + (n-3)b_{n-1} = 0.$$

The corresponding vertical array is:

$$\frac{b_0 \text{ arb.}}{}$$

$$b_1 = -\frac{(-2)b_0}{1 \cdot 1}$$

$$b_2 = -\frac{(-1)b_1}{2 \cdot 3}$$

$$b_3 = -\frac{(0)b_2}{3 \cdot 5}$$

$$\vdots$$

$$b_n = -\frac{(n-3)b_{n-1}}{n(2n-1)}.$$

Then $b_n = 0$ for $n \geq 3$ and, using $b_0 = 1$, $b_1$ and $b_2$ may be computed and found to have the values $b_1 = 2$ and $b_2 = \frac{1}{6}b_1 = \frac{1}{3}$. Therefore a second solution is

$$y_2 = 1 + 2x + \tfrac{1}{3}x^2. \tag{9}$$

Since the differential equation has no singular point, other than $x = 0$, in the finite plane, we conclude that the linearly independent solutions $y_1$ of (8) and $y_2$ of (9) are valid at least for $x > 0$. The validity of (9) is evident in this particular example because the series terminates.

The student should associate with each solution the region of validity guaranteed by the general theorem quoted in Section 110, though from now on the printed answers to the exercises will omit statement of the region of validity.

## Exercises

For each equation, obtain two linearly independent solutions valid near the origin for $x > 0$. Always state the region of validity of each solution that you obtain.

**1.** $2x(x + 1)y'' + 3(x + 1)y' - y = 0.$

ANS.   $y_1 = 1 + \sum\limits_{n=1}^{\infty} \dfrac{(-1)^{n+1}x^n}{4n^2 - 1}; y_2 = x^{-1/2} + x^{1/2}.$

**2.** $4x^2y'' + 4xy' + (4x^2 - 1)y = 0.$   ANS.   $y_1 = \sin x/\sqrt{x}; y_2 = \cos x/\sqrt{x}.$

**3.** $4x^2y'' + 4xy' - (4x^2 + 1)y = 0.$   ANS.   $y_1 = \sinh x/\sqrt{x}; y_2 = \cosh x/\sqrt{x}.$

**4.** $4xy'' + 3y' + 3y = 0.$   ANS.   $y_1 = x^{1/4} + \sum\limits_{n=1}^{\infty} \dfrac{(-3)^n x^{n+1/4}}{n!5 \cdot 9 \cdot 13 \cdots (4n+1)};$

$$y_2 = 1 + \sum\limits_{n=1}^{\infty} \dfrac{(-3)^n x^n}{n!3 \cdot 7 \cdot 11 \cdots (4n-1)}.$$

**5.** $2x^2(1 - x)y'' - x(1 + 7x)y' + y = 0.$

ANS.   $y_1 = x + \tfrac{1}{15}\sum\limits_{n=1}^{\infty}(2n + 3)(2n + 5)x^{n+1};$

$$y_2 = x^{1/2} + \tfrac{1}{2}\sum\limits_{n=1}^{\infty}(n + 1)(n + 2)x^{n+1/2}.$$

**6.** $2xy'' + 5(1 - 2x)y' - 5y = 0.$   ANS.   $y_1 = 1 + 3\sum\limits_{n=1}^{\infty} \dfrac{5^n x^n}{n!(2n + 1)(2n + 3)};$

$$y_2 = x^{-3/2} + 10x^{-1/2}.$$

**7.** $8x^2y'' + 10xy' - (1 + x)y = 0.$

ANS.   $y_1 = x^{1/4} + \sum\limits_{n=1}^{\infty} \dfrac{x^{n+1/4}}{2^n n!7 \cdot 11 \cdot 15 \cdots (4n+3)};$

$$y_2 = x^{-1/2} + \sum\limits_{n=1}^{\infty} \dfrac{x^{n-1/2}}{2^n n!1 \cdot 5 \cdot 9 \cdots (4n-3)}.$$

**8.** $3xy'' + (2 - x)y' - 2y = 0$.

ANS. $y_1 = \sum_{n=0}^{\infty} \frac{(3n + 4)x^{n+1/3}}{4 \cdot 3^n n!}$; $y_2 = 1 + \sum_{n=1}^{\infty} \frac{(n + 1)x^n}{2 \cdot 5 \cdot 8 \cdots (3n - 1)}$.

**9.** $2x(x + 3)y'' - 3(x + 1)y' + 2y = 0$.

ANS. $y_1 = x^{3/2} + \sum_{n=1}^{\infty} \frac{(-1)^{n+1}x^{n+3/2}}{3^{n-1}(2n - 1)(2n + 1)(2n + 3)}$; $y_2 = 1 + \frac{2}{3}x + \frac{1}{9}x^2$.

**10.** $2xy'' + (1 - 2x^2)y' - 4xy = 0$.

ANS. $y_1 = \sum_{k=0}^{\infty} \frac{x^{2k+1/2}}{2^k k!} = x^{1/2} e^{x^2/2}$; $y_2 = 1 + \sum_{k=1}^{\infty} \frac{2^k x^{2k}}{3 \cdot 7 \cdot 11 \cdots (4k - 1)}$.

**11.** $x(4 - x)y'' + (2 - x)y' + 4y = 0$.

ANS. $y_1 = x^{1/2} + \sum_{n=1}^{\infty} \frac{(2n + 3)[(-3)(-1) \cdot 1 \cdots (2n - 5)]x^{n+1/2}}{3 \cdot 2^{3n} n!}$;

$y_2 = 1 - 2x + \frac{1}{2}x^2$.

**12.** $3x^2 y'' + xy' - (1 + x)y = 0$. ANS. $y_1 = x + \sum_{n=1}^{\infty} \frac{x^{n+1}}{n! 7 \cdot 10 \cdot 13 \cdots (3n + 4)}$;

$y_2 = x^{-1/3} + \sum_{n=1}^{\infty} \frac{x^{n-1/3}}{n!(-1) \cdot 2 \cdot 5 \cdots (3n - 4)}$.

**13.** $2xy'' + (1 + 2x)y' + 4y = 0$.

ANS. $y_1 = \sum_{n=0}^{\infty} \frac{(-1)^n(2n + 3)x^{n+1/2}}{3 \cdot n!}$; $y_2 = 1 + \sum_{n=1}^{\infty} \frac{(-1)^n 2^n(n + 1)x^n}{1 \cdot 3 \cdot 5 \cdots (2n - 1)}$.

**14.** $2xy'' + (1 + 2x)y' - 5y = 0$.

ANS. $y_1 = x^{1/2} + \frac{4}{3}x^{3/2} + \frac{4}{15}x^{5/2}$; $y_2 = \sum_{n=0}^{\infty} \frac{15(-1)^{n+1} x^n}{n!(2n - 5)(2n - 3)(2n - 1)}$.

**15.** $2x^2 y'' - 3x(1 - x)y' + 2y = 0$. ANS. $y_1 = x^2 + \sum_{n=1}^{\infty} \frac{(-1)^n 3^n(n + 1)x^{n+2}}{5 \cdot 7 \cdot 9 \cdots (2n + 3)}$;

$y_2 = x^{1/2} + \sum_{n=1}^{\infty} \frac{(-1)^{n+1} 3^n(2n - 1)x^{n+1/2}}{2^n n!}$.

**16.** $2x^2 y'' + x(4x - 1)y' + 2(3x - 1)y = 0$.

ANS. $y_1 = \sum_{n=0}^{\infty} \frac{(-1)^n 2^n x^{n+2}}{n!} = x^2 e^{-2x}$;

$y_2 = x^{-1/2} + \sum_{n=1}^{\infty} \frac{(-1)^n 4^n x^{n-1/2}}{(-3)(-1) \cdot 1 \cdots (2n - 5)}$.

**17.** $2xy'' - (1 + 2x^2)y' - xy = 0$.

ANS. $y_1 = x^{3/2} + \sum_{k=1}^{\infty} \frac{2^k x^{2k+3/2}}{7 \cdot 11 \cdot 15 \cdots (4k + 3)}$; $y_2 = 1 + \sum_{k=1}^{\infty} \frac{x^{2k}}{2^k k!} = \exp\left(\frac{1}{2}x^2\right)$.

18. The equation of exercise 17 above has a particular solution $y_2 = \exp(\tfrac{1}{2}x^2)$ obtained by the series method. Make a change of dependent variable in the differential equation, using $y = v \exp(\tfrac{1}{2}x^2)$ (the device of Section 44), and thus obtain the general solution in "closed form."

$$\text{ANS.} \quad y = c_1 \exp(\tfrac{1}{2}x^2) + c_2 \exp(\tfrac{1}{2}x^2) \int_0^x \beta^{1/2} \exp(-\tfrac{1}{2}\beta^2) \, d\beta.$$

In exercises 19 to 22, use the power series method to find solutions valid for $x > 0$. What is causing the recurrence relations to degenerate into one-term relations?

19. $2x^2 y'' + xy' - y = 0$.          ANS.  $y_1 = x; y_2 = x^{-1/2}$.
20. $2x^2 y'' - 3xy' + 2y = 0$.        ANS.  $y_1 = x^2; y_2 = x^{1/2}$.
21. $9x^2 y'' + 2y = 0$.               ANS.  $y_1 = x^{2/3}; y_2 = x^{1/3}$.
22. $2x^2 y'' + 5xy' - 2y = 0$.        ANS.  $y_1 = x^{1/2}; y_2 = x^{-2}$.
23. Obtain $dy/dx$ and $d^2y/dx^2$ in terms of derivatives of $y$ with respect to a new independent variable $t$ related to $x$ by $t = \ln x$ for $x > 0$.

$$\text{ANS.} \quad \frac{dy}{dx} = e^{-t}\frac{dy}{dt}, \frac{d^2y}{dx^2} = e^{-2t}\left[\frac{d^2y}{dt^2} - \frac{dy}{dt}\right].$$

24. Use the result of exercise 23 above to show that the change of independent variable from $x$ to $t$, where $t = \ln x$, transforms the equation*

$$ax^2\frac{d^2y}{dx^2} + bx\frac{dy}{dx} + cy = 0,$$

$a, b, c$ constants, into a linear equation with constant coefficients.

Solve exercises 25 through 34 by the method implied by exercise 24 above; that is, by changing independent variable to $t = \ln x$ for $x > 0$.

25. Exercise 19.                          26. Exercise 20.
27. Exercise 21.                          28. Exercise 22.
29. $x^2 y'' + 2xy' - 12y = 0$.            ANS.  $y_1 = x^3; y_2 = x^{-4}$.
30. $x^2 y'' + xy' - 9y = 0$.              ANS.  $y_1 = x^3; y_2 = x^{-3}$.
31. $x^2 y'' - 3xy' + 4y = 0$.             ANS.  $y = x^2(c_1 + c_2 \ln x)$.
32. $x^2 y'' - 5xy' + 9y = 0$.             ANS.  $y = x^3(c_1 + c_2 \ln x)$.
33. $x^2 y'' + 5xy' + 5y = 0$.       ANS.  $y = x^{-2}[c_1 \cos(\ln x) + c_2 \sin(\ln x)]$.
34. $(x^3 D^3 + 4x^2 D^2 - 8xD + 8)y = 0$. You will need to extend the result of exercise 23 to the third derivative, obtaining

$$x^3\frac{d^3y}{dx^3} = \frac{d^3y}{dt^3} - 3\frac{d^2y}{dt^2} + 2\frac{dy}{dt}.$$          ANS.  $y = c_1 x + c_2 x^2 + c_3 x^{-4}$.

* An equation such as the one of this exercise, which contains only terms of the kind $cx^k D^k y$ with $c$ constant and $k = 0, 1, 2, 3, \ldots$, is called an equation of *Cauchy type*, or of *Euler type*.

## 112. Differentiation of a product of functions

It will soon prove necessary for us to differentiate efficiently a product of a number of functions. Suppose that

$$u = u_1 u_2 u_3 \cdots u_n, \tag{1}$$

each of the $u$'s being a function of the parameter $c$. Let differentiation with respect to $c$ be indicated by primes. Then from

$$\ln u = \ln u_1 + \ln u_2 + \ln u_3 + \cdots + \ln u_n$$

it follows that

$$\frac{u'}{u} = \frac{u'_1}{u_1} + \frac{u'_2}{u_2} + \frac{u'_3}{u_3} + \cdots + \frac{u'_n}{u_n}.$$

Hence

$$u' = u\left\{\frac{u'_1}{u_1} + \frac{u'_2}{u_2} + \frac{u'_3}{u_3} + \cdots + \frac{u'_n}{u_n}\right\}. \tag{2}$$

Thus to differentiate a product we may multiply the original product by a conversion factor (which converts the product into its derivative) consisting of the sum of the derivatives of the logarithms of the separate factors.

When the factors involved are themselves powers of polynomials, there is a convenient way of forming the conversion factor mentally. That factor is the sum of the conversion factors for the individual parts.

The way most of us learned to differentiate a power of a quantity is to multiply the exponent, the derivative of the original quantity, and the quantity with its exponent lowered by one. Thus, if

$$y = (ac + b)^k,$$

then

$$\frac{dy}{dc} = y\left\{\frac{ka}{ac + b}\right\},$$

the division by $(ac + b)$ converting $(ac + b)^k$ into $(ac + b)^{k-1}$.

EXAMPLE (a):   If

$$u = \frac{c^2(c + 1)}{(4c - 1)^3(7c + 2)^6},$$

then

$$\frac{du}{dc} = u\left\{\frac{2}{c} + \frac{1}{c + 1} - \frac{12}{4c - 1} - \frac{42}{7c + 2}\right\}.$$

Note that the denominator factors in the function $u$ are thought of as numerator factors with negative exponents.

**EXAMPLE (b):**   If

$$y = \frac{c + n}{c(c + 1)(c + 2)\cdots(c + n - 1)},$$

then

$$\frac{dy}{dc} = y\left\{\frac{1}{c + n} - \frac{1}{c} - \frac{1}{c + 1} - \frac{1}{c + 2} - \cdots - \frac{1}{c + n - 1}\right\}.$$

**EXAMPLE (c):**   If

$$w = \frac{2^n c^3}{[(c + 2)(c + 3)\cdots(c + n + 1)]^2},$$

then

$$\frac{dw}{dc} = w\left\{\frac{3}{c} - 2\left(\frac{1}{c + 2} + \frac{1}{c + 3} + \cdots + \frac{1}{c + n + 1}\right)\right\}.$$

## 113. Indicial equation with equal roots

When the indicial equation has equal roots, the method of Section 111 cannot yield two linearly independent solutions. The work with one value of $c$ would be a pure repetition of that with the other value of $c$. A new attack is needed.

Consider the problem of solving the equation

$$x^2 y'' + 3xy' + (1 - 2x)y = 0 \tag{1}$$

for $x > 0$. It will turn out that the roots of the indicial equation are equal, a fact that can be determined ahead of time by setting up the indicial equation as developed in the theory, page 351. Here

$$p(x) = \frac{3}{x}, \qquad q(x) = \frac{1 - 2x}{x^2},$$

so $p_0 = 3$ and $q_0 = 1$. The indicial equation is

$$c^2 + 2c + 1 = 0,$$

with roots $c_1 = c_2 = -1$.

Any attempt to obtain solutions by putting

$$y = \sum_{n=0}^{\infty} a_n x^{n+c} \tag{2}$$

into equation (1) is certain to force us to choose $c = -1$ and thus get only one solution. We know that we must not choose $c$ yet if we are to get two solutions. Hence let us put the $y$ of equation (2) into the left member of equation (1) and try to come as close as we can to making that left member zero without choosing $c$.

It is convenient to have a notation for the left member of equation (1); let us use

$$L(y) = x^2 y'' + 3xy' + (1 - 2x)y. \tag{3}$$

For the $y$ of equation (2) we find that

$$L(y) = \sum_{n=0}^{\infty} (n + c)(n + c - 1)a_n x^{n+c} + \sum_{n=0}^{\infty} 3(n + c)a_n x^{n+c}$$

$$+ \sum_{n=0}^{\infty} a_n x^{n+c} - \sum_{n=0}^{\infty} 2a_n x^{n+c+1},$$

from which

$$L(y) = \sum_{n=0}^{\infty} [(n + c)^2 + 2(n + c) + 1]a_n x^{n+c} - \sum_{n=0}^{\infty} 2a_n x^{n+c+1}.$$

The usual simplifications lead to

$$L(y) = \sum_{n=0}^{\infty} (n + c + 1)^2 a_n x^{n+c} - \sum_{n=1}^{\infty} 2a_{n-1} x^{n+c}. \tag{4}$$

Recalling that the indicial equation comes from setting the coefficient in the $n = 0$ term equal to zero, we purposely avoid trying to make that term vanish yet. But by choosing the $a$'s and leaving $c$ as a parameter we can make every term but that first one in $L(y)$ vanish. Therefore we set equal to zero each coefficient, except for the $n = 0$ term, of the various powers of $x$ on the right in equation (4); thus

$$n \geq 1: \quad (n + c + 1)^2 a_n - 2a_{n-1} = 0. \tag{5}$$

The successive application of the recurrence relation (5) will determine each $a_n$, $n \geq 1$, in terms of $a_0$ and $c$. Indeed, from the array

$$a_1 = \frac{2a_0}{(c + 2)^2}$$

$$a_2 = \frac{2a_1}{(c + 3)^2}$$

$$\vdots$$

$$a_n = \frac{2a_{n-1}}{(c + n + 1)^2},$$

it follows by the usual multiplication device that

$$n \geq 1: \qquad a_n = \frac{2^n a_0}{[(c + 2)(c + 3) \cdots (c + n + 1)]^2}.$$

To arrive at a specific solution, let us choose $a_0 = 1$.

Using the $a$'s determined above, we write a $y$ that is dependent upon both $x$ and $c$, namely,

$$y(x, c) = x^c + \sum_{n=1}^{\infty} a_n(c) x^{n+c}, \qquad x > 0, \tag{6}$$

in which

$$n \geq 1: \qquad a_n(c) = \frac{2^n}{[(c + 2)(c + 3) \cdots (c + n + 1)]^2}. \tag{7}$$

The $y$ of equation (6) has been so determined that for that $y$ the right member of equation (4) must reduce to a single term, the $n = 0$ term. That is, for the $y(x, c)$ of equation (6), we have

$$L[y(x, c)] = (c + 1)^2 x^c. \tag{8}$$

A solution of the original differential equation is a function $y$ for which $L(y) = 0$. Now we see why the choice $c = -1$ yields a solution; it makes the right member of equation (8) zero.

But the factor $(c + 1)$ occurs squared in equation (8), an automatic consequence of the equality of the roots of the indicial equation. We know from elementary calculus that, if a function contains a power of a certain factor dependent upon $c$, then the derivative with respect to $c$ of that function contains the same factor to a power one lower than in the original.

For equation (8), in particular, differentiation of each member with respect to $c$ yields

$$\frac{\partial}{\partial c} L[y(x, c)] = L\left[\frac{\partial y(x, c)}{\partial c}\right] = 2(c + 1)x^c + (c + 1)^2 x^c \ln x, \tag{9}$$

the right member containing the factor $(c + 1)$ to the first power, as we knew it must from the theorem quoted.

In (9), the order of differentiations with respect to $x$ and $c$ was interchanged. It is best to avoid the need for justifying such steps by verifying the solutions, (13) and (14) below, directly. The verification in this instance is straightforward but a bit lengthy and is omitted here.

From equations (8) and (9) it can be seen that two solutions of the equation $L(y) = 0$ are

$$y_1 = [y(x, c)]_{c=-1} = y(x, -1)$$

and

$$y_2 = \left[\frac{\partial y(x, c)}{\partial c}\right]_{c=-1}$$

because $c = -1$ makes the right member of each of equations (8) and (9) vanish. That $y_1$ and $y_2$ are linearly independent will be evident later.

We have

$$y(x, c) = x^c + \sum_{n=1}^{\infty} a_n(c)x^{n+c} \tag{6}$$

and we need $\partial y(x, c)/\partial c$. From (6) it follows that

$$\frac{\partial y(x, c)}{\partial c} = x^c \ln x + \sum_{n=1}^{\infty} a_n(c)x^{n+c} \ln x + \sum_{n=1}^{\infty} a'_n(c)x^{n+c},$$

which simplifies at once to the form

$$\frac{\partial y(x, c)}{\partial c} = y(x, c) \ln x + \sum_{n=1}^{\infty} a'_n(c)x^{n+c}. \tag{10}$$

The solutions $y_1$ and $y_2$ will be obtained by putting $c = -1$ in equations (6) and (10); that is,

$$y_1 = x^{-1} + \sum_{n=1}^{\infty} a_n(-1)x^{n-1}, \tag{11}$$

$$y_2 = y_1 \ln x + \sum_{n=1}^{\infty} a'_n(-1)x^{n-1}. \tag{12}$$

Therefore we need to evaluate $a_n(c)$ and $a'_n(c)$ at $c = -1$. We know that

$$a_n(c) = \frac{2^n}{[(c + 2)(c + 3)\cdots(c + n + 1)]^2},$$

from which, by the method of Section 112, we obtain immediately

$$a'_n(c) = -2a_n(c)\left\{\frac{1}{c + 2} + \frac{1}{c + 3} + \cdots + \frac{1}{c + n + 1}\right\}.$$

We now use $c = -1$ to obtain

$$a_n(-1) = \frac{2^n}{(n!)^2}$$

and

$$a'_n(-1) = -2 \cdot \frac{2^n}{(n!)^2}\left\{1 + \frac{1}{2} + \frac{1}{3} + \cdots + \frac{1}{n}\right\}.$$

A frequently used notation for a partial sum of the harmonic series is useful here. It is

$$H_n = 1 + \frac{1}{2} + \frac{1}{3} + \cdots + \frac{1}{n} = \sum_{k=1}^{n} \frac{1}{k}.$$

We can now write $a'_n(-1)$ more simply as

$$a'_n(-1) = -\frac{2^{n+1} H_n}{(n!)^2}.$$

Finally, the desired solutions can be written in the form

$$y_1 = x^{-1} + \sum_{n=1}^{\infty} \frac{2^n x^{n-1}}{(n!)^2} \tag{13}$$

and

$$y_2 = y_1 \ln x - \sum_{n=1}^{\infty} \frac{2^{n+1} H_n x^{n-1}}{(n!)^2}. \tag{14}$$

The general solution, valid for $x > 0$, is

$$y = Ay_1 + By_2,$$

with $A$ and $B$ arbitrary constants. The linear independence of $y_1$ and $y_2$ should be evident because of the presence of $\ln x$ in $y_2$. In detail, $xy_1$ has a power series expansion about $x = 0$ but $xy_2$ does not, so that one cannot be a constant times the other.

Examination of the procedure used in solving this differential equation shows that the method is in no way dependent upon the specific coefficients, except that the indicial equation has equal roots. That is, the success of the method is due to the fact that the $n = 0$ term in $L(y)$ contains a square factor.

### Exercises

Obtain two linearly independent solutions valid for $x > 0$ unless otherwise instructed.

**1.** $x^2 y'' - x(1 + x)y' + y = 0.$

ANS. $y_1 = \sum_{n=0}^{\infty} \frac{x^{n+1}}{n!} = x e^x; y_2 = y_1 \ln x - \sum_{n=1}^{\infty} \frac{H_n x^{n+1}}{n!}.$

**2.** $4x^2 y'' + (1 - 2x)y = 0.$

ANS. $y_1 = \sum_{n=0}^{\infty} \frac{x^{n+1/2}}{2^n (n!)^2}; y_2 = y_1 \ln x - \sum_{n=1}^{\infty} \frac{H_n x^{n+1/2}}{2^{n-1}(n!)^2}.$

**3.** $x^2 y'' + x(x - 3)y' + 4y = 0.$

ANS. $y_1 = \sum_{n=0}^{\infty} \frac{(-1)^n (n+1) x^{n+2}}{n!};$

$y_2 = y_1 \ln x + \sum_{n=1}^{\infty} \frac{(-1)^{n+1}[n + (n+1)H_n] x^{n+2}}{n!}.$

**4.** $x^2y'' + 3xy' + (1 + 4x^2)y = 0$.

$$\text{ANS.}\quad y_1 = \sum_{k=0}^{\infty} \frac{(-1)^k x^{2k-1}}{(k!)^2}; y_2 = y_1 \ln x - \sum_{k=1}^{\infty} \frac{(-1)^k H_k x^{2k-1}}{(k!)^2}.$$

**5.** $x(1 + x)y'' + (1 + 5x)y' + 3y = 0$.

$$\text{ANS.}\quad y_1 = 1 + \tfrac{1}{2} \sum_{n=1}^{\infty} (-1)^n (n + 1)(n + 2)x^n;$$

$$y_2 = y_1 \ln x - \tfrac{3}{2}(y_1 - 1) + \tfrac{1}{2} \sum_{n=1}^{\infty} (-1)^n(2n + 3)x^n.$$

**6.** $xy'' + y' + xy = 0$. This is known as Bessel's equation of index zero. It is widely encountered in both pure and applied mathematics. (See also Sections 124 and 125.)

$$\text{ANS.}\quad y_1 = \sum_{k=0}^{\infty} \frac{(-1)^k x^{2k}}{2^{2k}(k!)^2}; y_2 = y_1 \ln x - \sum_{k=1}^{\infty} \frac{(-1)^k H_k x^{2k}}{2^{2k}(k!)^2}.$$

**7.** $x^2y'' - x(1 + 3x)y' + (1 - 6x)y = 0$.     $\text{ANS.}\quad y_1 = \sum_{n=0}^{\infty} \frac{3^n(n + 1)(n + 2)x^{n+1}}{2 \cdot n!};$

$$y_2 = y_1 \ln x + \sum_{n=1}^{\infty} \frac{3^n(n + 1)(n + 2)(H_{n+2} - 2H_n - \tfrac{3}{2})x^{n+1}}{2 \cdot n!}.$$

**8.** $x^2y'' + x(x - 1)y' + (1 - x)y = 0$.

$$\text{ANS.}\quad y_1 = x; y_2 = y_1 \ln x + \sum_{n=1}^{\infty} \frac{(-1)^n x^{n+1}}{n \cdot n!}.$$

**9.** $x(x - 2)y'' + 2(x - 1)y' - 2y = 0$.

$$\text{ANS.}\quad y_1 = 1 - x; y_2 = y_1 \ln x + \tfrac{5}{2}x - \sum_{n=2}^{\infty} \frac{(n + 1)x^n}{2^n n(n - 1)}.$$

**10.** Solve the equation of exercise 9 about the point $x = 2$.

$$\text{ANS.}\quad y_1 = 1 + (x - 2);$$

$$y_2 = y_1 \ln (x - 2) - \tfrac{5}{2}(x - 2) - \sum_{n=2}^{\infty} \frac{(-1)^n(n + 1)(x - 2)^n}{2^n n(n - 1)}.$$

**11.** Solve about $x = 4$:   $4(x - 4)^2 y'' + (x - 4)(x - 8)y' + xy = 0$.

$$\text{ANS.}\quad y_1 = \sum_{n=0}^{\infty} \frac{(-1)^n(n + 1)(x - 4)^{n+1}}{2^{2n} n!};$$

$$y_2 = y_1 \ln (x - 4) + \sum_{n=1}^{\infty} \frac{(-1)^n(n + 1)(H_{n+1} - 2H_n - 1)(x - 4)^{n+1}}{2^{2n} n!}.$$

**12.** $xy'' + (1 - x^2)y' - xy = 0$.     $\text{ANS.}\quad y_1 = 1 + \sum_{k=1}^{\infty} \frac{1 \cdot 3 \cdot 5 \cdots (2k - 1)x^{2k}}{2^{2k}(k!)^2};$

$$y_2 = y_1 \ln x + \sum_{k=1}^{\infty} \frac{1 \cdot 3 \cdot 5 \cdots (2k-1)\{1 + \frac{1}{3} + \cdots + 1/(2k-1) - H_k\}x^{2k}}{2^{2k}(k!)^2}.$$

**13.** Show that

$$1 + \tfrac{1}{3} + \tfrac{1}{5} + \cdots + \frac{1}{2k-1} = H_{2k} - \tfrac{1}{2}H_k$$

and apply the result to simplification of the formula for $y_2$ in the answer to exercise 12.

**14.** $x^2 y'' + x(3 + 2x)y' + (1 + 3x)y = 0$. In simplfying $y_2$, use the formula given in exercise 13.

$$\text{ANS.} \quad y_1 = x^{-1} + \sum_{n=1}^{\infty} \frac{(-1)^n 1 \cdot 3 \cdot 5 \cdots (2n-1)x^{n-1}}{(n!)^2};$$

$$y_2 = y_1 \ln x + \sum_{n=1}^{\infty} \frac{(-1)^n 1 \cdot 3 \cdot 5 \cdots (2n-1)(2H_{2n} - 3H_n)x^{n-1}}{(n!)^2}.$$

**15.** $4x^2 y'' + 8x(x+1)y' + y = 0$.

$$\text{ANS.} \quad y_1 = x^{-1/2} + \sum_{n=1}^{\infty} \frac{(-1)^n[(-1) \cdot 1 \cdot 3 \cdot 5 \cdots (2n-3)]x^{n-1/2}}{(n!)^2};$$

$$y_2 = y_1 \ln x$$

$$+ \sum_{n=1}^{\infty} \frac{(-1)^n[(-1) \cdot 1 \cdot 3 \cdots (2n-3)](2H_{2n-2} - H_{n-1} - 2H_n - 2)x^{n-1/2}}{(n!)^2}.$$

**16.** $x^2 y'' + 3x(1+x)y' + (1-3x)y = 0$.

$$\text{ANS.} \quad y_1 = x^{-1} + 6 + \tfrac{9}{2}x; \, y_2 = y_1 \ln x - 15 - \tfrac{81}{4}x + \sum_{n=3}^{\infty} \frac{2(-3)^n x^{n-1}}{n!n(n-1)(n-2)}.$$

**17.** $xy'' + (1-x)y' - y = 0$.

$$\text{ANS.} \quad y_1 = 1 + \sum_{n=1}^{\infty} \frac{x^n}{n!} = e^x; \, y_2 = y_1 \ln x - \sum_{n=1}^{\infty} \frac{H_n x^n}{n!}.$$

**18.** Refer to exercise 17 above. There one solution was found to be $y_1 = e^x$. Use the change of dependent variable, $y = ve^x$, to obtain the general solution of the differential equation in the form

$$y = k_1 e^x \int_{k_2}^{x} \beta^{-1} e^{-\beta} \, d\beta.$$

## 114. Indicial equation with equal roots, an alternative

In Section 113 we saw that when the indicial equation has equal roots, $c_2 = c_1$, two linearly independent solutions always appear of the form

$$y_1 = x^{c_1} + \sum_{n=1}^{\infty} a_n x^{n+c_1}, \tag{1}$$

$$y_2 = y_1 \ln x + \sum_{n=1}^{\infty} b_n x^{n+c_1}, \tag{2}$$

where $c_1$, $a_n$, $b_n$ are dependent upon the coefficients in the particular equation being solved.

It is possible to avoid some of the computational difficulties encountered in Section 113 in computing the $b_n$ of (2) by first determining $c_1$ and $y_1$, then substituting the $y_2$ of (2) directly into the differential equation and finding a recurrence relation that must be satisfied by $b_n$. The resulting recurrence relation may well be difficult to solve in closed form, but at least we can successively produce as many of the $b_n$ as we choose.

EXAMPLE (a): For the differential equation of Section 113,

$$L(y) = x^2 y'' + 3xy' + (1 - 2x)y = 0, \tag{3}$$

we saw that the roots of the indicial equation were both $-1$ and that a nonlogarithmic solution was

$$y_1 = x^{-1} + \sum_{n=1}^{\infty} \frac{2^n x^{n-1}}{(n!)^2}. \tag{4}$$

We know that a logarithmic solution of the form

$$y_2 = y_1 \ln x + \sum_{n=1}^{\infty} b_n x^{n-1} \tag{5}$$

exists and we shall determine the $b_n$ by forcing this $y_2$ to be a solution of (3). We have

$$y_2' = y_1' \ln x + x^{-1} y_1 + \sum_{n=1}^{\infty} (n - 1)b_n x^{n-2},$$

$$y_2'' = y_1'' \ln x + 2x^{-1} y_1' - x^{-2} y_1 + \sum_{n=1}^{\infty} (n - 1)(n - 2)b_n x^{n-3},$$

so that

$$L(y_2) = L(y_1) \ln x + 2y_1 + 2xy_1' + \sum_{n=1}^{\infty} (n - 2)(n - 1)b_n x^{n-1}$$

$$+ \sum_{n=1}^{\infty} 3(n - 1)b_n x^{n-1} + \sum_{n=1}^{\infty} b_n x^{n-1} - 2 \sum_{n=1}^{\infty} b_n x^n,$$

from which

$$L(y_2) = 2y_1 + 2xy_1' + b_1 + \sum_{n=2}^{\infty} (n^2 b_n - 2b_{n-1})x^{n-1}.$$

The logarithmic term vanishes because $L(y_1) = 0$.
Substituting from (4) for $y_1$ yields

$$L(y_2) = 2x^{-1} + 2 \sum_{n=1}^{\infty} \frac{2^n x^{n-1}}{(n!)^2} - 2x^{-1} + 2 \sum_{n=1}^{\infty} \frac{2^n(n-1)x^{n-1}}{(n!)^2}$$

$$+ b_1 + \sum_{n=2}^{\infty} (n^2 b_n - 2b_{n-1})x^{n-1},$$

or

$$L(y_2) = b_1 + 4 + \sum_{n=2}^{\infty} \left[ n^2 b_n - 2b_{n-1} + \frac{n2^{n+1}}{(n!)^2} \right] x^{n-1}.$$

If $y_2$ is to be a solution of equation (3), then $b_1 = -4$ and $b_n$ must satisfy the recurrence relation

$$n^2 b_n - 2b_{n-1} + \frac{n2^{n+1}}{(n!)^2} = 0, \qquad n \geq 2. \tag{6}$$

A simple calculation yields $b_2 = -3$ and $b_3 = -22/27$, but a closed form for $b_n$ is difficult to obtain from (6).

In Section 113 we found the values of $b_n$ to be

$$b_n = \frac{-2^{n+1} H_n}{(n!)^2}. \tag{7}$$

It is not difficult to show that this expression satisfies the recurrence relation (6), but it is difficult to obtain the form (7) from (6). Even so, for computational purposes the alternative form for $y_2$ given by the series (5) and the recurrence relation (6) is often useful.

EXAMPLE (b):   Solve the differential equation

$$x^2 y'' - x(1 + x)y' + y = 0 \tag{8}$$

of exercise 1, Section 113.

The two roots of the indicial equation are found to be $c_1 = c_2 = 1$ and the nonlogarithmic solution is

$$y_1 = \sum_{n=0}^{\infty} \frac{x^{n+1}}{n!}. \tag{9}$$

We seek a second solution of the form

$$y_2 = y_1 \ln x + \sum_{n=1}^{\infty} b_n x^{n+1},$$

so that

$$y_2' = y_1' \ln x + x^{-1} y_1 + \sum_{n=1}^{\infty} (n + 1) b_n x^n,$$

$$y_2'' = y_1'' \ln x + 2x^{-1} y_1' - x^{-2} y_1 + \sum_{n=1}^{\infty} n(n + 1) b_n x^{n-1}.$$

Substitution of these expressions into (8) gives

$$2xy_1' - 2y_1 - xy_1 + \sum_{n=1}^{\infty} n^2 b_n x^{n+1} - \sum_{n=1}^{\infty} (n + 1) b_n x^{n+2} = 0,$$

or

$$2xy_1' - (2 + x)y_1 + b_1 x^2 + \sum_{n=2}^{\infty} (n^2 b_n - n b_{n-1}) x^{n+1} = 0.$$

Using the series (9) for $y_1$ gives

$$2x + \sum_{n=1}^{\infty} \frac{2(n + 1)}{n!} x^{n+1} - 2x - 2 \sum_{n=1}^{\infty} \frac{x^{n+1}}{n!} - x^2$$

$$- \sum_{n=1}^{\infty} \frac{x^{n+2}}{n!} + b_1 x^2 + \sum_{n=2}^{\infty} (n^2 b_n - n b_{n-1}) x^{n+1} = 0,$$

which may be written

$$(1 + b_1)x^2 + \sum_{n=2}^{\infty} \frac{x^{n+1}}{(n - 1)!} + \sum_{n=2}^{\infty} (n^2 b_n - n b_{n-1}) x^{n+1} = 0.$$

It follows that $b_1 = -1$ and

$$n^2 b_n - n b_{n-1} + \frac{1}{(n - 1)!} = 0, \qquad n \geq 2. \tag{10}$$

If this problem is solved by the method of Section 113, we obtain

$$b_n = \frac{-H_n}{n!}.$$

It is easy to show that this expression satisfies the recurrence relation (10).


## Exercises

For exercises 1 through 9 of Section 113, find the logarithmic solution by finding a recurrence relation for the $b_n$ of equation (5) of this section.

2. ANS.   $b_1 = -1$ and $2n^2 b_n - b_{n-1} + \dfrac{n}{2^{n-2}(n!)^2} = 0, n \geq 2$.

**3.** ANS. $b_1 = 3$ and $n^2b_n + (n + 1)b_{n-1} + \dfrac{(-1)^n(n + 2)}{(n - 1)!} = 0, n \geq 2$.

**4.** ANS. $b_{2k-1} = 0, b_2 = 1$, and $4k^2b_{2k} + 4b_{2k-2} + \dfrac{(-1)^k 4k}{(k!)^2} = 0, k \geq 2$.

**5.** ANS. $b_1 = 2$ and $nb_n + (n + 2)b_{n-1} + (-1)^n(n + 1) = 0, n \geq 2$.

**6.** ANS. $b_{2k-1} = 0, b_2 = \frac{1}{4}$ and $4k^2b_{2k} + b_{2k-2} + \dfrac{(-1)^k 4k}{2^{2k}(k!)^2} = 0, k \geq 2$.

**7.** ANS. $b_1 = -15$ and $n^2b_n - 3(n + 2)b_{n-1} + \dfrac{3^n n(n + 1)(n + 4)}{2 \cdot n!} = 0, n \geq 2$.

**8.** ANS. $b_1 = -1$ and $n^2b_n + (n - 1)b_{n-1} = 0, n \geq 2$.

**9.** ANS. $b_1 = \frac{5}{2}, b_2 = -\frac{3}{8}$, and $2n^2b_n = (n + 1)(n - 2)b_{n-1}, n \geq 3$.

## 115. Indicial equation with difference of roots a positive integer, nonlogarithmic case

Consider the equation

$$xy'' - (4 + x)y' + 2y = 0. \tag{1}$$

As usual, let $L(y)$ stand for the left member of (1) and put

$$y = \sum_{n=0}^{\infty} a_n x^{n+c}. \tag{2}$$

At once we find that for the $y$ of equation (2), the left member of equation (1) takes the form

$$L(y) = \sum_{n=0}^{\infty} [(n + c)(n + c - 1) - 4(n + c)]a_n x^{n+c-1}$$

$$- \sum_{n=0}^{\infty} (n + c - 2)a_n x^{n+c}$$

or

$$L(y) = \sum_{n=0}^{\infty} (n + c)(n + c - 5)a_n x^{n+c-1} - \sum_{n=1}^{\infty} (n + c - 3)a_{n-1} x^{n+c-1}.$$

The indicial equation is $c(c - 5) = 0$, so

$$c_1 = 5, \qquad c_2 = 0, \qquad s = c_1 - c_2 = 5.$$

We reason that we may hope for two power series solutions, one starting with an $x^0$ term and the other with an $x^5$ term. If we use the large root $c = 5$ and try a series

$$\sum_{n=0}^{\infty} a_n x^{n+5},$$

it is evident that we can get at most one solution; the $x^0$ term would never enter.

On the other hand, if we use the smaller root $c = 0$, then a trial solution of the form

$$\sum_{n=0}^{\infty} a_n x^{n+0}$$

has a chance of picking up both solutions because the $n = 5 \, (n = s)$ term does contain $x^5$.

If $s$ is a positive integer, we try a series of the form (2) using the smaller root $c_2$. If $a_0$ and $a_s$ both turn out to be arbitrary, we obtain the general solution by this method. Otherwise the relation that should determine $a_s$ will be impossible (with our usual assumption that $a_0 \neq 0$) and the general solution will involve a logarithm as it did in the case of equal roots. That logarithmic case will be treated in the next section.

Let us return to the numerical problem. Using the smaller root $c = 0$, we now know that for

$$y = \sum_{n=0}^{\infty} a_n x^n \tag{3}$$

we get

$$L(y) = \sum_{n=0}^{\infty} n(n-5)a_n x^{n-1} - \sum_{n=1}^{\infty} (n-3)a_{n-1} x^{n-1}.$$

Therefore, to make $L(y) = 0$, we must have

$$n = 0: \qquad 0 \cdot a_0 = 0, \ (a_0 \text{ arbitrary}),$$

$$n \geq 1: \qquad n(n-5)a_n - (n-3)a_{n-1} = 0.$$

Since division by $(n-5)$ cannot be accomplished until $n > 5$, it is best to write out the separate relations through the critical one for $a_5$. We thus obtain

$$n = 1: \qquad -4a_1 + 2a_0 = 0,$$

$$n = 2: \qquad -6a_2 + a_1 = 0,$$

$$n = 3: \qquad -6a_3 + 0 \cdot a_2 = 0,$$

$$n = 4: \qquad -4a_4 - a_3 = 0,$$

$$n = 5: \qquad 0 \cdot a_5 - 2a_4 = 0,$$

$$n \geqq 6: \qquad a_n = \frac{(n-3)a_{n-1}}{n(n-5)}.$$

It follows from these relations that

$$a_1 = \tfrac{1}{2}a_0,$$

$$a_2 = \tfrac{1}{6}a_1 = \tfrac{1}{12}a_0,$$

$$a_3 = 0,$$

$$a_4 = 0,$$

$$0 \cdot a_5 = 0,$$

so $a_5$ is arbitrary. Each $a_n, n > 5$, will be obtained from $a_5$. In the usual way it is found that

$$a_6 = \frac{3a_5}{6 \cdot 1},$$

$$a_7 = \frac{4a_6}{7 \cdot 2},$$

$$\vdots$$

$$a_n = \frac{(n-3)a_{n-1}}{n(n-5)},$$

from which

$$a_n = \frac{3 \cdot 4 \cdot 5 \cdots (n-3)a_5}{[6 \cdot 7 \cdot 8 \cdots n](n-5)!} = \frac{3 \cdot 4 \cdot 5 a_5}{(n-2)(n-1)n(n-5)!}.$$

Therefore, with $a_0$ and $a_5$ arbitrary, the general solution may be written

$$y = a_0(1 + \tfrac{1}{2}x + \tfrac{1}{12}x^2) + a_5 \left[ x^5 + \sum_{n=6}^{\infty} \frac{60x^n}{(n-5)!n(n-1)(n-2)} \right].$$

The coefficient of $a_5$ may also be written, with a shift of index, in the form shown below:

$$\sum_{n=0}^{\infty} \frac{60x^{n+5}}{n!(n+5)(n+4)(n+3)}.$$

Before proceeding to the exercises, let us examine an equation for which the fortunate circumstance $a_0$ and $a_5$ both arbitrary does not occur. For the equation

$$x^2 y'' + x(1 - x)y' - (1 + 3x)y = 0 \tag{4}$$

the trial

$$y = \sum_{n=0}^{\infty} a_n x^{n+c}$$

leads to

$$L(y) = \sum_{n=0}^{\infty} (n + c + 1)(n + c - 1)a_n x^{n+c} - \sum_{n=1}^{\infty} (n + c + 2)a_{n-1}x^{n+c}.$$

Since $c_1 = 1$ and $c_2 = -1$, we use $c = -1$ and find the relations

$$n \geq 1: \quad n(n - 2)a_n - (n + 1)a_{n-1} = 0,$$

with $a_0$ arbitrary. Let us write down the separate relations out to the critical one,

$$n = 1: \quad -a_1 - 2a_0 = 0,$$

$$n = 2: \quad 0 \cdot a_2 - 3a_1 = 0,$$

$$n \geq 3: \quad a_n = \frac{(n + 1)a_{n-1}}{n(n - 2)}.$$

It follows that

$$a_1 = -2a_0$$

$$0 \cdot a_2 = 3a_1 = -6a_0.$$

These relations cannot be satisfied except by choosing $a_0 = 0$. But if that is done, $a_2$ will be the only arbitrary constant and the only solution coming out of the work will be that corresponding to the large value of $c$, $c = 1$. This is an instance where a logarithmic solution is indicated and the equation will be solved in the next section.

A good way to waste time is to use $a_0 = 0$, $a_1 = 0$, $a_2$ arbitrary, and to determine $a_n$, $n \geq 3$, from the above recurrence relation. In that way extra work can be done to get a solution that will be reobtained automatically when solving the equation by the method of the next section.

## Exercises

Obtain the general solution near $x = 0$ except when instructed otherwise. State the region of validity of each solution.

1. $x^2 y'' + 2x(x - 2)y' + 2(2 - 3x)y = 0$.

$$\text{ANS.} \quad y = a_0(x - 2x^2 + 2x^3) + a_3 \left[ x^4 + \sum_{n=4}^{\infty} \frac{6(-2)^{n-3}x^{n+1}}{n!} \right].$$

**2.** $x^2(1 + 2x)y'' + 2x(1 + 6x)y' - 2y = 0.$

ANS.  $y = a_0(x^{-2} - 6x^{-1} + 24) + a_3\left[x + \frac{1}{20} \sum_{n=4}^{\infty} (-2)^{n-3}(n + 2)(n + 1)x^{n-2}\right].$

**3.** $x^2y'' + x(2 + 3x)y' - 2y = 0.$

ANS.  $y = a_0(x^{-2} - 3x^{-1} + \frac{9}{2}) + a_3\left[x + \sum_{n=4}^{\infty} \frac{2(-1)^{n-3}3^{n-2}x^{n-2}}{n!}\right].$

**4.** $xy'' - (3 + x)y' + 2y = 0.$

ANS.  $y = a_0(1 + \frac{2}{3}x + \frac{1}{6}x^2) + a_4 \sum_{n=4}^{\infty} \frac{24(n - 3)x^n}{n!}.$

**5.** $x(1 + x)y'' + (x + 5)y' - 4y = 0.$

ANS.  $y = a_0(x^{-4} + 4x^{-3} + 5x^{-2}) + a_4(1 + \frac{4}{5}x + \frac{1}{5}x^2).$

**6.** Solve the equation of exercise 5 about the point $x = -1.$

ANS.  $y = a_0[1 + (x + 1) + \frac{1}{2}(x + 1)^2] + \frac{a_5}{12} \sum_{n=5}^{\infty} (n - 4)(n - 3)(n + 1)(x + 1)^n.$

**7.** $x^2y'' + x^2y' - 2y = 0.$

ANS.  $y = a_0(x^{-1} - \frac{1}{2}) + 6a_3 \sum_{n=3}^{\infty} \frac{(-1)^{n+1}(n - 2)x^{n-1}}{n!}.$

**8.** $x(1 - x)y'' - 3y' + 2y = 0.$    ANS.  $y = a_0(1 + \frac{2}{3}x + \frac{1}{3}x^2) + a_4 \sum_{n=4}^{\infty} (n - 3)x^n.$

**9.** Solve the equation of exercise 8 about the point $x = 1.$

ANS.  $y = a_0[(x - 1)^{-2} + 4(x - 1)^{-1}] + a_2[1 + \frac{2}{3}(x - 1) + \frac{1}{6}(x - 1)^2].$

**10.** $xy'' + (4 + 3x)y' + 3y = 0.$

ANS.  $y = a_0(x^{-3} - 3x^{-2} + \frac{9}{2}x^{-1}) + 6a_3 \sum_{n=3}^{\infty} \frac{(-3)^{n-3}x^{n-3}}{n!}.$

**11.** $xy'' - 2(x + 2)y' + 4y = 0.$

ANS.  $y = a_0(1 + x + \frac{1}{3}x^2) + a_5 \sum_{n=5}^{\infty} \frac{60 \cdot 2^{n-5}x^n}{(n - 5)!n(n - 1)(n - 2)}.$

**12.** $xy'' + (3 + 2x)y' + 4y = 0.$    ANS.  $y = a_0x^{-2} + a_2 \sum_{n=2}^{\infty} \frac{(-1)^n 2^{n-1}x^{n-2}}{n \cdot (n - 2)!}.$

**13.** $x(x + 3)y'' - 9y' - 6y = 0.$

ANS.  $y = a_0(1 - \frac{2}{3}x + \frac{1}{3}x^2 - \frac{4}{27}x^3) + a_4\left[x^4 + \sum_{n=5}^{\infty} \frac{(-1)^n(n + 1)x^n}{5 \cdot 3^{n-4}}\right].$

**14.** $x(1 - 2x)y'' - 2(2 + x)y' + 8y = 0.$

ANS.  $y = a_0(1 + 2x + 2x^2) + \frac{1}{3}a_5 \sum_{n=5}^{\infty} 2^{n-7}(n - 4)(n - 3)(n + 1)x^n.$

**15.** $xy'' + (x^3 - 1)y' + x^2y = 0.$

**16.** $x^2(4x - 1)y'' + x(5x + 1)y' + 3y = 0$.

ANS.   $y = a_0(x^{-1} - 1) + a_4\left[x^3 + 12 \sum\limits_{n=5}^{\infty} \dfrac{13 \cdot 17 \cdot 21 \cdots (4n - 7)x^{n-1}}{(n - 4)! \cdot n(n - 1)}\right]$.

## 116. Indicial equation with difference of roots a positive integer, logarithmic case

In the preceding section we examined the equation

$$x^2 y'' + x(1 - x)y' - (1 + 3x)y = 0, \qquad x > 0, \tag{1}$$

and found that its indicial equation has roots $c_1 = 1, c_2 = -1$. Since there is no power series solution starting with $x^{c_2}$, we suspect the presence of a logarithmic term and start to treat the equation in the manner of the previous logarithmic case, that of equal roots.

From the assumed form

$$y = \sum_{n=0}^{\infty} a_n x^{n+c}$$

we easily determine the left member of equation (1) to be

$$L(y) = \sum_{n=0}^{\infty} (n + c + 1)(n + c - 1)a_n x^{n+c} - \sum_{n=0}^{\infty} (n + c + 3)a_n x^{n+c+1}$$

$$= \sum_{n=0}^{\infty} (n + c + 1)(n + c - 1)a_n x^{n+c} - \sum_{n=1}^{\infty} (n + c + 2)a_{n-1} x^{n+c}.$$

As usual, each term after the first one in the series for $L(y)$ can be made zero by choosing the $a_n, n \geq 1$, without choosing $c$. Let us put

$$n \geq 1: \qquad a_n = \frac{(n + c + 2)a_{n-1}}{(n + c + 1)(n + c - 1)},$$

from which it follows at once that

$$n \geq 1: \qquad a_n = \frac{(c + 3)(c + 4) \cdots (c + n + 2)a_0}{[(c + 2)(c + 3) \cdots (c + n + 1)][c(c + 1) \cdots (c + n - 1)]},$$

or

$$n \geq 1: \qquad a_n = \frac{(c + n + 2)a_0}{(c + 2)[c(c + 1) \cdots (c + n - 1)]}.$$

From the $a_n$ obtained above, all terms after the first one in the power series for $L(y)$ have been made to vanish, so with

$$y = a_0 x^c + \sum_{n=1}^{\infty} \frac{(c + n + 2)a_0 x^{n+c}}{(c + 2)[c(c + 1) \cdots (c + n - 1)]} \tag{2}$$

it must follow that

$$L(y) = (c + 1)(c - 1)a_0 x^c. \tag{3}$$

From the larger root, $c = 1$, only one solution can be obtained. From the smaller root, $c = -1$, two solutions would be available, following the technique of using $y(x, c)$ and $\partial y(x, c)/\partial c$, as in the case of equal roots if the right member of (3) contained the factor $(c + 1)^2$ instead of just $(c + 1)$ to the first power. But $a_0$ is still arbitrary, so we take

$$a_0 = (c + 1)$$

to get the desired square factor on the right in equation (3).

Another way of seeing that it is desirable to choose $a_0 = (c + 1)$ is as follows. We know that eventually it is going to be necessary to use $c = -1$ in equation (2). But within the series, the denominator contains the factor $(c + 1)$ for all terms after the $n = 1$ term. As equation (2) stands now, the terms $n \geq 2$ would not exist with $c = -1$. Therefore we remove the troublesome factor $(c + 1)$ from the denominator by choosing $a_0 = (c + 1)$.

With $a_0 = (c + 1)$ we have

$$y(x, c) = (c + 1)x^c + \sum_{n=1}^{\infty} \frac{(c + 1)(c + n + 2)x^{n+c}}{(c + 2)[c(c + 1) \cdots (c + n - 1)]} \tag{4}$$

for which

$$L[y(x, c)] = (c + 1)^2(c - 1)x^c. \tag{5}$$

The same argument as the one used when the indicial equation had equal roots shows that the two linearly independent solutions being sought may be obtained as

$$y_1 = y(x, -1), \tag{6}$$

and

$$y_2 = \left(\frac{\partial y(x, c)}{\partial c}\right)_{c=-1}. \tag{7}$$

Naturally, it is wise to cancel the factor $(c + 1)$ from the numerator and denominator in the terms of the series in (4). But the factor $(c + 1)$ does not enter the denominator until the term $n = 2$. Therefore it seems best to write out the terms that far separately. We rewrite equation (4) as

$$y(x, c) = (c + 1)x^c + \frac{(c + 1)(c + 3)x^{1+c}}{(c + 2)c} + \frac{(c + 4)x^{2+c}}{(c + 2)c}$$

$$+ \sum_{n=3}^{\infty} \frac{(c + n + 2)x^{n+c}}{(c + 2)c[(c + 2)(c + 3) \cdots (c + n - 1)]}. \tag{8}$$

Differentiation with respect to $c$ of the members of equation (8) yields

$$\frac{\partial y(x, c)}{\partial c} = y(x, c) \ln x + x^c$$

$$+ \frac{(c + 1)(c + 3)x^{1+c}}{(c + 2)c} \left\{ \frac{1}{c + 1} + \frac{1}{c + 3} - \frac{1}{c + 2} - \frac{1}{c} \right\}$$

$$+ \frac{(c + 4)x^{2+c}}{(c + 2)c} \left\{ \frac{1}{c + 4} - \frac{1}{c + 2} - \frac{1}{c} \right\}$$

$$+ \sum_{n=3}^{\infty} \frac{(c + n + 2)x^{n+c}\left\{ \dfrac{1}{c + n + 2} - \dfrac{1}{c + 2} - \dfrac{1}{c} - \left( \dfrac{1}{c + 2} + \dfrac{1}{c + 3} + \cdots + \dfrac{1}{c + n - 1} \right) \right\}}{(c + 2)c[(c + 2)(c + 3) \cdots (c + n - 1)]}.$$

$\hspace{10cm}$ (9)

All that remains to be done is to get $y_1$ and $y_2$ by using $c = -1$ in the expressions above for $y(x, c)$ and $\partial y(x, c)/\partial c$. In the third term on the right in equation (9), we first (mentally) insert the factor $(c + 1)$ throughout the quantity in the curly brackets.

The desired solutions are thus found to be

$$y_1 = 0 \cdot x^{-1} + 0 \cdot x^0 - 3x + \sum_{n=3}^{\infty} \frac{(n + 1)x^{n-1}}{(-1)[1 \cdot 2 \cdots (n - 2)]}$$

and

$$y_2 = y_1 \ln x + x^{-1} - 2x^0 - 3x\{\tfrac{1}{3} - 1 + 1\}$$

$$+ \sum_{n=3}^{\infty} \frac{(n + 1)x^{n-1}\left\{ \dfrac{1}{n + 1} - 1 + 1 - \left( 1 + \tfrac{1}{2} + \cdots + \dfrac{1}{n - 2} \right) \right\}}{(-1)[1 \cdot 2 \cdots (n - 2)]}.$$

These results can be written more compactly as

$$y_1 = -3x - \sum_{n=3}^{\infty} \frac{(n + 1)x^{n-1}}{(n - 2)!}$$

$\hspace{10cm}$ (10)

and

$$y_2 = y_1 \ln x + x^{-1} - 2 - x - \sum_{n=3}^{\infty} \frac{[1 - (n + 1)H_{n-2}]x^{n-1}}{(n - 2)!}.$$

$\hspace{10cm}$ (11)

It is also possible to absorb one more term into the summation and to improve the appearance of these results. The student can show that

$$y_1 = -\sum_{n=0}^{\infty} \frac{(n + 3)x^{n+1}}{n!}$$

and

$$y_2 = y_1 \ln x + x^{-1} - 2 - \sum_{n=0}^{\infty} \frac{[1 - (n + 3)H_n]x^{n+1}}{n!}$$

as long as the common conventions (definitions) $H_0 = 0$ and $0! = 1$ are used.
The general solution of the original differential equation is

$$y = Ay_1 + By_2,$$

and it is valid for all $x > 0$, since the differential equation has no other singular points in the finite plane.

The step taken in passing from equation (4) to equation (8) should be used regularly in this type of solution. Without its use indeterminate forms that may cause confusion will be encountered.

An essential point in this method is the choice $a_0 = (c - c_2)$, where $c_2$ is the smaller root of the indicial equation.

## Exercises

Find two linearly independent solutions, valid for $x > 0$, unless otherwise instructed.

**1.** $xy'' + y = 0$.
    ANS. $y_1 = \sum_{n=1}^{\infty} \frac{(-1)^n x^n}{n!(n-1)!}$;

$$y_2 = y_1 \ln x + 1 + x - \sum_{n=2}^{\infty} \frac{(-1)^n (H_n + H_{n-1})x^n}{n!(n-1)!}.$$

**2.** $x^2 y'' - 3xy' + (3 + 4x)y = 0$.
    ANS. $y_1 = \sum_{n=2}^{\infty} \frac{(-1)^{n+1} 4^n x^{n+1}}{n!(n-2)!}$;

$$y_2 = y_1 \ln x + y_1 + x + 4x^2 + \sum_{n=2}^{\infty} \frac{(-4)^n (H_n + H_{n-2})x^{n+1}}{n!(n-2)!}.$$

**3.** $2xy'' + 6y' + y = 0$.
    ANS. $y_1 = \sum_{n=2}^{\infty} \frac{(-1)^{n+1} x^{n-2}}{2^n n!(n-2)!}$;

$$y_2 = y_1 \ln x + y_1 + x^{-2} + \tfrac{1}{2}x^{-1} + \sum_{n=2}^{\infty} \frac{(-1)^n (H_n + H_{n-2})x^{n-2}}{2^n n!(n-2)!}.$$

**4.** $4x^2 y'' + 2x(2 - x)y' - (1 + 3x)y = 0$.
    ANS. $y_1 = \sum_{n=1}^{\infty} \frac{x^{n-1/2}}{2^{n-1}(n-1)!}$;

$$y_2 = y_1 \ln x + 2x^{-1/2} - \sum_{n=2}^{\infty} \frac{H_{n-1} x^{n-1/2}}{2^{n-1}(n-1)!}.$$

**5.** $x^2 y'' - x(6 + x)y' + 10y = 0$.
    ANS. $y_1 = \sum_{n=3}^{\infty} \frac{(n+1)x^{n+2}}{2(n-3)!}$;

$$y_2 = y_1 \ln x + \tfrac{1}{2}y_1 + x^2 - x^3 + \tfrac{3}{2}x^4 + \sum_{n=3}^{\infty} \frac{[1 - (n+1)H_{n-3}]x^{n+2}}{2(n-3)!}.$$

**6.** $x^2y'' + xy' + (x^2 - 1)y = 0$. This is Bessel's equation of index one. See also Sections 124 and 125.

ANS. $y_1 = \sum_{k=1}^{\infty} \frac{(-1)^k x^{2k-1}}{2^{2k-1}k!(k-1)!};$

$$y_2 = y_1 \ln x + x^{-1} - \sum_{k=1}^{\infty} \frac{(-1)^k(H_k + H_{k-1})x^{2k-1}}{2^{2k}k!(k-1)!}.$$

**7.** $xy'' + (3 + 2x)y' + 8y = 0$.

ANS. $y_1 = \sum_{n=2}^{\infty} \frac{(-1)^{n+1}2^n(n + 1)x^{n-2}}{(n - 2)!};$

$$y_2 = y_1 \ln x + x^{-2} + 4x^{-1} + \sum_{n=2}^{\infty} \frac{(-2)^n[(n + 1)H_{n-2} - 1]x^{n-2}}{(n - 2)!}.$$

**8.** $x(1 - x)y'' + 2(1 - x)y' + 2y = 0$.

ANS. $y_1 = -2 + 2x;$

$$y_2 = y_1 \ln x + x^{-1} + 1 - 5x + \sum_{n=3}^{\infty} \frac{2x^{n-1}}{(n - 1)(n - 2)}.$$

**9.** Show that the answers to exercise 8 may be replaced by

$$y_3 = 1 - x; y_4 = y_3 \ln x - \tfrac{1}{2}x^{-1} + 2x - \sum_{n=3}^{\infty} \frac{x^{n-1}}{(n - 1)(n - 2)}.$$

**10.** Solve the equation of exercise 8 near the point $x = 1$.

ANS. $y_1 = 2(x - 1);$

$$y_2 = y_1 \ln (x - 1) + 1 - 3(x - 1) + \sum_{n=2}^{\infty} \frac{(-1)^{n+1}(n + 1)(x - 1)^n}{n - 1}.$$

**11.** $x^2y'' - 5xy' + (8 + 5x)y = 0$.

ANS. $y_1 = \sum_{n=2}^{\infty} \frac{(-1)^{n+1}5^n x^{n+2}}{n!(n - 2)!};$

$$y_2 = y_1 \ln x + y_1 + x^2 + 5x^3 + \sum_{n=2}^{\infty} \frac{(-5)^n(H_n + H_{n-2})x^{n+2}}{n!(n - 2)!}.$$

**12.** $xy'' + (3 - x)y' - 5y = 0$.

ANS. $y_1 = -\sum_{n=2}^{\infty} \frac{(n + 1)(n + 2)x^{n-2}}{2(n - 2)!};$

$$y_2 = y_1 \ln x - \tfrac{1}{2}y_1 + x^{-2} - 3x^{-1}$$

$$- \sum_{n=2}^{\infty} \frac{(n + 1)(n + 2)(H_{n+2} - H_n - H_{n-2})x^{n-2}}{2(n - 2)!}.$$

**13.** $9x^2y'' - 15xy' + 7(1 + x)y = 0$.

ANS. $y_1 = \sum_{n=2}^{\infty} \frac{(-1)^{n+1}7^n x^{n+1/3}}{3^{2n-1}n!(n - 2)!};$

$$y_2 = y_1 \ln x + 3x^{1/3} + \tfrac{7}{3}x^{4/3} + \sum_{n=2}^{\infty} \frac{(-7)^n(H_n + H_{n-2} - 1)x^{n+1/3}}{3^{2n-1}n!(n - 2)!}.$$

**14.** $x^2y'' + x(1 - 2x)y' - (x + 1)y = 0$.

ANS. $y_1 = \sum_{n=2}^{\infty} \frac{1 \cdot 3 \cdot 5 \cdots (2n - 3)x^{n-1}}{n!(n - 2)!};$

$$y_2 = y_1 \ln x + x^{-1} + 1 - y_1$$

$$+ \sum_{n=2}^{\infty} \frac{[1 \cdot 3 \cdot 5 \cdots (2n - 3)](2H_{2n-2} - H_n - H_{n-1} - H_{n-2})x^{n-1}}{n!(n - 2)!}.$$

## 117. Solution for large $x$

The power series solutions that have been studied up to this stage converge in regions surrounding some point $x = x_0$, usually the origin. Such solutions, although they may converge for large values of $x$, are apt to do so with discouraging slowness. For this reason, and for others of a more theoretical nature, we shall investigate the problem of obtaining solutions particularly useful for large $x$.

Consider the equation

$$b_0(x)y'' + b_1(x)y' + b_2(x)y = 0 \tag{1}$$

with polynomial coefficients. Let us put

$$w = \frac{1}{x}. \tag{2}$$

Then

$$\frac{dy}{dx} = \frac{dw}{dx}\frac{dy}{dw} = -\frac{1}{x^2}\frac{dy}{dw} = -w^2\frac{dy}{dw}$$

and

$$\frac{d^2y}{dx^2} = \frac{dw}{dx}\frac{d}{dw}\left(-w^2\frac{dy}{dw}\right) = -w^2\left(-w^2\frac{d^2y}{dw^2} - 2w\frac{dy}{dw}\right) = w^4\frac{d^2y}{dw^2} + 2w^3\frac{dy}{dw}.$$

Thus equation (1) is transformed into the following equation in $y$ and $w$:

$$b_0\left(\frac{1}{w}\right)w^4\frac{d^2y}{dw^2} + \left[2w^3b_0\left(\frac{1}{w}\right) - w^2b_1\left(\frac{1}{w}\right)\right]\frac{dy}{dw} + b_2\left(\frac{1}{w}\right)y = 0. \tag{3}$$

Since $b_0, b_1$, and $b_2$ are polynomials, equation (3) is readily converted into an equation with polynomial coefficients.

If the point $w = 0$ is an ordinary point or a regular singular point of equation (3), then our previous methods of attack will yield solutions valid for small $w$. But $w = 1/x$, so small $w$ means large $x$.

As a matter of terminology, whatever is true about equation (3) at $w = 0$ is said to be true about equation (1) "at the point at infinity." For instance, if the transformed equation has $w = 0$ as an ordinary point, then we say that equation (1) has an ordinary point at infinity. (See exercises 1 through 6, page 381.)

EXAMPLE: Obtain solutions valid for large $x$ for the equation

$$x^2 y'' + (3x - 1)y' + y = 0. \tag{4}$$

This equation has an irregular singular point at the origin and has no other singular points in the finite plane. To investigate the nature of equation (4) for large $x$, put $x = 1/w$. We have already found that

$$\frac{dy}{dx} = -w^2 \frac{dy}{dw} \tag{5}$$

and

$$\frac{d^2 y}{dx^2} = w^4 \frac{d^2 y}{dw^2} + 2w^3 \frac{dy}{dw}. \tag{6}$$

With the aid of (5) and (6) we see that equation (4) becomes

$$\frac{1}{w^2}\left( w^4 \frac{d^2 y}{dw^2} + 2w^3 \frac{dy}{dw} \right) + \left( \frac{3}{w} - 1 \right)\left( -w^2 \frac{dy}{dw} \right) + y = 0,$$

or

$$w^2 \frac{d^2 y}{dw^2} - w(1 - w)\frac{dy}{dw} + y = 0, \tag{7}$$

an equation that we wish to solve about $w = 0$.

Since $w = 0$ is a regular point of equation (7), the point at infinity is a regular singular point of equation (4).

From the assumed form

$$y = \sum_{n=0}^{\infty} a_n w^{n+c},$$

it follows from equation (7) by our usual methods that

$$L(y) = \sum_{n=0}^{\infty} (n + c - 1)^2 a_n w^{n+c} + \sum_{n=1}^{\infty} (n + c - 1)a_{n-1} w^{n+c},$$

in which $L(y)$ now represents the left member of equation (7). The indicial equation has equal roots, $c = 1, 1$. Let us set up $y(w, c)$ as usual. From the recurrence relation

$$n \geq 1: \qquad a_n = -\frac{a_{n-1}}{n + c - 1},$$

it follows that

$$a_n = \frac{(-1)^n a_0}{c(c + 1)\cdots(c + n - 1)}.$$

Hence we choose

$$y(w, c) = w^c + \sum_{n=1}^{\infty} \frac{(-1)^n w^{n+c}}{c(c+1)\cdots(c+n-1)},$$

and then find that

$$\frac{\partial y(w, c)}{\partial c} = y(w, c) \ln w - \sum_{n=1}^{\infty} \frac{(-1)^n w^{n+c} \left\{ \dfrac{1}{c} + \dfrac{1}{c+1} + \cdots + \dfrac{1}{c+n-1} \right\}}{c(c+1)\cdots(c+n-1)}.$$

Employing the root $c = 1$, we arrive at the solutions

$$y_1 = w + \sum_{n=1}^{\infty} \frac{(-1)^n w^{n+1}}{n!}$$

and

$$y_2 = y_1 \ln w - \sum_{n=1}^{\infty} \frac{(-1)^n H_n w^{n+1}}{n!}.$$

Therefore the original differential equation has the two linearly independent solutions

$$y_1 = \sum_{n=0}^{\infty} \frac{(-1)^n x^{-n-1}}{n!} = x^{-1} e^{-1/x} \tag{8}$$

and

$$y_2 = y_1 \ln (1/x) - \sum_{n=1}^{\infty} \frac{(-1)^n H_n x^{-n-1}}{n!}. \tag{9}$$

These solutions are valid for all $x > 0$.


## Exercises

In exercises 1 through 6, the singular points in the finite plane have already been located and classified. For each equation, determine whether the point at infinity is an ordinary point (O.P.), a regular singular point (R.S.P.), or an irregular singular point (I.S.P.). Do not solve the equations.

1. $x^3(x-1)y'' + (x-1)y' + 4xy = 0.$ (exercise 1, p. 349.)        ANS.   R.S.P.
2. $x^2(x^2-4)y'' + 2x^3 y' + 3y = 0.$ (exercise 2, p. 349.)        ANS.   O.P.
3. $y'' + xy = 0.$ (exercise 3, p. 349.)        ANS.   I.S.P.
4. $x^2 y'' + y = 0.$ (exercise 4, p. 349.)        ANS.   R.S.P.
5. $x^4 y'' + y = 0.$ (exercise 5, p. 349.)        ANS.   R.S.P.
6. $x^4 y'' + 2x^3 y' + 4y = 0.$ (exercise 18, p. 349.)        ANS.   O.P.

In exercises 7 through 19, find solutions valid for large positive $x$ unless otherwise instructed.

**7.** $x^4y'' + x(1 + 2x^2)y' + 5y = 0.$

ANS. $y = a_0 \sum\limits_{k=0}^{\infty} \dfrac{-15x^{-2k}}{2^k k!(2k - 1)(2k - 3)(2k - 5)} + a_1[x^{-1} - \frac{2}{3}x^{-3} + \frac{1}{15}x^{-5}].$

**8.** $2x^3y'' - x(2 - 5x)y' + y = 0.$

ANS. $y_1 = x^{-3/2} + \sum\limits_{n=1}^{\infty} \dfrac{(-2)^n(n + 1)x^{-n-3/2}}{5 \cdot 7 \cdot 9 \cdots (2n + 3)} ; y_2 = \sum\limits_{n=0}^{\infty} \dfrac{(-1)^{n+1}(2n - 1)x^{-n}}{n!}.$

**9.** $x(1 - x)y'' - 3y' + 2y = 0,$ the equation of exercise 8, page 373.

ANS. $y = a_0(x^2 + 2x + 3) + \frac{1}{4}a_3 \sum\limits_{n=0}^{\infty} (n + 4)x^{-n-1}.$

**10.** $x^3y'' + x(2 - 3x)y' - (5 - 4x)y = 0.$ See exercise 13, page 365.

ANS. $y_1 = x^2 + \sum\limits_{n=1}^{\infty} \dfrac{1 \cdot 3 \cdot 5 \cdots (2n - 1)x^{-n+2}}{(n!)^2} ;$

$y_2 = y_1 \ln(1/x) + \sum\limits_{n=1}^{\infty} \dfrac{1 \cdot 3 \cdot 5 \cdots (2n - 1)(2H_{2n} - 3H_n)x^{-n+2}}{(n!)^2}.$

**11.** $2x^2(x - 1)y'' + x(5x - 3)y' + (x + 1)y = 0.$

ANS. $y_1 = x^{-1}; y_2 = - \sum\limits_{n=0}^{\infty} \dfrac{x^{-n-1/2}}{2n - 1}.$

**12.** Solve the equation of exercise 11 about the point $x = 0.$

ANS. $y_3 = \sum\limits_{n=0}^{\infty} \dfrac{3x^{n+1/2}}{2n + 3} ; y_4 = x^{-1}.$

**13.** $2x^2(1 - x)y'' - 5x(1 + x)y' + (5 - x)y = 0.$

ANS. $y_1 = \frac{1}{15} \sum\limits_{n=0}^{\infty} (n + 1)(2n + 3)(2n + 5)x^{-n-1} ;$

$y_2 = \frac{1}{2} \sum\limits_{n=0}^{\infty} (n + 1)(n + 2)(2n + 1)x^{-n-1/2}.$

**14.** Solve the equation of exercise 13 about the point $x = 0.$

ANS. $y_3 = \frac{1}{10} \sum\limits_{n=0}^{\infty} (n + 1)(n + 2)(2n + 5)x^{n+5/2} ;$

$y_4 = - \sum\limits_{n=0}^{\infty} (n + 1)(2n - 1)(2n + 1)x^{n+1}.$

**15.** $x(1 + x)y'' + (1 + 5x)y' + 3y = 0,$ the equation of exercise 5, page 364.

ANS. $y_1 = \sum\limits_{n=2}^{\infty} (-1)^{n+1}n(n - 1)x^{-n-1} ;$

$$y_2 = y_1 \ln(1/x) + x^{-1} + x^{-2} + \sum_{n=2}^{\infty} (-1)^{n+1}(n^2 + n - 1)x^{-n-1}.$$

**16.** $x^2(4 + x^2)y'' + 2x(4 + x^2)y' + y = 0.$

$$\text{ANS.} \quad y_1 = 1 + \sum_{k=1}^{\infty} \frac{(-1)^k[(-1) \cdot 3 \cdot 7 \cdots (4k - 5)]^2 x^{-2k}}{(2k)!};$$

$$y_2 = x^{-1} + \sum_{k=1}^{\infty} \frac{(-1)^k[1 \cdot 5 \cdot 9 \cdots (4k - 3)]^2 x^{-2k-1}}{(2k + 1)!}.$$

**17.** $x(1 - x)y'' + (1 - 4x)y' - 2y = 0,$ the equation of exercise 18, page 390.

$$\text{ANS.} \quad y_1 = \sum_{n=1}^{\infty} nx^{-n-1}; \, y_2 = y_1 \ln(1/x) + \sum_{n=0}^{\infty} x^{-n-1}.$$

**18.** $x(1 + 4x)y'' + (1 + 8x)y' + y = 0,$ the equation of exercise 49, page 391.

$$\text{ANS.} \quad y_1 = x^{-1/2} + \sum_{n=1}^{\infty} \frac{(-1)^n[1 \cdot 3 \cdot 5 \cdots (2n - 1)]^2 x^{-n-1/2}}{2^{4n}(n!)^2};$$

$$y_2 = y_1 \ln(1/x) + \sum_{n=1}^{\infty} \frac{(-1)^n[1 \cdot 3 \cdot 5 \cdots (2n - 1)]^2(H_{2n} - H_n)x^{-n-1/2}}{2^{4n-2}(n!)^2}.$$

**19.** The equation of exercise 6 above.       ANS.   $y_1 = \cos(2x^{-1}); y_2 = \sin(2x^{-1}).$

## 118. Many-term recurrence relations

In solving an equation near a regular singular point it will sometimes happen that a many-term recurrence relation is encountered. In nonlogarithmic cases, the methods developed in Chapter 18 are easily applied and no complications result except that usually no explicit formula will be obtained for the coefficients.

In logarithmic cases, the methods introduced in Chapter 18, constructing $y(x, c)$ and $\partial y(x, c)/\partial c$, can become awkward when a many-term recurrence relation is present. There is another attack which has its good points.

Consider the problem of solving the equation

$$L(y) = x^2 y'' + x(3 + x)y' + (1 + x + x^2)y = 0 \tag{1}$$

for $x > 0$. From

$$y = \sum_{n=0}^{\infty} a_n x^{n+c} \tag{2}$$

it is easily shown that

$$L(y) = \sum_{n=0}^{\infty} (n + c + 1)^2 a_n x^{n+c} + \sum_{n=1}^{\infty} (n + c)a_{n-1}x^{n+c} + \sum_{n=2}^{\infty} a_{n-2}x^{n+c}. \tag{3}$$

Therefore the indicial equation is $(c + 1)^2 = 0$.

Since the roots of the indicial equation are equal, $c = -1, -1$, it follows that there exist the solutions

$$y_1 = \sum_{n=0}^{\infty} a_n x^{n-1}, \tag{4}$$

$$y_2 = y_1 \ln x + \sum_{n=1}^{\infty} b_n x^{n-1}, \tag{5}$$

valid for $x > 0$. The region of validity is obtained from the differential equation; the form of the solutions can be seen by the reasoning in Section 113.

We shall determine the $a_n$, $n > 0$, by requiring that $L(y_1) = 0$. Then the $b_n$, $n \geq 1$, will be determined in terms of the $a_n$ by requiring that $L(y_2) = 0$.

From $L(y_1) = 0$ it follows that

$$\sum_{n=0}^{\infty} n^2 a_n x^{n-1} + \sum_{n=1}^{\infty} (n-1) a_{n-1} x^{n-1} + \sum_{n=2}^{\infty} a_{n-2} x^{n-1} = 0.$$

Let us choose $a_0 = 1$. Then the rest of the $a$'s are determined by

$$n = 1: \qquad a_1 + 0 \cdot a_0 = 0,$$

$$n \geq 2: \qquad n^2 a_n + (n-1) a_{n-1} + a_{n-2} = 0.$$

Therefore one solution of the differential equation (1) is

$$y_1 = \sum_{n=0}^{\infty} a_n x^{n-1}, \tag{6}$$

in which $a_0 = 1$, $a_1 = 0$,

$$n \geq 2: \qquad a_n = -\frac{(n-1) a_{n-1} + a_{n-2}}{n^2}.$$

Next we wish to require that $L(y_2) = 0$. From

$$y_2 = y_1 \ln x + \sum_{n=1}^{\infty} b_n x^{n-1} \tag{5}$$

it follows that

$$y_2' = y_1' \ln x + x^{-1} y_1 + \sum_{n=1}^{\infty} (n-1) b_n x^{n-2}$$

and

$$y_2'' = y_1'' \ln x + 2x^{-1} y_1' - x^{-2} y_1 + \sum_{n=1}^{\infty} (n-1)(n-2) b_n x^{n-3}.$$

Now direct computation of $L(y_2)$ yields

$$L(y_2) = L(y_1) \ln x + 2xy_1' - y_1 + xy_1 + 3y_1 + \sum_{n=1}^{\infty} n^2 b_n x^{n-1}$$

$$+ \sum_{n=2}^{\infty} (n-1)b_{n-1}x^{n-1} + \sum_{n=3}^{\infty} b_{n-2}x^{n-1}.$$

Since $L(y_1) = 0$, the requirement $L(y_2) = 0$ leads to the equation

$$\sum_{n=1}^{\infty} n^2 b_n x^{n-1} + \sum_{n=2}^{\infty} (n-1)b_{n-1}x^{n-1} + \sum_{n=3}^{\infty} b_{n-2}x^{n-1}$$

$$= -2xy_1' - 2y_1 - xy_1$$

$$= -\sum_{n=1}^{\infty} 2na_n x^{n-1} - \sum_{n=1}^{\infty} a_{n-1}x^{n-1}, \tag{7}$$

in which the right member has been simplified by using equation (6). From the identity (7), relations for the determination of the $b_n$ from the $a_n$ follow. They are

$$n = 1: \quad b_1 = -2a_1 - a_0,$$

$$n = 2: \quad 4b_2 + b_1 = -4a_2 - a_1,$$

$$n \geq 3: \quad n^2 b_n + (n-1)b_{n-1} + b_{n-2} = -2na_n - a_{n-1}.$$

Therefore the original differential equation has the two linearly independent solutions given by the $y_1$ of equation (6) and by

$$y_2 = y_1 \ln x + \sum_{n=1}^{\infty} b_n x^{n-1}, \tag{8}$$

in which $b_1 = -1, b_2 = \frac{1}{2}$,

$$n \geq 3: \quad b_n = -\frac{(n-1)b_{n-1} + b_{n-2}}{n^2} - \frac{2a_n}{n} - \frac{a_{n-1}}{n^2}.$$

If the indicial equation has roots that differ by a positive integer and if a logarithmic solution exists, then the two solutions will have the form

$$y_1 = \sum_{n=0}^{\infty} a_n x^{n+c_1},$$

$$y_2 = y_1 \ln x + \sum_{n=0}^{\infty} b_n x^{n+c_2},$$

where $c_1$ is the larger and $c_2$ the smaller root of the indicial equation. The $a_n$ and $b_n$ can still be determined by the procedure used in this section.

## Exercises

Solve each equation for $x > 0$ unless otherwise instructed.

**1.** $x^2 y'' + 3xy' + (1 + x + x^3)y = 0$.

ANS.   $y_1 = \sum\limits_{n=0}^{\infty} a_n x^{n-1}$ in which $a_0 = 1, a_1 = -1, a_2 = \frac{1}{4}$,

$$n \geq 3: a_n = -\frac{a_{n-1} + a_{n-3}}{n^2};$$

$$y_2 = y_1 \ln x + \sum\limits_{n=1}^{\infty} b_n x^{n-1}, \text{ in which } b_1 = 2, b_2 = -\frac{3}{4}, b_3 = \frac{19}{108},$$

$$n \geq 4: b_n = -\frac{b_{n-1} + b_{n-3}}{n^2} - \frac{2a_n}{n}.$$

**2.** $2x(1 - x)y'' + (1 - 2x)y' + (2 + x)y = 0$.

ANS.   $y_1 = \sum\limits_{n=0}^{\infty} a_n x^{n+1/2}$, in which $a_0 = 1, a_1 = -\frac{1}{2}$,

$$n \geq 2: a_n = \frac{(2n + 1)(2n - 3)a_{n-1} - 2a_{n-2}}{2n(2n + 1)};$$

$$y_2 = \sum\limits_{n=0}^{\infty} b_n x^n, \text{ in which } b_0 = 1, b_1 = -2,$$

$$n \geq 2: b_n = \frac{2n(n - 2)b_{n-1} - b_{n-2}}{n(2n - 1)}.$$

**3.** $xy'' + y' + x(1 + x)y = 0$.

ANS.   $y_1 = \sum\limits_{n=0}^{\infty} a_n x^n$, in which $a_0 = 1, a_1 = 0, a_2 = -\frac{1}{4}$,

$$n \geq 3: a_n = -\frac{a_{n-2} + a_{n-3}}{n^2};$$

$$y_2 = y_1 \ln x + \sum\limits_{n=1}^{\infty} b_n x^n, \text{ in which } b_1 = 0, b_2 = \frac{1}{4}, b_3 = \frac{2}{27},$$

$$n \geq 4: b_n = -\frac{b_{n-2} + b_{n-3}}{n^2} - \frac{2a_n}{n}.$$

**4.** $x^2 y'' + x(1 + x)y' - (1 - 3x + 6x^2)y = 0$.

ANS.   $y = \sum\limits_{n=0}^{\infty} a_n x^{n-1}$, in which $a_0$ is arbitrary, $a_1 = 2a_0$, $a_2$ is arbitrary,

$$n \geq 3: a_n = -\frac{(n + 1)a_{n-1} - 6a_{n-2}}{n(n - 2)}.$$

**5.** Show that the series in the answer to exercise 4 start out as follows:

$$y = a_0(x^{-1} + 2 + 4x^2 - \tfrac{5}{2}x^3 + \tfrac{13}{5}x^4 - \tfrac{83}{60}x^2 + \cdots)$$

$$+ a_2(x - \tfrac{4}{3}x^2 + \tfrac{19}{12}x^3 - \tfrac{7}{6}x^4 + \tfrac{53}{72}x^5 + \cdots).$$

**6.** $xy'' + xy' + (1 + x^4)y = 0$. Here the indicial equation has roots $c = 0$, $c = 1$, and an attempt to get a complete solution without $\ln x$ fails. Then we put

$$y_1 = \sum_{n=0}^{\infty} a_n x^{n+1},$$

$$y_2 = y_1 \ln x + \sum_{n=0}^{\infty} b_n x^n.$$

The coefficient $b_1 (s = 1)$ turns out to be arbitrary and we choose it to be zero. Show that the indicated $y_1$ and $y_2$ are solutions if

$$a_0 = 1, a_1 = -1, a_2 = \tfrac{1}{2}, a_3 = -\tfrac{1}{6}, a_4 = \tfrac{1}{24},$$

$$n \geq 5 : a_n = -\frac{(n+1)a_{n-1} + a_{n-5}}{n(n+1)},$$

and if the $b$'s are given by

$$b_0 = -1, b_1 = 0 \text{ (so chosen)}, b_2 = 1, b_3 = -\tfrac{3}{4}, b_4 = \tfrac{11}{36},$$

$$n \geq 5 : b_n = -\frac{nb_{n-1} + b_{n-5}}{n(n-1)} - \frac{(2n-1)a_{n-1} + a_{n-2}}{n(n-1)}.$$

**7.** For the $y_1$ in exercise 6, prove that the $a_n$ alternate in sign. Also compute the terms of $y_1$ out to the $x^7$ term.

ANS.   $y_1 = x - x^2 + \tfrac{1}{2}x^3 - \tfrac{1}{6}x^4 + \tfrac{1}{24}x^5 - \tfrac{1}{24}x^6 + \tfrac{31}{1008}x^7 + \cdots.$

**8.** $x(x-2)^2 y'' - 2(x-2)y' + 2y = 0.$                    ANS.   $y_1 = 1 - \tfrac{1}{2}x;$

$$y_2 = y_1 \ln x + \sum_{n=1}^{\infty} b_n x^n, \text{ in which } b_1 = \tfrac{1}{2}, b_2 = -\tfrac{1}{8}, b_3 = -\tfrac{1}{48},$$

$$n \geq 4 : b_n = \frac{n-2}{4n^2}[2(2n-1)b_{n-1} - (n-3)b_{n-2}].$$

**9.** Solve the equation of exercise 8 for $x > 2$.                    ANS.   $y_1 = (x - 2);$

$$y_2 = y_1 \ln (x - 2) + \sum_{n=1}^{\infty} \frac{(-1)^n (x-2)^{n+1}}{2^n n}.$$

**10.** $2xy'' + (1 - x)y' - (1 + x)y = 0.$

ANS.   $y_1 = \sum_{n=0}^{\infty} a_n x^{n+1/2}$ in which $a_0 = 1, a_1 = \tfrac{1}{2},$

$$n \geq 2 : a_n = \frac{(2n+1)a_{n-1} + 2a_{n-2}}{2n(2n+1)};$$

$$y_2 = \sum_{n=0}^{\infty} b_n x^n \text{ in which } b_0 = 1, b_1 = 1,$$

$$n \geq 2 : b_n = \frac{nb_{n-1} + b_{n-2}}{n(2n-1)}.$$

**11.** Show that the answers to exercise 10 are also given by

$$y_2 = e^x, \quad y_1 = e^x \int_0^{\sqrt{x}} \exp\left(-\tfrac{3}{2}\beta^2\right) d\beta.$$

## 119. Summary

Confronted by a linear equation

$$L(y) = 0, \tag{1}$$

we first determine the location and nature of the singular points of the equation. In practice, the use to which the results will be put will dictate that solutions are desired near a certain point or points. In seeking solutions valid about the point $x = x_0$, always first translate the origin, putting $x - x_0 = v$.

Solutions valid near an ordinary point $x = 0$ of equation (1) take the form

$$y = \sum_{n=0}^{\infty} a_n x^n \tag{2}$$

with $a_0$ and $a_1$ arbitrary, if equation (1) is of second order.

If $x = 0$ is a regular singular point of equation (1) and we wish to get solutions valid for $x > 0$, we first put

$$y = \sum_{n=0}^{\infty} a_n x^{n+c}. \tag{3}$$

For the $y$ of (3), we obtain the series for $L(y)$ by substitution. From the $n = 0$ term of that series, the indicial equation may be written. When the difference of the roots of the indicial equation is not an integer, or if the roots are equal, the technique is straightforward following the method of Section 111 or of Section 113.

When the roots differ by a nonzero integer, then the solution may, or may not, involve $\ln x$. The recurrence relation for $n = s$, where $s$ is the difference of the roots, is the critical one. We must then determine whether the relations for $n = 1, 2, \cdots, s$ leave $a_0$ and $a_s$ both arbitrary. If they do, two power series of the form (3) will be solutions of the differential equation. If $a_0$ and $a_s$ are not both arbitrary, the case is a logarithmic one. Then the device of Section 116 may be used.

The technique can be varied, if desired, by always choosing the $a_n$ in terms of $c$ so the series for $L(y)$ reduces to a single term. Thus a series of the form

$$y(x, c) = a_0 \left[ x^c + \sum_{n=1}^{\infty} f_n(c) x^{n+c} \right] \tag{4}$$

will be determined for which

$$L[y(x, c)] = a_0(c - c_1)(c - c_2)x^{c-k}, \tag{5}$$

where $k = 0$ or $1$ for the equations being treated here and $c_1$ and $c_2$ are the roots of the indicial equation. Then it can be determined from the actual coefficients in (4) whether the use of $c = c_1$ and $c = c_2$ will result in two solutions of the differential equation. If $c_1 = c_2$, the results would be identical and the use of $\partial y(x, c)/\partial c$ is indicated. The other logarithmic case will be identified by the fact that some one or more of the coefficients $f_n(c)$ will not exist when $c = c_2$, the smaller root. Then again the differentiation process is needed, after the introduction of $a_0 = c - c_2$.

The method sketched above has a disadvantage in that it seems to tempt the user into automatic application of rules, always a dangerous procedure in mathematics. When a student thoroughly understands what is happening in each of the four possible cases, this method may safely be used and saves some labor.

Extension of the methods of this and the preceding chapter to linear equations of higher order is direct. As an example, a fourth-order equation whose indicial equation has roots $c = 2, 2, 2, \frac{1}{2}$ would be treated as follows. A series would be determined for $y(x, c)$,

$$y(x, c) = a_0 \left[ x^c + \sum_{n=1}^{\infty} f_n(c)x^{n+c} \right]$$

for which the left member of the original equation reduces to one term, such as

$$L[y(x, c)] = (c - 2)^3(2c - 1)a_0 x^c.$$

Then four linearly independent solutions could be obtained:

$$y_1 = y(x, 2); \qquad y_2 = \left[ \frac{\partial y(x, c)}{\partial c} \right]_{c=2}; \qquad y_3 = \left[ \frac{\partial^2 y(x, c)}{\partial c^2} \right]_{c=2};$$

$$y_4 = y(x, \tfrac{1}{2}).$$

## Miscellaneous Exercises

In each exercise, obtain solutions valid for $x > 0$.

1. $xy'' - (2 + x)y' - y = 0$.
2. $x^2 y'' + 2x^2 y' - 2y = 0$.
3. $x^2(1 + x^2)y'' + 2x(3 + x^2)y' + 6y = 0$.
4. $2xy'' + (1 + 2x)y' - 3y = 0$.
5. $x(1 - x^2)y'' - (7 + x^2)y' + 4xy = 0$.
6. $4x^2 y'' - 2x(2 + x)y' + (3 + x)y = 0$.
7. $2xy'' + y' + y = 0$.
8. $4x^2 y'' - x^2 y' + y = 0$.

9. $2x^2y'' - x(1 + 2x)y' + (1 + 3x)y = 0.$
10. $4x^2y'' + 3x^2y' + (1 + 3x)y = 0.$
11. $4x^2y'' + 2x^2y' - (x + 3)y = 0.$
12. $x^2y'' + x(3 + x)y' + (1 + 2x)y = 0.$
13. $x(1 - 2x)y'' - 2(2 + x)y' + 18y = 0.$
14. $x^2y'' - 3xy' + 4(1 + x)y = 0.$
15. $4x^2y'' + 2x(x - 4)y' + (5 - 3x)y = 0.$
16. $x(1 - x)y'' - (4 + x)y' + 4y = 0.$

$$\text{ANS.} \quad y = a_0(1 + x + \tfrac{1}{2}x^2) + \frac{a_5}{12}\sum_{n=5}^{\infty}(n - 4)(n - 3)(n + 1)x^n.$$

17. Solve the equation of exercise 16 about the point $x = 1$.
$$\text{ANS.} \quad y = a_0[(x - 1)^{-4} + 4(x - 1)^{-3} + 5(x - 1)^{-2}]$$
$$+ a_4[1 + \tfrac{4}{5}(x - 1) + \tfrac{1}{5}(x - 1)^2].$$

18. $x(1 - x)y'' + (1 - 4x)y' - 2y = 0.$
19. Show that the solutions of exercise 18 may be written in the form
$$y_1 = (1 - x)^{-2},$$
$$y_2 = (1 - x)^{-2}(\ln x - x).$$

20. $xy'' + (1 - x)y' + 3y = 0.$ $\quad\quad\quad$ ANS. $\quad y_1 = 1 - 3x + \tfrac{3}{2}x^2 - \tfrac{1}{6}x^3;$

$$y_2 = y_1 \ln x + 7x - \tfrac{23}{4}x^2 + \tfrac{11}{12}x^3 - 6\sum_{n=4}^{\infty}\frac{x^n}{n!n(n - 1)(n - 2)(n - 3)}.$$

21. $xy'' - (2 + x)y' - 2y = 0.$
22. $2x^2y'' - x(2x + 7)y' + 2(x + 5)y = 0.$
23. $(1 - x^2)y'' - 10xy' - 18y = 0.$
24. $y'' + 2xy' - 8y = 0.$
25. $2x(1 - x)y'' + (1 - 2x)y' + 8y = 0.$
26. $2x^2y'' - x(1 + 2x)y' + (1 + 4x)y = 0.$
27. $x^2y'' - x(1 + x^2)y' + (1 - x^2)y = 0.$
28. $x^2y'' + x(x^2 - 3)y' + 4y = 0.$
29. $(1 + x^2)y'' - 2y = 0.$
30. $x^2y'' - 3x(1 + x)y' + 4(1 - x)y = 0.$
31. $y''' + xy = 0.$
32. $xy'' + (1 - x^2)y' + 2xy = 0.$
33. $x(1 - x^2)y'' + 5(1 - x^2)y' - 4xy = 0.$
34. $x^2y'' + xy' - (x^2 + 4)y = 0.$
35. $2xy'' + (3 - x)y' - 3y = 0.$
36. $xy'' + (2 - x)y' - y = 0.$
37. $y'' - 2xy' + 6y = 0.$
38. $x^2y'' - x(3 + 2x)y' + (3 - x)y = 0.$
39. $4x^2y'' + 2x(x + 2)y' + (5x - 1)y = 0.$
40. $xy'' + 3y' - y = 0.$
41. $4x^2y'' + (3x + 1)y = 0.$
42. $x^2y'' + x(3x - 1)y' + (3x + 1)y = 0.$

**43.** $x^2y'' + x(4x - 3)y' + (8x + 3)y = 0.$
**44.** $2(1 + x^2)y'' + 7xy' + 2y = 0.$
**45.** $3xy'' + 2(1 - x)y' - 2y = 0.$
**46.** $xy'' - (1 + 3x)y' - 4y = 0.$
**47.** $x^2y'' - x(3 + 2x)y' + (4 - x)y = 0.$
**48.** $2x(1 - x)y'' + y' + 4y = 0.$
**49.** $x(1 + 4x)y'' + (1 + 8x)y' + y = 0.$
**50.** $xy'' + (3 - 2x)y' + 4y = 0.$
**51.** $x^2y'' + x(2x - 3)y' + (4x + 3)y = 0.$
**52.** $(1 - x^2)y'' - 2xy' + 12y = 0.$
**53.** $x^2(1 + x)y'' + x(3 + 5x)y' + (1 + 4x)y = 0.$
**54.** $x^2y'' + x^2y' + (3x - 2)y = 0.$
**55.** $2x^2y'' + 3xy' - (1 + x)y = 0.$

# 19

# Equations of Hypergeometric Type

## 120. Equations to be treated in this chapter

With the methods studied in Chapters 17 and 18, we are able to solve many equations that appear frequently in modern physics and engineering as well as in pure mathematics. We shall consider briefly the hypergeometric equation, Bessel's equation, and the equations that lead to the study of Laguerre, Legendre, and Hermite polynomials. There are, in mathematical literature, thousands of research papers devoted entirely or in part to the study of the functions that are solutions of the equations to be studied in this chapter. Here we do no more than call to the attention of the student the existence of these special functions, which are of such great value to theoretical physicists, engineers, and many mathematicians. An introduction to the properties of these and other special functions can be found in Rainville's *Special Functions* (New York: Macmillan Publishing Co., Inc., 1960).

## 121. The factorial function

It will be convenient for us to employ a notation that is widely encountered in advanced mathematics. We define the factorial function $(a)_n$ for $n$ equal to zero or a positive integer by

$$(a)_n = a(a+1)(a+2)\cdots(a+n-1), \qquad \text{for } n \geqq 1; \qquad (1)$$

$$(a)_0 = 1, \qquad \text{for } a \neq 0.$$

Thus the symbol $(a)_n$ denotes a product of $n$ factors starting with the factor $a$, each factor being one larger than the factor before it. For instance,

$$(7)_4 = 7\cdot 8\cdot 9\cdot 10,$$

$$(-5)_3 = (-5)(-4)(-3),$$

$$(-\tfrac{1}{2})_3 = (-\tfrac{1}{2})(\tfrac{1}{2})(\tfrac{3}{2}).$$

The factorial function is a generalization of the ordinary factorial. Indeed,

$$(1)_n = 1\cdot 2\cdot 3\cdots n = n!. \qquad (2)$$

In our study of the gamma function in Section 64, we derived the functional relation

$$\Gamma(x+1) = x\Gamma(x), \qquad \text{for } x > 0. \qquad (3)$$

By repeated use of the relation (3), we find that, if $n$ is an integer,

$$\Gamma(a+n) = (a+n-1)\Gamma(a+n-1)$$

$$= (a+n-1)(a+n-2)\Gamma(a+n-2)$$

$$\vdots$$

$$= (a+n-1)(a+n-2)\cdots(a)\Gamma(a)$$

$$= (a)_n\Gamma(a).$$

Therefore the factorial function and the Gamma function are related by

$$(a)_n = \frac{\Gamma(a+n)}{\Gamma(a)}, \qquad n \text{ integer}, \ n > 0, \text{ and } a > 0. \qquad (4)$$

Actually (4) can be shown to be valid for any complex $a$ except zero or a negative integer.

## 122. The hypergeometric equation

Let us now consider any second-order linear differential equation that has only three singular points (one could be at infinity). Suppose that each of these singularities is regular. It can be shown* that such an equation can be trans-

_____

* See, for instance, E. D. Rainville, *Intermediate Differential Equations*, 2nd ed. (New York: Macmillan Publishing Co., Inc., 1964), Chapter 6.

formed by change of variables into the hypergeometric equation

$$x(1 - x)y'' + [c - (a + b + 1)x]y' - aby = 0, \tag{1}$$

in which $a, b, c$ are fixed parameters.

Let us solve equation (1) about the regular singular point $x = 0$. For the moment let $c$ be not an integer. For (1), the indicial equation has roots zero and $(1 - c)$. We put

$$y = \sum_{n=0}^{\infty} e_n x^n$$

in equation (1) and thus arrive, after the usual simplifications, at

$$\sum_{n=0}^{\infty} n(n + c - 1)e_n x^{n-1} - \sum_{n=0}^{\infty} (n + a)(n + b)e_n x^n = 0. \tag{2}$$

Shift index in (2) to get

$$\sum_{n=0}^{\infty} n(n + c - 1)\,e_n x^{n-1} - \sum_{n=1}^{\infty} (n + a - 1)(n + b - 1)\,e_{n-1}x^{n-1} = 0. \tag{3}$$

We thus find that $e_0$ is arbitrary and, for $n \geqq 1$,

$$e_n = \frac{(n + a - 1)(n + b - 1)}{n(n + c - 1)} e_{n-1}. \tag{4}$$

The recurrence relation (4) may be solved by our customary device. The result is, for $n \geqq 1$,

$$e_n = \frac{a(a + 1)(a + 2)\cdots(a + n - 1) \cdot b(b + 1)(b + 2)\cdots(b + n - 1)e_0}{n!\,c(c + 1)(c + 2)\cdots(c + n - 1)}. \tag{5}$$

But (5) is greatly simplified by use of the factorial function. We rewrite (5) as

$$e_n = \frac{(a)_n(b)_n}{n!\,(c)_n} e_0. \tag{6}$$

Let us choose $e_0 = 1$ and write our first solution of the hypergeometric equation as

$$y_1 = 1 + \sum_{n=1}^{\infty} \frac{(a)_n(b)_n x^n}{(c)_n n!}. \tag{7}$$

The particular solution $y_1$ in (7) is called the hypergeometric function and for it a common symbol is $F(a, b; c; x)$. That is,

$$F(a, b; c; x) = 1 + \sum_{n=1}^{\infty} \frac{(a)_n(b)_n x^n}{(c)_n n!},$$

and $y_1 = F(a, b; c; x)$ is a solution of equation (1).

The other root of the indicial equation is $(1 - c)$. We may put

$$y = \sum_{n=0}^{\infty} f_n x^{n+1-c}$$

into equation (1), determine $f_n$ in the usual manner, and arrive at a second solution

$$y_2 = x^{1-c} + \sum_{n=1}^{\infty} \frac{(a + 1 - c)_n (b + 1 - c)_n x^{n+1-c}}{(2 - c)_n n!}. \tag{8}$$

In the hypergeometric notation this second solution (8) may be written

$$y_2 = x^{1-c} F(a + 1 - c, b + 1 - c; 2 - c; x),$$

which means exactly the same as (8). The solutions (7) and (8) are valid in $0 < x < 1$, a region extending to the nearest other singular point of the differential equation (1).

If $c$ is an integer, one of the solutions (7) or (8) is correct, but the other involves a zero denominator. For example, if $c = 5$, then in (8), $(2 - c)_n = (-3)_n$ and as soon as $n \geq 4$, $(-3)_n = 0$. For,

$$(-3)_4 = (-3)(-2)(-1)(0) = 0.$$

If $c$ is an integer but $a$ and $b$ are nonintegral, one of the solutions about $x = 0$ of the hypergeometric equation is of logarithmic type. If $c$ and one or both of $a$ and $b$ are integers, the solution may or may not involve a logarithm. To save space we omit logarithmic solutions of the hypergeometric equation.

## 123. Laguerre polynomials

The equation

$$xy'' + (1 - x)y' + ny = 0 \tag{1}$$

is called Laguerre's equation. If $n$ is a nonnegative integer, one solution of equation (1) is a polynomial.

Consider the solution of (1) about the regular singular point $x = 0$. The indicial equation has equal roots $c = 0, 0$. Hence one solution will involve a logarithm. We seek the nonlogarithmic solution.

Let us put

$$y = \sum_{k=0}^{\infty} a_k x^k$$

into (1) and obtain, in the usual way,

$$\sum_{k=0}^{\infty} k^2 a_k x^{k-1} - \sum_{k=1}^{\infty} (k - 1 - n)a_{k-1} x^{k-1} = 0. \tag{2}$$

From (2) we find that

$$k \geqq 1: \qquad a_k = \frac{(k - 1 - n)a_{k-1}}{k^2}$$

$$= \frac{(-n)(-n + 1) \cdots (-n + k - 1)a_0}{(k!)^2}$$

$$= \frac{(-n)_k}{(k!)^2} a_0.$$

If $n$ is a nonnegative integer, $(-n)_k = 0$ for $k > n$. Therefore, with $a_0$ chosen equal to unity, one solution of equation (1) is

$$y_1 = \sum_{k=0}^{n} \frac{(-n)_k x^k}{(k!)^2}. \qquad (3)$$

The right member of (3) is called the Laguerre polynomial and is usually denoted by $L_n(x)$:

$$L_n(x) = \sum_{k=0}^{n} \frac{(-n)_k x^k}{(k!)^2} = \sum_{k=0}^{n} \frac{(-1)^k n! x^k}{(k!)^2 (n - k)!}. \qquad (4)$$

The student should prove the equivalence of the two summations in (4) by showing that

$$(-n)_k = \frac{(-1)^k n!}{(n - k)!}.$$

One solution of (1) is $y_1 = L_n(x)$. The associated logarithmic solution may, after considerable simplification, be put in the form

$$y_2 = L_n(x) \ln x + \sum_{k=1}^{n} \frac{(-n)_k (H_{n-k} - H_n - 2H_k) x^k}{(k!)^2}$$

$$+ \sum_{k=1}^{\infty} \frac{(-1)^n n! (k - 1)! x^{k+n}}{[(k + n)!]^2}. \qquad (5)$$

The solution (3) is valid for all finite $x$; the solution (5) is valid for $x > 0$.

## 124. Bessel's equation with index not an integer

The equation

$$x^2 y'' + xy' + (x^2 - n^2)y = 0 \qquad (1)$$

is called Bessel's equation of index $n$. Equation (1) has a regular singular point at $x = 0$, but no other singular points in the finite plane. At $x = 0$ the roots of the indicial equation are $c_1 = n$, $c_2 = -n$. In this section we assume that $n$ is not an integer.

It is a simple exercise in the methods of Chapter 18 to show that, if $n$ is not an integer, two linearly independent solutions of (1) are

$$y_1 = \sum_{k=0}^{\infty} \frac{(-1)^k x^{2k+n}}{2^{2k} k!(1+n)_k}, \tag{2}$$

$$y_2 = \sum_{k=0}^{\infty} \frac{(-1)^k x^{2k-n}}{2^{2k} k!(1-n)_k}, \tag{3}$$

valid for $x > 0$.
The function

$$y_3 = \frac{1}{2^n \Gamma(1+n)} y_1 = \sum_{k=0}^{\infty} \frac{(-1)^k x^{2k+n}}{2^{2k+n} k! \Gamma(k+n+1)},$$

also a solution of equation (1), is called $J_n(x)$, the Bessel function of the first kind and of index $n$. Thus

$$y_3 = J_n(x) = \sum_{k=0}^{\infty} \frac{(-1)^k x^{2k+n}}{2^{2k+n} k! \Gamma(k+n+1)} \tag{4}$$

is a solution of (1) and the general solution of (1) may be written

$$y = A J_n(x) + B J_{-n}(x), \qquad n \neq \text{an integer.} \tag{5}$$

That $J_{-n}(x)$ is a solution of the differential equation (1) should be evident from the fact that the parameter $n$ enters (1) only in the term $n^2$. It is also true that

$$J_{-n}(x) = \frac{1}{2^{-n} \Gamma(1-n)} y_2.$$

## 125. Bessel's equation with index an integer

In Bessel's equation

$$x^2 y'' + x y' + (x^2 - n^2) y = 0, \tag{1}$$

let $n$ now be zero or a positive integer. Then

$$y_1 = J_n(x) = \sum_{k=0}^{\infty} \frac{(-1)^k x^{2k+n}}{2^{2k+n} k! \Gamma(k+n+1)} \tag{2}$$

is one solution of equation (1). Any solution linearly independent of (2) must contain ln $x$. We have already solved (1) for $n = 0$ in exercise 6, page 364, and for $n = 1$ in exercise 6, page 378.

For $n$ an integer $\geq 2$, put

$$y = \sum_{j=0}^{\infty} a_j x^{j+c},$$

proceed with the technique of Section 116, determine $y(x, c)$ and $\dfrac{\partial}{\partial c} y(x, c)$, and then use $c = -n$ to obtain two solutions:

$$y_2 = \sum_{k=n}^{\infty} \frac{(-1)^k x^{2k-n}}{2^{2k-1}(1-n)_{n-1}(k-n)!\,k!} \tag{3}$$

and

$$y_3 = y_2 \ln x + x^{-n} + \sum_{k=1}^{n-1} \frac{(-1)^k x^{2k-n}}{2^{2k}(1-n)_k k!}$$

$$+ \sum_{k=n}^{\infty} \frac{(-1)^{k+1}(H_{k-n} + H_k - H_{n-1})x^{2k-n}}{2^{2k-1}(1-n)_{n-1}(k-n)!\,k!}. \tag{4}$$

A shift of index in (3) from $k$ to $(k + n)$ yields

$$y_2 = \sum_{k=0}^{\infty} \frac{(-1)^{k+n} x^{2k+n}}{2^{2k+2n-1}(1-n)_{n-1}k!(k+n)!}.$$

But for $n \geq 2$, $(1 - n)_{n-1} = (-1)^{n-1}(n - 1)!$, so

$$y_2 = \frac{-1}{2^{n-1}(n-1)!} J_n(x).$$

We can therefore replace solution (3) with

$$y_1 = J_n(x). \tag{5}$$

By similar manipulations, we replace solution (4) with

$$y_4 = J_n(x) \ln x + \sum_{k=0}^{n-1} \frac{(-1)^{k+1}(n-1)!\,x^{2k-n}}{2^{2k+1-n}k!(1-n)_k}$$

$$+ \tfrac{1}{2} \sum_{k=0}^{\infty} \frac{(-1)^{k+1}(H_k + H_{k+n})x^{2k+n}}{2^{2k+n}k!(k+n)!}. \tag{6}$$

For $n$ an integer $> 1$, equations (5) and (6) can be used as the fundamental pair of linearly independent solutions of Bessel's equation (1) for $x > 0$.

## 126. Hermite polynomials

The equation

$$y'' - 2xy' + 2ny = 0 \qquad (1)$$

is called Hermite's equation. Since equation (1) has no singular points in the finite plane, $x = 0$ is an ordinary point of the equation. We put

$$y = \sum_{j=0}^{\infty} a_j x^j$$

and employ the methods of Chapter 17 to obtain the general solution

$$y = a_0 \left[ 1 + \sum_{k=1}^{\infty} \frac{2^k(-n)(-n+2)\cdots(-n+2k-2)x^{2k}}{(2k)!} \right]$$
$$+ a_1 \left[ x + \sum_{k=1}^{\infty} \frac{2^k(1-n)(1-n+2)\cdots(1-n+2k-2)x^{2k+1}}{(2k+1)!} \right], \quad (2)$$

valid for all finite $x$ and with $a_0$ and $a_1$ arbitrary.

Interest in equation (1) is greatest when $n$ is a positive integer or zero. If $n$ is an even integer, the coefficient of $a_0$ in (2) terminates, each term for $k \geq \frac{1}{2}(n+2)$ being zero. If $n$ is an odd integer, the coefficient of $a_1$ in (2) terminates, each term for $k \geq \frac{1}{2}(n+1)$ being zero. Thus Hermite's equation always has a polynomial solution, of degree $n$, for $n$ zero or a positive integer. It is elementary but tedious to obtain from (2) a single expression for this polynomial solution. The result is

$$H_n(x) = \sum_{k=0}^{[\frac{1}{2}n]} \frac{(-1)^k n!(2x)^{n-2k}}{k!(n-2k)!}, \qquad (3)$$

in which $[\frac{1}{2}n]$ stands for the greatest integer $\leq \frac{1}{2}n$.

The polynomial $H_n(x)$ of (3) is the Hermite polynomial; $y = H_n(x)$ is a solution of equation (1).

## 127. Legendre polynomials

The equation

$$(1 - x^2)y'' - 2xy' + n(n+1)y = 0 \qquad (1)$$

is called Legendre's equation. Let us solve (1) about the regular singular point $x = 1$. We put $x - 1 = v$ and obtain the transformed equation

$$v(v + 2)\frac{d^2y}{dv^2} + 2(v + 1)\frac{dy}{dv} - n(n + 1)y = 0. \tag{2}$$

At $v = 0$, equation (2) has, as roots of its indicial equation, $c = 0, 0$. Hence one solution is logarithmic. We are interested here only in the non-logarithmic solution.

Following the methods of Chapter 18, put

$$y = \sum_{k=0}^{\infty} a_k v^k$$

into equation (2) and thus arrive at the results: $a_0$ is arbitrary and

$$k \geqq 1: \qquad a_k = \frac{-(k - n - 1)(k + n)a_{k-1}}{2k^2}. \tag{3}$$

Solve the recurrence relation (3) and thus obtain

$$a_k = \frac{(-1)^k(-n)_k(1 + n)_k a_0}{2^k(k!)^2},$$

with the factorial notation of Section 121.

We may now write one solution of equation (1) in the form

$$y_1 = 1 + \sum_{k=1}^{\infty} \frac{(-1)^k(-n)_k(n + 1)_k(x - 1)^k}{2^k(k!)^2}. \tag{4}$$

Since $k! = (1)_k$, we may put (4) into the form

$$y_1 = 1 + \sum_{k=1}^{\infty} \frac{(-n)_k(n + 1)_k}{(1)_k k!}\left(\frac{1 - x}{2}\right)^k. \tag{5}$$

The right member of equation (5) is an example of the hypergeometric function that we met in Section 122. In fact,

$$y_1 = F\left(-n, n + 1; 1; \frac{1 - x}{2}\right). \tag{6}$$

If $n$ is a positive integer or zero, the series in (4), (5), or (6) terminates. It is then called the Legendre polynomial and designated $P_n(x)$. We write our nonlogarithmic solution of Legendre's equation as

$$y_1 = P_n(x) = F\left(-n, n + 1; 1; \frac{1 - x}{2}\right). \tag{7}$$

## 128. The confluent* hypergeometric equation

The equation
$$xy'' + (c - x)y' - ay = 0 \tag{1}$$
has a regular singular point at $x = 0$ with zero and $(1 - c)$ as roots of the indicial equation there. Equation (1) is called the confluent hypergeometric equation. If $c$ is not an integer, there is no logarithmic solution of (1) about $x = 0$, so we restrict ourselves here to that simple situation.

In the usual manner we put
$$y = \sum_{n=0}^{\infty} b_n x^n$$
into (1) and thus find that $b_0$ is arbitrary and
$$n \geq 1: \qquad b_n = \frac{(n - 1 + a)b_{n-1}}{n(n - 1 + c)}.$$
The recurrence relation yields
$$n \geq 1: \qquad b_n = \frac{(a)_n b_0}{n!(c)_n}$$
in the notation of the factorial function of Section 121.

Therefore equation (1) has a solution
$$y_1 = 1 + \sum_{n=1}^{\infty} \frac{(a)_n x^n}{(c)_n n!}, \tag{2}$$
valid for all finite $x$. Note how much the right member of (2) resembles the hypergeometric function
$$F(a, b; c; x) = 1 + \sum_{n=1}^{\infty} \frac{(a)_n (b)_n x^n}{(c)_n n!} \tag{3}$$
of Section 122. In (2), the series has only one numerator parameter, $a$, and one denominator parameter, $c$. In (3) there are two numerator parameters, $a$ and $b$, and one denominator parameter, $c$. It is therefore customary to use a notation like that in (3) for the right member of (2). We write
$$_1F_1(a; c; x) = 1 + \sum_{n=1}^{\infty} \frac{(a)_n x^n}{(c)_n n!}, \tag{4}$$
with the subscripts before and after the $F$ denoting the number of numerator

---

* The concept of confluence of singularities is treated in E. D. Rainville, *Intermediate Differential Equations*, 2nd ed. (New York: Macmillan Publishing Co., Inc., 1964), Chapter 10.

and denominator parameters, respectively. When it is thought desirable, the function symbol on the left in (3) is similarly written $_2F_1(a, b; c; x)$. Functions of hypergeometric type with any number of numerator and denominator parameters have been studied for many years.

The subscripts on the $F$ play a useful role when the nature of the function, but not its specific parameters, is under discussion. For instance, we say, "Any $_0F_1$ is essentially a Bessel function of the first kind," and "The Laguerre polynomial is a terminating $_1F_1$." The detailed statements are as follows.

$$J_n(x) = \frac{(x/2)^n}{\Gamma(n + 1)} {}_0F_1\left(-; n + 1; -\frac{x^2}{4}\right), \tag{5}$$

$$L_n(x) = {}_1F_1(-n; 1; x). \tag{6}$$

We have seen that the differential equation (1) has one solution, as indicated in (2),

$$y_1 = {}_1F_1(a; c; x). \tag{7}$$

The student can show that another solution, linearly independent of (7), is

$$y_2 = x^{1-c} {}_1F_1(a + 1 - c; 2 - c; x), \tag{8}$$

as long as $c$ is not an integer.

Again we omit discussion of the logarithmic solutions, which may enter if $c$ is integral.

# 20

## Numerical Methods

### 129. General remarks

There is no general method for obtaining an explicit formula for the solution of a differential equation. Specific equations do occur for which no known attack yields a solution or for which the explicit forms of solution are not well adapted to computation. For these reasons, systematic, efficient methods for the numerical approximation to solutions are important. Unfortunately, a clear grasp of good numerical methods requires time-consuming practice and often also the availability of modern computing machines.

This chapter is restricted to a fragmentary discussion of some simple and moderately useful methods. The purpose here is to give the student a concept of the fundamental principles of numerical approximations to solutions. We shall take one problem, which does not yield to the methods developed earlier, and apply to it several simple numerical processes.

## 130. The increment method

We seek to obtain that solution of the differential equation

$$y' = y^2 - x^2 \tag{1}$$

for which $y = 1$ when $x = 0$. We wish to know the solution $y = y(x)$ in the range $0 \leqq x \leqq \frac{1}{2}$.

Equation (1) may be written in differential form as

$$dy = (y^2 - x^2)\, dx. \tag{2}$$

Figure 44 shows the geometrical significance of the differential $dy$ and of $\Delta y$, the actual change in $y$, as induced by an increment $dx$ (or $\Delta x$) applied to $x$. In calculus it is shown that, near a point where the derivative exists, $dy$ can be made to approximate $\Delta y$ as closely as desired by taking $\Delta x$ sufficiently small.

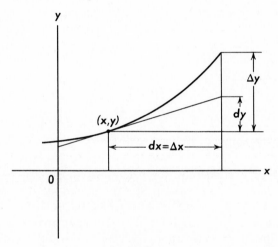

FIGURE 44

We know the value of $y$ at $x = 0$; we wish to compute $y$ for $0 \leqq x \leqq \frac{1}{2}$. Suppose we choose $\Delta x = 0.1$; then $dy$ can be computed from

$$dy = (y^2 - x^2)\, \Delta x.$$

Indeed, $dy = (1 - 0)(0.1) = 0.1$. Thus for $x = 0 + 0.1$, the approximate value of $y$ is $1 + 0.1$. Now we have $x = 0.1$, $y = 1.1$. Let us choose $\Delta x = 0.1$ again. Then

$$dy = [(1.1)^2 - (0.1)^2]\, \Delta x,$$

so $dy = 0.12$. Hence at $x = 0.2$, the approximate value of $y$ is 1.22. The complete computation using $\Delta x = 0.1$ is shown in Table 1.

TABLE 1

$\Delta x = 0.1$

| $x$ | $y$ | $y^2$ | $x^2$ | $(y^2 - x^2)$ | $dy$ |
|-----|-----|-------|-------|---------------|------|
| 0.0 | 1.00 | 1.00 | 0.00 | 1.00 | 0.10 |
| 0.1 | 1.10 | 1.21 | 0.01 | 1.20 | 0.12 |
| 0.2 | 1.22 | 1.49 | 0.04 | 1.45 | 0.14 |
| 0.3 | 1.36 | 1.85 | 0.09 | 1.76 | 0.18 |
| 0.4 | 1.54 | 2.37 | 0.16 | 2.21 | 0.22 |
| 0.5 | 1.76 | | | | |

The increment $\Delta x$ need not be constant throughout the interval. Where the slope is larger it pays to take a smaller increment. For simplicity in computations, equal increments are used here.

It is helpful to repeat the computation with a smaller increment and to note the changes that result in the approximate values of $y$. Table 2 shows a computation with $\Delta x = 0.05$ throughout.

TABLE 2

$\Delta x = 0.05$

| $x$ | $y$ | $y^2$ | $x^2$ | $(y^2 - x^2)$ | $dy$ |
|-----|-----|-------|-------|---------------|------|
| 0.00 | 1.000 | 1.000 | 0.000 | 1.000 | 0.050 |
| 0.05 | 1.050 | 1.102 | 0.002 | 1.100 | 0.055 |
| 0.10 | 1.105 | 1.221 | 0.010 | 1.211 | 0.061 |
| 0.15 | 1.166 | 1.360 | 0.022 | 1.338 | 0.067 |
| 0.20 | 1.233 | 1.520 | 0.040 | 1.480 | 0.074 |
| 0.25 | 1.307 | 1.708 | 0.062 | 1.646 | 0.082 |
| 0.30 | 1.389 | 1.929 | 0.090 | 1.839 | 0.092 |
| 0.35 | 1.481 | 2.193 | 0.122 | 2.071 | 0.104 |
| 0.40 | 1.585 | 2.512 | 0.160 | 2.352 | 0.118 |
| 0.45 | 1.703 | 2.900 | 0.202 | 2.698 | 0.135 |
| 0.50 | 1.838 | | | | |

In Table 3 the value of $y$ obtained from the computations in Tables 1 and 2 and also the values of $y$ obtained by using $\Delta x = 0.01$ (computation not shown) are exhibited beside the values of $y$ correct to two decimal places. The correct values are obtained by the method of Section 133. Their availability is in a sense accidental. Frequently, we know of no way to obtain the $y$ value correct to a specified degree of accuracy. In such instances it is customary to resort to decreasing the size of the increment until the $y$ values show changes no larger than the errors we are willing to permit. Then it is *hoped* that the steadying down of the $y$ values is due to our being close to the correct solution rather than (as is quite possible) to the slowness of convergence of the process used.

TABLE 3

| When: $x$ | $\Delta x = 0.1$ $y$ | $\Delta x = 0.05$ $y$ | $\Delta x = 0.01$ $y$ | Correct $y$ |
|---|---|---|---|---|
| 0.0 | 1.00 | 1.00 | 1.00 | 1.00 |
| 0.1 | 1.10 | 1.11 | 1.11 | 1.11 |
| 0.2 | 1.22 | 1.23 | 1.24 | 1.25 |
| 0.3 | 1.36 | 1.39 | 1.41 | 1.42 |
| 0.4 | 1.54 | 1.58 | 1.62 | 1.64 |
| 0.5 | 1.76 | 1.84 | 1.91 | 1.93 |

### Exercises

In each of the following exercises, use the increment method with the prescribed $\Delta x$ to approximate the solution of the initial value problem in the given interval. In exercises 1 through 6, solve the problem by elementary methods and compare the approximate values of $y$ with the correct values.

1. $y' = x + y$;     when $x = 0$, $y = 1$;     $\Delta x = 0.1$ and $0 \le x \le 1$.
2. Use $\Delta x = 0.05$ in exercise 1.
3. $y' = x + y$;     when $x = 0$, $y = 2$;     $\Delta x = 0.1$ and $0 \le x \le 1$.
4. $y' = x + y$;     when $x = 1$, $y = 1$;     $\Delta x = 0.1$ and $1 \le x \le 2$.
5. $y' = x + y$;     when $x = 2$, $y = -1$;     $\Delta x = 0.1$ and $2 \le x \le 3$.
6. $y' = 2x - 3y$;   when $x = 0$, $y = 2$;     $\Delta x = 0.1$ and $0 \le x \le 1$.
7. $y' = e^{-xy}$;      when $x = 0$, $y = 0$;     $\Delta x = 0.2$ and $0 \le x \le 2$.
8. Use $\Delta x = 0.1$ in exercise 7.
9. $y' = (1 + x^2 + y^2)^{-1}$;     when $x = 0$, $y = 0$;   $\Delta x = 0.2$ and $0 \le x \le 2$.
10. Use $\Delta x = 0.2$ in exercise 9.
11. $y' = (\cos x + \sin y)^{1/2}$;     when $x = 0$, $y = 1$;   $\Delta x = 0.2$ and $0 \le x \le 2$.
12. Use $\Delta x = 0.1$ in exercise 11.

13. $y' = \dfrac{x^2 + y^2}{x^2 - y^2 + 2}$;     when $x = 0$, $y = 0$;   $\Delta x = 0.2$ and $0 \le x \le 2$.

## 131. A method of successive approximation

Next let us attack the same problem as before,

$$y' = y^2 - x^2; \qquad x = 0, \, y = 1, \tag{1}$$

with $y$ desired in the interval $0 \le x \le \frac{1}{2}$, by the method suggested in the discussion of the existence theorem in Chapter 15. Applying the statements made in that discussion, we conclude that the desired solution is $y = y(x)$ where

$$y(x) = \lim_{n \to \infty} y_n(x)$$

and the sequence of approximations $y_n(x)$ is given by $y_0(x) = 1$ and, for $n \ge 1$,

$$y_n(x) = 1 + \int_0^x [y_{n-1}^2(t) - t^2] \, dt. \tag{2}$$

For the problem at hand

$$y_1(x) = 1 + \int_0^x (1 - t^2) \, dt,$$

$$y_1(x) = 1 + x - \tfrac{1}{3}x^3.$$

Next we obtain a second approximation, finding $y_2(x)$ from $y_1(x)$ by means of (2). Thus we find that

$$y_2(x) = 1 + \int_0^x [(1 + t - \tfrac{1}{3}t^3)^2 - t^2] \, dt,$$

$$y_2(x) = 1 + x + x^2 - \tfrac{1}{6}x^4 - \tfrac{2}{15}x^5 + \tfrac{1}{63}x^7.$$

Then $y_3(x), y_4(x), \ldots,$ can be obtained in a similar manner, each from the preceding element of the sequence $y_n(x)$.

In Table 4 the values taken on by $y_1(x)$, $y_2(x)$, and $y_3(x)$ at intervals of 0.1 in $x$ are shown beside the corresponding values of $y(x)$, correct to two decimal places, as obtained in Section 133.

TABLE 4

| $x$ | $y_1(x)$ | $y_2(x)$ | $y_3(x)$ | $y(x)$ |
|-----|----------|----------|----------|--------|
| 0.0 | 1.00 | 1.00 | 1.00 | 1.00 |
| 0.1 | 1.10 | 1.11 | 1.11 | 1.11 |
| 0.2 | 1.20 | 1.24 | 1.25 | 1.25 |
| 0.3 | 1.29 | 1.39 | 1.41 | 1.42 |
| 0.4 | 1.38 | 1.56 | 1.62 | 1.64 |
| 0.5 | 1.46 | 1.74 | 1.87 | 1.93 |

It must be realized that the usefulness of this method is not dependent upon our being able to carry out the integrations in a formal sense. It may be best to perform the integrations mechanically with a planimeter or by some numerical process such as Simpson's rule.

## Exercises

**1.** Apply the method of this section to the problem

$$y' = x + y; \quad \text{when } x = 0, y = 1,$$

(exercise 1, Section 130). Obtain $y_1(x)$, $y_2(x)$, and $y_3(x)$.

ANS.   $y_1(x) = 1 + x + \frac{1}{2}x^2$; $y_2(x) = 1 + x + x^2 + \frac{1}{6}x^3$;
$$y_3(x) = 1 + x + x^2 + \frac{1}{3}x^3 + \frac{1}{24}x^4.$$

**2.** Compute a table of values to two decimal places of $y_1$, $y_2$, $y_3$ of exercise 1 for $x = 0$ to $x = 1$ at intervals of 0.1. Tabulate also the correct values of $y$ obtained from the elementary solution to the problem.

**3.** Obtain $y_1(x)$, $y_2(x)$, and $y_3(x)$ for the initial value problem (exercise 4, Section 130).

$$y' = x + y; \quad \text{when } x = 1, y = 1.$$

[*Hint:* Express the integrand of the integral in equation (2) in powers of $t - 1$ before integrating.]

ANS.   $y_1(x) = 1 + 2(x - 1) + \frac{1}{2}(x - 1)^2$;
$y_2(x) = 1 + 2(x - 1) + \frac{3}{2}(x - 1)^2 + \frac{1}{6}(x - 1)^3$;
$y_3(x) = 1 + 2(x - 1) + \frac{3}{2}(x - 1)^2 + \frac{1}{2}(x - 1)^3 + \frac{1}{24}(x - 1)^4.$

**4.** Compute a table of values to two decimal places of the $y_1$, $y_2$, $y_3$ in exercise 3 for $x = 1$ to $x = 2$ at intervals of 0.1. Also tabulate the correct values of $y$ obtained from the elementary solution to the problem.

## 132. An improvement on the preceding method

In the method used in Section 131, each of the $y_n(x)$, where $n = 0, 1, 2, \ldots$, yields an approximation to the solution $y = y(x)$. It is plausible that, usually, the more nearly correct a particular approximation $y_k(x)$, the better will be its successor $y_{k+1}(x)$.

The initial value problem we are treating is

$$y' = y^2 - x^2; \quad x = 0, y = 1$$

and it tells us at once that at $x = 0$, the slope is $y' = 1$. But in Section 131, by blindly following the suggestion in Chapter 15, we started out with $y_0(x) = 1$, a line that does not have the correct slope at $x = 0$.

It is therefore reasonable to alter our initial approximation by choosing $y_0(x)$ to have the correct slope at $x = 0$, still making it also pass through the desired point $x = 0$, $y = 1$. Hence we choose

$$y_0(x) = 1 + x$$

and proceed to compute $y_1(x)$, $y_2(x)$, $\cdots$, as before. The successive stages of approximation to $y(x)$ now become

$$y_1(x) = 1 + \int_0^x [(1 + t)^2 - t^2] \, dt,$$

$$y_1(x) = 1 + x + x^2;$$

and

$$y_2(x) = 1 + \int_0^x [(1 + t + t^2)^2 - t^2] \, dt,$$

$$y_2(x) = 1 + x + x^2 + \tfrac{2}{3}x^3 + \tfrac{1}{2}x^4 + \tfrac{1}{5}x^5;$$

and so on.

In Table 5 the values of $y_1$, $y_2$, $y_3$ obtained by this method are shown beside the correct values of $y$.

TABLE 5

| $x$ | $y_1(x)$ | $y_2(x)$ | $y_3(x)$ | $y(x)$ |
|-----|----------|----------|----------|--------|
| 0.0 | 1.00 | 1.00 | 1.00 | 1.00 |
| 0.1 | 1.11 | 1.11 | 1.11 | 1.11 |
| 0.2 | 1.24 | 1.25 | 1.25 | 1.25 |
| 0.3 | 1.39 | 1.41 | 1.42 | 1.42 |
| 0.4 | 1.56 | 1.62 | 1.64 | 1.64 |
| 0.5 | 1.75 | 1.87 | 1.92 | 1.93 |

## Exercises

1. Apply the method of this section to obtain approximations $y_1$, $y_2$, $y_3$ for the problem of exercise 1 of Section 130.

    ANS.   $y_1(x) = 1 + x + x^2$; $y_2(x) = 1 + x + x^2 + \tfrac{1}{3}x^3$;
    $y_3(x) = 1 + x + x^2 + \tfrac{1}{3}x^3 + \tfrac{1}{12}x^4$.

2. Tabulate to two decimal places $y_1$, $y_2$, $y_3$ of exercise 1 beside the corresponding values of the exact solution $y(x) = 2e^x - 1 - x$.

3. Apply the method of this section to obtain the approximations $y_1$, $y_2$, $y_3$ for the problem of exercise 3 of Section 131.

    ANS.   $y_3(x) = 1 + 2(x - 1) + \tfrac{3}{2}(x - 1)^2 + \tfrac{1}{2}(x - 1)^3 + \tfrac{1}{8}(x - 1)^4$.

4. Compare the $y_1$, $y_2$, $y_3$ of exercise 3 with the Taylor series in powers of $x - 1$ for the exact solution

$$y(x) = 3 \exp(x - 1) - (x - 1) - 2.$$

## 133. The use of Taylor's theorem

For students familiar with the elementary calculus, the most natural approach to the approximation of solutions is to make use of Taylor's theorem. If we consider the initial value problem

$$y' = F(x, y); \qquad x = x_0, y = y_0, \tag{1}$$

we may be able to compute successive derivatives of the solution $y = y(x)$ at $x = x_0$ by using equation (1). We adopt the notation

$$y_0' = y'(x_0), \qquad y_0'' = y''(x_0), \qquad \cdots, \qquad y_0^{(n)} = y^{(n)}(x_0),$$

and recall that Taylor's theorem suggests the approximation formula

$$y \cong y_0 + y_0'(x - x_0) + \frac{y_0''}{2!}(x - x_0)^2 + \cdots + \frac{y_0^{(n)}}{n!}(x - x_0)^n. \tag{2}$$

One advantage in using the approximation in (2) is that we may be able to estimate the error in our calculation by examining the value of the remainder term in Taylor's theorem. For relation (2) this remainder takes the form

$$\frac{y^{(n+1)}(c)}{(n+1)!}(x - x_0)^{n+1}, \tag{3}$$

where $c$ is some number between $x$ and $x_0$.

It should be clear that the practicality of this technique for approximating solutions will depend greatly on the difficulty of obtaining values of the derivatives involved. If the function $F(x, y)$ is at all complicated, there may be a great deal of computation necessary to produce a reasonable approximation using Taylor's theorem.

For the example

$$y' = y^2 - x^2; \qquad x_0 = 0, y_0 = 1, \tag{4}$$

it is relatively easy to show that

$$\begin{aligned}
y'' &= 2yy' - 2x, \\
y''' &= 2yy'' + 2(y')^2 - 2, \\
y^{(4)} &= 2yy''' + 6y'y'', \\
y^{(5)} &= 2yy^{(4)} + 8y'y''' + 6(y'')^2.
\end{aligned} \tag{5}$$

Thus we can obtain the values

$$y_0 = 1, \quad y_0' = 1, \quad y_0'' = 2, \quad y_0''' = 4, \quad y_0^{(4)} = 20, \quad \text{and} \quad y_0^{(5)} = 96.$$

Equation (2) thus becomes

$$y \cong 1 + x + x^2 + \tfrac{2}{3}x^3 + \tfrac{5}{6}x^4. \tag{6}$$

Some indication of the accuracy of equation (6) is given in Table 6. The values of $y$ obtained from equation (6) for several values of $x$ are exhibited beside the values of $y$ correct to two decimal places.

TABLE 6

| x | y | Correct y |
|---|---|---|
| 0.0 | 1.00 | 1.00 |
| 0.1 | 1.11 | 1.11 |
| 0.2 | 1.25 | 1.25 |
| 0.3 | 1.41 | 1.42 |
| 0.4 | 1.62 | 1.64 |
| 0.5 | 1.89 | 1.93 |

Another indication of the error in using equation (6) for estimating the value of $y$ for $x = 0.5$ can be obtained by examining the next term in the Taylor's series expansion. That is, we may compute the value of $(96/5!)x^5$ at $x = 0.5$ and find that the error is at least as big as 0.02.

A more careful study of the remainder term given in equation (3) could be used to get a better estimate of the error in our results. In practice, however, it is extremely difficult to make these error estimates because of the complexity of the derivative formulas involved and the fact that the value of $c$ is unknown to us.

### Exercises

For each of the following initial value problems, use Taylor's theorem retaining powers of $x - x_0$ sufficiently large to approximate the values of $y$ accurately to two decimal places on the given interval using the prescribed increments in $x$. In exercises 1 through 6, compare the estimated values with the correct values obtained by solving the problem exactly using elementary methods.

1. Exercise 1, page 406.      ANS.   $y_5(x) = 1 + x + x^2 + \frac{1}{3}x^3 + \frac{1}{12}x^4 + \frac{1}{60}x^5$.
2. Exercise 3, page 406.      ANS.   $y_5(x) = 2 + 2x + \frac{3}{2}x^2 + \frac{1}{2}x^3 + \frac{1}{8}x^4 + \frac{1}{40}x^5$.
3. Exercise 4, page 406.
     ANS.   $y_5(x) = 1 + 2(x - 1) + \frac{3}{2}(x - 1)^2 + \frac{1}{2}(x - 1)^3 + \frac{1}{8}(x - 1)^4 + \frac{1}{40}(x - 1)^5$.
4. Exercise 5, page 406.
     ANS.   $y_5(x) = -1 + (x - 2) + (x - 2)^2 + \frac{1}{3}(x - 2)^3 + \frac{1}{12}(x - 2)^4 + \frac{1}{60}(x - 2)^5$.
5. Exercise 6, page 406.
6. $y' = y^2 - x^2$;     when $x = 0, y = 1$;   $\Delta x = 0.1$;   $0 \le x \le 0.5$.
7. $y' = y^2 + x^2$;     when $x = 0, y = 1$;   $\Delta x = 0.1$;   $0 \le x \le 0.5$.
8. Use Taylor's series to determine to three places the value of the solution of the problem

$$y' = -xy^2; \qquad \text{when } x = 0, y = 1,$$

for $x = 0.1, 0.2$, and 0.3. Compare your results with the values obtained by solving the problem by elementary means.

## 134. The Runge-Kutta method

From a computational point of view, the major drawback in using Taylor's series to estimate the values of solutions of differential equations is that each coefficient in the series involves a different derivative function. Thus each approximation requires computations of the values of several different functions. We now consider a widely used technique that requires the computation of a single function at several different points rather than the computation of several different functions at a single point.

We consider the initial value problem

$$y' = F(x, y); \qquad \text{when } x = x_n, \, y = y_n, \tag{1}$$

and for convenience adopt the notation

$$F_n = F(x_n, y_n). \tag{2}$$

Let us begin by considering the tangent line to the solution curve at the point $(x_n, y_n)$. The equation of this line is given by

$$y = y_n + F_n(x - x_n). \tag{3}$$

The value of $y$ for this tangent line at $x = x_n + h$ is thus $y = y_n + hF_n$. If we define $K_1 = F_n$ and compute $F$ at the point $(x_n + h, \, y_n + hK_1)$ we obtain $K_2 = F(x_n + h, \, y_n + hK_1)$. Thus $K_1$ and $K_2$ represent the values of $y'$ at two points, the two endpoints of a segment of the tangent line. If we consider the arithmetic mean of these values of $y'$, namely $\frac{1}{2}(K_1 + K_2)$, and replace the tangent line with a new line through $(x_n, y_n)$ having this slope, we obtain

$$y = y_n + \tfrac{1}{2}(K_1 + K_2)(x - x_n).$$

For $x = x_n + h$, this line has a point whose $y$ coordinate is

$$y = y_n + \frac{h}{2}(K_1 + K_2), \tag{4}$$

where

$$K_1 = F_n, \tag{5}$$

and

$$K_2 = F(x_n + h, \, y_n + hK_1). \tag{6}$$

The generalizations of the idea above are the basis for the method of Runge-Kutta. Instead of choosing the tangent line as a means for approximating the value of $y$ at $x = x_n + h$, we choose a line whose slope is an average of the values of $y'$ at several carefully chosen points. When only two points are used, as just shown, the idea can be pictured as in Figure 45.

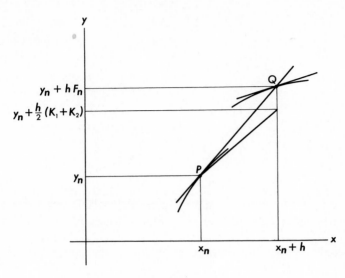

FIGURE 45

We now describe the intuitive idea behind the more elaborate scheme. Again we define $K_1 = F_n$ to be the slope at the point $P$. This time, we define $K_2$ to be the slope at the midpoint $M$ of the segment of the tangent line $PQ$. From equation (3) we find that $M$ is $(x_n + \frac{1}{2}h, y_n + \frac{1}{2}hK_1)$ and thus $K_2 = F(x_n + \frac{1}{2}h, y_n + \frac{1}{2}hK_1)$.

The line through $P$ with slope $K_2$ has equation

$$y = y_n + K_2(x - x_n),$$

and for $x = x_n + \frac{1}{2}h$, we obtain a point on this second line, namely $(x_n + \frac{1}{2}h, y_n + \frac{1}{2}hK_2)$. Now, defining $K_3 = F(x_n + \frac{1}{2}h, y_n + \frac{1}{2}hK_2)$, we consider a third line through $P$, this one having slope $K_3$. Its equation is

$$y = y_n + K_3(x - x_n).$$

The value of $y$ for this third line at $x = x_n + h$ is $y_n + hK_3$. We define $K_4 = F(x_n + h, y_n + hK_3)$ as the value of $y'$ at the fourth point.

Thus the numbers $K_1, K_2, K_3$, and $K_4$ represent the values of $y'$ at four points, one with $x = x_n$, two with $x = x_n + \frac{1}{2}h$, and one with $x = x_n + h$. We now determine the weighted mean of these four numbers

$$K = \tfrac{1}{6}(K_1 + 2K_2 + 2K_3 + K_4),$$

and consider a line through $P$ with slope $K$. Its equation is

$$y = y_n + K(x - x_n).$$

The value of $y$ for this fourth line at $x = x_n + h$ is

$$y_{n+1} = y_n + hK, \tag{7}$$

where

$$K = \tfrac{1}{6}(K_1 + 2K_2 + 2K_3 + K_4), \tag{8}$$

$$K_1 = F_n, \tag{9}$$

$$K_2 = F(x_n + \tfrac{1}{2}h, y_n + \tfrac{1}{2}hK_1), \tag{10}$$

$$K_3 = F(x_n + \tfrac{1}{2}h, y_n + \tfrac{1}{2}hK_2), \tag{11}$$

$$K_4 = F(x_n + h, y_n + hK_3). \tag{12}$$

The formulas (7) to (12) are due to Runge (1856–1927) and Kutta (1867–1944). The particular weighting factors assigned to the $K_1, K_2, K_3$, and $K_4$ in equation (8) are chosen so that the value of $y_{n+1}$ computed by the Runge-Kutta method and the value computed by a five-term Taylor formula,

$$y_{n+1} = y_n + hy'_n + \frac{h^2 y''_n}{2!} + \frac{h^3 y'''_n}{3!} + \frac{h^4 y_n^{(4)}}{4!}$$

differ by an amount that is proportional to $h^5$. The proof of this fact will not be given here but can be found in various texts on numerical analysis. We notice in passing, however, that if $F(x, y)$ does not explicitly involve the variable $y$, then the Runge-Kutta formulas reduce to the familiar Simpson's rule of elementary calculus. (See exercise 4 below.)

EXAMPLE: Solve the example of Section 130 by the Runge-Kutta method.

We present in Table 7 the results of the computation for $x = 0.1$ and $x = 0.2$ and leave the remaining computations for the exercises.

TABLE 7

| $x$ | 0 | 0.1 | 0.2 |
|---|---|---|---|
| $y$ | 1.00 | 1.11 | 1.25 |
| $K_1$ | 1.00 | 1.22 | |
| $x + \tfrac{1}{2}h$ | 0.05 | 0.15 | |
| $y + \tfrac{1}{2}hK_1$ | 1.05 | 1.17 | |
| $K_2$ | 1.10 | 1.35 | |
| $y + \tfrac{1}{2}hK_2$ | 1.06 | 1.18 | |
| $K_3$ | 1.12 | 1.37 | |
| $x + h$ | 0.10 | 0.20 | |
| $y + hK_3$ | 1.11 | 1.25 | |
| $K_4$ | 1.22 | 1.52 | |
| $K$ | 1.11 | 1.36 | |

### Exercises

In each of the following exercises, use the Runge-Kutta method to approximate the solution of the initial value problem in the given interval. In exercises 2 through 6, compare the approximate values with correct values obtained by elementary methods.

1. Continue the computation of the example above to obtain approximate values of $y$ for $x = 0.3, 0.4$, and $0.5$.
2. Exercise 1, page 406.
3. Exercise 2, page 406.
4. Exercise 3, page 406.
5. Exercise 4, page 406.
6. Exercise 5, page 406.
7. Exercise 6, page 406.
8. Exercise 7, page 406.
9. Exercise 8, page 406.
10. Exercise 10, page 406.
11. Exercise 11, page 406.
12. Exercise 12, page 406.
13. Exercise 13, page 406.
14. Exercise 8, page 411.
15. Show that if the function $F(x, y)$ in equation (1) of this section does not explicitly involve the variable $y$, then the Runge-Kutta formulas (7) to (12) reduce to a special case of Simpson's rule.

## 135. A continuing method

The methods used in the previous sections of this chapter may be called "starting" methods for finding approximations of solutions of the problem

$$y' = F(x, y); \qquad x = x_0, y = y_0. \tag{1}$$

By this we mean that no additional information is known other than that given in the problem (1) itself. Once an approximate value of $y_1$ has been obtained for $x_1 = x_0 + h$, we have then used $y_1$ to compute $y_2$, and so on. We shall now describe a "continuing" method developed by Milne (1890–1971).*

Suppose we know the values of $y_n, y_{n-1}, y_{n-2}$, and $y_{n-3}$. Then we can compute the values of $F_n, F_{n-1}, F_{n-2}, F_{n-3}$ from equation (1). Next we approximate $y'(x)$ by a cubic polynomial that passes through the four points $(x_n, F_n), (x_{n-1}, F_{n-1}), (x_{n-2}, F_{n-2})$, and $(x_{n-3}, F_{n-3})$. It can be proved that this can be done and that the polynomial thus obtained is unique.

---

* See for instance W. E. Milne, *Numerical Solutions of Differential Equations* (New York: John Wiley & Sons, Inc., 1953), Chapters 3 and 4.

Using this polynomial in place of $y'(x)$ in the integral

$$y_{n+1} - y_{n-3} = \int_{x_{n-3}}^{x_{n+1}} y'(x)\, dx, \tag{2}$$

performing the integration and simplifying gives an approximation for $y_{n+1}$. The result is

$$y_{n+1}^{(1)} = y_{n-3} + \frac{4h}{3}(2F_n - F_{n-1} + 2F_{n-2}). \tag{3}$$

The details of this derivation are discussed in exercise 4 below.

The problem of estimating the error in our approximations and of designing programs for reducing or correcting for errors is of course crucial to any method we may use. In the method of Milne, (3) is called a predictor formula and the value $y_{n+1}^{(1)}$ obtained from it is then used to find a corrected value for $y_{n+1}$.

A derivation of the correction formula is suggested in exercise 5 below. The result is

$$y_{n+1}^{(2)} = y_{n-1} + \frac{h}{3}(F_{n-1} + 4F_n + F_{n+1}), \tag{4}$$

where the value of $F_{n+1}$ is calculated by using the $y_{n+1}^{(1)}$ obtained from the predictor formula.

To illustrate the procedure outlined previously we use Milne's method to find the value of $y$ at $x = 0.4$ for the problem

$$y' = y^2 - x^2; \qquad x = 0, \ y = 1.$$

For starting points we take the values of $y_1$, $y_2$, and $y_3$, which were computed using the Runge-Kutta method. These numbers are presented in Table 8.

TABLE 8

| $n$ | $x_n$ | $y_n$ | $F_n$ |
|-----|-------|-------|-------|
| 0 | 0.0 | 1.00 | 1.00 |
| 1 | 0.1 | 1.11 | 1.22 |
| 2 | 0.2 | 1.25 | 1.52 |
| 3 | 0.3 | 1.42 | 1.92 |

Now using the predictor formula (3), we obtain

$$y_4^{(1)} = y_0 + \frac{4h}{3}(2F_3 - F_2 - 2F_1)$$

$$= 1.00 + \frac{4(0.1)}{3}[2(1.92) - 1.52 + 2(1.22)]$$

$$= 1.63.$$

Using this value to compute $F_4$ and applying the corrector formula (4), we have

$$y_4^{(2)} = y_2 + \frac{h}{3}(F_2 + 4F_3 + F_4)$$

$$= 1.25 + \frac{0.1}{3}[1.52 + 4(1.92) + 2.50]$$

$$= 1.64.$$

## *Exercises*

In exercises 2 through 5, use the Runge-Kutta method to obtain estimated values for $y_1$, $y_2$, $y_3$ and then compute approximations for $y_4$ and $y_5$ by Milne's method.

1. Continue the problem of this section to estimate values of $y$ at 0.5 and 0.6.
2. Exercise 1, page 406.
3. Exercise 3, page 406.
4. Exercise 4, page 406.
5. Exercise 6, page 406.
6. Exercise 7, page 406.
7. Exercise 9, page 406.
8. Exercise 11, page 406.
9. Exercise 13, page 406.
10. Let

$$\nabla F_n = F_n - F_{n-1},$$

$$\nabla^2 F_n = \nabla(\nabla F_n) = \nabla(F_n - F_{n-1}) = F_n - 2F_{n-1} + F_{n-2},$$

$$\nabla^3 F_n = \nabla(\nabla^2 F_n).$$

(a) Verify that the graph of

$$y = F_n + \frac{\nabla F_n}{1!h}(x - x_n) + \frac{\nabla^2 F_n}{2!h^2}(x - x_n)(x - x_{n-1})$$

$$+ \frac{\nabla^3 F_n}{3!h^3}(x - x_n)(x - x_{n-1})(x - x_{n-2})$$

passes through the points $(x_{n-3}, F_{n-3})$, $(x_{n-2}, F_{n-2})$, $(x_{n-1}, F_{n-1})$, and $(x_n, F_n)$.

(b) Using the above polynomial as a replacement for the integrand in equation (2) above, derive Milne's formula (3).

11. Suppose the value $y_{n+1}^{(1)}$ of equation (3) above is used to estimate $F_{n+1}$. By substitution into the differential equation and the use of Simpson's rule, show that a recalculation of $y_{n+1}$ gives the result of formula (4) above.

# Partial Differential Equations

## 136. Remarks on partial differential equations

A partial differential equation is an equation that contains one or more partial derivatives. Such equations occur frequently in applications of mathematics. The subject of partial differential equations offers sufficient ramifications and difficulties to be of interest for its own sake.

In this book we shall devote the space allotted to partial differential equations almost entirely to a kind of boundary value problem that enters modern applied mathematics at every turn.

Partial differential equations can have solutions involving arbitrary functions and solutions involving an unlimited number of arbitrary constants. The general solution of a linear partial differential equation of order $n$ may involve $n$ arbitrary functions.

The general solution of a partial differential equation is almost never (the wave equation is one of the few exceptions) of any practical use in solving boundary value problems associated with that equation.

### 137. Some partial differential equations of applied mathematics

Certain partial differential equations enter applied mathematics so frequently and in so many connections that their study is remarkably remunerative. A sufficiently thorough study of these equations would lead the student eventually into every phase of classical mathematics and, in particular, would lead almost at once to contact with special functions that are widely used in modern quantum theory and elsewhere in theoretical physics and engineering.

The derivation of the differential equations to be listed here is beyond the scope of this book. These derivations can be found (not all in each book) in the books listed for reference at the end of this section.

Some of the ways in which these equations are useful will appear in the detailed applications in Chapters 24 and 26.

Let $x, y, z$ be rectangular coordinates in ordinary space. Then the equation

$$\frac{\partial^2 V}{\partial x^2} + \frac{\partial^2 V}{\partial y^2} + \frac{\partial^2 V}{\partial z^2} = 0 \tag{1}$$

is called Laplace's equation. It enters problems in steady-state temperature, electrostatic potential, fluid flow of the steady-state variety, and so on.

If a problem involving equation (1) is such that a physical object in the problem is a circular cylinder, then it is possible that cylindrical coordinates will facilitate solution of the problem. We shall encounter such a problem later. It is possible to change equation (1) into an equation in which the independent variables are cylindrical coordinates $r, \theta, z$, related to the $x, y, z$ of equation (1) by the equations

$$x = r \cos \theta, \qquad y = r \sin \theta, \qquad z = z.$$

The resulting equation, Laplace's equation in cylindrical coordinates, is

$$\frac{\partial^2 V}{\partial r^2} + \frac{1}{r}\frac{\partial V}{\partial r} + \frac{1}{r^2}\frac{\partial^2 V}{\partial \theta^2} + \frac{\partial^2 V}{\partial z^2} = 0. \tag{2}$$

Note that the use of $z$ in both coordinate systems above is safe in making the change of variables only because $z$ is not involved in the equations with other variables. That is, in a change of independent variables such as

$$x = x_1 + y_1 + z_1, \qquad y = x_1 - y_1, \qquad z = z_1,$$

or its equivalent

$$x_1 = \tfrac{1}{2}(x + y - z), \qquad y_1 = \tfrac{1}{2}(x - y - z), \qquad z_1 = z,$$

incorrect conclusions would often result from any attempt to drop the subscript on the $z_1$, even though $z = z_1$. For instance, from the change of variables under discussion, it follows that

$$\frac{\partial V}{\partial z} = -\frac{1}{2}\frac{\partial V}{\partial x_1} - \frac{1}{2}\frac{\partial V}{\partial y_1} + \frac{\partial V}{\partial z_1}.$$

Hence

$$\frac{\partial V}{\partial z} \neq \frac{\partial V}{\partial z_1}$$

even though $z = z_1$.

Let us return to Laplace's equation. In spherical coordinates $\rho$, $\theta$, $\phi$, related to $x, y, z$ by the equations

$$x = \rho \sin \phi \cos \theta, \qquad y = \rho \sin \phi \sin \theta, \qquad z = \rho \cos \phi,$$

Laplace's equation is

$$\frac{\partial^2 V}{\partial \rho^2} = \frac{2}{\rho}\frac{\partial V}{\partial \rho} + \frac{1}{\rho^2}\frac{\partial^2 V}{\partial \phi^2} + \frac{\cot \phi}{\rho^2}\frac{\partial V}{\partial \phi} + \frac{\csc^2 \phi}{\rho^2}\frac{\partial^2 V}{\partial \theta^2} = 0. \tag{3}$$

With an additional independent variable $t$ representing time, and with a constant denoted by $a$, we can write the wave equation in rectangular coordinates,

$$\frac{\partial^2 V}{\partial t^2} = a^2\left(\frac{\partial^2 V}{\partial x^2} + \frac{\partial^2 V}{\partial y^2} + \frac{\partial^2 V}{\partial z^2}\right). \tag{4}$$

Equation (4) occurs in problems involving wave motions. We shall meet it later in the problem of the vibrating string.

Whenever the physical problem suggests a different choice for a coordinate system, usually by the shape of the objects involved, the pertinent partial differential equation can be transformed into one with the desired new independent variables.

Suppose that for some solid under consideration, $u$ represents the temperature at a point with rectangular coordinates $x, y, z$ and at time $t$. The origin of coordinates and the initial time $t = 0$ may be assigned at our convenience. If there are no heat sources present, the temperature $u$ must satisfy the heat equation

$$\frac{\partial u}{\partial t} = h^2\left(\frac{\partial^2 u}{\partial x^2} + \frac{\partial^2 u}{\partial y^2} + \frac{\partial^2 u}{\partial z^2}\right), \tag{5}$$

in which $h^2$ is a physical constant called thermal diffusivity. Equation (5) is derived under the assumption that the density, specific heat, and thermal conductivity are constant for the solid being studied. More comment on the question of the validity of equation (5) will be made in Section 154.

Equation (5) is the equation that pertains in many types of diffusion, not only when heat is being diffused. It is often called the equation of diffusion.

In the subject of elasticity, certain problems in plane stress can be solved with the aid of Airy's stress function $\phi$, which must satisfy the partial differential equation

$$\frac{\partial^4 \phi}{\partial x^4} + 2\frac{\partial^4 \phi}{\partial x^2 \, \partial y^2} + \frac{\partial^4 \phi}{\partial y^4} = 0. \tag{6}$$

Numerous other partial differential equations occur in applications, though not with the dominating insistency of equations (1), (4), and (5).

In this book two methods for solving boundary value problems in partial differential equations will be examined. The Laplace transform, which was studied in Chapters 11 through 14, is a useful tool for certain kinds of boundary value problems. The transform technique will be further developed in Chapter 25 and then used in Chapter 26.

A second method, the classical one of separation of variables, will be discussed in the remainder of this chapter. We shall find that two other topics, orthogonal sets and Fourier series, need to be treated before we can proceed, in Chapter 24, to use separation of variables efficiently to solve problems involving partial differential equations.

### References

CHURCHILL, R. V. *Operational Mathematics*, 3rd ed. New York: McGraw-Hill Book Company, 1972.

CHURCHILL, R. V., and J. W. BROWN. *Fourier Series and Boundary Value Problems*, 3rd ed. New York: McGraw-Hill Book Company, 1978.

HOPF, L. *Differential Equations of Physics*. New York: Dover Publications, Inc., 1948.

JEFFREYS, H., and B. S. JEFFREYS. *Methods of Mathematical Physics*, 3rd ed. Cambridge: Cambridge University Press, 1962.

## 138. Method of separation of variables

Before attacking an actual boundary value problem in partial differential equations, it is wise to become somewhat proficient in getting solutions of the differential equations. When we have acquired some facility in obtaining solutions, we can then tackle the tougher problem of fitting them together to satisfy stipulated boundary conditions.

The device to be exhibited here is particularly useful in connection with linear equations, although it does not always apply to such equations.

Consider the equation

$$\frac{\partial u}{\partial t} = h^2 \frac{\partial^2 u}{\partial x^2} \tag{1}$$

with $h$ constant. A solution of equation (1) will in general be a function of the two independent variables $t$ and $x$ and of the parameter $h$.

Let us seek a solution that is a product of a function of $t$ alone by a function of $x$ alone. We put

$$u = f(t)v(x)$$

in equation (1) and arrive at

$$f'(t)v(x) = h^2 f(t)v''(x), \tag{2}$$

where primes denote derivatives with respect to the indicated argument.

Dividing each member of (2) by the product $f(t)v(x)$, we get

$$\frac{f'(t)}{f(t)} = \frac{h^2 v''(x)}{v(x)}. \tag{3}$$

Now equation (3) is said to have its variables (the independent variables) separated; that is, the left member of equation (3) is a function of $t$ alone and the right member of equation (3) is a function of $x$ alone.

Since $x$ and $t$ are independent variables, the only way in which a function of $x$ alone can equal a function of $t$ alone is for each function to be constant. Thus from (3) it follows at once that

$$\frac{f'(t)}{f(t)} = k, \tag{4}$$

$$\frac{h^2 v''(x)}{v(x)} = k, \tag{5}$$

in which $k$ is arbitrary.

Another way of obtaining equations (4) and (5) is this: Differentiate each member of equation (3) with respect to $t$ (either independent variable could be used) and thus get

$$\frac{d}{dt}\frac{f'(t)}{f(t)} = 0,$$

since the right member of equation (3) is independent of $t$. First we obtain equation (4) by integration and then equation (5) follows from (4) and (3).

Equation (4) may be rewritten

$$\frac{df}{dt} = kf,$$

from which its general solution

$$f = c_1 e^{kt}$$

follows immediately.

Before going further into the solution of our problem, we attempt to choose a convenient form for the arbitrary constant introduced in equations (4) and (5). Equation (5) suggests that $k$ be taken as a multiple of $h^2$.

Let us then return to equations (4) and (5) and put $k = h^2\beta^2$, so we have

$$\frac{f'(t)}{f(t)} = h^2\beta^2 \tag{6}$$

and

$$\frac{h^2 v''(x)}{v(x)} = h^2\beta^2. \tag{7}$$

Using real $\beta$ and the choice $k = h^2\beta^2$, we are implying that the constant $k$ is positive. Later we shall obtain solutions corresponding to the choice of a negative constant.

From equations (6) and (7) we find at once that

$$f(t) = c_1 \exp(h^2\beta^2 t) \tag{8}$$

and

$$v(x) = c_2 \cosh \beta x + c_3 \sinh \beta x. \tag{9}$$

Since $u = f(t)v(x)$, we are led to the result that the partial differential equation

$$\frac{\partial u}{\partial t} = h^2 \frac{\partial^2 u}{\partial x^2} \tag{1}$$

has solutions

$$u = \exp(h^2\beta^2 t)[a \cosh \beta x + b \sinh \beta x], \tag{10}$$

in which $\beta$, $a$, and $b$ are arbitrary constants. The $a$ and $b$ of equation (10) are respectively given by $a = c_1 c_2$ and $b = c_1 c_3$ in terms of the constants of equations (8) and (9).

If we return to equations (4) and (5) with the choice $k = -h^2\alpha^2$, so $k$ is taken to be a negative constant, then we find that the partial differential equation (1) has the solutions

$$u = \exp(-h^2\alpha^2 t)[A \cos \alpha x + B \sin \alpha x], \tag{11}$$

in which $\alpha$, $A$, and $B$ are arbitrary constants.

Finally, let the constant $k$ be zero. It is straightforward to determine that the corresponding solutions of the differential equation (1) are

$$u = C_1 + C_2 x, \tag{12}$$

in which $C_1$ and $C_2$ are constants.

Direct verification that equations (10), (11), and (12) are actually solutions of equation (1) is simple.

Since the partial differential equation (1) is linear, we may construct solutions by forming linear combinations of solutions. Thus from (10), (11), and (12) with varying choices of $\alpha, \beta, A, B, a, b, C_1, C_2$, we can construct as many solutions of (1) as we wish.

The distinction between equations (10) and (11) is dependent upon the parameters and variables remaining real. Our aim is to develop tools for solving physical problems; hence we do intend to keep things real.

## Exercises

Except where other instructions are given, use the method of separation of variables to obtain solutions in real form for each differential equation.

1. $\dfrac{\partial^2 u}{\partial t^2} = a^2 \dfrac{\partial^2 u}{\partial x^2}$.

ANS.  $u = (A_1 \cos a\beta t + A_2 \sin a\beta t)(B_1 \cos \beta x + B_2 \sin \beta x)$,
$u = (A_3 + A_4 t)(B_3 + B_4 x)$, and
$u = (A_5 \cosh a\beta t + A_6 \sinh a\beta t)(B_5 \cosh \beta x + B_6 \sinh \beta x)$,
in which $\beta$ and the $A$'s and $B$'s are arbitrary constants.

2. $\dfrac{\partial^2 v}{\partial x^2} + \dfrac{\partial^2 v}{\partial y^2} = 0$.

ANS.  $v = (A_1 e^{\beta x} + A_2 e^{-\beta x})(B_1 \cos \beta y + B_2 \sin \beta y)$, and
$v = (A_3 + A_4 x)(B_3 + B_4 y)$,
in which $\beta$ and the $A$'s and $B$'s are arbitrary constants; the roles played by $x$ and $y$ in the former solution can be interchanged.

3. $\dfrac{\partial^2 u}{\partial t^2} + 2b\dfrac{\partial u}{\partial t} = a^2 \dfrac{\partial^2 u}{\partial x^2}$. Show that this equation has the solutions:

$$u = g(t)(B_1 \cos kx + B_2 \sin kx),$$

where $g(t)$ can assume any one of the forms:

$$e^{-bt}(A_1 e^{\gamma t} + A_2 e^{-\gamma t}), \quad \gamma^2 = b^2 - a^2 k^2, \quad \text{if } b^2 - a^2 k^2 > 0,$$

or

$$e^{-bt}(A_3 \cos \delta t + A_4 \sin \delta t), \quad \delta^2 = a^2 k^2 - b^2, \quad \text{if } a^2 k^2 - b^2 > 0,$$

or

$$e^{-bt}(A_5 + A_6 t), \quad \text{if } a^2 k^2 - b^2 = 0,$$

and the solutions

$$u = (A_7 + A_8 e^{-2bt})(B_3 + B_4 x).$$

Find also solutions containing $e^{kx}$ and $e^{-kx}$.

4. $\dfrac{\partial w}{\partial y} = y\dfrac{\partial w}{\partial x}$.

ANS.  $w = A \exp [k(2x + y^2)], k$ and $A$ arbitrary.

5. $x\dfrac{\partial w}{\partial x} = w + y\dfrac{\partial w}{\partial y}.$    ANS.   $w = Ax^k y^{k-1}$, $k$ and $A$ arbitrary.

6. Subject the partial differential equation of exercise 5 to the change of dependent variable $w = v/y$ and show that the resultant equation for $v$ is

$$x\frac{\partial v}{\partial x} = y\frac{\partial v}{\partial y}.$$

7. Show that the method of separation of variables does not succeed, without modifications, for the equation

$$\frac{\partial^2 u}{\partial x^2} + 4\frac{\partial^2 u}{\partial x\,\partial t} + 5\frac{\partial^2 u}{\partial t^2} = 0.$$

8. For the equation of exercise 7, seek a solution of the form

$$u = e^{kt} f(x)$$

and thus obtain the solutions

$$u = \exp\,[k(t - 2x)][A_1 \cos kx + A_2 \sin kx],$$

where $k$, $A_1$, $A_2$ are arbitrary. Show also that the equation of exercise 7 has the solutions $u = Ax + B$ and $u = Ct + D$, with $A, B, C, D$ arbitrary.

9. For the equation

$$\frac{\partial^2 u}{\partial x^2} + 4x\frac{\partial u}{\partial x} + \frac{\partial^2 u}{\partial y^2} = 0,$$

put $u = f(x)g(y)$ and thus obtain solutions

$$u = [A_1 e^{2ky} + A_2 e^{-2ky}][B_1 f_1(x) + B_2 f_2(x)],$$

in which $f_1(x)$ and $f_2(x)$ are any two linearly independent solutions of the equation

$$f'' + 4xf' + 4k^2 f = 0.$$

Obtain also the solutions that correspond to $k = 0$ in the above. For $k \neq 0$, the functions $f_1$ and $f_2$ should be obtained by the method of Chapter 17. They may be found in the form

$$f_1(x) = 1 + \sum_{m=1}^{\infty} \frac{(-4)^m(k^2)(k^2 + 2)(k^2 + 4)\cdots(k^2 + 2m - 2)x^{2m}}{(2m)!},$$

$$f_2(x) = x + \sum_{m=1}^{\infty} \frac{(-4)^m(k^2 + 1)(k^2 + 3)(k^2 + 5)\cdots(k^2 + 2m - 1)x^{2m+1}}{(2m + 1)!}.$$

Find similar solutions involving $\cos 2ky$ and $\sin 2ky$.

10. Use the change of variable $u = e^{\beta t} g(x)$ to find solutions of the equation

$$\frac{\partial^2 u}{\partial x^2} + 2\frac{\partial^2 u}{\partial x\,\partial t} + \frac{\partial^2 u}{\partial t^2} = 0.$$

ANS.   $u = \exp\,[\beta(t - x)][A_1 + A_2 x].$

11. Show by direct computation that if $f_1(y)$ and $f_2(y)$ are any functions with continuous second derivatives, $f_1''(y)$ and $f_2''(y)$, then

$$u = f_1(x - at) + f_2(x + at)$$

satisfies the simple wave equation (exercise 1),

$$\frac{\partial^2 u}{\partial t^2} = a^2 \frac{\partial^2 u}{\partial x^2}.$$

12. Show that $v = f(xy)$ is a solution of the equation of exercise 6.
13. Show that $w = f(2x + y^2)$ is a solution of the equation of exercise 4.
14. Show that $u = f_1(t - x) + xf_2(t - x)$ is a solution of the equation of exercise 10.

## 139. A problem on the conduction of heat in a slab

Among the equations of applied mathematics already stated is the heat equation in rectangular coordinates,

$$\frac{\partial u}{\partial t} = h^2\left(\frac{\partial^2 u}{\partial x^2} + \frac{\partial^2 u}{\partial y^2} + \frac{\partial^2 u}{\partial z^2}\right), \qquad (1)$$

in which

$x, y, z$ = rectangular space coordinates,
$t$ = time coordinate,
$h^2$ = thermal diffusivity,
$u$ = temperature.

The constant $h^2$ and the variables $x, y, z, t, u$ may be in any consistent set of units. For instance, we may measure $x, y, z$ in feet, $t$ in hours, $u$ in degrees Fahrenheit, and $h^2$ in square feet per hour. The thermal diffusivity (assumed constant in our work) can be defined by

$$h^2 = \frac{K}{\sigma\delta},$$

in terms of quantities of elementary physics,
$K$ = thermal conductivity,
$\sigma$ = specific heat,
$\delta$ = density,

all pertaining to the material composing the solid whose temperature we seek.

For our first boundary value problem in partial differential equations it seems wise to set up as simple a problem as possible. We now construct a temperature problem that is set in such a way that the temperature is independent of two space variables, say $y$ and $z$. For such a problem, $u$ will be a function of only two independent variables ($x$ and $t$), which is the smallest number of independent variables possible in a partial differential equation.

Consider a huge flat slab of concrete, or some other material reasonably near homogeneity in texture. Let the thickness of the slab be $c$ units of length. Choose the origin of coordinates on a face of the slab as indicated in Figure 46 and assume that the slab extends very far in the $y$ and $z$ directions.

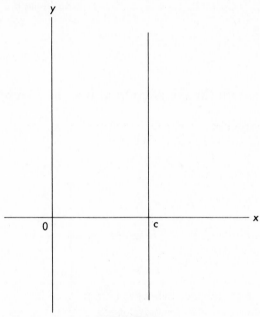

FIGURE 46

Let the initial ($t = 0$) temperature of the slab be $f(x)$, a function of $x$ alone, and let the surfaces $x = 0$, $x = c$ be kept at zero temperature for all $t > 0$. If the slab is considered infinite in the $y$ and $z$ directions, or more specifically if we treat only cross sections nearby (far from the distant surface of the slab), then the temperature $u$ at any time $t$ and position $x$ is determined by the boundary value problem:

$$\frac{\partial u}{\partial t} = h^2 \frac{\partial^2 u}{\partial x^2}, \qquad \text{for } 0 < t, 0 < x < c; \tag{2}$$

$$\text{As } t \to 0^+, \quad u \to f(x), \qquad \text{for } 0 < x < c; \tag{3}$$

$$\text{As } x \to 0^+, \quad u \to 0, \qquad \text{for } 0 < t; \tag{4}$$

$$\text{As } x \to c^-, \quad u \to 0, \qquad \text{for } 0 < t. \tag{5}$$

In the boundary value problem (2) through (5), the zero on the temperature scale has been chosen as the temperature at which the surfaces of the slab are held. Then the $f(x)$ is really the difference between the actual initial temperature and the subsequent constant boundary temperature.

Such a set of symbols as $t \to 0^+$ means that $t$ approaches zero through values greater than zero. Similarly, $x \to c^-$ means that $x$ approaches $c$ through values less than $c$, "$c$ minus something" at each stage during the approach. We may say, for instance, that $x$ approaches $c$ from the left.

It is particularly to be noted that we do not require such a condition as that the function $u(x, t)$ be $f(x)$ when $t = 0$. We require only that as $t \to 0^+$, then $u \to f(x)$ for each $x$ in the range $0 < x < c$.

The question of precisely how many boundary conditions, and of what nature, are to be associated with a given partial differential equation to assure the existence and uniqueness of a solution is a question of considerable difficulty. In this book we shall use the most popular practical guide, physical intuition. We seek a solution and, if we find one, we know it exists; if we do not find one, it is hardly balm for our wounds to be told that it exists. To the serious problem of uniqueness we close our eyes entirely.

Let us now attempt to solve the boundary value problem, that is, find a function $u(x, t)$ that will satisfy the partial differential equation (2) and that will also satisfy the conditions (3), (4), and (5).

We already know how to get some solutions of the differential equation (2). Indeed, on pages 422–425 the method of separation of variables was used to arrive at the solutions

$$u = \exp{(h^2 \beta^2 t)}[a \cosh \beta x + b \sinh \beta x], \tag{6}$$

with $a, b, \beta$ arbitrary constants, and the solutions

$$u = \exp{(-h^2 \alpha^2 t)}[A \cos \alpha x + B \sin \alpha x] \tag{7}$$

with $A, B, \alpha$ arbitrary constants.

It is now necessary to attempt to adjust the solutions (6) or (7) to satisfy the boundary conditions (3), (4), and (5). Trial shows quickly that it is simpler to satisfy conditions (4) and (5) first and then to tackle (3).

Let us try to satisfy (4) and (5) with solutions in the form of equation (6) above. Now condition (4) requires that when we let $x \to 0^+$, then $u \to 0$ for all positive $t$. Letting $x \to 0^+$ in equation (6) we conclude that

$$0 = \exp{(h^2 \beta^2 t)}[a + 0], \qquad \text{for } 0 < t.$$

Thus we are forced to conclude that $a = 0$, so the solution (6) becomes

$$u = b \exp{(h^2 \beta^2 t)} \sinh \beta x. \tag{8}$$

By condition (5) we must require that as $x \to c^-$, then again $u \to 0$ for all positive $t$; that is, from equation (8) and condition (5) we get

$$b \exp(h^2\beta^2 t) \sinh \beta c = 0, \qquad \text{for } 0 < t.$$

The exponential $\exp(h^2\beta^2 t)$ cannot vanish. For real values of $\beta$ and $c$ the function $\sinh \beta c$ is zero only when $\beta c = 0$. Hence it follows that $\beta = 0$ or $b = 0$, so $u \equiv 0$ and we have no chance of satisfying the remaining condition, (3). Let us therefore abandon equation (6) and concentrate upon the solutions

$$u = \exp(-h^2\alpha^2 t)[A \cos \alpha x + B \sin \alpha x]. \tag{7}$$

Let us impose conditions (4) and (5) upon the $u$ of equation (7). First we let $x \to 0^+$ and conclude that

$$0 = \exp(-h^2\alpha^2 t)[A + 0], \qquad \text{for } 0 < t,$$

so we must choose $A = 0$. Then (7) reduces to

$$u = B \exp(-h^2\alpha^2 t) \sin \alpha x. \tag{9}$$

Next we impose condition (5), that $u \to 0$ when $x \to c^-$, so

$$0 = B \exp(-h^2\alpha^2 t) \sin \alpha c, \qquad \text{for } 0 < t.$$

We must not choose $B = 0$, if we are to have any success in satisfying the additional condition (3). The function $\exp(-h^2\alpha^2 t)$ does not vanish. Therefore, for the $u$ of (9) to satisfy condition (5), it is necessary that

$$\sin \alpha c = 0. \tag{10}$$

The sine function is zero when, and only when, its argument is an integral multiple of $\pi$; that is, $\sin z$ is zero when $z = 0, \pm\pi, \pm 2\pi, \ldots, \pm n\pi, \ldots$. Therefore from (10) it follows that

$$\alpha c = n\pi, n \text{ integral.} \tag{11}$$

Since $c$ is given, equation (11) serves to determine the arbitrary parameter $\alpha$—rather, to restrict the values of $\alpha$ to those given by (11).

With $\alpha = n\pi/c$, the solutions (9) above become

$$u = B \exp\left[-\left(\frac{n\pi h}{c}\right)^2 t\right] \sin \frac{n\pi x}{c}$$

with $n$ integral and $B$ arbitrary. Since we need not use the same arbitrary constant $B$ for different values of $n$, it is wiser to write solutions

$$u_n = B_n \exp\left[-\left(\frac{n\pi h}{c}\right)^2 t\right] \sin \frac{n\pi x}{c}, \qquad n \text{ integral.} \tag{12}$$

We lose nothing by restricting the $n$ in equation (12) to the positive integers $1, 2, 3, \ldots$, for $n = 0$ leads to the trivial solution $u \equiv 0$ and the negative integral values for $n$ lead to essentially the same solutions as do the positive integral values.

Let us see where we stand at present. Each of the functions $u_n$ defined by equation (12) is a solution of the differential equation

$$\frac{\partial u}{\partial t} = h^2 \frac{\partial^2 u}{\partial x^2}, \qquad \text{for } 0 < x < c, \, 0 < t, \tag{2}$$

and each of those functions satisfies the two conditions

$$\text{As } x \to 0^+, \quad u \to 0, \qquad \text{for } 0 < t, \tag{4}$$

and

$$\text{As } x \to c^-, \quad u \to 0, \qquad \text{for } 0 < t. \tag{5}$$

It remains to find from the solutions $u_n$ a solution $u(x, t)$ that will also satisfy the boundary (or initial) condition

$$\text{As } t \to 0^+, \quad u \to f(x), \qquad \text{for } 0 < x < c. \tag{3}$$

Since the partial differential equation involved is linear and homogeneous in $u$ and its derivatives, a sum of solutions is also a solution. From the known solutions $u_1, u_2, u_3, \ldots, u_n, \ldots$, we may thus construct others. With sufficiently strong convergence conditions it is true that even the infinite series

$$u = \sum_{n=1}^{\infty} u_n$$

or

$$u(x, t) = \sum_{n=1}^{\infty} B_n \exp\left[ -\left( \frac{n\pi h}{c} \right)^2 t \right] \sin \frac{n\pi x}{c} \tag{13}$$

is also a solution of the differential equation (2).

The $u(x, t)$ of equation (13) satisfies equation (2) and the boundary conditions (4) and (5). If $u(x, t)$ is to satisfy condition (3), then for each $x$ in the interval $0 < x < c$ the right member of equation (13) should approach $f(x)$ as $t \to 0^+$. We assume that we may interchange the order of limit (as $t \to 0^+$) and summation and conclude that condition (3) formally requires that

$$f(x) = \sum_{n=1}^{\infty} B_n \sin \frac{n\pi x}{c}, \qquad \text{for } 0 < x < c. \tag{14}$$

Thus we can solve the problem under consideration if we can choose the constants $B_n$ so that the infinite series on the right in (14) has $f(x)$ for its sum for each $x$ in the interval $0 < x < c$. That there exist such coefficients $B_n$ is far from evident. For a large class of functions $f(x)$, an expansion of the type of equation (14) does exist, as will be seen in Chapter 23. Once the $B_n$ are known, they are to be inserted on the right in equation (13), which is then the final solution of the boundary value problem consisting of equation (2) and conditions (3), (4), and (5).

To be able to complete the solution of boundary value problems of the kind under consideration here, we need to acquire a knowledge of methods of expansion of functions into trigonometric series. Chapter 23 is devoted to the development of that type of expansion, providing us with tools for solving numerous boundary value problems in Chapter 24.

# Orthogonal Sets

## 140. Orthogonality

A set of functions $\{f_0(x), f_1(x), f_2(x), \ldots, f_n(x), \ldots\}$, is said to be *an orthogonal set with respect to the weight function $w(x)$ over the interval $a \leqq x \leqq b$ if*

$$\int_a^b w(x)f_n(x)f_m(x)\,dx = 0 \qquad \text{for } m \neq n,$$

$$\neq 0 \qquad \text{for } m = n.$$

Orthogonality is a property widely encountered in certain branches of mathematics. Much use is made of the representation of functions in series of the form

$$\sum_{n=0}^{\infty} c_n f_n(x)$$

in which the $c_n$ are numerical coefficients and $\{f_n(x)\}$ is an orthogonal set.

A tremendous literature exists regarding orthogonal sets of functions. A student who wishes to pursue the subject beyond the material in this chapter can get a thorough introduction to it in courses and books on orthogonal polynomials, Fourier series, Fourier analysis, and so on. A short list of references is given at the end of Chapter 23.

One simple version of a theorem in the subject of orthogonal functions may be stated as follows:

Given a set of functions

$$\{\phi_0(x), \phi_1(x), \phi_2(x), \ldots, \phi_n(x), \ldots\},$$

linearly independent (Section 25) and continuous in the interval

$$a \leqq x \leqq b,$$

and given a weight function $w(x)$ positive and continuous in that same interval, then a set of functions

$$\{f_0(x), f_1(x), f_2(x), \ldots, f_n(x), \ldots\}$$

exists with the properties that:

(a) Each $f_n(x)$ is a linear combination of $\phi$'s.
(b) The $f_n(x)$ are linearly independent on the interval $a \leqq x \leqq b$.
(c) $\{f_n(x)\}$ is an orthogonal set with respect to the weight function $w(x)$ over the interval $a \leqq x \leqq b$.

We already know that the functions $1, x, x^2, \ldots, x^n, \ldots$, are linearly independent and are continuous over any finite interval. Linear combinations of powers of $x$ are polynomials. Hence, given an interval and a proper weight function, there exists in particular a set of polynomials orthogonal with respect to that weight function over that interval. With added conditions on the weight function the restriction to a finite interval can be removed.

## 141. Simple sets of polynomials

A set of polynomials $\{f_n(x)\}$, $n = 0, 1, 2, 3, \ldots$, is called a simple set if $f_n(x)$ is of degree precisely $n$. The set then contains one polynomial of each degree, $0, 1, 2, \ldots, n, \ldots$.

An important property of simple sets is that if $g_m(x)$ is any polynomial of degree $m$ and $\{f_n(x)\}$ is a simple set of polynomials, then constants $c_k$ exist such that

$$g_m(x) = \sum_{k=0}^{m} c_k f_k(x). \tag{1}$$

To show that this is true, let the highest degree term in $g_m(x)$ be $a_m x^m$ and the highest degree term in $f_m(x)$ be $b_m x^m$. Define $c_m = a_m/b_m$, noting that

$b_m \neq 0$. Then the polynomial

$$g_m(x) - c_m f_m(x)$$

is of degree at most $(m - 1)$. On this polynomial use the same procedure as was used on $g_m(x)$. It follows that $c_{m-1}$ exists so that

$$g_m(x) - c_m f_m(x) - c_{m-1} f_{m-1}(x)$$

is of degree at most $(m - 2)$. Iteration of the process yields equation (1) in $(m + 1)$ steps. Note that any $c_k$ except $c_m$ may be zero.

## 142. Orthogonal polynomials

We next obtain for polynomials a condition equivalent to our definition (Section 140) of orthogonality.

THEOREM 32:    *If $\{f_n(x)\}$ is a simple set of polynomials, a necessary and suffi-cient condition that $\{f_n(x)\}$ be orthogonal with respect to $w(x)$ over the interval $a \leq x \leq b$ is that*

$$\int_a^b w(x) x^k f_n(x)\, dx = 0, \qquad k = 0, 1, 2, \ldots, (n - 1), \tag{1}$$

$$\neq 0, \qquad k = n.$$

PROOF.    Suppose that (1) is satisfied. Since $\{x^k\}$ is a simple set, we can write

$$f_m(x) = \sum_{k=0}^m a_k x^k. \tag{2}$$

If $m < n$, it follows that

$$\int_a^b w(x) f_m(x) f_n(x)\, dx = \sum_{k=0}^m a_k \int_a^b w(x) x^k f_n(x)\, dx = 0$$

by (1), since each $k$ involved is less than $n$. If $m > n$, interchange the roles of $m$ and $n$, and repeat the argument. If $m = n$ in (2), then $a_n \neq 0$, and we have

$$\int_a^b w(x) f_n^2(x)\, dx = \sum_{k=0}^n a_k \int_a^b w(x) x^k f_n(x)\, dx$$

$$= a_n \int_a^b w(x) x^n f_n(x)\, dx \neq 0.$$

Thus we see that the condition (1) is sufficient for orthogonality of the set $\{f_n(x)\}$.

Next suppose the $f_n(x)$ satisfy the condition for orthogonality as laid down in Section 140. Since $\{f_n(x)\}$ is a simple set, we can write

$$x^k = \sum_{m=0}^{k} b_m f_m(x). \tag{3}$$

If $k < n$, it follows that

$$\int_a^b w(x)x^k f_n(x)\,dx = \sum_{m=0}^{k} b_m \int_a^b w(x)f_m(x)f_n(x)\,dx = 0,$$

since no $m$ can equal $n$. If $k = n$ in (3), then $b_n \neq 0$, and we have

$$\int_a^b w(x)x^n f_n(x)\,dx = \sum_{m=0}^{n} b_m \int_a^b w(x)f_m(x)f_n(x)\,dx$$

$$= b_n \int_a^b w(x)f_n^2(x)\,dx \neq 0.$$

This completes the proof of Theorem 32.


## 143. Zeros of orthogonal polynomials

We shall show that real orthogonal polynomials have all their zeros real, distinct, and in the interval of orthogonality.

THEOREM 33:   *If $\{f_n(x)\}$ is a simple set of real polynomials orthogonal with respect to $w(x)$ over the interval $a \leq x \leq b$, and if $w(x) > 0$ over $a < x < b$, then the zeros of $f_n(x)$ are distinct and all lie in the open interval $a < x < b$.*

PROOF.   For $n \geq 1$, the polynomial $f_n(x)$ does change sign in the interval $a < x < b$ because, by Theorem 32 (with $k = 0$),

$$\int_a^b w(x)f_n(x)\,dx = 0$$

and $w(x)$ cannot change sign in $a < x < b$.

Let $f_n(x)$ change sign in $a < x < b$ at precisely the distinct points $x = \alpha_1$, $\alpha_2, \ldots, \alpha_s$. The $\alpha$'s are precisely the zeros of odd multiplicity of $f_n(x)$ in the interval. Since $f_n(x)$ is of degree $n$, it has $n$ zeros, multiplicity counted. Thus we know that $s \leq n$.

Form the function

$$\psi(x) = (x - \alpha_1)(x - \alpha_2)\cdots(x - \alpha_s). \tag{1}$$

Then, in $a < x < b$, $\psi(x)$ changes sign at $x = \alpha_1, \alpha_2, \ldots, \alpha_s$ and nowhere else.

If $s < n$, $\psi(x)$ is of degree less than $n$, so

$$\int_a^b w(x)f_n(x)\psi(x)\,dx = 0 \tag{2}$$

by the application of Theorem 32 to each term in the expanded form of $\psi(x)$. But the integrand in (2) does not change sign anywhere in the interval of integration, because $w(x) > 0$ and the functions $f_n(x)$ and $\psi(x)$ change sign at precisely the same points. Therefore the integral in (2) cannot vanish and the assumption $s < n$ has led us to a contradiction.

Thus we have $s = n$. That is, among the $n$ zeros of $f_n(x)$ there are precisely $n$ of odd multiplicity in the open interval $a < x < b$. Therefore each zero is of multiplicity one; the zeros are distinct. The proof of Theorem 33 is complete.

## 144. Orthogonality of Legendre polynomials

The Legendre polynomials

$$P_n(x) = F\left(-n, n + 1; 1; \frac{1 - x}{2}\right) \tag{1}$$

of Section 127 were obtained by solving the differential equation

$$(1 - x^2)y'' - 2xy' + n(n + 1)y = 0. \tag{2}$$

The $P_n(x)$ form a simple set of polynomials for which we now obtain an orthogonality property.

From (2) we have

$$(1 - x^2)P_n''(x) - 2xP_n'(x) + n(n + 1)P_n(x) = 0,$$

$$D[(1 - x^2)P_n'(x)] + n(n + 1)P_n(x) = 0; \qquad D = \frac{d}{dx}. \tag{3}$$

For index $m$ we have

$$D[(1 - x^2)P_m'(x)] + m(m + 1)P_m(x) = 0. \tag{4}$$

We are interested in the product $P_m(x)P_n(x)$. Hence we multiply (3) throughout by $P_m(x)$, (4) throughout by $P_n(x)$, and subtract to obtain

$$P_m(x)D[(1 - x^2)P_n'(x)] - P_n(x)D[(1 - x^2)P_m'(x)]$$

$$+ [n(n + 1) - m(m + 1)]P_m(x)P_n(x) = 0.$$

The equation above may be rewritten

$$(n^2 - m^2 + n - m)P_m(x)P_n(x)$$
$$= P_n(x)D[(1 - x^2)P'_m(x)] - P_m(x)D[(1 - x^2)P'_n(x)]. \tag{5}$$

Now, by the formula for differentiating a product, we get

$$D[(1 - x^2)P_n(x)P'_m(x)] = P_n(x)D[(1 - x^2)P'_m(x)] + (1 - x^2)P'_n(x)P'_m(x)$$

and

$$D[(1 - x^2)P'_n(x)P_m(x)] = P_m(x)D[(1 - x^2)P'_n(x)] + (1 - x^2)P'_n(x)P'_m(x).$$

Hence

$$D[(1 - x^2)\{P_n(x)P'_m(x) - P'_n(x)P_m(x)\}]$$
$$= P_n(x)D[(1 - x^2)P'_m(x)] - P_m(x)D[(1 - x^2)P'_n(x)].$$

Furthermore, $n^2 - m^2 + n - m = (n - m)(n + m + 1)$. Therefore we can write (5) as

$$(n - m)(n + m + 1)P_m(x)P_n(x)$$
$$= D[(1 - x^2)\{P_n(x)P'_m(x) - P'_n(x)P_m(x)\}]. \tag{6}$$

We have now expressed the product of any two Legendre polynomials as a derivative. Derivatives are easy to integrate. Equation (6) yields

$$(n - m)(n + m + 1)\int_a^b P_m(x)P_n(x)\,dx$$
$$= [(1 - x^2)\{P_n(x)P'_m(x) - P'_n(x)P_m(x)\}]_a^b. \tag{7}$$

We may choose any $a$ and $b$ that we wish. Since $(1 - x^2)$ is zero at $x = -1$ and $x = 1$, we conclude that

$$(n - m)(n + m + 1)\int_{-1}^1 P_m(x)P_n(x)\,dx = 0. \tag{8}$$

Since $n$ and $m$ are to be nonnegative integers, $n + m + 1 \neq 0$. Hence if $m \neq n, n - m \neq 0$ and (8) yields

$$\int_{-1}^1 P_m(x)P_n(x)\,dx = 0. \tag{9}$$

The Legendre polynomials are real, so $\int_{-1}^1 P_n^2(x)\,dx \neq 0$.

We have shown that the Legendre polynomials $P_n(x)$ form an orthogonal set with respect to the weight function $w(x) = 1$ over the interval $-1 < x < 1$. Since $\{P_n(x)\}$ is a simple set of real polynomials, the theorems of Sections 142 and 143 apply to it.

Further study of $P_n(x)$ would occupy more space than seems appropriate in an elementary differential equations text. We now list a few of the simplest from among the hundreds of known properties of these interesting polynomials:

$$(1 - 2xt + t^2)^{-1/2} = \sum_{n=0}^{\infty} P_n(x)t^n, \tag{10}$$

$$\int_{-1}^{1} P_n^2(x)\, dx = \frac{2}{2n + 1}, \tag{11}$$

$$P_n(x) = \frac{1}{2^n n!} D^n(x^2 - 1)^n; \qquad D = \frac{d}{dx}, \tag{12}$$

$$xP_n'(x) = nP_n(x) + P_{n-1}'(x), \tag{13}$$

$$(x^2 - 1)P_n'(x) = nxP_n(x) - nP_{n-1}(x), \tag{14}$$

$$nP_n(x) = (2n - 1)xP_{n-1}(x) - (n - 1)P_{n-2}(x). \tag{15}$$

## 145. Other orthogonal sets

In Chapter 19 we solved several differential equations of hypergeometric type. In Section 123 we encountered the Laguerre polynomial

$$L_n(x) = \sum_{k=0}^{n} \frac{(-n)_k x^k}{(k!)^2} = \sum_{k=0}^{n} \frac{(-1)^k n! x^k}{(k!)^2 (n - k)!} \tag{1}$$

as a solution of the differential equation

$$xL_n''(x) + (1 - x)L_n'(x) + nL_n(x) = 0. \tag{2}$$

Equation (2) can be put in the form

$$D[x\, e^{-x}L_n'(x)] + n\, e^{-x}L_n(x) = 0, \tag{3}$$

from which the orthogonality of the set of Laguerre polynomials follows. (See exercise 1 below.)

The Hermite polynomial of Section 126,

$$H_n(x) = \sum_{k=0}^{[n/2]} \frac{(-1)^k n! (2x)^{n-2k}}{k! (n - 2k)!}, \tag{4}$$

satisfies the differential equation

$$H_n''(x) - 2xH_n'(x) + 2nH_n(x) = 0. \tag{5}$$

Equation (5) can be put in the form

$$D[\exp(-x^2)H_n'(x)] + 2n \exp(-x^2)H_n(x) = 0, \tag{6}$$

from which the orthogonality of the set of Hermite polynomials follows. (See exercise 3 below.)

The Bessel function $J_n(x)$ of Section 125 can be shown to have orthogonality properties also, but they are beyond the scope of this book. See, for example, R. V. Churchill and J. W. Brown, *Fourier Series and Boundary Value Problems*, 3rd ed. (New York: McGraw-Hill Book Company, 1978).

## *Exercises*

1. Use equation (3) above and the method of Section 144 to show that the set of Laguerre polynomials is orthogonal with respect to the weight function $e^{-x}$ over the interval $0 \leqq x < \infty$.
2. Show, with the aid of exercise 1, that the zeros of the Laguerre polynomial $L_n(x)$ are distinct and positive.
3. Use equation (6) above and the method of Section 144 to show that the set of Hermite polynomials is orthogonal with respect to the weight function $e^{-x^2}$ over the interval $-\infty < x < \infty$.
4. Show, with the aid of exercise 3, that the zeros of the Hermite polynomial $H_n(x)$ are real and distinct.

# Fourier Series

## 146. Orthogonality of a set of sines and cosines

The functions $\sin \alpha x$ and $\cos \alpha x$ occur in the formal solution of certain boundary value problems in partial differential equations, as was indicated in Chapter 21. We shall now obtain an orthogonality property for a set of such functions with $\alpha$ specified. An interval must be involved; let the origin be chosen at the center of the interval so the latter appears in the symmetric form: $-c \leqq x \leqq c$.

We shall show that the set of functions,

$$\left\{ \begin{array}{ll} \sin (n\pi x/c), & n = 1, 2, 3, \ldots \\ \cos (n\pi x/c), & n = 0, 1, 2, \ldots \end{array} \right\}, \tag{A}$$

or

$$\left\{ \begin{array}{l} \sin (\pi x/c), \sin (2\pi x/c), \sin (3\pi x/c), \cdots, \sin (n\pi x/c), \ldots \\ 1, \cos (\pi x/c), \cos (2\pi x/c), \cos (3\pi x/c), \cdots, \cos (n\pi x/c), \ldots \end{array} \right\}, \tag{A}$$

*is orthogonal with respect to the weight function* $w(x) = 1$ *over the interval*

$-c \leqq x \leqq c$. That is, we shall prove that the integral from $x = -c$ to $x = +c$ of the product of any two different members of the set (A) is zero.

First consider the integral of the product of any of the sine functions in (A) and any of the cosine functions in (A). The result

$$I_1 = \int_{-c}^{c} \sin \frac{n\pi x}{c} \cos \frac{k\pi x}{c} dx = 0$$

follows at once from the fact that the integrand is an odd function of $x$; in this instance the result does not depend upon the fact that $k$ and $n$ are integers.

Next consider the integral of the product of two different sine functions from the set (A),

$$I_2 = \int_{-c}^{c} \sin \frac{n\pi x}{c} \sin \frac{k\pi x}{c} dx, \qquad k \neq n.$$

Let us introduce a new variable of integration for simplicity in writing; put

$$\frac{\pi x}{c} = \beta,$$

from which

$$dx = \frac{c}{\pi} d\beta.$$

Then $I_2$ can be written

$$I_2 = \frac{c}{\pi} \int_{-\pi}^{\pi} \sin n\beta \sin k\beta \, d\beta.$$

Now from trigonometry we get the formula

$$\sin n\beta \sin k\beta = \tfrac{1}{2}[\cos (n - k)\beta - \cos (n + k)\beta]$$

which is useful in performing the desired integration. Thus it follows that the integral becomes

$$I_2 = \frac{c}{2\pi} \int_{-\pi}^{\pi} [\cos (n - k)\beta - \cos (n + k)\beta] \, d\beta$$

$$= \frac{c}{2\pi} \left[ \frac{\sin (n - k)\beta}{n - k} - \frac{\sin (n + k)\beta}{n + k} \right]_{-\pi}^{\pi},$$

since neither $(n - k)$ nor $(n + k)$ can be zero. Because $n$ and $k$ are positive integers, $\sin (n - k)\beta$ and $\sin (n + k)\beta$ each vanish at $\beta = \pi$ and $\beta = -\pi$; then

$$I_2 = 0$$

for $n, k = 1, 2, 3, \ldots,$ and $k \neq n$.

Finally, consider the integral of the product of two different cosine functions from the set (A),

$$I_3 = \int_{-c}^{c} \cos \frac{n\pi x}{c} \cos \frac{k\pi x}{c} \, dx,$$

where $n, k = 0, 1, 2, 3, \ldots; k \neq n$. The method used on $I_2$ works equally well here to yield

$$I_3 = \frac{c}{2\pi} \left[ \frac{\sin (n - k)\beta}{n - k} + \frac{\sin (n + k)\beta}{n + k} \right]_{-\pi}^{\pi} = 0.$$

It is easy to see that the integral of the square of any function from the set (A) will not vanish—its integrand is positive except at an occasional point. The values of those integrals are readily obtained. The integral

$$I_4 = \int_{-c}^{c} \sin^2 \frac{n\pi x}{c} \, dx$$

has an even integrand. Hence it can be written as

$$I_4 = 2 \int_{0}^{c} \sin^2 \frac{n\pi x}{c} \, dx.$$

Elementary methods of integration yield

$$I_4 = \int_{0}^{c} \left( 1 - \cos \frac{2n\pi x}{c} \right) dx$$

$$= \left[ x - \frac{c}{2n\pi} \sin \frac{2n\pi x}{c} \right]_{0}^{c} = c.$$

Therefore

$$\int_{-c}^{c} \sin^2 \frac{n\pi x}{c} \, dx = c, \qquad \text{for } n = 1, 2, 3, \ldots.$$

In the same way it follows that, for $n > 0$, $n$ integral,

$$I_5 = \int_{-c}^{c} \cos^2 \frac{n\pi x}{c} \, dx$$

$$= \left[ x + \frac{c}{2n\pi} \sin \frac{2n\pi x}{c} \right]_{0}^{c} = c.$$

For $n = 0$ the integral $I_5$ becomes

$$I_6 = \int_{-c}^{c} 1 \cdot dx = 2c.$$

Thus

$$\int_{-c}^{c} \cos^2 \frac{n\pi x}{c}\, dx = c, \qquad \text{for } n = 1, 2, 3, \ldots,$$

$$= 2c, \qquad \text{for } n = 0.$$

We have shown that the set

$$\begin{Bmatrix} \sin(n\pi x/c), & n = 1, 2, 3, \ldots \\ \cos(m\pi x/c), & m = 0, 1, 2, \ldots \end{Bmatrix}, \tag{A}$$

is orthogonal with respect to the weight function $w(x) = 1$ over the interval $-c \leqq x \leqq c$. We have also evaluated the integrals of the squares of the functions of the set (A).

## 147. Fourier series: an expansion theorem

With the assumption that there exists a series expansion of the type

$$f(x) = \frac{1}{2}a_0 + \sum_{n=1}^{\infty} \left( a_n \cos \frac{n\pi x}{c} + b_n \sin \frac{n\pi x}{c} \right), \tag{1}*$$

valid in the interval $-c \leqq x \leqq c$, it is a simple matter to determine the coefficients, $a_n$ and $b_n$. Indeed, disregarding the question of validity of interchange of order of summation and integration, we proceed as follows.

Multiply each term of equation (1) by $\sin(k\pi x/c)\, dx$, where $k$ is a positive integer, and then integrate each term from $-c$ to $+c$, thus arriving at

$$\int_{-c}^{c} f(x) \sin \frac{k\pi x}{c}\, dx = \frac{1}{2}a_0 \int_{-c}^{c} \sin \frac{k\pi x}{c}\, dx$$

$$+ \sum_{n=1}^{\infty} \left[ a_n \int_{-c}^{c} \cos \frac{n\pi x}{c} \sin \frac{k\pi x}{c}\, dx + b_n \int_{-c}^{c} \sin \frac{n\pi x}{c} \sin \frac{k\pi x}{c}\, dx \right]. \tag{2}$$

As seen earlier,

$$\int_{-c}^{c} \cos \frac{n\pi x}{c} \sin \frac{k\pi x}{c}\, dx = 0, \qquad \text{for all } k \text{ and } n, \tag{3}$$

and

$$\int_{-c}^{c} \sin \frac{n\pi x}{c} \sin \frac{k\pi x}{c}\, dx = 0, \qquad \text{for } k \neq n; k, n = 1, 2, 3, \ldots. \tag{4}$$

---

* A reason for the apparently peculiar notation, $\frac{1}{2}a_0$, for the constant term will be seen quite soon, page 446.

Therefore each term on the right-hand side of equation (2) is zero except for the term $n = k$. Thus equation (2) reduces to

$$\int_{-c}^{c} f(x) \sin \frac{k\pi x}{c} \, dx = b_k \int_{-c}^{c} \sin^2 \frac{k\pi x}{c} \, dx. \tag{5}$$

Since

$$\int_{-c}^{c} \sin^2 \frac{k\pi x}{c} \, dx = c,$$

we have

$$b_k = \frac{1}{c} \int_{-c}^{c} f(x) \sin \frac{k\pi x}{c} \, dx, \qquad k = 1, 2, 3, \ldots,$$

from which the coefficients $b_n$ in equation (1) follow by mere replacement of $k$ with $n$; that is,

$$b_n = \frac{1}{c} \int_{-c}^{c} f(x) \sin \frac{n\pi x}{c} \, dx, \qquad n = 1, 2, 3, \ldots. \tag{6}$$

Let us obtain the $a_n$ in a like manner. Using the multiplier $\cos(k\pi x/c) \, dx$ throughout equation (1) and then integrating term by term from $x = -c$ to $x = +c$, we get

$$\int_{-c}^{c} f(x) \cos \frac{k\pi x}{c} \, dx = \frac{1}{2} a_0 \int_{-c}^{c} \cos \frac{k\pi x}{c} \, dx$$

$$+ \sum_{n=1}^{\infty} \left[ a_n \int_{-c}^{c} \cos \frac{n\pi x}{c} \cos \frac{k\pi x}{c} \, dx + b_n \int_{-c}^{c} \sin \frac{n\pi x}{c} \cos \frac{k\pi x}{c} \, dx \right]. \tag{7}$$

The coefficient of $b_n$ in (7) is zero for all $n$ and $k$. If $k \neq 0$, we know that

$$\int_{-c}^{c} \cos \frac{n\pi x}{c} \cos \frac{k\pi x}{c} \, dx = 0, \qquad \text{for } n \neq k,$$

$$= c, \qquad \text{for } n = k,$$

and also the coefficient of $\frac{1}{2} a_0$ is zero. Thus, for $k \neq 0$, equation (7) reduces to

$$\int_{-c}^{c} f(x) \cos \frac{k\pi x}{c} \, dx = a_k \int_{-c}^{c} \cos^2 \frac{k\pi x}{c} \, dx,$$

from which $a_k$, and therefore $a_n$, can be found in the way $b_k$ was determined.

Thus we get

$$a_n = \frac{1}{c} \int_{-c}^{c} f(x) \cos \frac{n\pi x}{c} \, dx, \qquad n = 1, 2, 3, \ldots. \tag{8}$$

Next let us determine $a_0$. Suppose $k = 0$ in equation (7) so we have the equation

$$\int_{-c}^{c} f(x) \, dx = \frac{1}{2} a_0 \int_{-c}^{c} dx + \sum_{n=1}^{\infty} \left[ a_n \int_{-c}^{c} \cos \frac{n\pi x}{c} \, dx + b_n \int_{-c}^{c} \sin \frac{n\pi x}{c} \, dx \right].$$

The terms involving $n \geq 1$ are each zero. Hence

$$\int_{-c}^{c} f(x) \, dx = \tfrac{1}{2} a_0 (2c),$$

from which we obtain

$$a_0 = \frac{1}{c} \int_{-c}^{c} f(x) \, dx. \tag{9}$$

Equation (9) fits in with equation (8) as the special case $n = 0$. Had the factor $\frac{1}{2}$ not been inserted in equation (1), a separate formula would have been needed. As it is, we may write the formal expansion as follows:

$$f(x) = \frac{1}{2} a_0 + \sum_{n=1}^{\infty} \left( a_n \cos \frac{n\pi x}{c} + b_n \sin \frac{n\pi x}{c} \right) \tag{10}$$

with

$$a_n = \frac{1}{c} \int_{-c}^{c} f(x) \cos \frac{n\pi x}{c} \, dx, \qquad n = 0, 1, 2, \ldots, \tag{11}$$

$$b_n = \frac{1}{c} \int_{-c}^{c} f(x) \sin \frac{n\pi x}{c} \, dx, \qquad n = 1, 2, 3, \ldots. \tag{12}$$

Before proceeding to specific examples and applications it behooves us to state conditions under which the equality in (10) makes sense.

When $a_n$ and $b_n$ are given by (11) and (12) above, then the right-hand member of equation (10) is called the *Fourier series, over the interval* $-c \leq x \leq c$, *for the function* $f(x)$. A statement of conditions sufficient to insure that the Fourier series in (10) represents the function $f(x)$ in a reasonably meaningful manner follows.

Let $f(x)$ be continuous and differentiable at every point in the interval $-c \leq x \leq c$ except for, at most, a finite number of points, and at those points let $f(x)$ and $f'(x)$ have right- and left-hand limits. Such a function is exhibited in Figure 47.

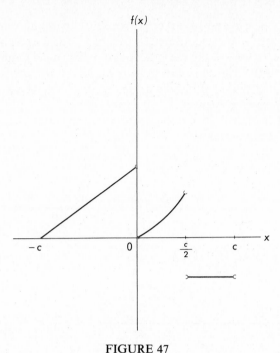

FIGURE 47

THEOREM 34: *Under the stipulations of the preceding paragraph, the Fourier series for $f(x)$, namely the series on the right in equation (10) with coefficients given by equations (11) and (12), converges to the value $f(x)$ at each point of continuity of $f(x)$; at each point of discontinuity of $f(x)$ the Fourier series converges to the arithmetic mean of the values approached by $f(x)$ from the right and the left.*

Since the Fourier series for $f(x)$ may not converge to the value $f(x)$ everywhere (for instance, at discontinuities of the function), it is customary to replace the equals sign in equation (10) by the symbol $\sim$, which may be read "has for its Fourier series." We write

$$f(x) \sim \frac{1}{2}a_0 + \sum_{n=1}^{\infty}\left(a_n \cos\frac{n\pi x}{c} + b_n \sin\frac{n\pi x}{c}\right), \qquad (13)$$

with $a_n$ and $b_n$ given by equations (11) and (12).

An interesting fact and one often useful as a check in numerical problems is that $\frac{1}{2}a_0$ is the average value of $f(x)$ over the interval $-c < x < c$.

The sine and cosine functions are periodic with period $2\pi$, so the terms in the Fourier series (13) for $f(x)$ are periodic with period $2c$. Therefore the series represents (converges to) a function that is as described above for the interval $-c < x < c$ and repeats that structure over and over outside that interval. For the function exhibited in Figure 47, the corresponding Fourier series would converge to the periodic function shown in Figure 48. Note the convergence to the average value at discontinuities, the periodicity, and the way in which the two together determine the value to which the series converges at $x = c$ and $x = -c$.

These statements will be amply illustrated in the numerical examples and exercises of the next section.

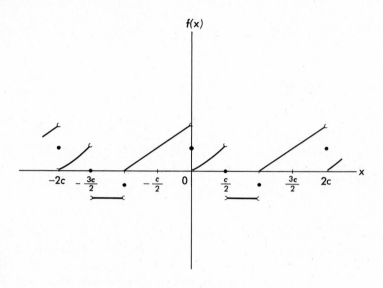

FIGURE 48

## 148. Numerical examples of Fourier series

We shall now construct the Fourier series for two specific functions.

EXAMPLE (a):  Construct the Fourier series over the interval

$$-2 \leqq x \leqq 2$$

for the function defined by

$$f(x) = 2, \qquad -2 < x \leqq 0,$$
$$= x, \qquad 0 < x < 2, \qquad (1)$$

and sketch the graph of the function to which the series converges.

First we sketch $f(x)$ itself, the result being exhibited in Figure 49. Note that $f(x)$ is undefined except for $x$ between $x = -2$ and $x = +2$.

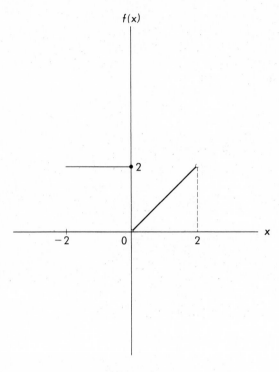

FIGURE 49

For the function described in (1),

$$f(x) \sim \frac{1}{2}a_0 + \sum_{n=1}^{\infty} \left( a_n \cos\frac{n\pi x}{2} + b_n \sin\frac{n\pi x}{2} \right),$$

in which

$$a_n = \frac{1}{2}\int_{-2}^{2} f(x) \cos\frac{n\pi x}{2}\, dx; \qquad n = 0, 1, 2, \ldots, \tag{2}$$

and

$$b_n = \frac{1}{2}\int_{-2}^{2} f(x) \sin\frac{n\pi x}{2}\, dx; \qquad n = 1, 2, 3, \ldots. \tag{3}$$

Since in the description of $f(x)$ different formulas were used in the two intervals $-2 < x < 0$ and $0 < x < 2$, it is convenient to separate the integrals in (2) and (3) into corresponding parts. Thus, inserting the $f(x)$ of (1) into the integral (2), leads us to the form

$$a_n = \frac{1}{2} \int_{-2}^{0} 2 \cos \frac{n\pi x}{2} \, dx + \frac{1}{2} \int_{0}^{2} x \cos \frac{n\pi x}{2} \, dx. \tag{4}$$

For these integrals, the method of integration will differ according to whether $n = 0$ or $n \neq 0$.

If $n \neq 0$, then

$$a_n = \frac{2}{n\pi} \left[ \sin \frac{n\pi x}{2} \right]_{-2}^{0} + \frac{1}{2} \left[ \frac{2}{n\pi} x \sin \frac{n\pi x}{2} + \left( \frac{2}{n\pi} \right)^2 \cos \frac{n\pi x}{2} \right]_{0}^{2},$$

or

$$a_n = \frac{2}{n\pi} [0 - 0] + \frac{1}{2} \left[ 0 + \left( \frac{2}{n\pi} \right)^2 \cos n\pi - 0 - \left( \frac{2}{n\pi} \right)^2 \right].$$

Hence for $n \neq 0$, the $a_n$ are given by the formula

$$a_n = \frac{-2(1 - \cos n\pi)}{n^2 \pi^2}, \qquad n = 1, 2, 3, \ldots . \tag{5}$$

For $n = 0$ the above integrations are not valid (division by $n$), but we return to (4), put $n = 0$, and get

$$a_0 = \frac{1}{2} \int_{-2}^{0} 2 \, dx + \frac{1}{2} \int_{0}^{2} x \, dx,$$

from which

$$a_0 = [x]_{-2}^{0} + \tfrac{1}{4}[x^2]_0^2 = 2 + 1 = 3.$$

The $b_n$ may be obtained in a like manner. From (3) and (1) it follows that

$$b_n = \frac{1}{2} \int_{-2}^{0} 2 \sin \frac{n\pi x}{2} \, dx + \frac{1}{2} \int_{0}^{2} x \sin \frac{n\pi x}{2} \, dx.$$

Thus

$$b_n = \frac{2}{n\pi} \left[ - \cos \frac{n\pi x}{2} \right]_{-2}^{0} + \frac{1}{2} \left[ -\left( \frac{2}{n\pi} \right) x \cos \frac{n\pi x}{2} + \left( \frac{2}{n\pi} \right)^2 \sin \frac{n\pi x}{2} \right]_{0}^{2},$$

from which, since $\cos(-n\pi) = \cos n\pi$,

$$b_n = \frac{2}{n\pi} [-1 + \cos n\pi] + \frac{1}{2} \left[ -\frac{2}{n\pi} \cdot 2 \cos n\pi + 0 + 0 - 0 \right],$$

or

$$b_n = -\frac{2}{n\pi}, \qquad n = 1, 2, 3, \ldots . \tag{6}$$

For integral $n$, $\cos n\pi = (-1)^n$, as is seen by examining both sides for even and odd $n$. Therefore the formula (5) above can also be written

$$a_n = \frac{-2[1 - (-1)^n]}{n^2\pi^2}, \qquad n = 1, 2, 3, \ldots. \tag{7}$$

We can now write the Fourier series, over the interval $-2 < x < 2$ for the $f(x)$ of this example,

$$f(x) \sim \frac{3}{2} - 2 \sum_{n=1}^{\infty} \left[ \frac{1 - (-1)^n}{n^2\pi^2} \cos\frac{n\pi x}{2} + \frac{1}{n\pi} \sin\frac{n\pi x}{2} \right]. \tag{8}$$

Several pertinent remarks can be made about (8). The right-hand member of (8) converges to the function shown in the sketch in Figure 50. It converges to $f(x)$ at each point where $f(x)$ is defined except at the discontinuity at $x = 0$. Though $f(0) = 2$, the series converges to unity at $x = 0$.

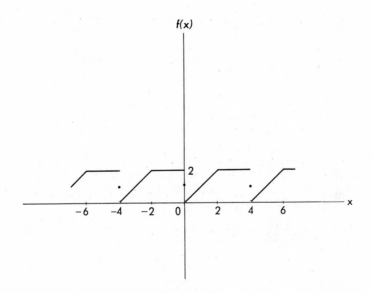

FIGURE 50

We may therefore write

$$f(x) = \frac{3}{2} - 2 \sum_{n=1}^{\infty} \left[ \frac{1 - (-1)^n}{n^2\pi^2} \cos\frac{n\pi x}{2} + \frac{1}{n\pi} \sin\frac{n\pi x}{2} \right] \tag{9}$$

for $-2 < x < 0$ and for $0 < x < 2$.

It is sometimes desirable to define a new function $\phi(x)$ as follows:

$$\phi(x) = f(x), \qquad -2 < x < 0,$$
$$= 1, \qquad\qquad x = 0,$$
$$= f(x), \qquad 0 < x < 2,$$

and

$$\phi(x + 4) = \phi(x).$$

This $\phi(x)$ is the function exhibited in Figure 50. If $\phi(x)$ is put in the place of $f(x)$ in (8) above, then the symbol $\sim$ may be replaced by the symbol $=$ for all $x$.

Because $[1 - (-1)^n]$ is zero for even $n$, the Fourier series on the right in (8) may be written in the somewhat more compact form

$$f(x) \sim \frac{3}{2} - \frac{4}{\pi^2} \sum_{k=0}^{\infty} \frac{\cos{[(2k + 1)\pi x/2]}}{(2k + 1)^2} - \frac{2}{\pi} \sum_{n=1}^{\infty} \frac{\sin{(n\pi x/2)}}{n}. \tag{10}$$

This is one instance in which an infinite rearrangement in the order of terms, passing from (8) to (10), is easily justified. Consider (10) again after studying the sections on Fourier sine series and Fourier cosine series.

Let us next use the expansion in (8) or (10) to sum two numerical series. For instance, if we put $x = 0$ in (10), then the series has the sum unity as indicated above. Hence

$$1 = \frac{3}{2} - \frac{4}{\pi^2} \sum_{k=0}^{\infty} \frac{1}{(2k + 1)^2} - \frac{2}{\pi} \sum_{n=1}^{\infty} \frac{0}{n},$$

or

$$\sum_{k=0}^{\infty} \frac{1}{(2k + 1)^2} = \frac{\pi^2}{8}. \tag{11}$$

For $x = 1$, the series in (10) has the sum unity again. Using $x = 1$ in (10) we are led to

$$1 = \frac{3}{2} - \frac{4}{\pi^2} \sum_{k=0}^{\infty} \frac{\cos{[(2k + 1)\pi/2]}}{(2k + 1)^2} - \frac{2}{\pi} \sum_{n=1}^{\infty} \frac{\sin{(n\pi/2)}}{n}.$$

Now $\cos{[(2k + 1)\pi/2]} = 0$ and $\sin{(n\pi/2)}$ may be obtained as follows. For even $n$, $n = 2k$, we get

$$\sin\frac{2k\pi}{2} = \sin{k\pi} = 0.$$

For odd $n$, $n = 2k + 1$,

$$\sin\frac{(2k + 1)\pi}{2} = \sin{(k\pi + \tfrac{1}{2}\pi)} = \cos{k\pi} = (-1)^k.$$

Thus we arrive at the equation

$$1 = \frac{3}{2} - \frac{2}{\pi} \sum_{k=0}^{\infty} \frac{(-1)^k}{2k+1},$$

or

$$\sum_{k=0}^{\infty} \frac{(-1)^k}{2k+1} = \frac{\pi}{4}, \tag{12}$$

which can be verified also by the fact that the left-hand member represents arc tan 1.

EXAMPLE (b):  Obtain the Fourier series over the interval $-\pi$ to $\pi$ for the function $x^2$. We know that

$$x^2 \sim \frac{1}{2}a_0 + \sum_{n=1}^{\infty} [a_n \cos nx + b_n \sin nx] \tag{13}$$

for $-\pi < x < \pi$, where

$$a_n = \frac{1}{\pi} \int_{-\pi}^{\pi} x^2 \cos nx \, dx; \qquad n = 0, 1, 2, \ldots, \tag{14}$$

$$b_n = \frac{1}{\pi} \int_{-\pi}^{\pi} x^2 \sin nx \, dx; \qquad n = 1, 2, 3, \ldots. \tag{15}$$

Now $x^2$ is an even function of $x$ and $\sin nx$ is an odd function of $x$, so the product $x^2 \sin nx$ is an odd function of $x$. Therefore $b_n = 0$ for every $n$. Since $x^2 \cos nx$ is an even function of $x$,

$$a_n = \frac{2}{\pi} \int_{0}^{\pi} x^2 \cos nx \, dx; \qquad n = 0, 1, 2, \ldots. \tag{16}$$

For $n \neq 0$,

$$a_n = \frac{2}{\pi} \left[ \frac{x^2 \sin nx}{n} + \frac{2x \cos nx}{n^2} - \frac{2 \sin nx}{n^3} \right]_0^{\pi},$$

from which

$$a_n = \frac{2}{\pi} \left[ \frac{2\pi \cos n\pi}{n^2} \right] = \frac{4(-1)^n}{n^2}, \qquad n = 1, 2, 3, \ldots.$$

A separate integration is needed for $a_0$. We get

$$a_0 = \frac{2}{\pi} \int_{0}^{\pi} x^2 \, dx = \frac{2}{\pi} \cdot \frac{\pi^3}{3} = \frac{2\pi^2}{3}.$$

Therefore, in the interval $-\pi < x < \pi$,

$$x^2 \sim \frac{\pi^2}{3} + 4 \sum_{n=1}^{\infty} \frac{(-1)^n \cos nx}{n^2}.$$

Indeed, because of continuity of the function involved, we may write

$$x^2 = \frac{\pi^2}{3} + 4 \sum_{n=1}^{\infty} \frac{(-1)^n \cos nx}{n^2}, \qquad \text{for } -\pi \leq x \leq \pi. \tag{17}$$

Beyond the indicated interval, the series on the right in equation (17) represents the periodic extension of the original function. The sum of the series is sketched in Figure 51.

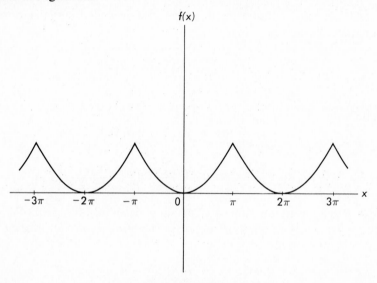

**FIGURE 51**

## Exercises

In exercises 1 through 22, obtain the Fourier series over the indicated interval for the given function. Always sketch the function that is the sum of the series obtained.

**1.** Interval, $-c < x < c$; function, $f(x) = 0, \qquad -c < x < 0,$

$$= c - x, \qquad 0 < x < c.$$

$$\text{ANS.} \quad f(x) \sim \frac{c}{4} + \frac{c}{\pi^2} \sum_{n=1}^{\infty} \frac{1}{n^2}\left[\{1 - (-1)^n\} \cos \frac{n\pi x}{c} + n\pi \sin \frac{n\pi x}{c}\right].$$

**2.** Interval, $-c < x < c$; function, $f(x) = x$.

$$\text{ANS.} \quad f(x) \sim \frac{2c}{\pi} \sum_{n=1}^{\infty} \frac{(-1)^{n+1} \sin (n\pi x/c)}{n}.$$

3. Interval, $-c < x < c$; function, $f(x) = x^2$. Check your answer with that in Example (b) in the text.

ANS.   $f(x) \sim \dfrac{c^2}{3} + \dfrac{4c^2}{\pi^2} \displaystyle\sum_{n=1}^{\infty} \dfrac{(-1)^n \cos(n\pi x/c)}{n^2}$.

4. Interval, $-c < x < c$; function, $f(x) = 0$,        $-c < x < 0$,

                             $= (c - x)^2$,        $0 < x < c$.

ANS.   $f(x) \sim \dfrac{c^2}{6} + \dfrac{c^2}{\pi^3} \displaystyle\sum_{n=1}^{\infty} \dfrac{1}{n^3}\left[ 2n\pi \cos\dfrac{n\pi x}{c} + \{n^2\pi^2 - 2 + 2(-1)^n\} \sin\dfrac{n\pi x}{c} \right]$.

5. Interval, $-c < x < c$; function, $f(x) = 0$,      $-c < x < 0$,

                             $= 1$,          $0 < x < c$.

ANS.   $f(x) \sim \dfrac{1}{2} + \displaystyle\sum_{n=1}^{\infty} \dfrac{1 - (-1)^n}{n\pi} \sin\dfrac{n\pi x}{c}$,

or

         $f(x) \sim \dfrac{1}{2} + \dfrac{2}{\pi} \displaystyle\sum_{k=0}^{\infty} \dfrac{\sin[(2k + 1)\pi x/c]}{2k + 1}$.

6. Interval, $-c < x < c$; function, $f(x) = x^3$.

ANS.   $f(x) \sim \dfrac{2c^3}{\pi^3} \displaystyle\sum_{n=1}^{\infty} (-1)^{n+1} \dfrac{(n^2\pi^2 - 6) \sin(n\pi x/c)}{n^3}$.

7. Interval, $-\pi < x < \pi$; function, $f(x) = 3\pi + 2x$,      $-\pi < x < 0$,

                             $= \pi + 2x$,       $0 < x < \pi$.

ANS.   $f(x) \sim 2\pi - 2 \displaystyle\sum_{k=1}^{\infty} \dfrac{\sin 2kx}{k}$.

8. Interval, $-c < x < c$; function, $f(x) = x(c + x)$,     $-c < x < 0$,

                             $= (c - x)^2$,      $0 < x < c$.

ANS.   $f(x) \sim \dfrac{c^2}{12} + \dfrac{c^2}{\pi^2} \displaystyle\sum_{n=1}^{\infty} \dfrac{1}{n^2}\left[ \{3 + (-1)^n\} \cos\dfrac{n\pi x}{c} + n\pi \sin\dfrac{n\pi x}{c} \right]$.

9. Interval, $-2 < x < 2$; function, $f(x) = x + 1$,      $-2 < x < 0$,

                             $= 1$,           $0 \leqq x < 2$.

ANS.   $f(x) \sim \dfrac{1}{2} + \dfrac{2}{\pi^2} \displaystyle\sum_{n=1}^{\infty} \dfrac{1}{n^2}[\{1 - (-1)^n\} \cos\tfrac{1}{2}n\pi x + n\pi(-1)^{n+1} \sin\tfrac{1}{2}n\pi x]$.

10. Interval, $-1 < x < 1$; function, $f(x) = 0$,      $-1 < x < 0$,

                             $= 1$,        $0 < x < \tfrac{1}{2}$,

                             $= 0$,        $\tfrac{1}{2} < x < 1$.

ANS.   $f(x) \sim \dfrac{1}{4} + \dfrac{1}{\pi} \displaystyle\sum_{n=1}^{\infty} \dfrac{1}{n}\left[ \sin\dfrac{n\pi}{2} \cos n\pi x + \left(1 - \cos\dfrac{n\pi}{2}\right) \sin n\pi x \right]$.

**11.** Interval, $-\pi < x < \pi$; function, $f(x) = 0,$  $-\pi < x < 0,$

$$= x^2, \qquad 0 < x < \pi.$$

ANS. $f(x) \sim \dfrac{\pi^2}{6} + 2 \displaystyle\sum_{n=1}^{\infty} \dfrac{(-1)^n \cos nx}{n^2}$

$$+ \frac{1}{\pi} \sum_{n=1}^{\infty} \frac{1}{n^3}[(-1)^{n+1} n^2 \pi^2 - 2 + 2(-1)^n] \sin nx.$$

**12.** Interval, $-\pi < x < \pi$; function, $f(x) = \cos 2x.$  ANS. $f(x) \sim \cos 2x.$

**13.** Interval, $-\pi < x < \pi$; function, $f(x) = \cos (x/2).$

ANS. $f(x) \sim \dfrac{2}{\pi} + \dfrac{4}{\pi} \displaystyle\sum_{n=1}^{\infty} \dfrac{(-1)^{n+1} \cos nx}{(2n-1)(2n+1)}.$

**14.** Interval, $-\pi < x < \pi$; function, $f(x) = \sin^2 x.$

ANS. $\sin^2 x \sim \frac{1}{2} - \frac{1}{2} \cos 2x.$

**15.** Interval, $-c < x < c$; function, $f(x) = e^x.$

ANS. $f(x) \sim \dfrac{\sinh c}{c} + \displaystyle\sum_{n=1}^{\infty} \dfrac{2(-1)^n \sinh c[c \cos (n\pi x/c) - n\pi \sin (n\pi x/c)]}{c^2 + n^2 \pi^2}.$

**16.** Interval, $-c < x < c$; function, $f(x) = 0,$  $-c < x < 0,$

$$= e^{-x}, \qquad 0 < x < c.$$

ANS. $f(x) \sim \dfrac{1 - e^{-c}}{2c} + \displaystyle\sum_{n=1}^{\infty} \dfrac{1 - (-1)^n e^{-c}}{c^2 + n^2 \pi^2} \left( c \cos \dfrac{n\pi x}{c} + n\pi \sin \dfrac{n\pi x}{c} \right).$

**17.** Interval, $-c < x < c$; function, $f(x) = 0,$  $-c < x < \frac{1}{2}c,$

$$= 1, \qquad \tfrac{1}{2}c < x < c.$$

ANS. $f(x) \sim \dfrac{1}{4} - \dfrac{1}{\pi} \displaystyle\sum_{n=1}^{\infty} \dfrac{1}{n} \left[ \sin \tfrac{1}{2} n\pi \cos \dfrac{n\pi x}{c} + (\cos n\pi - \cos \tfrac{1}{2} n\pi) \sin \dfrac{n\pi x}{c} \right].$

**18.** Interval, $-c < x < c$; function, $f(x) = 0,$  $-c < x < 0,$

$$= x, \qquad 0 < x < c.$$

ANS. $f(x) \sim \dfrac{1}{4}c - \dfrac{c}{\pi^2} \displaystyle\sum_{n=1}^{\infty} \dfrac{1}{n^2} \left[ \{1 - (-1)^n\} \cos \dfrac{n\pi x}{c} + n\pi(-1)^n \sin \dfrac{n\pi x}{c} \right].$

**19.** Interval, $-4 < x < 4$; function, $f(x) = 1,$  $-4 < x < 2,$

$$= 0, \qquad 2 < x < 4.$$

**20.** Interval, $-c < x < c$; function, $f(x) = 0,$  $-c < x < 0,$

$$= x(c - x), \qquad 0 < x < c.$$

**21.** Interval, $-c < x < c$; function, $f(x) = c + x,$  $-c < x < 0,$

$$= 0, \qquad 0 < x < c.$$

**22.** Interval, $-c < x < c$; function, $f(x) = x^4.$

ANS. $f(x) \sim \dfrac{c^4}{5} + 8c^4 \displaystyle\sum_{n=1}^{\infty} (-1)^n \dfrac{n^2 \pi^2 - 6}{n^4 \pi^4} \cos \dfrac{n\pi x}{c}.$

**23.** Use the answer to exercise 3 to show that $\displaystyle\sum_{n=1}^{\infty} \frac{(-1)^{n+1}}{n^2} = \frac{\pi^2}{12}$.

**24.** Use the answer to exercise 8 to show that $\displaystyle\sum_{n=1}^{\infty} \frac{1}{n^2} = \frac{\pi^2}{6}$.

**25.** Use the answer to exercise 22 to show that $\displaystyle\sum_{n=1}^{\infty} \frac{1}{n^4} = \frac{\pi^4}{90}$.

**26.** Use $x = 0$ in the answer to exercise 15 to sum the series $\displaystyle\sum_{n=1}^{\infty} \frac{(-1)^n}{c^2 + n^2\pi^2}$.

ANS. $\displaystyle\frac{c - \sinh c}{2c^2 \sinh c}$.

**27.** Let $c \to 0$ in the result of exercise 26 and check with exercise 23.

## 149. Fourier sine series

On page 431 we found it desirable to have an expansion of a function $f(x)$ in a series involving only sine functions, the expansion to represent the original $f(x)$ in an interval $0 < x < c$. With the notation we have been using, the Fourier series

$$\frac{1}{2}a_0 + \sum_{n=1}^{\infty} \left( a_n \cos\frac{n\pi x}{c} + b_n \sin\frac{n\pi x}{c} \right)$$

will reduce to a series with each term containing a sine function if somehow the $a_n$, where $n = 0, 1, 2, \ldots$, can be made to be zero. Examining the formula for $a_n$, page 446, reveals that the $a_n$ will vanish if the function being expanded is an odd function over the interval $-c < x < c$.

Therefore, to get a sine series for $f(x)$ we introduce a new function $g(x)$ defined to equal $f(x)$ in the interval $0 < x < c$ and to be the odd extension of that function in the remaining interval, $-c < x < 0$. That is, we define $g(x)$ by

$$g(x) = f(x), \qquad\qquad 0 < x < c,$$
$$= -f(-x), \qquad -c < x < 0.$$

Then $g(x)$ is an odd function over the interval $-c < x < c$. Hence from

$$g(x) \sim \frac{1}{2}a_0 + \sum_{n=1}^{\infty} \left( a_n \cos\frac{n\pi x}{c} + b_n \sin\frac{n\pi x}{c} \right)$$

it follows that

$$a_n = \frac{1}{c}\int_{-c}^{c} g(x) \cos\frac{n\pi x}{c}\, dx = 0, \qquad n = 0, 1, 2, \ldots,$$

and that

$$b_n = \frac{1}{c} \int_{-c}^{c} g(x) \sin \frac{n\pi x}{c} dx = \frac{2}{c} \int_{0}^{c} f(x) \sin \frac{n\pi x}{c} dx.$$

The resultant series represents $f(x)$ in the interval $0 < x < c$, because $g(x)$ and $f(x)$ are identical over that portion of the whole interval.

Thus we have

$$f(x) \sim \sum_{n=1}^{\infty} b_n \sin \frac{n\pi x}{c}, \qquad 0 < x < c, \tag{1}$$

in which

$$b_n = \frac{2}{c} \int_{0}^{c} f(x) \sin \frac{n\pi x}{c} dx, \qquad n = 1, 2, 3, \ldots. \tag{2}$$

The representation (1) is called the *Fourier sine series* for $f(x)$ over the interval $0 < x < c$.

It should be realized that the device of introducing the function $g(x)$ was a tool for arriving at (1) and (2); there is no need to repeat it in specific problems. Those we handle by direct use of (1) and (2) above.

EXAMPLE:   Expand $f(x) = x^2$ in a Fourier sine series over the interval $0 < x < 1$.

At once we may write, for $0 < x < 1$,

$$x^2 \sim \sum_{n=1}^{\infty} b_n \sin n\pi x, \tag{3}$$

in which

$$b_n = 2 \int_{0}^{1} x^2 \sin n\pi x \, dx$$

$$= 2 \left[ -\frac{x^2 \cos n\pi x}{n\pi} + \frac{2x \sin n\pi x}{(n\pi)^2} + \frac{2 \cos n\pi x}{(n\pi)^3} \right]_{0}^{1}$$

$$= 2 \left[ -\frac{\cos n\pi}{n\pi} + \frac{2 \cos n\pi}{n^3 \pi^3} - \frac{2}{n^3 \pi^3} \right]. \tag{4}$$

Hence the Fourier sine series, over $0 < x < 1$, for $x^2$ is

$$x^2 \sim 2 \sum_{n=1}^{\infty} \left[ \frac{(-1)^{n+1}}{n\pi} - \frac{2\{1 - (-1)^n\}}{n^3 \pi^3} \right] \sin n\pi x. \tag{5}$$

The series on the right in (5) converges to the function exhibited in Figure 52, that function being called the odd periodic extension, with period 2, of the function

$$f(x) = x^2, \qquad 0 < x < 1.$$

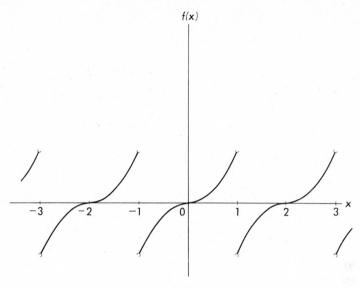

**FIGURE 52**

## *Exercises*

In each exercise, obtain the Fourier sine series over the stipulated interval for the function given. Sketch the function that is the sum of the series obtained.

**1.** Interval, $0 < x < c$; function, $f(x) = 1$.

$$\text{ANS.} \quad f(x) \sim \frac{4}{\pi} \sum_{k=0}^{\infty} \frac{\sin\left[(2k+1)\pi x/c\right]}{2k+1}.$$

**2.** Interval, $0 < x < c$; function, $f(x) = x$. Compare your result with that in exercise 2, page 431.

**3.** Interval, $0 < x < c$; function, $f(x) = x^2$. Check your answer with that for the example in the text above.

$$\text{ANS.} \quad f(x) \sim 2c^2 \sum_{n=1}^{\infty} \left[ \frac{(-1)^{n+1}}{n\pi} - \frac{2\{1 - (-1)^n\}}{n^3\pi^3} \right] \sin \frac{n\pi x}{c}.$$

**4.** Interval, $0 < x < c$; function, $f(x) = c - x$. $\quad$ ANS. $\quad f(x) \sim \dfrac{2c}{\pi} \sum_{n=1}^{\infty} \dfrac{1}{n} \sin \dfrac{n\pi x}{c}.$

**5.** Interval, $0 < x < 2c$; function, $f(x) = c - x$. $\quad$ ANS. $\quad$ Same as in exercise 4.

**6.** Interval, $0 < x < 4c$; function, $f(x) = c - x$. Compare with exercises 4 and 5.

$$\text{ANS.} \quad f(x) \sim \frac{2c}{\pi} \sum_{n=1}^{\infty} \frac{1 + 3(-1)^n}{n} \sin \frac{n\pi x}{4c}.$$

**7.** Interval, $0 < x < c$; function, $f(x) = x(c - x)$.

$$\text{ANS.} \quad f(x) \sim \frac{8c^2}{\pi^3} \sum_{k=0}^{\infty} \frac{\sin\left[(2k+1)\pi x/c\right]}{(2k+1)^3}.$$

8. Interval, $0 < x < 2$; function, $f(x) = x,$     $0 < x < 1,$

$= 2 - x,$     $1 < x < 2.$

ANS.   $f(x) \sim \dfrac{8}{\pi^2} \displaystyle\sum_{k=0}^{\infty} \dfrac{(-1)^k}{(2k+1)^2} \sin{[(2k+1)\pi x/2]}.$

9. Interval, $0 < t < t_1$; function, $f(t) = 1,$     $0 < t < t_0,$

$= 0,$     $t_0 < t < t_1.$

ANS.   $f(t) \sim \dfrac{2}{\pi} \displaystyle\sum_{n=1}^{\infty} \dfrac{1}{n}\left(1 - \cos\dfrac{n\pi t_0}{t_1}\right)\sin\dfrac{n\pi t}{t_1}.$

10. Interval, $0 < x < 1$; function, $f(x) = 0,$     $0 < x < \frac{1}{2},$

$= 1,$     $\frac{1}{2} < x < 1.$

ANS.   $f(x) \sim \dfrac{2}{\pi} \displaystyle\sum_{n=1}^{\infty} \dfrac{1}{n}\left(\cos\dfrac{n\pi}{2} - \cos n\pi\right)\sin n\pi x.$

11. Interval, $0 < x < 1$; function, $f(x) = 0,$     $0 < x < \frac{1}{2},$

$= x - \frac{1}{2},$     $\frac{1}{2} < x < 1.$

ANS.   $f(x) \sim \displaystyle\sum_{n=1}^{\infty} \left[\dfrac{(-1)^{n+1}}{n\pi} - \dfrac{2\sin{(n\pi/2)}}{n^2\pi^2}\right]\sin n\pi x.$

12. Interval, $0 < x < \pi$; function, $f(x) = \sin 3x.$     ANS.   $f(x) \sim \sin 3x.$

13. Interval, $0 < x < \pi$; function, $f(x) = \cos 2x.$ Note the special treatment necessary for the evaluation of $b_2$.

ANS.   $f(x) \sim \dfrac{4}{\pi} \displaystyle\sum_{k=0}^{\infty} \dfrac{(2k+1)\sin{[(2k+1)x]}}{(2k-1)(2k+3)}.$

14. Interval, $0 < x < \pi$; function, $f(x) = \cos x.$     ANS.   $f(x) \sim \dfrac{8}{\pi} \displaystyle\sum_{k=1}^{\infty} \dfrac{k\sin 2kx}{4k^2 - 1}.$

15. Interval, $0 < x < c$; function, $f(x) = e^{-x}.$

ANS.   $f(x) \sim \displaystyle\sum_{n=1}^{\infty} \dfrac{2n\pi[1 - (-1)^n\, e^{-c}]\sin{(n\pi x/c)}}{c^2 + n^2\pi^2}.$

16. Interval, $0 < x < c$; function, $f(x) = \sinh kx.$

ANS.   $f(x) \sim \sinh kc \displaystyle\sum_{n=1}^{\infty} \dfrac{(-1)^{n+1}2n\pi}{(kc)^2 + (n\pi)^2}\sin\dfrac{n\pi x}{c}.$

17. Interval, $0 < x < c$; function, $f(x) = \cosh kx.$

ANS.   $f(x) \sim \displaystyle\sum_{n=1}^{\infty} \dfrac{2n\pi[1 + (-1)^{n+1}\cosh kc]}{(kc)^2 + (n\pi)^2}\sin\dfrac{n\pi x}{c}.$

18. Interval, $0 < x < c$; function, $f(x) = x^3.$     ANS.   See exercise 6, p. 455.

19. Interval, $0 < x < c$; function, $f(x) = x^4.$

ANS.   $f(x) \sim \dfrac{2c^4}{\pi} \displaystyle\sum_{n=1}^{\infty} \left[(-1)^{n+1}\left\{\dfrac{1}{n} - \dfrac{12}{\pi^2 n^3} + \dfrac{24}{\pi^4 n^5}\right\} + \dfrac{24}{\pi^4 n^5}\right]\sin\dfrac{n\pi x}{c}.$

**20.** Interval, $0 < x < c$; function, $f(x) = x,$ $\quad 0 < x < \tfrac{1}{2}c,$

$\qquad\qquad\qquad\qquad\qquad = 0, \quad \tfrac{1}{2}c < x < c.$

ANS. $\quad f(x) \sim \dfrac{c}{\pi^2} \displaystyle\sum_{n=1}^{\infty} \left( \dfrac{2}{n^2} \sin \dfrac{n\pi}{2} - \dfrac{\pi}{n} \cos \dfrac{n\pi}{2} \right) \sin \dfrac{n\pi x}{c}.$

**21.** Interval, $0 < x < 1$; function, $f(x) = (x-1)^2$.

ANS. $\quad f(x) \sim \dfrac{2}{\pi^3} \displaystyle\sum_{n=1}^{\infty} \dfrac{1}{n^3} [n^2\pi^2 - 2 + 2(-1)^n] \sin n\pi x.$

## 150. Fourier cosine series

In a manner entirely similar to that used to obtain the Fourier sine series, it is possible to obtain a series of cosine terms, including a constant term, for a function defined over the interval $0 < x < c$. Indeed, given $f(x)$ defined over the interval $0 < x < c$ and satisfying there the conditions stipulated in Section 147, we may define an auxiliary function $h(x)$ by

$$h(x) = f(x), \qquad 0 < x < c,$$
$$= f(-x), \qquad -c < x < 0.$$

Then $h(x)$ is an even function of $x$ and, of course, it is equal to $f(x)$ over the interval where $f(x)$ was defined. Since $h(x)$ is even, it follows that, in its ordinary Fourier expansion over the interval $-c < x < c$, the $b_n$ are all zero,

$$b_n = \frac{1}{c} \int_{-c}^{c} h(x) \sin \frac{n\pi x}{c} \, dx = 0,$$

because of the oddness of the integrand. Furthermore, since $h(x)$ is even, $h(x) \cos(n\pi x/c)$ is also even and

$$a_n = \frac{2}{c} \int_{0}^{c} h(x) \cos \frac{n\pi x}{c} \, dx = \frac{2}{c} \int_{0}^{c} f(x) \cos \frac{n\pi x}{c} \, dx.$$

Since $h(x)$ and $f(x)$ are identical over the interval $0 < x < c$, we may write what is customarily called the *Fourier cosine series* for $f(x)$ over that interval, namely,

$$f(x) \sim \frac{1}{2}a_0 + \sum_{n=1}^{\infty} a_n \cos \frac{n\pi x}{c}, \quad 0 < x < c, \qquad (1)$$

in which

$$a_n = \frac{2}{c} \int_{0}^{c} f(x) \cos \frac{n\pi x}{c} \, dx. \qquad (2)$$

EXAMPLE : Find the Fourier cosine series over the interval $0 < x < c$ for the function $f(x) = x$.

At once we have

$$f(x) \sim \frac{1}{2}a_0 + \sum_{n=1}^{\infty} a_n \cos \frac{n\pi x}{c},$$

in which

$$a_n = \frac{2}{c} \int_0^c x \cos \frac{n\pi x}{c} \, dx.$$

For $n \neq 0$, the $a_n$ may be evaluated as follows:

$$a_n = \frac{2}{c} \left[ \frac{c}{n\pi} x \sin \frac{n\pi x}{c} + \left( \frac{c}{n\pi} \right)^2 \cos \frac{n\pi x}{c} \right]_0^c$$

$$= \frac{2}{c} \left[ \left( \frac{c}{n\pi} \right)^2 \cos n\pi - \left( \frac{c}{n\pi} \right)^2 \right]$$

$$= -\frac{2c}{n^2\pi^2}(1 - \cos n\pi), \qquad n \neq 0.$$

The remaining coefficient $a_0$ is readily obtained:

$$a_0 = \frac{2}{c} \int_0^c x \, dx = \frac{2}{c} \cdot \frac{c^2}{2} = c.$$

Thus the Fourier cosine series over the interval $0 < x < c$ for the function $f(x) = x$ is

$$f(x) \sim \frac{1}{2}c - \frac{2c}{\pi^2} \sum_{n=1}^{\infty} \frac{1 - (-1)^n}{n^2} \cos \frac{n\pi x}{c},$$

which may also be written in the form

$$f(x) \sim \frac{1}{2}c - \frac{4c}{\pi^2} \sum_{k=0}^{\infty} \frac{\cos [(2k + 1)\pi x/c]}{(2k + 1)^2}. \tag{3}$$

The infinite series on the right in (3) converges to a function which is often called the even periodic extension of the function $f(x) = x$. The graph of this extension is exhibited in Figure 53.

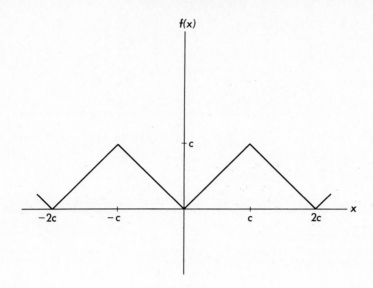

FIGURE 53

## Exercises

In each exercise, obtain the Fourier cosine series for the given function over the stipulated interval and sketch the function to which the series converges.

**1.** Interval, $0 < x < 2$; function, $f(x) = x,$      $0 < x < 1,$
                                    $= 2 - x,$    $1 < x < 2.$

ANS.   $f(x) \sim \dfrac{1}{2} + \dfrac{4}{\pi^2} \sum_{n=1}^{\infty} \dfrac{1}{n^2}\left[2\cos\dfrac{n\pi}{2} - 1 - (-1)^n\right]\cos\dfrac{n\pi x}{2}.$

**2.** Interval, $0 < t < t_1$; function, $f(t) = 1,$     $0 < t < t_0,$
                                 $= 0,$      $t_0 < t < t_1.$

ANS.   $f(t) \sim \dfrac{t_0}{t_1} + \dfrac{2}{\pi} \sum_{n=1}^{\infty} \dfrac{1}{n}\sin\dfrac{n\pi t_0}{t_1}\cos\dfrac{n\pi t}{t_1}.$

**3.** Interval, $0 < x < 1$; function, $f(x) = (x - 1)^2.$

ANS.   $f(x) \sim \dfrac{1}{3} + \dfrac{4}{\pi^2} \sum_{n=1}^{\infty} \dfrac{\cos n\pi x}{n^2}.$

**4.** Interval, $0 < x < c$; function, $f(x) = x(c - x).$

ANS.   $f(x) \sim \dfrac{c^2}{6} - \dfrac{c^2}{\pi^2} \sum_{k=1}^{\infty} \dfrac{\cos(2k\pi x/c)}{k^2}.$

**5.** Interval, $0 < x < c$; function, $f(x) = c - x.$

ANS.   $f(x) \sim \dfrac{1}{2}c + \dfrac{4c}{\pi^2} \sum_{k=0}^{\infty} \dfrac{\cos[(2k + 1)\pi x/c]}{(2k + 1)^2}.$

**6.** Interval, $0 < x < 1$; function, $f(x) = 0$, $\quad 0 < x < \frac{1}{2}$,
$$= 1, \quad \frac{1}{2} < x < 1.$$

ANS. $f(x) \sim \dfrac{1}{2} - \dfrac{2}{\pi} \displaystyle\sum_{k=0}^{\infty} \dfrac{(-1)^k \cos\left[(2k+1)\pi x\right]}{2k+1}$.

**7.** Interval, $0 < x < 1$; function, $f(x) = 0$, $\quad 0 < x < \frac{1}{2}$,
$$= x - \tfrac{1}{2}, \quad \frac{1}{2} < x < 1.$$

ANS. $f(x) \sim \dfrac{1}{8} + \dfrac{2}{\pi^2} \displaystyle\sum_{n=1}^{\infty} \dfrac{1}{n^2}\left(\cos n\pi - \cos \dfrac{n\pi}{2}\right) \cos n\pi x$.

**8.** Interval, $0 < x < 1$; function, $f(x) = \frac{1}{2} - x$, $\quad 0 < x < \frac{1}{2}$,
$$= 0, \quad \frac{1}{2} < x < 1.$$

ANS. $f(x) \sim \dfrac{1}{8} + \dfrac{2}{\pi^2} \displaystyle\sum_{n=1}^{\infty} \dfrac{1}{n^2}\left(1 - \cos \dfrac{n\pi}{2}\right) \cos n\pi x$.

**9.** Interval, $0 < x < \pi$; function, $f(x) = \cos 2x$. $\qquad$ ANS. $f(x) \sim \cos 2x$.

**10.** Interval, $0 < x < \pi$; function, $f(x) = \sin 2x$.

ANS. $f(x) \sim -\dfrac{8}{\pi} \displaystyle\sum_{k=0}^{\infty} \dfrac{\cos\left[(2k+1)x\right]}{(2k-1)(2k+3)}$.

**11.** Interval, $0 < x < c$; function, $f(x) = x$, $\quad 0 < x < \frac{1}{2}c$,
$$= 0, \quad \frac{1}{2}c < x < c.$$

ANS. $f(x) \sim \dfrac{c}{8} + \dfrac{c}{\pi^2} \displaystyle\sum_{n=1}^{\infty} \dfrac{1}{n^2}[n\pi \sin \tfrac{1}{2}n\pi - 2(1 - \cos \tfrac{1}{2}n\pi)] \cos \dfrac{n\pi x}{c}$.

**12.** Interval, $0 < x < c$; function $f(x) = e^{-x}$. Notice how the $a_0$ term fits in with the others this time, making separate integration unnecessary.

ANS. $f(x) \sim \dfrac{1 - e^{-c}}{c} + 2c \displaystyle\sum_{n=1}^{\infty} \dfrac{1 - (-1)^n e^{-c}}{c^2 + n^2\pi^2} \cos \dfrac{n\pi x}{c}$.

**13.** Interval, $0 < x < c$; function, $f(x) = \cosh kx$.

ANS. $f(x) \sim \dfrac{\sinh kc}{kc} + \sinh kc \displaystyle\sum_{n=1}^{\infty} \dfrac{2kc(-1)^n}{(kc)^2 + (n\pi)^2} \cos \dfrac{n\pi x}{c}$.

**14.** Interval, $0 < x < c$; function, $f(x) = \sinh kx$.

ANS. $f(x) \sim \dfrac{\cosh kc - 1}{kc} + \displaystyle\sum_{n=1}^{\infty} \dfrac{2kc[(-1)^n \cosh kc - 1]}{(kc)^2 + (n\pi)^2} \cos \dfrac{n\pi x}{c}$.

**15.** Interval, $0 < x < c$; function, $f(x) = x^3$.

ANS. $f(x) \sim \dfrac{c^3}{4} + \dfrac{6c^3}{\pi^2} \displaystyle\sum_{n=1}^{\infty} \left[\dfrac{(-1)^n}{n^2} + \dfrac{2}{\pi^2} \cdot \dfrac{1 - (-1)^n}{n^4}\right] \cos \dfrac{n\pi x}{c}$.

**16.** Interval, $0 < x < c$; function, $f(x) = x^4$. $\qquad$ ANS. See exercise 22, p. 456.

## 151. Numerical Fourier analysis

In the preceding sections and in the applications in the next chapter, the functions for which Fourier series are required are expressed by means of formulas, as for example

$$f(x) = x, \qquad 0 < x < 1,$$

$$= 2 - x, \qquad 1 < x < 2.$$

Then the Fourier coefficients, $a_n$, $b_n$, are obtained by formal integrations.

In practice it often happens that a function will, in the first place, be described only by a graph or by a table of numerical values. Then the Fourier coefficients should be determined by performing the appropriate integrations by some numerical, mechanical, or graphical method. For instance, in the heat-conduction problem studied in Section 139, page 427, the initial temperature distribution $f(x)$ might well consist of a table of initial temperature readings for points at various distances from one surface of the slab. At the end of Section 139, on page 431, it is seen that the solution

$$u(x, t) = \sum_{n=1}^{\infty} B_n \exp\left[ -\left(\frac{n\pi h}{c}\right)^2 t \right] \sin \frac{n\pi x}{c} \tag{1}$$

of the temperature problem involves the coefficients $B_n$, which are to be chosen so that

$$f(x) = \sum_{n=1}^{\infty} B_n \sin \frac{n\pi x}{c}, \qquad \text{for } 0 < x < c. \tag{2}$$

We can see now that (2) is to be the Fourier sine series expansion of $f(x)$. Hence the $B_n$ are given by

$$B_n = \frac{2}{c} \int_0^c f(x) \sin \frac{n\pi x}{c} dx$$

from which $B_n$ is to be found numerically and then inserted in (1).

It is natural that in this book there is a marked tendency to consider each topic encountered only in the light of its bearing on differential equations, or even on a particular phase of the subject of differential equations. In all fairness it must be mentioned that Fourier series are involved in many other ways in mathematics and in other sciences. For contact with the subject of curve-fitting and the method of least squares, see Churchill and Brown's *Fourier Series and Boundary Value Problems*, cited in the References, page 422.

## 152. Improvement in rapidity of convergence

In practical problems, trigonometric series occur sometimes without the sum function being known in any other form. Computations with such series can be irksome unless the series converge with reasonable rapidity.

Suppose that it is desired to compute, at several points in the interval $0 < x < \pi$, the sum of the series

$$\sum_{n=1}^{\infty} \frac{(-1)^n n \cos nx}{n^3 + 7}. \tag{1}$$

The series in (1) converges absolutely since its general term is less in absolute value than $1/n^2$ and $\sum\limits_{n=1}^{\infty} 1/n^2$ converges. Let the sum of the series in (1) be denoted by $\phi(x)$.

For large $n$ the coefficients in the series (1) are well approximated by $(-1)^n/n^2$. But we know the sum of the corresponding series with those coefficients; on pages 453–454 we showed that

$$x^2 = \frac{\pi^2}{3} + 4 \sum_{n=1}^{\infty} \frac{(-1)^n \cos nx}{n^2}, \qquad \text{for } -\pi \leqq x \leqq \pi. \tag{2}$$

Therefore,

$$\frac{1}{4}\left(x^2 - \frac{\pi^2}{3}\right) = \sum_{n=1}^{\infty} \frac{(-1)^n \cos nx}{n^2}, \qquad -\pi \leqq x \leqq \pi. \tag{3}$$

Since the coefficients in the series in (3) and those in the series for $\phi(x)$,

$$\phi(x) = \sum_{n=1}^{\infty} \frac{(-1)^n n \cos nx}{n^3 + 7}, \qquad 0 < x < \pi, \tag{4}$$

are nearly equal for large $n$, it follows that the difference of those coefficients should be small. So we subtract the members of equation (3) from the corresponding ones of equation (4) and get

$$\phi(x) - \frac{1}{4}\left(x^2 - \frac{\pi^2}{3}\right) = \sum_{n=1}^{\infty} (-1)^n \left[\frac{n}{n^3 + 7} - \frac{1}{n^2}\right] \cos nx$$

$$= \sum_{n=1}^{\infty} \frac{7(-1)^{n+1} \cos nx}{n^2(n^3 + 7)}.$$

Thus we obtain for $\phi(x)$ the formula

$$\phi(x) = \frac{1}{4}\left(x^2 - \frac{\pi^2}{3}\right) + 7 \sum_{n=1}^{\infty} \frac{(-1)^{n+1} \cos nx}{n^2(n^3 + 7)}, \qquad 0 < x < \pi, \tag{5}$$

with which computation of $\phi(x)$ is simplified because the coefficients of cos $nx$ in (5) get small more rapidly than those in (4) as $n$ increases.

The device illustrated above is worth keeping in mind when computing with infinite series—whether trigonometric or not. The method is largely dependent upon the presence of a collection of series for which the sum is known.

## References on Fourier series

CHURCHILL, R. V., and J. W. BROWN. *Fourier Series and Boundary Value Problems*, 3rd ed. New York: McGraw-Hill Book Company, 1978.

JACKSON, DUNHAM. *Fourier Series and Orthogonal Polynomials*. Carus Mathematical Monograph No. 6. Menasha, Wis.: Mathematical Association of America, 1941.

LANGER, R. E. *Fourier Series, The Genesis and Evolution of a Theory*. The first Slaught memorial paper, published as a supplement to the *Amer. Math. Mon.*, **54,** 1947.

# Boundary Value Problems

### 153. The one-dimensional heat equation

The equation that governs the conduction of heat,

$$\frac{\partial u}{\partial t} = h^2 \left( \frac{\partial^2 u}{\partial x^2} + \frac{\partial^2 u}{\partial y^2} + \frac{\partial^2 u}{\partial z^2} \right), \tag{1}$$

was introduced on page 421. The symbols in it and a set of consistent units often employed in engineering practice are described below:

$x, y, z$ = rectangular space coordinates (ft),
$\quad t$ = time (hr),
$\quad u$ = temperature (°F),
$\quad h^2$ = thermal diffusivity (ft$^2$/hr).

Another frequently used set of units for the above quantities replaces feet by centimeters, hours by seconds, and degrees Fahrenheit by degrees centigrade.

It has already been indicated in Section 139 that under proper physical conditions it is reasonable to study a certain special case of equation (1), the one-dimensional heat equation

$$\frac{\partial u}{\partial t} = h^2 \frac{\partial^2 u}{\partial x^2}.$$

In Section 139 we obtained from the boundary value problem

$$\frac{\partial u}{\partial t} = h^2 \frac{\partial^2 u}{\partial x^2}, \qquad \text{for } 0 < t, 0 < x < c; \tag{2}$$

$$\text{As } t \to 0^+, u \to f(x), \qquad \text{for } 0 < x < c; \tag{3}$$

$$\text{As } x \to 0^+, u \to 0, \qquad \text{for } 0 < t; \tag{4}$$

$$\text{As } x \to c^-, u \to 0, \qquad \text{for } 0 < t, \tag{5}$$

the relation

$$u(x, t) = \sum_{n=1}^{\infty} B_n \exp\left[ -\left(\frac{n\pi h}{c}\right)^2 t \right] \sin \frac{n\pi x}{c}, \tag{6}$$

where the $B_n$ were to be determined so that

$$f(x) = \sum_{n=1}^{\infty} B_n \sin \frac{n\pi x}{c}, \qquad \text{for } 0 < x < c. \tag{7}$$

Then in Chapter 23 we found that equation (7) suggests that the series on the right be the Fourier sine series for $f(x)$ over the interval $0 < x < c$, and therefore that

$$B_n = \frac{2}{c} \int_0^c f(x) \sin \frac{n\pi x}{c} \, dx. \tag{8}$$

It is not difficult, but requires material beyond this course, to verify that (6) with coefficients $B_n$ given by (8) is actually a solution; that is, that (6) possesses, for properly chosen $f(x)$, the required convergence properties in addition to its formal satisfaction of the differential equation (2) and the boundary conditions (3), (4), and (5).

The amount of heat that flows across an element of surface in a specified time is proportional to the rate of change of temperature in the direction normal (perpendicular) to that surface. Thus the flux of heat in the $x$ direction (across a surface normal to the $x$ direction) is taken to be

$$-K \frac{\partial u}{\partial x},$$

the constant of proportionality being $K$, the thermal conductivity of the material involved. The significance of the negative sign can be seen by considering an example in which the temperature increases with increasing $x$. Then $\partial u/\partial x$ is positive, but heat flows toward negative $x$, from the warmer portion to the colder portion; hence the flux is taken to be negative.

For us the expression for flux of heat will be used most often in forming boundary conditions involving insulation. If there is total insulation at a surface normal to the $x$ direction, then there is no flux of heat across that surface, so

$$\frac{\partial u}{\partial x} = 0$$

at that surface.

EXAMPLE: Find the temperature in a flat slab of unit width such that:

   (a) Its initial temperature varies uniformly from zero at one face to $u_0$ at the other.
   (b) The temperature of the face initially at zero remains at zero for $t > 0$.
   (c) The face initially at temperature $u_0$ is insulated for $t > 0$.

If $x$ is measured from the face at zero temperature, the problem may be written

$$\frac{\partial u}{\partial t} = h^2 \frac{\partial^2 u}{\partial x^2}, \qquad \text{for } 0 < x < 1, 0 < t; \tag{9}$$

$$\text{As } t \to 0^+, u \to u_0 x, \qquad \text{for } 0 < x < 1; \tag{10}$$

$$\text{As } x \to 0^+, u \to 0, \qquad \text{for } 0 < t; \tag{11}$$

$$\text{As } x \to 1^-, \frac{\partial u}{\partial x} \to 0, \qquad \text{for } 0 < t. \tag{12}$$

First we seek functions that satisfy the differential equation (9), using the technique of separating the independent variables. As before we get

$$u = \exp(-h^2\alpha^2 t)[A \cos \alpha x + B \sin \alpha x] \tag{13}$$

with $\alpha$, $A$, and $B$ arbitrary. Condition (11) demands that

$$0 = A \exp(-h^2\alpha^2 t), \qquad \text{for } 0 < t,$$

so we must take $A = 0$. We now have

$$u = B \exp(-h^2\alpha^2 t) \sin \alpha x, \tag{14}$$

which satisfies (9) and (11). From (14) it follows that

$$\frac{\partial u}{\partial x} = \alpha B \exp(-h^2\alpha^2 t) \cos \alpha x,$$

so condition (12) requires that

$$0 = \alpha B \exp(-h^2\alpha^2 t) \cos \alpha, \qquad \text{for } 0 < t.$$

We must not choose $\alpha = 0$ or $B = 0$ because then (14) would yield $u = 0$, which cannot satisfy the remaining condition (10). The factor $\exp(-h^2\alpha^2 t)$ cannot vanish for any $t$, much less for all positive $t$. Thus we conclude that

$$\cos \alpha = 0. \tag{15}$$

From (15) it follows that

$$\alpha = (2k + 1)\pi/2; \qquad k = 0, 1, 2, \ldots.$$

We now have the functions

$$u = B_k \exp\left[-\tfrac{1}{4}h^2(2k + 1)^2\pi^2 t\right] \sin\left[(2k + 1)\pi x/2\right]; \qquad k = 0, 1, 2, \ldots,$$

each of which satisfies (9), (11), and (12). To attack the condition (10), we form the series

$$u(x, t) = \sum_{k=0}^{\infty} B_k \exp\left[-\tfrac{1}{4}h^2(2k + 1)^2\pi^2 t\right] \sin\left[(2k + 1)\pi x/2\right] \tag{16}$$

and require, because of (10), that

$$u_0 x = \sum_{k=0}^{\infty} B_k \sin\left[(2k + 1)\pi x/2\right], \qquad \text{for } 0 < x < 1. \tag{17}$$

Comparison of the right member of (17) with the general Fourier sine series expansion for the interval $0 < x < c$ shows that the series in (17) is an expansion over the interval $0 < x < 2$ and that its even-numbered terms are missing. That is, we seek a Fourier sine series expansion

$$f(x) \sim \sum_{n=1}^{\infty} b_n \sin \frac{n\pi x}{2}, \qquad \text{for } 0 < x < 2 \tag{18}$$

where

$$f(x) = u_0 x, \qquad \text{for } 0 < x < 1,$$

and $f(x)$ is so chosen for $1 < x < 2$ that in (18) the terms with even $n$ will drop out.

Physically, it is not difficult to see that we wish to extend the slab its own width beyond $x = 1$ in some way to prevent the heat from flowing across the insulated face $x = 1$. Once that fact is realized it soon follows that we need all temperature conditions to be symmetric with respect to that insulated face $x = 1$.

The initial temperature $f(x)$ of our original slab is shown in Figure 54. Let us prescribe $f(x)$ over $1 < x < 2$ to be the reflection through $x = 1$ of the initial temperature, so in $0 < x < 2$ the initial temperature of the extended slab is as shown in Figure 55.

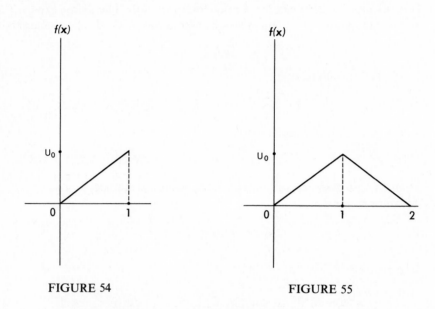

FIGURE 54                    FIGURE 55

The boundary value problem (9) through (12) can now be replaced by a new one with slab width 2, initial temperature as exhibited in Figure 55, and with faces $x = 0$ and $x = 2$ held at zero temperature for $t > 0$. The solution to the old problem is the same as that to the new problem, except that it is to be used only for $0 < x < 1$.

An alternative procedure is to revert to equation (18) with

$$f(x) = u_0 x, \qquad \text{for } 0 < x < 1,$$
$$= u_0(2 - x), \qquad \text{for } 1 < x < 2,$$

and thus to obtain $b_n$ from which the $B_k$ of (16) and (17) follows.

One more method deserves mention. After justification, such as in exercise 10 below, it is permissible to obtain the $B_k$ directly from equation (17) without open recourse to any devices such as the ones above.

The student can show in one or more of these ways that the problem (9) through (12) has for a solution

$$u(x, t) = \frac{8u_0}{\pi^2} \sum_{k=0}^{\infty} \frac{(-1)^k}{(2k + 1)^2} \exp[-\tfrac{1}{4}h^2(2k + 1)^2\pi^2 t] \sin\frac{(2k + 1)\pi x}{2}. \quad (19)$$

## Exercises

1. Use the method, not the formulas, of this section to solve the problem of a slab initially at a constant temperature $u_0$ throughout and having its faces $x = 0$ and $x = c$ held at zero temperature for $t > 0$.

$$\text{ANS.} \quad u = \frac{4u_0}{\pi} \sum_{k=0}^{\infty} \frac{1}{2k+1} \exp\left[\frac{-h^2\pi^2(2k+1)^2 t}{c^2}\right] \sin\frac{(2k+1)\pi x}{c}.$$

2. Obtain the average temperature across the slab of exercise 1 for $t > 0$.

$$\text{ANS.} \quad u_m(t) = \frac{8u_0}{\pi^2} \sum_{k=0}^{\infty} \frac{1}{(2k+1)^2} \exp\left[-\frac{h^2\pi^2(2k+1)^2 t}{c^2}\right].$$

3. For the one-dimensional heat equation (2) above, find a solution $u$ such that $u$ is independent of $t$, $u = A$ for $x = 0$, and $u = 0$ for $x = c$; $A$ is constant.

$$\text{ANS.} \quad u = A(c - x)/c.$$

4. With the aid of the result of exercise 3, solve the problem of a slab of width $c$, with initial temperature zero throughout, and with faces $x = 0$ and $x = c$ held at temperatures $A$ and zero respectively for $t > 0$.

$$\text{ANS.} \quad u = A(c - x)/c - \frac{2A}{\pi} \sum_{n=1}^{\infty} \frac{1}{n} \exp\left[-\left(\frac{hn\pi}{c}\right)^2 t\right] \sin\frac{n\pi x}{c}.$$

5. Combine the result of exercise 4 with the material of this section to solve the problem of the slab such that

$$\begin{array}{lll} \text{As } t \to 0^+, u \to f(x), & \text{for } 0 < x < c; \\ \text{As } x \to 0^+, u \to A, & \text{for } 0 < t; \\ \text{As } x \to c^-, u \to 0, & \text{for } 0 < t. \end{array}$$

6. For a particular concrete, the thermal diffusivity $h^2$ is about 0.04 (ft$^2$/hr), so we may reasonably choose $h^2\pi^2 = 0.4$. A slab 20 ft thick is initially at temperature 130°F and has its surfaces held at 60°F for $t > 0$. Show that the temperature in degrees Fahrenheit at the center of the slab is given by the formula

$$u = 60 + \frac{280}{\pi} \sum_{k=0}^{\infty} \frac{(-1)^k}{2k+1} \exp\left[-\frac{(2k+1)^2 t}{1000}\right].$$

7. Two slabs of the concrete of exercise 6 (with $h^2\pi^2 = 0.4$ ft$^2$/hr), one slab 15 ft thick and the other 5 ft thick, are placed side by side. The thicker slab is initially at temperature 120°F, the thinner one at 30°F. The outside faces are to be held at 30°F for $t > 0$. Find the temperature throughout the slab for $t > 0$. Measure $x$ from the outer face of the thicker slab.

$$\text{ANS.} \quad u = 30 + \frac{180}{\pi} \sum_{n=1}^{\infty} \frac{1 - \cos(3n\pi/4)}{n} \exp\left(-\frac{n^2 t}{1000}\right) \sin\frac{n\pi x}{20}.$$

8. Two slabs of the same material, one 2 ft thick and the other 1 ft thick, are to be placed side by side. The thicker slab is initially at temperature $A$, the thinner one initially at zero. The outside faces are to be held at zero temperature for $t > 0$.

Find the temperature at the center of the 2-ft slab.

$$\text{ANS.} \quad \frac{4A}{\pi} \sum_{n=1}^{\infty} \frac{1}{n} \sin^3 \frac{n\pi}{3} \exp\left[ -\left(\frac{hn\pi}{3}\right)^2 t \right].$$

9. Knowing that the temperature function that is the answer to exercise 8 has the value $A$ at $t = 0$, show that

$$\sum_{k=0}^{\infty} \frac{(-1)^k (2k+1)}{(3k+1)(3k+2)} = \frac{2\pi}{9\sqrt{3}}.$$

10. By extending $f(x)$ in a proper way (see the example of this section), prove that, with an $f(x)$ defined in $0 < x < c$ and satisfying the conditions of the convergence theorem stated in Section 147, the right member in the expansion

$$f(x) \sim \sum_{k=0}^{\infty} B_k \sin \frac{(2k+1)\pi x}{2c}, \qquad \text{for } 0 < x < c,$$

in which

$$B_k = \frac{2}{c} \int_0^c f(x) \sin \frac{(2k+1)\pi x}{2c} \, dx,$$

represents the extended function in the sense of Section 147.

11. Interpret as a heat conduction problem and solve equation (2) of this section with the conditions that

As $x \to 0^+$, $\partial u / \partial x \to 0$,     for $0 < t$;
As $x \to c^-$, $u \to 0$,     for $0 < t$;
As $t \to 0^+$, $u \to f(x)$,     for $0 < x < c$.

12. Two metal rods of the same material, each of length $L$, have their sides insulated so that heat can flow only longitudinally. One rod is at temperature $A$, the other at temperature zero. At time $t = 0$ the rods are placed end to end as in Figure 56. The exposed end of the first rod is then insulated; the exposed end of the second rod is thereafter held at temperature $B$. Determine the temperature at the juncture of the rods for $t > 0$.

0             $L$             $2L$

FIGURE 56

## 154. Experimental verification of the validity of the heat equation

It is reassuring to have our mathematical formulas supported in comparison with observed phenomena and pleasant to see those same formulas

being of practical* value. Both experiences are encountered in the study of
the conduction of heat in concrete dams.

When concrete is poured a chemical reaction causes heat to be generated
in the material. Exposure to air temperatures cools the concrete, the inner
portions cooling more slowly than those near the surface. The temperature
differences create stresses and cause expansions and contractions. Because
of these facts it is customary when building a large dam to leave contraction
joints, openings to facilitate the safe expansion and contraction of the
concrete. After the concrete has lost most of its heat of setting, the dam is
grouted (the contraction joints filled) and the dam is ready for use so far as
the temperature problem is concerned.

The question of when the dam will be ready for grouting is a serious one
for the designer. If it is known that the waiting period would be extremely
long without special procedures, the concrete may be cooled as it is poured.
This was done with Boulder Dam (Hoover Dam), which was designed by
the United States Bureau of Reclamation. For Boulder Dam the waiting
period would have been one-and-a-half centuries; it is a large dam.

The temperature problem needs extensive idealization to bring it down to
the level for which decently computable solutions are known. Figure 57

FIGURE 57

shows a typical dam cross section; on it is indicated the thickness $c$ at a
random elevation. The designing engineers sometimes proceed to determine
the temperatures to be expected at various elevations by replacing the

* There is a story, perhaps a legend, that H. J. S. Smith, a mathematician of no mean standing,
once proposed the toast: "To pure mathematics, may it never be of use to anyone!". Many
mathematicians regard as pure any part of mathematics that seems to us worthy of study
primarily because of its inherent beauty, regardless of applicability to mundane affairs.

FIGURE 58

temperature problem for Figure 57 by that for the flat slab in Figure 58. The
width $c$ may be varied to approximate the thickness at various elevations.

The designer knows what the initial temperature of the concrete will be
(laboratory tests) and he knows what air temperatures to expect (United
States Weather Bureau). The solution of the heat problem for Figure 58, with
known initial temperature and variable surface temperatures, may be handled
by superposition of solutions (see the next section), using, for instance, the
successive mean monthly anticipated air temperatures. The designer can then
predict concrete temperatures across the dam at various elevations and from
them conclude when it would be safe to grout the dam.

Replacing Figure 58 by a wedge-shaped cross section to get closer to the
appearance of Figure 57 seems to make a negligible difference in the end
result, the predicted time for grouting.

During the construction of some large dams, thermocouples are installed
to permit future readings of temperature at various points in the dam. Years
later computations are made by the method described above and the computed temperature curves are compared to the observed ones. The results
are pleasant to behold. Frequently, the observed temperature history and the
predicted temperature agree within 2 or 3°F for years at a time, through the
gradual dissipation of the heat of setting and the nearly periodic fluctuations
due to seasonal changes in air temperatures.

## 155. Surface temperature varying with time

As indicated in the preceding section, a practical problem may force us to consider variable surface temperatures. A commonly encountered example is that of a slab initially at constant temperature $A_0$ and with surfaces maintained thereafter at the variable air temperature $A(t)$. We are thus led to consider the simple heat equation

$$\frac{\partial u}{\partial t} = h^2 \frac{\partial^2 u}{\partial x^2}, \qquad \text{for } 0 < t,\, 0 < x < c, \tag{1}$$

with conditions

$$\text{As } t \to 0^+,\, u \to A_0, \qquad \text{for } 0 < x < c; \tag{2}$$

$$\text{As } x \to 0^+,\, u \to A(t), \qquad \text{for } 0 < t; \tag{3}$$

$$\text{As } x \to c^-,\, u \to A(t), \qquad \text{for } 0 < t. \tag{4}$$

Here $A(t)$ represents surface (air) temperature as a function of time. We are concerned only with $A(t)$ of the form exhibited in Figure 59. It may

FIGURE 59

help to think of $A(t)$ as giving mean monthly predicted air temperature, $A_1$ over the first month

$$0 < t < t_1,$$

$A_2$ over the second month

$$t_1 < t < t_2,$$

and so on.

The function $A(t)$ may be expressed, using $t_0 = 0$, by

$$A(t) = A_n, \qquad \text{for } t_{n-1} < t < t_n;\, n = 1, 2, 3, \ldots. \tag{5}$$

For our purpose there is no need for the time intervals to be equal, but it is necessary that $A(t)$ be constant within each time interval. This problem is also amenable to treatment by Laplace transform methods, but will be solved by only one method in this book.

Since equation (1) is linear and homogeneous in $u$, any linear combination of solutions of (1) is also a solution. Since $t$ enters (1) only in $\partial u/\partial t$, any solution remains a solution under a translation of the time origin; that is, if $u(x, t)$ satisfies equation (1), then $u(x, t - t_n)$ satisfies (1) for any constant $t_n$.

The fundamental solution on which we base our treatment of the problem (1) through (4) is

$$F(x, t) = 1 - \frac{4}{\pi} \sum_{k=0}^{\infty} \frac{1}{2k + 1} \exp\left[-\frac{(2k + 1)^2 h^2 \pi^2 t}{c^2}\right] \sin \frac{(2k + 1)\pi x}{c}, \quad (6)$$

for $0 < t, 0 < x < c$. Directly, or by comparison with exercise 1, page 459, the function $F(x, t)$ of (6) can be seen to be a solution of the heat equation (1) and to have the properties:

$$\text{As } t \to 0^+, F(x, t) \to 0, \qquad \text{for } 0 < x < c; \quad (7)$$

$$\text{As } x \to 0^+, F(x, t) \to 1, \qquad \text{for } 0 < t; \quad (8)$$

$$\text{As } x \to c^-, F(x, t) \to 1, \qquad \text{for } 0 < t. \quad (9)$$

Since $F(x, t)$ in (6) was undefined for $t \leq 0$, let us take the liberty of defining it as identically zero whenever its second argument $t$ is nonpositive.

$$F(x, t) \equiv 0, \qquad \text{for } t \leq 0. \quad (10)$$

Then the desired solution of the problem (1) through (4) can be written at once as

$$u(x, t) = A_0 + \sum_{n=1}^{\infty} (A_n - A_{n-1})F(x, t - t_{n-1}), \qquad \text{for } 0 < t, 0 < x < c. \quad (11)$$

Note that the series on the right in (11) terminates for any specific $t$, since sooner or later (as $n$ increases) the argument $(t - t_{n-1})$ becomes and remains negative.

The $u$ of (11) is a linear combination of solutions of (1); it satisfies (2) because of (7). As $x \to 0^+$ or $x \to c^-$, in any range $t_{k-1} < t \leq t_k$,

$$u \to A_0 + \sum_{n=1}^{k} (A_n - A_{n-1}) = A_k,$$

because of (8) and (9) and the fact that $F(x, t - t_{n-1}) \equiv 0$ for $t \leq t_{n-1}$.

In actual practice, computations with the solution (11) are greatly simplified because the $F(x, t)$ of (6) is essentially the same for all diffusivities and all slab widths. In equation (6), put $h^2 \pi^2 t/c^2 = \tau$ and $\pi x/c = \zeta$. The manner in which the new arguments $\tau$ and $\zeta$ are chosen will be discussed in Section 166.

We may now write

$$F(x, t) = 1 - \frac{4}{\pi} \sum_{k=0}^{\infty} \frac{\exp\left[-(2k+1)^2\tau\right]\sin(2k+1)\zeta}{2k+1} = \phi(\zeta, \tau),$$

for $0 < \tau, 0 < \zeta < \pi$. The function $\phi$ can be tabulated at intervals of $\zeta$ and $\tau$. For a particular slab problem, values of $h^2$, $c$, $t$, and $x$ are used to compute the pertinent $\zeta$ and $\tau$ and the values of $\phi$ are read from the table.

## 156. Heat conduction in a sphere

Consider a solid sphere initially at a known temperature that depends only upon distance from the center of the sphere. Let the surface of the sphere be held at zero temperature for $t > 0$. We shall determine the temperature in the sphere for positive $t$ under the assumption that the heat equation

$$\frac{\partial u}{\partial t} = h^2\left(\frac{\partial^2 u}{\partial x^2} + \frac{\partial^2 u}{\partial y^2} + \frac{\partial^2 u}{\partial z^2}\right) \tag{1}$$

is valid.

Since the object under study is a sphere, we choose the origin at the center of the sphere and introduce spherical coordinates, related to $x$, $y$, $z$, by

$$x = \rho \sin \phi \cos \theta, \qquad y = \rho \sin \phi \sin \theta, \qquad z = \rho \cos \phi.$$

Then the heat equation (1) becomes

$$\frac{\partial u}{\partial t} = h^2\left(\frac{\partial^2 u}{\partial \rho^2} + \frac{2}{\rho}\frac{\partial u}{\partial \rho} + \frac{1}{\rho^2}\frac{\partial^2 u}{\partial \phi^2} + \frac{\cot\phi}{\rho^2}\frac{\partial u}{\partial \phi} + \frac{\csc^2\phi}{\rho^2}\frac{\partial^2 u}{\partial \theta^2}\right) \tag{2}$$

For the problem we wish to solve, the temperature is independent of the coordinates $\theta$ and $\phi$, so equation (2) reduces to

$$\frac{\partial u}{\partial t} = h^2\left(\frac{\partial^2 u}{\partial \rho^2} + \frac{2}{\rho}\frac{\partial u}{\partial \rho}\right). \tag{3}$$

Let $R$ be the radius of the sphere and $f(\rho)$ be the initial temperature. Then the problem confronting us is

$$\frac{\partial u}{\partial t} = h^2\left(\frac{\partial^2 u}{\partial \rho^2} + \frac{2}{\rho}\frac{\partial u}{\partial \rho}\right), \qquad \text{for } 0 < t, 0 < \rho < R; \tag{4}$$

$$\text{As } \rho \to R^-, u \to 0, \qquad \text{for } 0 < t; \tag{5}$$

$$\text{As } t \to 0^+, u \to f(\rho), \qquad \text{for } 0 \leqq \rho < R. \tag{6}$$

The student can easily show that the change of dependent variable,

$$u = \frac{v}{\rho}, \tag{7}$$

transforms the problem (4) through (6) into the problem

$$\frac{\partial v}{\partial t} = h^2 \frac{\partial^2 v}{\partial \rho^2}, \qquad \text{for } 0 < t, 0 < \rho < R; \tag{8}$$

$$\text{As } \rho \to R^-, v \to 0, \qquad \text{for } 0 < t; \tag{9}$$

$$\text{As } \rho \to 0^+, \frac{v}{\rho} \to \text{a limit}, \qquad \text{for } 0 < t; \tag{10}$$

$$\text{As } t \to 0^+, v \to \rho f(\rho), \qquad \text{for } 0 < \rho < R. \tag{11}$$

The added condition (10) is a reflection of the fact that the temperature $u$ is to exist at $\rho = 0$ in spite of relation (7). The new problem (8) through (11) is much like those treated at the beginning of this chapter. Its solution is left as an exercise.

The corresponding problem of finding the temperatures in a solid cylinder is less elementary and involves series of Bessel functions. It may be found worked out in many books.*

## Exercises

**1.** Solve the problem (4) through (6) by the method outlined above.

ANS.  $u = \dfrac{1}{\rho} \displaystyle\sum_{n=1}^{\infty} b_n \exp\left[-\left(\dfrac{hn\pi}{R}\right)^2 t\right] \sin\dfrac{n\pi\rho}{R}$, in which

$$b_n = \frac{2}{R} \int_0^R \rho f(\rho) \sin\frac{n\pi\rho}{R}\, d\rho.$$

**2.** A sphere of radius $R$ is initially at a constant temperature $u_0$ throughout, then has its surface held at temperature $u_1$ for $t > 0$. Find the temperature throughout the sphere for $t > 0$ and in particular the temperature $u_c$ at the center of the sphere.

ANS.  $u = u_1 + \dfrac{2R(u_0 - u_1)}{\rho\pi} \displaystyle\sum_{n=1}^{\infty} \dfrac{(-1)^{n+1}}{n} \exp\left[-\left(\dfrac{hn\pi}{R}\right)^2 t\right] \sin\dfrac{n\pi\rho}{R};$

$u_c = u_1 + 2(u_0 - u_1) \displaystyle\sum_{n=1}^{\infty} (-1)^{n+1} \exp\left[-\left(\dfrac{hn\pi}{R}\right)^2 t\right]$, for $t > 0$.

---

* See, for example, E. D. Rainville, *Intermediate Differential Equations*, 2nd ed., (New York: Macmillan Publishing Co., Inc., 1964), p. 279, or R. V. Churchill and J. W. Brown, *Fourier Series and Boundary Value Problems*, 3rd ed. (New York: McGraw-Hill Book Company, 1978).

## 157. The simple wave equation

If an elastic string held fixed at two points is taut and then is displaced from its equilibrium position and released, the subsequent displacements from the position of equilibrium may be determined by solving a boundary value problem. Figure 60 shows a representative displacement of the string,

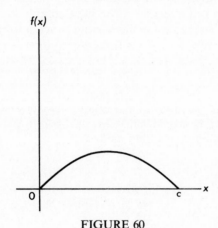

FIGURE 60

which is to be held fixed at $x = 0$ and $x = c$. The displacement $y$ for $0 < x < c$ and $0 < t$ is to be found from the known initial displacement $f(x)$, the initial velocity $\phi(x)$, and the fact that $y$ must satisfy the one-dimensional wave equation

$$\frac{\partial^2 y}{\partial t^2} = a^2 \frac{\partial^2 y}{\partial x^2}, \tag{1}$$

in which the parameter $a$ is a constant that depends upon the physical properties of the string.

The boundary value problem to be solved is

$$\frac{\partial^2 y}{\partial t^2} = a^2 \frac{\partial^2 y}{\partial x^2}, \qquad \text{for } 0 < x < c, 0 < t; \tag{2}$$

$$\text{As } x \to 0^+, y \to 0, \qquad \text{for } 0 < t; \tag{3}$$

$$\text{As } x \to c^-, y \to 0, \qquad \text{for } 0 < t; \tag{4}$$

$$\text{As } t \to 0^+, y \to f(x), \qquad \text{for } 0 < x < c; \tag{5}$$

$$\text{As } t \to 0^+, \frac{\partial y}{\partial t} \to \phi(x), \text{ for } 0 < x < c. \tag{6}$$

It is inherent in the string problem that $f(x)$ be continuous and that $f(0) = f(c) = 0$. Either $f(x)$ or $\phi(x)$ may be zero throughout the interval. Indeed, the boundary value problem (2) through (6) can always be replaced by two problems, one with $f(x)$ replaced by zero, the other with $\phi(x)$ replaced by zero. The sum of the solutions of those two problems is the solution of the problem with both an initial velocity and an initial displacement.

The solution of problems such as (2) through (6) with various $f(x)$ and $\phi(x)$ can be accomplished by the method of separation of variables and use of Fourier series, as was done with the heat conduction problems earlier. That work is left for exercises for the student, since it involves no new technique. Note the usefulness of the solutions determined in exercise 1, page 425.

## Exercises

In exercises 1 through 5, find the displacement for $t > 0$ of the vibrating string problem of this section under the condition that the initial velocity is to be zero and that the initial displacement is given by the $f(x)$ described.

**1.** $f(x) = x,$      for $\quad 0 \leq x \leq c/2,$
     $= c - x,$      for $c/2 \leq x \leq c.$

ANS.   $y = \dfrac{4c}{\pi^2} \sum\limits_{k=0}^{\infty} \dfrac{(-1)^k \cos\left[(2k+1)a\pi t/c\right] \sin\left[(2k+1)\pi x/c\right]}{(2k+1)^2}.$

**2.** $f(x) = x(c - x)/c.$    ANS.   $y = \dfrac{8c}{\pi^3} \sum\limits_{k=0}^{\infty} \dfrac{\cos\left[(2k+1)a\pi t/c\right] \sin\left[(2k+1)\pi x/c\right]}{(2k+1)^3}.$

**3.** $f(x) = x,$      for $\quad 0 \leq x \leq c/4,$
     $= c/4,$      for $\quad c/4 \leq x \leq 3c/4,$
     $= c - x,$      for $3c/4 \leq x \leq c.$    See Figure 61.

ANS.   $y = \dfrac{2c}{\pi^2} \sum\limits_{n=1}^{\infty} \dfrac{1}{n^2}\left(\sin\dfrac{n\pi}{4} + \sin\dfrac{3n\pi}{4}\right)\cos\dfrac{na\pi t}{c}\sin\dfrac{n\pi x}{c}.$

**4.** $f(x) = x,$      for $\quad 0 \leq x \leq \tfrac{1}{4}c,$
     $= \tfrac{1}{2}c - x,$      for $\quad \tfrac{1}{4}c \leq x \leq 3c/4,$
     $= x - c,$      for $3c/4 \leq x \leq c.$ See Figure 62.

ANS.   $y = \dfrac{2c}{\pi^2} \sum\limits_{k=0}^{\infty} \dfrac{(-1)^k \cos\left[(4k+2)\pi at/c\right] \sin\left[(4k+2)\pi x/c\right]}{(2k+1)^2}.$

**5.** $f(x) = x,$      for $0 \leq x \leq \tfrac{1}{4}c,$
     $= \tfrac{1}{2}c - x,$      for $\tfrac{1}{4}c \leq x \leq \tfrac{1}{2}c,$
     $= 0,$      for $\tfrac{1}{2}c \leq x \leq c.$

ANS.   $y = \dfrac{2c}{\pi^2} \sum\limits_{n=1}^{\infty} \dfrac{1}{n^2}(2\sin\tfrac{1}{4}n\pi - \sin\tfrac{1}{2}n\pi)\cos\dfrac{n\pi at}{c}\sin\dfrac{n\pi x}{c}.$

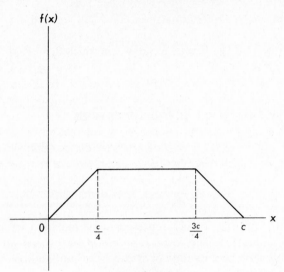

FIGURE 61

**6.** Find the displacement of the string of this section if the initial displacement is zero and the initial velocity is given by $\phi(x) = ax(c - x)/(4c^2)$.

ANS. $y = \dfrac{2c}{\pi^4} \sum\limits_{k=0}^{\infty} \dfrac{\sin\left[(2k + 1)a\pi t/c\right]\sin\left[(2k + 1)\pi x/c\right]}{(2k + 1)^4}$.

**7.** Find the displacement of the string of this section if the initial displacement is zero and the initial velocity is given by

$$\phi(x) = 0, \qquad \text{for} \quad 0 \leq x \leq c/3,$$
$$= v_0, \qquad \text{for} \quad c/3 \leq x \leq 2c/3,$$
$$= 0, \qquad \text{for} \quad 2c/3 \leq x \leq c.$$

ANS. $y = \dfrac{2v_0 c}{\pi^2 a} \sum\limits_{n=1}^{\infty} \dfrac{\left[\cos\left(n\pi/3\right) - \cos\left(2n\pi/3\right)\right]\sin\left(n\pi a t/c\right)\sin\left(n\pi x/c\right)}{n^2}$.

**8.** Solve the problem (2) through (6) of this section with $\phi(x) = 0$.

FIGURE 62

**9.** Solve the problem (2) through (6) of this section with $f(x) = 0$.

ANS.    $y = \sum_{n=1}^{\infty} B_n \sin \dfrac{n\pi at}{c} \sin \dfrac{n\pi x}{c}$, in which $B_n = \dfrac{2}{n\pi a} \int_0^c \phi(x) \sin \dfrac{n\pi x}{c} \, dx.$

## 158. Laplace's equation in two dimensions

We conclude this chapter with a discussion of Laplace's equation, in particular of the two-dimensional case

$$\frac{\partial^2 u}{\partial x^2} + \frac{\partial^2 u}{\partial y^2} = 0. \tag{1}$$

The dependent variable $u$ may represent any one of various quantities, steady-state temperature, electrostatic potential, and so on, although in this section we shall use the language of steady-state temperature problems for simplicity in working and visualization.

The Fourier series method as it is being utilized in this chapter is particularly well adapted to solving steady-state temperature problems for a flat rectangular plate. Let the two faces of the plate be insulated; let no heat flow in the direction normal to them. Then the problem is two-dimensional. Each edge of the plate may be either insulated or held at a known temperature.

Consider a flat rectangular plate with edges of length $a$ and $b$. Let three edges be kept at zero temperature and the remaining one, one of those of length $a$, be kept at a specified temperature, a function of distance along that edge. Let us choose a coordinate system and accompanying notation as shown in Figure 63.

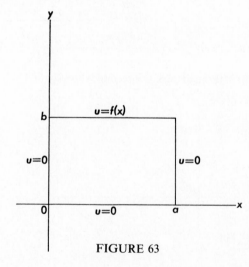

FIGURE 63

The steady-state temperature problem associated with Figure 63 may be written

$$\frac{\partial^2 u}{\partial x^2} + \frac{\partial^2 u}{\partial y^2} = 0, \qquad \text{for } 0 < x < a, 0 < y < b; \tag{2}$$

$$\text{As } x \to 0^+, u \to 0, \qquad \text{for } 0 < y < b; \tag{3}$$

$$\text{As } x \to a^-, u \to 0, \qquad \text{for } 0 < y < b; \tag{4}$$

$$\text{As } y \to 0^+, u \to 0, \qquad \text{for } 0 < x < a; \tag{5}$$

$$\text{As } y \to b^-, u \to f(x), \qquad \text{for } 0 < x < a. \tag{6}$$

The solution of the problem (2) through (6) is found by the same method as was used on boundary value problems for the one-dimensional heat equation. Results appear in the exercises below.

When a rectangular plate has nonzero temperatures at more than one edge, the problem can be separated into two or more problems of the type of (2) through (6). Insulation of an edge can be treated by doubling the size of the plate, with the insulated edge becoming the new center line and then creating temperature conditions symmetric about that center line so no heat can flow across it. See the methods used in Section 153 at the start of this chapter.

Next let us consider a flat circular plate of radius $R$ subjected to assigned fixed temperatures at its edge and left until it reaches the steady-state condition. Using cylindrical coordinates $r$, $\theta$, $z$ and conditions that permit no heat flow normal to the faces of the plate, we arrive at a two-dimensional problem,

$$\frac{\partial^2 u}{\partial r^2} + \frac{1}{r}\frac{\partial u}{\partial r} + \frac{1}{r^2}\frac{\partial^2 u}{\partial \theta^2} = 0, \qquad \text{for } 0 < r < R, 0 < \theta < 2\pi; \tag{7}$$

$$\text{As } r \to R^-, u \to f(\theta), \qquad \text{for } 0 \leqq \theta < 2\pi; \tag{8}$$

$$\lim_{r \to 0^+} u \text{ exists}, \qquad \text{for } 0 \leqq \theta < 2\pi; \tag{9}$$

$$\lim_{\theta \to 0^+} u = \lim_{\theta \to 2\pi^-} u, \qquad \text{for } 0 \leqq r < R. \tag{10}$$

If we are willing to define the edge temperature $f(\theta)$ for all $\theta$ and make it have the period $2\pi$, then the condition (10) can be replaced by the requirement that $u$ be a periodic function of $\theta$ with period $2\pi$.

The corresponding problems of a wedge, or of a portion of a wedge with concentric circular edges, do not involve any requirement of periodicity in their physical nature.

Separation of variables in equation (7) leads to a need for solving an ordinary equation of Euler–Cauchy type,

$$r^2\psi''(r) + r\psi'(r) - \alpha^2\psi(r) = 0, \tag{11}$$

where $\alpha$ is constant. See exercises 19 through 34, page 357.

## *Exercises*

Each of exercises 1 through 7 refers to the steady-state temperature problem for the rectangular plate of Figure 63, but with edge conditions as described in the individual exercise.

**1.** Edges $x = 0$, $x = a$, $y = 0$ held at zero temperature; the edge $y = b$ held at temperature $f(x)$.

ANS.   $u = \displaystyle\sum_{n=1}^{\infty} c_n \sinh \frac{n\pi y}{a} \sin \frac{n\pi x}{a}$, in which $c_n \sinh \dfrac{n\pi b}{a} = \dfrac{2}{a} \displaystyle\int_0^a f(x) \sin \frac{n\pi x}{a}\, dx$.

**2.** Edges $x = 0$, $x = a$, $y = 0$ held at zero; the edge $y = b$ held at temperature unity. Solve the problem directly or by using the result of exercise 1.

ANS.   $u = \dfrac{4}{\pi} \displaystyle\sum_{k=0}^{\infty} \frac{\sinh\left[(2k+1)\pi y/a\right] \sin\left[(2k+1)\pi x/a\right]}{(2k+1)\sinh\left[(2k+1)\pi b/a\right]}.$

**3.** Exercise 2 with the change that the edge $y = b$ is to be held at unity for $0 < x < a/2$ and at zero for $a/2 < x < a$.

ANS.   $u = \dfrac{2}{\pi} \displaystyle\sum_{n=1}^{\infty} \frac{\left[1 - \cos\left(n\pi/2\right)\right] \sinh\left(n\pi y/a\right) \sin\left(n\pi x/a\right)}{n \sinh\left(n\pi b/a\right)}.$

**4.** Edges $x = 0$ and $x = a$ held at zero; edge $y = 0$ insulated; edge $y = b$ held at temperature $f(x)$.

**5.** Edge $x = 0$ insulated; edges $x = a$ and $y = 0$ held at zero; edge $y = b$ held at temperature unity.

ANS.   $u = \dfrac{4}{\pi} \displaystyle\sum_{k=0}^{\infty} \frac{(-1)^k \sinh\left[(2k+1)\pi y/(2a)\right] \cos\left[(2k+1)\pi x/(2a)\right]}{(2k+1)\sinh\left[(2k+1)\pi b/(2a)\right]}.$

**6.** Edges $x = 0$ and $y = 0$ held at zero; edge $x = a$ insulated; edge $y = b$ held at temperature unity.

ANS.   $u = \dfrac{4}{\pi} \displaystyle\sum_{k=0}^{\infty} \frac{\sinh\left[(2k+1)\pi y/(2a)\right] \sin\left[(2k+1)\pi x/(2a)\right]}{(2k+1)\sinh\left[(2k+1)\pi b/(2a)\right]}.$

**7.** Edges $x = 0$ and $y = 0$ insulated; edge $x = a$ held at zero; edge $y = b$ held at temperature unity.

ANS.   $u = \dfrac{4}{\pi} \displaystyle\sum_{k=0}^{\infty} \frac{(-1)^k \cosh\left[(2k+1)\pi y/(2a)\right] \cos\left[(2k+1)\pi x/(2a)\right]}{(2k+1)\cosh\left[(2k+1)\pi b/(2a)\right]}.$

**8.** Show that the temperature at the center of the plate of exercise 2 above is

$$\frac{2}{\pi} \sum_{k=0}^{\infty} \frac{(-1)^k \operatorname{sech}\left[(2k+1)\pi b/(2a)\right]}{2k+1}.$$

**9.** For a square plate, show by superposition of solutions, without obtaining any solution explicitly, that when one face is held at temperature unity and the others are held at zero, then the temperature at the center is $\frac{1}{4}$. Then, by comparing your result with exercise 8, using $b = a$, conclude that

$$\sum_{k=0}^{\infty} \frac{(-1)^k \operatorname{sech}\left[(2k+1)\pi/2\right]}{2k+1} = \frac{\pi}{8}.$$

**10.** A circular plate has radius $R$. The edge $r = R$ of the plate is held at temperature unity for $0 < \theta < \pi$, at zero temperature for $\pi < \theta < 2\pi$. Find the temperature throughout the plate.

ANS.  $u = \dfrac{1}{2} + \dfrac{2}{\pi} \displaystyle\sum_{k=0}^{\infty} \left(\dfrac{r}{R}\right)^{2k+1} \dfrac{\sin(2k+1)\theta}{2k+1}.$

**11.** A plate with concentric circular boundaries $r = a$ and $r = b$, $0 < a < b$, has its inner boundary held at temperature $A$, its outer one at temperature $B$. Find the temperature throughout.

ANS.  $u = \dfrac{B \ln(r/a) - A \ln(r/b)}{\ln(b/a)}.$

**12.** The plate of exercise 11 has its inner edge $r = a$ insulated, its outer edge held at temperature unity for $0 < \theta < \pi$ and held at temperature zero for $\pi < \theta < 2\pi$. Find the temperature throughout.

ANS.  $u = \dfrac{1}{2} + \dfrac{2}{\pi} \displaystyle\sum_{k=0}^{\infty} \dfrac{(r/b)^{2k+1} + (a/r)^{2k+1}(a/b)^{2k+1}}{1 + (a/b)^{4k+2}} \cdot \dfrac{\sin(2k+1)\theta}{2k+1}.$

**13.** A flat wedge is defined in polar coordinates by the region $0 < r < R$, $0 < \theta < \beta$. Find the temperature throughout if the edges $\theta = 0$ and $\theta = \beta$ are held at temperature zero and the curved edge $r = R$ is held at temperature unity.

ANS.  $u = \dfrac{4}{\pi} \displaystyle\sum_{k=0}^{\infty} \left(\dfrac{r}{R}\right)^{(2k+1)\pi/\beta} \dfrac{\sin\left[(2k+1)\pi\theta/\beta\right]}{2k+1}.$

**14.** For the flat wedges of exercise 13, find the temperature if the edge $\theta = 0$ is held at zero, the edge $\theta = \beta$ is held at temperature unity, and the curved edge $r = R$ is held at temperature $f(\theta)$ for $0 < \theta < \beta$.

ANS.  $u = \theta/\beta + \displaystyle\sum_{n=1}^{\infty} A_n \left(\dfrac{r}{R}\right)^{n\pi/\beta} \sin\dfrac{n\pi\theta}{\beta},$  in which

$$A_n = \frac{2}{\beta} \int_0^{\beta} [f(\theta) - \theta/\beta] \sin\frac{n\pi\theta}{\beta}\, d\theta.$$

# Additional Properties of the Laplace Transform

## 159. Power series and inverse transforms

To use the Laplace transform on boundary value problems involving partial differential equations, we need certain transforms and inverse transforms that we did not obtain in Chapters 11 and 12.

Before proceeding to illustrative examples, we list certain elementary power series expansions for easy reference.

$$\frac{1}{1-x} = \sum_{n=0}^{\infty} x^n, \qquad |x| < 1; \tag{1}$$

$$e^x = \sum_{n=0}^{\infty} \frac{x^n}{n!}, \qquad \text{all } x; \tag{2}$$

$$\cos x = \sum_{n=0}^{\infty} \frac{(-1)^n x^{2n}}{(2n)!}, \qquad \text{all } x; \tag{3}$$

$$\sin x = \sum_{n=0}^{\infty} \frac{(-1)^n x^{2n+1}}{(2n+1)!}, \qquad \text{all } x; \tag{4}$$

$$\cosh x = \sum_{n=0}^{\infty} \frac{x^{2n}}{(2n)!}, \qquad \text{all } x; \tag{5}$$

$$\sinh x = \sum_{n=0}^{\infty} \frac{x^{2n+1}}{(2n+1)!}, \qquad \text{all } x; \tag{6}$$

$$\arctan x = \sum_{n=0}^{\infty} \frac{(-1)^n x^{2n+1}}{2n+1}, \qquad |x| < 1; \tag{7}$$

$$\frac{1}{(1-x)^m} = 1 + \sum_{n=1}^{\infty} \frac{m(m+1)\cdots(m+n-1)x^n}{n!}, \qquad |x| < 1; \tag{8}$$

$$\ln(1+x) = \sum_{n=0}^{\infty} \frac{(-1)^n x^{n+1}}{n+1}, \qquad |x| < 1; \tag{9}$$

$$\ln \frac{1+x}{1-x} = 2 \sum_{n=0}^{\infty} \frac{x^{2n+1}}{2n+1}, \qquad |x| < 1. \tag{10}$$

In seeking the Laplace transform, or the inverse transform, of a given function, we may find it inconvenient, difficult, or even beyond us to obtain the desired result by direct use of the theorems of Chapters 11 and 12 in a finite number of steps. Then we frequently turn to infinite series. If we can expand our function into a series such that we know how to obtain the desired transform or inverse transform of each term, we can thus solve our original problem.

EXAMPLE (a): Given that $L^{-1}\{f(s)\} = F(t)$, evaluate

$$L^{-1}\left\{ \frac{f(s)}{\sinh(cs)} \right\}.$$

We know that $\sinh z = \frac{1}{2}(e^z - e^{-z})$. Then

$$\frac{f(s)}{\sinh(cs)} = \frac{2f(s)}{e^{cs} - e^{-cs}}. \tag{11}$$

For $h > 0$, $s > 0$, we know how to evaluate $L^{-1}\{e^{-hs}f(s)\}$ by Theorem 22, page 208. Indeed,

$$L^{-1}\{e^{-hs}f(s)\} = F(t-h)\alpha(t-h), \qquad h > 0, s > 0. \tag{12}$$

We therefore rewrite (11) as

$$\frac{f(s)}{\sinh(cs)} = \frac{2f(s)e^{-cs}}{1 - e^{-2cs}} \tag{13}$$

because we can use the power series (1) to expand $(1 - e^{-2cs})^{-1}$ in a series of exponentials with negative arguments. From (1) we get

$$\frac{1}{1 - e^{-2cs}} = \sum_{n=0}^{\infty} \exp(-2ncs),$$

so, by (13),

$$\frac{f(s)}{\sinh(cs)} = 2 \sum_{n=0}^{\infty} f(s) \exp(-2ncs - cs). \tag{14}$$

We now use (12) to obtain, for $c > 0$, $s > 0$,

$$L^{-1}\frac{f(s)}{\sinh(cs)} = 2 \sum_{n=0}^{\infty} F(t - 2nc - c)\alpha(t - 2nc - c). \tag{15}$$

It is important to realize that the series on the right in (15) is a finite series. No matter how large the value of $t$ nor how small the (positive) $c$, the argument of the $\alpha$ function will become negative for sufficiently large $n$ and for all succeeding $n$ values. Thus each term of the series will be zero for all $n$ such that $(2n + 1)c > t$.

The procedure used in this example is of value to us in applications involving boundary value problems in partial differential equations, to be discussed in Chapter 26.

**EXAMPLE (b):** Evaluate $L\left\{\dfrac{1 - e^{-t}}{t}\right\}$.

By (2) we obtain

$$e^{-t} = \sum_{n=0}^{\infty} \frac{(-1)^n t^n}{n!} = 1 + \sum_{n=1}^{\infty} \frac{(-1)^n t^n}{n!}.$$

Therefore we may write

$$\frac{1 - e^{-t}}{t} = \sum_{n=1}^{\infty} \frac{(-1)^{n+1} t^{n-1}}{n!}.$$

A shift in index from $n$ to $(n + 1)$ yields

$$\frac{1 - e^{-t}}{t} = \sum_{n=0}^{\infty} \frac{(-1)^n t^n}{(n + 1)!}.$$

We know that $L\left\{\dfrac{t^n}{n!}\right\} = \dfrac{1}{s^{n+1}}$. Hence

$$L\left\{\frac{1 - e^{-t}}{t}\right\} = \sum_{n=0}^{\infty} \frac{(-1)^n}{(n + 1)s^{n+1}},$$

so comparison with (9) above yields

$$L\left\{\frac{1 - e^{-t}}{t}\right\} = \ln\left(1 + \frac{1}{s}\right), \qquad s > 0. \tag{16}$$

The restriction $s > 0$ may be obtained by examining the integral definition of the left member of (16). Note also the connection with exercise 19, page 193.

**EXAMPLE (c):** Evaluate $L^{-1}\left\{\ln\dfrac{s + 1}{s - 1}\right\}$.

From (10) we have

$$\ln\frac{s + 1}{s - 1} = \ln\frac{1 + \dfrac{1}{s}}{1 - \dfrac{1}{s}} = 2\sum_{n=0}^{\infty}\frac{1}{(2n + 1)s^{2n+1}}.$$

Now $L^{-1}\left\{\dfrac{1}{s^{2n+1}}\right\} = \dfrac{t^{2n}}{(2n)!}$. Hence

$$L^{-1}\left\{\ln\frac{s + 1}{s - 1}\right\} = 2\sum_{n=0}^{\infty}\frac{t^{2n}}{(2n + 1)!},$$

which, with the aid of (6), yields

$$L^{-1}\left\{\ln\frac{s + 1}{s - 1}\right\} = \frac{2}{t}\sinh t. \tag{17}$$

## Exercises

1. Evaluate $L\left\{\dfrac{\sin kt}{t}\right\}$.

ANS. $\arctan\dfrac{k}{s}, s > 0.$

2. Evaluate $L\left\{\dfrac{1 - \cos kt}{t}\right\}$.

ANS. $\frac{1}{2}\ln\left(1 + \dfrac{k^2}{s^2}\right), s > k > 0.$

3. Evaluate $L\left\{\dfrac{\sinh(kt)}{t}\right\}$.

ANS. $\frac{1}{2}\ln\dfrac{s + k}{s - k}, s > k > 0.$

4. Evaluate $L\left\{\dfrac{1 - \cosh(kt)}{t}\right\}$.

ANS. $\frac{1}{2}\ln\left(1 - \dfrac{k^2}{s^2}\right), s > k > 0.$

5. Evaluate $F(t) = L^{-1}\left\{\dfrac{1}{s^3(1 - e^{-2s})}\right\}$ and compute $F(5)$.

ANS. $F(t) = \frac{1}{2}\sum_{n=0}^{\infty}(t - 2n)^2\alpha(t - 2n); F(5) = 17.5.$

**6.** Evaluate $F(t) = L^{-1}\left\{\dfrac{1}{s^3 \cosh(2s)}\right\}$ and compute $F(12)$.

> ANS.   $F(t) = \displaystyle\sum_{n=0}^{\infty} (-1)^n (t - 4n - 2)^2 \alpha(t - 4n - 2);\ F(12) = 68.$

**7.** Let $\phi(t) = L^{-1}\left\{\dfrac{3}{s^4 \sinh(3s)}\right\}$. Compute $\phi(10)$.          ANS.   344.

**8.** Let $c > 0, s > 0$, and let $L^{-1}\{f(s)\} = F(t)$. Prove that

$$L^{-1}\left\{\frac{f(s)}{\cosh(cs)}\right\} = 2 \sum_{n=0}^{\infty} (-1)^n F(t - 2nc - c)\alpha(t - 2nc - c).$$

**9.** Let $c > 0, s > 0$, and let $L^{-1}\{f(s)\} = F(t)$. Prove that

$$L^{-1}\{f(s)\tanh(cs)\} = F(t) + 2 \sum_{n=1}^{\infty} (-1)^n F(t - 2nc)\alpha(t - 2nc).$$

**10.** Let $0 < x < 1$, where $x$ does not depend on $s$.
Find the inverse transform $y(x, t)$ of

$$\frac{4\, e^{xs}}{s^3(e^s + e^{-s})}$$

and then compute $y(\tfrac{1}{2}, 5)$, assuming continuity of $y$.          ANS.   $y(\tfrac{1}{2}, 5) = 28.5.$

**11.** In exercise 4, page 294, replace the alternating current element $E \sin \omega t$ by $EQ(t, c)$ in which $Q$ is the square-wave function of Figure 21, page 191.

> ANS.   $I(t) = \dfrac{E}{R} \exp\left(-\dfrac{t}{RC}\right) + \dfrac{2E}{R} \displaystyle\sum_{n=1}^{\infty} (-1)^n \exp\left(-\dfrac{t - nc}{RC}\right)\alpha(t - nc).$

**12.** In exercise 4, page 294, replace $E \sin \omega t$ by $EF(t)$ in which $F(t)$ is the half-wave rectification of $\sin \omega t$ as described in exercise 17, page 193.
ANS.   $I(t) =$

$$\frac{E}{\omega CZ^2} \sum_{n=0}^{\infty} \left[(-1)^n(\cos \omega t + \omega RC \sin \omega t) - \exp\left(-\frac{\omega t - n\pi}{\omega RC}\right)\right]\alpha\left(t - \frac{n\pi}{\omega}\right).$$

## 160.  The error function

The error function, abbreviated "erf," which was mentioned briefly in Section 23, page 81, is defined by

$$\text{erf } x = \frac{2}{\sqrt{\pi}} \int_0^x \exp(-\beta^2)\, d\beta. \tag{1}$$

This function arises in many ways. It is sometimes studied in elementary courses. We also encounter erf $x$ in evaluating inverse transforms of certain simple functions of $s$.

We know that $L^{-1}\{s^{-1/2}\} = (\pi t)^{-1/2}$ and therefore that

$$L^{-1}\left\{\frac{1}{\sqrt{s+1}}\right\} = \frac{e^{-t}}{\sqrt{\pi t}}.$$

Then the convolution theorem yields

$$L^{-1}\left\{\frac{1}{s\sqrt{s+1}}\right\} = \int_0^t 1 \cdot \frac{e^{-\beta}}{\sqrt{\pi\beta}}\, d\beta. \qquad (2)$$

On the right in (2) put $\sqrt{\beta} = \gamma$. Then $\beta^{-1/2}\, d\beta = 2\, d\gamma$ and we obtain

$$L^{-1}\left\{\frac{1}{s\sqrt{s+1}}\right\} = \frac{2}{\sqrt{\pi}}\int_0^{\sqrt{t}} \exp(-\gamma^2)\, d\gamma.$$

That is,

$$L^{-1}\left\{\frac{1}{s\sqrt{s+1}}\right\} = \operatorname{erf}(\sqrt{t}). \qquad (3)$$

A few basic properties of $\operatorname{erf} x$ are useful in our work and will now be obtained. Directly from the definition (1) it follows that the derivative of $\operatorname{erf} x$ is given by

$$\frac{d}{dx}\operatorname{erf} x = \frac{2}{\sqrt{\pi}}\exp(-x^2). \qquad (4)$$

From (1) and the power series for $\exp(-\beta^2)$ we get

$$\operatorname{erf} x = \frac{2}{\sqrt{\pi}}\sum_{n=0}^{\infty}\frac{(-1)^n x^{2n+1}}{(2n+1)n!}. \qquad (5)$$

In elementary calculus we found that

$$\int_0^{\infty}\exp(-\beta^2)\, d\beta = \frac{\sqrt{\pi}}{2}. \qquad (6)$$

From (6) we get

$$\lim_{x\to\infty}\operatorname{erf} x = 1. \qquad (7)$$

The values of $\operatorname{erf} x$ are easily computed for small $x$ from (5) above and for larger $x$ from the asymptotic expansion*

$$\operatorname{erf} x \sim 1 - \frac{\exp(-x^2)}{\sqrt{\pi}}\sum_{n=0}^{\infty}\frac{(-1)^n[1\cdot 3\cdot 5\cdots(2n-1)]}{2^n x^{2n+1}}. \qquad (8)$$

---

* See, for example, E. D. Rainville, *Special Functions* (New York: Macmillan Publishing Co., Inc., 1960), pp. 36–38. The function $\operatorname{erf} x$ is tabulated under the name "The Probability Integral," in B. O. Peirce and R. M. Foster, *A Short Table of Integrals*, 4th ed. (Lexington, Mass.: Ginn and Co., 1956), pp. 128–132.

It is convenient in our work to use what is called the complementary error function, denoted by erfc $x$ and defined by

$$\text{erfc } x = 1 - \text{erf } x, \tag{9}$$

which means also that

$$\text{erfc } x = \frac{2}{\sqrt{\pi}} \int_x^\infty \exp\left(-\beta^2\right) d\beta. \tag{10}$$

The properties of erf $x$ are readily converted to properties of erfc $x$. It is important that for any fixed $m$,

$$\lim_{x \to \infty} x^m \text{ erfc } x = 0, \tag{11}$$

which the student can demonstrate by considering the indeterminate form

$$\frac{\text{erfc } x}{x^{-m}}$$

and using the derivative of erfc $x$ as obtained from (4) above. See the exercises at the end of this section for other properties of erf $x$ and erfc $x$.

A transform that is important in certain applications (Sections 165–168) is

$$L\left\{\text{erfc}\left(\frac{k}{\sqrt{t}}\right)\right\}$$

in which $k$ is to be independent of $t$ and $k > 0$.

By the definition of erfc $x$ we have

$$\text{erfc}\left(\frac{k}{\sqrt{t}}\right) = \frac{2}{\sqrt{\pi}} \int_{k/\sqrt{t}}^\infty \exp\left(-\beta^2\right) d\beta. \tag{12}$$

In (12) put $\beta = k/\sqrt{v}$ so that the limits of integration become $v = t$ to $v = 0$. Since $d\beta = -\frac{1}{2}kv^{-3/2}\, dv$, we obtain (using the minus sign to reverse the order of integration)

$$\text{erfc}\left(\frac{k}{\sqrt{t}}\right) = \frac{k}{\sqrt{\pi}} \int_0^t v^{-3/2} \exp\left(-\frac{k^2}{v}\right) dv. \tag{13}$$

The integral on the right in (13) is a convolution integral. Hence

$$L\left\{\text{erfc}\left(\frac{k}{\sqrt{t}}\right)\right\} = \frac{k}{\sqrt{\pi}} L\{1\} \cdot L\left\{t^{-3/2} \exp\left(-\frac{k^2}{t}\right)\right\},$$

or

$$L\left\{\text{erfc}\left(\frac{k}{\sqrt{t}}\right)\right\} = \frac{k}{s\sqrt{\pi}} L\left\{t^{-3/2} \exp\left(-\frac{k^2}{t}\right)\right\}. \tag{14}$$

Now let

$$A(s) = L\left\{t^{-3/2}\exp\left(-\frac{k^2}{t}\right)\right\}. \tag{15}$$

Note that the functions $t^m \exp(-k^2/t)$ are of class $A$, page 181, for each $m$. From (15) it follows, by Theorem 16, page 186, that

$$\frac{dA}{ds} = L\left\{-t^{-1/2}\exp\left(-\frac{k^2}{t}\right)\right\} \tag{16}$$

and

$$\frac{d^2A}{ds^2} = L\left\{t^{1/2}\exp\left(-\frac{k^2}{t}\right)\right\}. \tag{17}$$

But also, by Theorem 12, page 184,

$$L\left\{\frac{d}{dt}t^{1/2}\exp\left(-\frac{k^2}{t}\right)\right\} = sL\left\{t^{1/2}\exp\left(-\frac{k^2}{t}\right)\right\} - \lim_{t\to 0^+}\left[t^{1/2}\exp\left(-\frac{k^2}{t}\right)\right],$$

or

$$L\left\{\tfrac{1}{2}t^{-1/2}\exp\left(-\frac{k^2}{t}\right) + k^2 t^{-3/2}\exp\left(-\frac{k^2}{t}\right)\right\} = sL\left\{t^{1/2}\exp\left(-\frac{k^2}{t}\right)\right\} - 0. \tag{18}$$

Because of (15), (16), and (17), equation (18) may be written

$$-\frac{1}{2}\frac{dA}{ds} + k^2 A = s\frac{d^2A}{ds^2}.$$

Therefore the desired function $A(s)$ is a solution of the differential equation

$$s\frac{d^2A}{ds^2} + \frac{1}{2}\frac{dA}{ds} - k^2 A = 0. \tag{19}$$

We need two boundary conditions to go with equation (19). We know that as $s \to \infty$, $A \to 0$. Now consider what happens as $s \to 0^+$.

By (15),

$$\lim_{s\to 0^+} A(s) = \lim_{s\to 0^+}\int_0^\infty e^{-st}t^{-3/2}\exp\left(-\frac{k^2}{t}\right)dt$$

$$= \int_0^\infty t^{-3/2}\exp\left(-\frac{k^2}{t}\right)dt.$$

Equation (13) yields (with $y$ replacing $t$)

$$\int_0^y v^{-3/2}\exp\left(-\frac{k^2}{v}\right)dv = \frac{\sqrt{\pi}}{k}\operatorname{erfc}\left(\frac{k}{\sqrt{y}}\right). \tag{20}$$

Therefore

$$\lim_{s \to 0^+} A(s) = \frac{\sqrt{\pi}}{k} \lim_{y \to \infty} \operatorname{erfc}\left(\frac{k}{\sqrt{y}}\right) = \frac{\sqrt{\pi}}{k} \operatorname{erfc} 0 = \frac{\sqrt{\pi}}{k}.$$

To get the general solution of the differential equation (19), we change independent variable* from $s$ to $z = \sqrt{s}$. Now by the chain rule of elementary calculus,

$$\frac{dA}{ds} = \frac{dz}{ds}\frac{dA}{dz} = \frac{1}{2\sqrt{s}}\frac{dA}{dz} = \frac{1}{2z}\frac{dA}{dz},$$

and

$$\frac{d^2A}{ds^2} = \frac{1}{4s}\frac{d^2A}{dz^2} - \frac{1}{4s\sqrt{s}}\frac{dA}{dz}.$$

Thus

$$s\frac{d^2A}{ds^2} = \frac{1}{4}\frac{d^2A}{dz^2} - \frac{1}{4z}\frac{dA}{dz}$$

and equation (19) becomes

$$\frac{d^2A}{dz^2} - 4k^2A = 0. \tag{21}$$

The general solution of (21) is

$$A = b_1 \exp(-2kz) + b_2 \exp(2kz),$$

so the general solution of (19) is

$$A = b_1 \exp(-2k\sqrt{s}) + b_2 \exp(2k\sqrt{s}). \tag{22}$$

We must determine the constants $b_1$ and $b_2$ from the conditions that $A \to 0$ as $s \to \infty$ and $A \to \sqrt{\pi}/k$ as $s \to 0^+$. As $s \to \infty$, $A$ will not approach a limit unless $b_2 = 0$. Then, letting $s \to 0^+$, we get

$$\frac{\sqrt{\pi}}{k} = b_1.$$

Therefore

$$A(s) = L\left\{t^{-3/2} \exp\left(-\frac{k^2}{t}\right)\right\} = \frac{\sqrt{\pi}}{k} \exp(-2k\sqrt{s}).$$

---

* Such a change of variable is dictated by the test on page 16 of E. D. Rainville, *Intermediate Differential Equations*, 2nd ed. (New York: Macmillan Publishing Co., Inc., 1964).

We return to (14) to write the desired transform

$$L\left\{\text{erfc}\left(\frac{k}{\sqrt{t}}\right)\right\} = \frac{1}{s}\exp\left(-2k\sqrt{s}\right), \qquad k > 0, s > 0. \tag{23}$$

We shall use (23) in the form

$$L^{-1}\left\{\frac{1}{s}\exp\left(-2k\sqrt{s}\right)\right\} = \text{erfc}\left(\frac{k}{\sqrt{t}}\right), \qquad k > 0, s > 0. \tag{24}$$

In Chapter 26 it will be important to combine the use of equation (24) and the series methods of Section 159.

Consider the problem of obtaining

$$L^{-1}\left\{\frac{\sinh\left(x\sqrt{s}\right)}{s\sinh\sqrt{s}}\right\}, \qquad 0 < x < 1, s > 0. \tag{25}$$

If $x$ were greater than unity, the inverse in (25) would not exist because of the behavior of $\sinh\left(x\sqrt{s}\right)/\sinh\sqrt{s}$ as $s \to \infty$.

Because we know (24), it is wise to turn to exponentials. We write

$$\frac{\sinh\left(x\sqrt{s}\right)}{\sinh\sqrt{s}} = \frac{\exp\left(x\sqrt{s}\right) - \exp\left(-x\sqrt{s}\right)}{\exp\left(\sqrt{s}\right) - \exp\left(-\sqrt{s}\right)}. \tag{26}$$

As in Section 159 we seek a series involving exponentials of negative argument. We therefore multiply numerator and denominator on the right in (26) by $\exp\left(-\sqrt{s}\right)$ and find that

$$\frac{\sinh\left(x\sqrt{s}\right)}{\sinh\sqrt{s}} = \frac{\exp\left[-(1-x)\sqrt{s}\right] - \exp\left[-(1+x)\sqrt{s}\right]}{1 - \exp\left(-2\sqrt{s}\right)}. \tag{27}$$

Now

$$\frac{1}{1 - \exp\left(-2\sqrt{s}\right)} = \sum_{n=0}^{\infty}\exp\left(-2n\sqrt{s}\right). \tag{28}$$

Therefore,

$$\frac{\sinh\left(x\sqrt{s}\right)}{s\sinh\sqrt{s}} = \sum_{n=0}^{\infty}\frac{1}{s}\{\exp\left[-(1-x+2n)\sqrt{s}\right] - \exp\left[-(1+x+2n)\sqrt{s}\right]\}.$$

For $0 < x < 1$ the exponentials have negative arguments and we may use (24) to conclude that

$$L^{-1}\left\{\frac{\sinh\left(x\sqrt{s}\right)}{s\sinh\sqrt{s}}\right\} = \sum_{n=0}^{\infty}\left[\text{erfc}\left(\frac{1-x+2n}{2\sqrt{t}}\right) - \text{erfc}\left(\frac{1+x+2n}{2\sqrt{t}}\right)\right]. \tag{29}$$

## Exercises

1. Show that for all real $x$, $|\text{erf } x| < 1$.

2. Show that erf $x$ is an odd function of $x$.

3. Show that $\displaystyle\lim_{x\to 0}\frac{\text{erf } x}{x} = \frac{2}{\sqrt{\pi}}$.

4. Use integration by parts to show that

$$\int_0^x \text{erf } y \, dy = x \, \text{erf } x - \frac{1}{\sqrt{\pi}}[1 - \exp(-x^2)].$$

5. Obtain equation (11), page 494.

6. Start with the power series for erf $x$, equation (5), page 493, and show that

$$L\{t^{-1/2}\,\text{erf}\,(\sqrt{t})\} = \frac{2}{\sqrt{\pi s}}\,\text{arc tan}\,\frac{1}{\sqrt{s}}, \qquad s > 0.$$

7. Use the fact that

$$\frac{1}{1 + \sqrt{1 + s}} = \frac{1 - \sqrt{1 + s}}{1 - (1 + s)} = -\frac{1}{s} + \frac{\sqrt{1 + s}}{s} = -\frac{1}{s} + \frac{1 + s}{s\sqrt{1 + s}}$$

and equation (3), page 493, to show that

$$L^{-1}\left\{\frac{1}{1 + \sqrt{1 + s}}\right\} = -1 + \text{erf}\,(\sqrt{t}) + \frac{e^{-t}}{\sqrt{\pi t}} = \frac{e^{-t}}{\sqrt{\pi t}} - \text{erfc}\,(\sqrt{t}).$$

8. Use equation (3), page 493, to conclude that

$$L^{-1}\left\{\frac{1}{(s - 1)\sqrt{s}}\right\} = e^t\,\text{erf}\,(\sqrt{t})$$

and therefore that

$$L^{-1}\left\{\frac{1}{\sqrt{s}(\sqrt{s} + 1)}\right\} = e^t\,\text{erfc}\,(\sqrt{t}).$$

9. Evaluate $L^{-1}\left\{\dfrac{1}{\sqrt{s} + 1}\right\}$.        ANS.  $\dfrac{1}{\sqrt{\pi t}} - e^t\,\text{erfc}\,(\sqrt{t})$.

10. Evaluate $L^{-1}\left\{\dfrac{1}{\sqrt{s} - 1}\right\}$.        ANS.  $\dfrac{1}{\sqrt{\pi t}} + e^t + e^t\,\text{erf}\,(\sqrt{t})$.

11. Define the function $\phi(t)$ by

$$\phi(t) = L^{-1}\left\{\text{erf}\,\frac{1}{s}\right\}.$$

Prove that

$$L\{\phi(\sqrt{t})\} = \frac{2}{\sqrt{\pi s}}\,\sin\frac{1}{\sqrt{s}}.$$

**12.** Show that for $x > 0$,

$$L^{-1}\left\{\frac{\operatorname{sech} x\sqrt{s}}{s}\right\} = 2\sum_{n=0}^{\infty} (-1)^n \operatorname{erfc}\left[\frac{(2n+1)x}{2\sqrt{t}}\right].$$

**13.** Show that for $x > 0$,

$$L^{-1}\left\{\frac{\operatorname{csch} x\sqrt{s}}{s}\right\} = 2\sum_{n=0}^{\infty} \operatorname{erfc}\left[\frac{(2n+1)x}{2\sqrt{t}}\right].$$

**14.** Derive the result

$$A(s) = L\left\{t^{-3/2}\exp\left(-\frac{k^2}{t}\right)\right\} = \frac{\sqrt{\pi}}{k}\exp(-2k\sqrt{s}), \qquad k > 0, s > 0$$

directly from the definition of a transform. In the integral

$$A(s) = \int_0^\infty \exp(-st - k^2 t^{-1}) t^{-3/2}\, dt$$

put $\beta = \sqrt{t}$ to get

$$A(s) = 2\int_0^\infty \beta^{-2}\exp(-s\beta^2 - k^2\beta^{-2})\, d\beta$$

or

$$A(s) = 2\exp(-2k\sqrt{s})\int_0^\infty \beta^{-2}\exp[-(\beta\sqrt{s} - k\beta^{-1})^2]\, d\beta.$$

Show that

$$\frac{dA}{ds} = -2\int_0^\infty \exp(-s\beta^2 - k^2\beta^{-2})\, d\beta$$

$$= -2\exp(-2k\sqrt{s})\int_0^\infty \exp[-(\beta\sqrt{s} - k\beta^{-1})^2]\, d\beta.$$

Thus arrive at the differential equation

$$\sqrt{s}\frac{dA}{ds} - kA = -2\sqrt{\pi}\exp(-2k\sqrt{s})$$

and from it obtain the desired function $A(s)$.

## 161. Bessel functions

The Bessel function

$$J_n(z) = \sum_{k=0}^{\infty} \frac{(-1)^k(\tfrac{1}{2}z)^{2k+n}}{k!\,\Gamma(k+n+1)}, \tag{1}$$

of the first kind and of index $n$, appeared in Sections 124 and 125. We meet $J_n(z)$ in a simple application of the series technique of Section 159. If we can expand a given function of $s$ in negative powers of $s$, surely we can get the inverse transform term by term. A simple example is the following:

$$\frac{1}{s}\exp\left(-\frac{x}{s}\right) = \sum_{k=0}^{\infty} \frac{(-1)^k x^k}{k!s^{k+1}},$$

which leads immediately to

$$L^{-1}\left\{\frac{1}{s}\exp\left(-\frac{x}{s}\right)\right\} = \sum_{k=0}^{\infty} \frac{(-1)^k x^k t^k}{k!k!}. \tag{2}$$

When $n = 0$ in (1) we get, since $\Gamma(k + 1) = k!$,

$$J_0(z) = \sum_{k=0}^{\infty} \frac{(-1)^k (\frac{1}{2}z)^{2k}}{k!k!}. \tag{3}$$

By comparing (2) with (3) we get

$$L^{-1}\left\{\frac{1}{s}\exp\left(-\frac{x}{s}\right)\right\} = J_0(2\sqrt{xt}); \qquad x > 0, s > 0. \tag{4}$$

From

$$\frac{1}{s^{n+1}}\exp\left(-\frac{x}{s}\right) = \sum_{k=0}^{\infty} \frac{(-1)^k x^k}{k!s^{k+n+1}}$$

we get

$$L^{-1}\left\{\frac{1}{s^{n+1}}\exp\left(-\frac{x}{s}\right)\right\} = \sum_{k=0}^{\infty} \frac{(-1)^k x^k t^{k+n}}{k!\Gamma(k + n + 1)}$$

$$= x^{-\frac{1}{2}n}t^{\frac{1}{2}n} \sum_{k=0}^{\infty} \frac{(-1)^k(\sqrt{xt})^{2k+n}}{k!\Gamma(k + n + 1)}.$$

Therefore, at least for $n \geqq 0$,

$$L^{-1}\left\{\frac{1}{s^{n+1}}\exp\left(-\frac{x}{s}\right)\right\} = \left(\frac{t}{x}\right)^{\frac{1}{2}n} J_n(2\sqrt{xt}), \qquad s > 0, x > 0. \tag{5}$$

With more knowledge of the Gamma function, we could use series methods to obtain the transform of $J_n(xt)$ for general $n$. Here we restrict ourselves to $n = 0$ for simplicity.

From (1) we obtain

$$J_0(xt) = \sum_{k=0}^{\infty} \frac{(-1)^k(\frac{1}{2}x)^{2k}t^{2k}}{k!k!}.$$

Then

$$L\{J_0(xt)\} = \sum_{k=0}^{\infty} \frac{(-1)^k(\frac{1}{2}x)^{2k}(2k)!}{k!k!s^{2k+1}}.$$

But $(2k)! = 2^k k![1 \cdot 3 \cdot 5 \cdots (2k - 1)]$. Hence

$$L\{J_0(xt)\} = \frac{1}{s}\left[1 + \sum_{k=1}^{\infty} \frac{(-1)^k[1 \cdot 3 \cdot 5 \cdots (2k-1)]x^{2k}}{2^k \cdot k!s^{2k}}\right],$$

or

$$L\{J_0(xt)\} = \frac{1}{s}\left(1 + \frac{x^2}{s^2}\right)^{-1/2}.$$

Therefore

$$L\{J_0(xt)\} = \frac{1}{\sqrt{s^2 + x^2}}. \tag{6}$$

From (1) it is easy to conclude that

$$\frac{d}{dz}J_0(z) = -J_1(z).$$

Then

$$\frac{d}{dt}J_0(xt) = -xJ_1(xt)$$

and we obtain

$$L\{-xJ_1(xt)\} = L\left\{\frac{d}{dt}J_0(xt)\right\}$$

$$= sL\{J_0(xt)\} - J_0(0).$$

But $J_0(0) = 1$, so

$$L\{-xJ_1(xt)\} = \frac{s}{\sqrt{s^2 + x^2}} - 1,$$

or

$$L\{J_1(xt)\} = \frac{\sqrt{s^2 + x^2} - s}{x\sqrt{s^2 + x^2}}. \tag{7}$$

## Exercise

1. The modified Bessel function of the first kind and of index $n$ is

$$I_n(z) = \sum_{k=0}^{\infty} \frac{(\frac{1}{2}z)^{2k+n}}{k!\Gamma(k + n + 1)}.$$

Show that

$$L^{-1}\left\{\frac{1}{s^{n+1}}\exp\left(\frac{x}{s}\right)\right\} = \left(\frac{t}{x}\right)^{\frac{1}{2}n} I_n(2\sqrt{xt}).$$

## 162. Differential equations with variable coefficients

Any reader who has become overly optimistic about the efficacy of the Laplace transform as a tool in treating linear differential equations should keep in mind that we have restricted our work so far to equations with constant coefficients.

Suppose that we are confronted with an initial value problem involving the equation

$$F''(t) + t^2 F(t) = 0. \tag{1}$$

Let $L\{F(t)\} = f(s)$ and put $F(0) = A$, $F'(0) = B$. Then application of the operator $L$ transforms equation (1) into

$$s^2 f(s) - sA - B + \frac{d^2}{ds^2} f(s) = 0,$$

or

$$f''(s) + s^2 f(s) = As + B. \tag{2}$$

The problem of getting the complementary function for equation (2) is the same as it is for equation (1); no progress has been made. The left member of (1) remained essentially unchanged under the Laplace transformation.

The behavior of (1) under $L$ is not unique. Indeed, the differential equations with polynomial coefficients that remain invariant under the Laplace transformation have been classified.*

Since $L\{t^n F(t)\} = (-1)^n (d^n/ds^n) f(s)$, it follows that the operator $L$ can be used to transform one differential equation with polynomial coefficients into another differential equation with polynomial coefficients and that the order of the new equation will equal the maximum degree of the polynomial coefficients in the original equation. The Laplace transform is simply not the proper tool for attacking differential equations with variable coefficients. For such a purpose, the classical method of solution by power series is a good tool to use.

---

*E. D. Rainville, Linear differential invariance under an operator related to the Laplace transformation, *Amer. J. Math.*, **62**:391–405 (1940).

# Partial Differential Equations; Transform Methods

### 163. Boundary value problems

For some boundary value problems involving partial differential equations, the Laplace transform provides an effective method of attack; for other problems the transform method contributes additional information even when the older techniques, such as separation of variables and Fourier series, may be easier to use. There remain problems for which the Laplace transform method contributes nothing but complications.

In this chapter we present a few applications and a detailed study of the solution of some simple problems. Our goal is to give the student sufficient background to enable him to use the Laplace transform on problems he encounters in practice and to give him some criteria to use in deciding whether the transform method is an appropriate tool for a given problem.

We first solve some artificial problems that have been constructed to exhibit the technique and underlying ideas without introducing the complexities common to many physical applications. The student who fully understands and can execute the solutions of such simple problems will find no difficulty, other than an increase in amount of labor, in solving corresponding problems arising in physical situations.

EXAMPLE:    Solve the problem consisting of the equation

$$\frac{\partial^2 y}{\partial x^2} = 16\frac{\partial^2 y}{\partial t^2}, \qquad \text{for } t > 0, x > 0; \tag{1}$$

with the conditions

$$t \to 0^+, y \to 0, \qquad \text{for } x > 0; \tag{2}$$

$$t \to 0^+, \frac{\partial y}{\partial t} \to -1, \qquad \text{for } x > 0; \tag{3}$$

$$x \to 0^+, y \to t^2, \qquad \text{for } t > 0; \tag{4}$$

$$\lim_{x \to \infty} y(x, t) \text{ exists}, \qquad \text{for fixed } t > 0. \tag{5}$$

The characteristics of the problem that suggest it is worthwhile to try the Laplace transform technique are:

(a) The differential equation is linear (necessary).
(b) The equation has constant coefficients (highly desirable).
(c) At least one independent variable has the range 0 to $\infty$ (highly desirable).
(d) There are appropriate initial ($t = 0$) conditions involving the independent variable in (c) above (desirable).

In this problem the independent variable $x$ also has the range 0 to $\infty$, but there is only one condition at $x = 0$; two conditions are needed in transforming a second derivative. We shall therefore attack this problem with Laplace transforms with respect to the variable $t$.
    Let

$$L\{y(x, t)\} = w(x, s), \tag{6}$$

in which $x$ is treated as a constant (parameter) as far as the Laplace transformation is concerned. Since we shall verify our solution, there is no risk in assuming that the operations of differentiations with respect to $x$ and Laplace transforms with respect to $t$ are commutative.
    Because (1) has constant coefficients, derivatives with respect to the transform variable $s$ will not appear. The partial differential equation (1) will be transformed into an ordinary differential equation with independent variable $x$ and with $s$ involved as a parameter. In view of (6), application of the operator $L$ transforms (1), (2), and (3) into

$$\frac{d^2 w}{dx^2} = 16(s^2 w + 1), \qquad x > 0. \tag{7}$$

The conditions (4) and (5) become

$$x \to 0^+, \ w \to \frac{2}{s^3}, \tag{8}$$

$$\lim_{x \to \infty} w(x, s) \text{ exists.} \tag{9}$$

We now solve the new problem, (7), (8), and (9), for $w(x, s)$ and then obtain $y(x, t)$ as the inverse transform of $w$. Let us rewrite (7) in the form

$$\frac{d^2 w}{dx^2} - 16s^2 w = 16 \tag{10}$$

and keep in mind that $x$ is the independent variable and $s$ is a parameter. When we get the general solution of (10), the arbitrary constants in it may well be functions of $s$; they must not involve $x$.

The general solution of (10) should be found by inspection. It is

$$w = -\frac{1}{s^2} + c_1(s) \exp(-4sx) + c_2(s) \exp(4sx), \qquad x > 0, s > 0. \tag{11}$$

Because of (9), the $w$ of (11) is to approach a limit as $x \to \infty$. The first two terms on the right in (11) approach limits as $x \to \infty$, but the term with the positive exponent, $\exp(4sx)$, will not do so unless

$$c_2(s) \equiv 0. \tag{12}$$

That is, (9) forces (12) upon us. The $w$ of (11) then becomes

$$w = -\frac{1}{s^2} + c_1(s) \exp(-4sx), \qquad x > 0, s > 0. \tag{13}$$

Application of condition (8) to the $w$ of (13) yields

$$\frac{2}{s^3} = c_1(s) - \frac{1}{s^2}; \qquad c_1(s) = \frac{2}{s^3} + \frac{1}{s^2}.$$

Thus we find that

$$w(x, s) = -\frac{1}{s^2} + \left( \frac{2}{s^3} + \frac{1}{s^2} \right) \exp(-4sx), \qquad x > 0, s > 0. \tag{14}$$

We already know that if

$$L^{-1}\{f(s)\} = F(t),$$

$$L^{-1}\{e^{-cs}f(s)\} = F(t - c)\alpha(t - c). \tag{15}$$

Therefore the application of the operator $L^{-1}$ throughout (14) gives us

$$y(x, t) = -t + [(t - 4x)^2 + (t - 4x)]\alpha(t - 4x), \qquad x > 0, t > 0. \tag{16}$$

It is our contention that the $y$ of (16) satisfies the boundary value problem (1) through (5). Let us now verify the solution in detail.

From (16) it follows that

$$\frac{\partial y}{\partial t} = -1 + [2(t - 4x) + 1]\alpha(t - 4x), \qquad x > 0, t > 0, t \neq 4x. \qquad (17)$$

Note the discontinuity in the derivative for $t = 4x$. This is forcing us into the admission that we obtain a solution of the problem only on each side of the line $t = 4x$ in the first quadrant of the $xt$ plane. Our $y$ will not satisfy the differential equation along that line because the second derivative cannot exist there. This is a reflection of the fact that (1) is a "hyperbolic differential equation." Whether the "solution" does or does not satisfy the differential equation along what are called the characteristic lines of the equation depends upon the specific boundary conditions. We shall treat each problem individually with no attempt to examine the general situation.

From (17) we obtain

$$\frac{\partial^2 y}{\partial t^2} = 2\alpha(t - 4x), \qquad x > 0, t > 0, t \neq 4x. \qquad (18)$$

Equation (16) also yields

$$\frac{\partial y}{\partial x} = [-8(t - 4x) - 4]\alpha(t - 4x), \qquad x > 0, t > 0, t \neq 4x, \qquad (19)$$

and

$$\frac{\partial^2 y}{\partial x^2} = 32\alpha(t - 4x), \qquad x > 0, t > 0, t \neq 4x. \qquad (20)$$

Equations (18) and (20) combine to show that the $y$ of (16) is a solution of the differential equation (1) in the $xt$ region desired, except along the line $t = 4x$, where the second derivatives do not exist.

Next we verify that our $y$ satisfies the boundary conditions. To see whether $y$ satisfies condition (2), we must hold $x$ fixed, but positive, and then let $t$ approach zero through positive values. As

$$t \to 0^+, \quad y \to 0 + [(-4x)^2 + (-4x)]\alpha(-4x) = 0, \qquad \text{for } x > 0.$$

Thus (2) is satisfied. Note that $\alpha(-4x)$ would not have been zero for negative $x$.

From (17), with $x$ fixed and positive, it follows that as

$$t \to 0^+, \quad \frac{\partial y}{\partial t} \to -1 + [2(-4x) + 1]\alpha(-4x) = -1, \qquad \text{for } x > 0.$$

Thus (3) is satisfied. Once more the fact that $x$ is positive plays an important role in the verification.

Consider condition (4). In it we must hold $t$ fixed and positive. Then, by (16), as

$$x \to 0^+, \quad y \to -t + (t^2 + t)\alpha(t) = -t + t^2 + t = t^2, \qquad \text{for } t > 0.$$

Then (4) is satisfied.

Finally, the $y$ of (16) satisfies condition (5), since

$$\lim_{x \to \infty} y(x, t) = -t + 0 = -t, \qquad \text{for } t > 0,$$

because for sufficiently large $x$ and fixed $t$, $(t - 4x)$ is negative and therefore $\alpha(t - 4x) = 0$. This completes the verification of the solution (16).

## Exercises

In each exercise, solve the problem and verify your solution completely.

**1.** $\dfrac{\partial y}{\partial x} + 4\dfrac{\partial y}{\partial t} = -8t,$      for $t > 0, x > 0$;

$t \to 0^+, y \to 0,$      for $x > 0$;
$x \to 0^+, y \to 2t^2,$      for $t > 0$.

           ANS.   $y(x, t) = -t^2 + 3(t - 4x)^2 \alpha(t - 4x).$

**2.** $\dfrac{\partial y}{\partial x} + 2\dfrac{\partial y}{\partial t} = 4t,$      for $t > 0, x > 0$;

$t \to 0^+, y \to 0,$      for $x > 0$;
$x \to 0^+, y \to 2t^3,$      for $t > 0$.

           ANS.   $y(x, t) = t^2 + [2(t - 2x)^3 - (t - 2x)^2]\alpha(t - 2x).$

**3.** Solve exercise 1 with the condition as $t \to 0^+$ replaced by $t \to 0^+, y \to x$.

           ANS.   $y(x, t) = x - \tfrac{1}{4}t - t^2 + [3(t - 4x)^2 + \tfrac{1}{4}(t - 4x)]\alpha(t - 4x).$

**4.** Solve exercise 2 with the condition as $t \to 0^+$ replaced by $t \to 0^+, y \to 2x$.

           ANS.   $y(x, t) = 2x - t + t^2 + [2(t - 2x)^3 - (t - 2x)^2 + (t - 2x)]\alpha(t - 2x).$

**5.** $\dfrac{\partial^2 y}{\partial x^2} = 16\dfrac{\partial^2 y}{\partial t^2},$      for $t > 0, x > 0$;

$t \to 0^+, y \to 0,$      for $x > 0$;

$t \to 0^+, \dfrac{\partial y}{\partial t} \to -2,$      for $x > 0$;

$x \to 0^+, y \to t,$      for $t > 0$;

$\lim_{x \to \infty} y(x, t)$ exists,      for $t > 0$.            ANS.   $y = 3(t - 4x)\alpha(t - 4x) - 2t.$

**6.** $\dfrac{\partial^2 y}{\partial t^2} = 4\dfrac{\partial^2 y}{\partial x^2},$      for $t > 0, x > 0$;

$t \to 0^+, y \to 0,$      for $x > 0$;

$t \to 0^+, \dfrac{\partial y}{\partial t} \to 2,$      for $x > 0$;

$x \to 0^+, y \to \sin t,$      for $t > 0$;

$\lim_{x \to \infty} y(x, t)$ exists,      for $t > 0$.

           ANS.   $y = 2t + [\sin(t - \tfrac{1}{2}x) - 2(t - \tfrac{1}{2}x)]\alpha(t - \tfrac{1}{2}x).$

## 164.  The wave equation

The transverse displacement $y$ of an elastic string must satisfy the one-dimensional wave equation

$$\frac{\partial^2 y}{\partial t^2} = a^2 \frac{\partial^2 y}{\partial x^2}$$

of Section 157, in which the positive constant $a$ has the dimensions of a velocity, centimeters per second, and so on.

Suppose a long elastic string is initially taut and at rest so that we may take, at $t = 0$,

$$y = 0 \quad \text{and} \quad \frac{\partial y}{\partial t} = 0, \qquad \text{for } x \geqq 0.$$

We assume the string long enough that the assumption that it extends from $x = 0$ to $\infty$ introduces no appreciable error over the time interval in which we are interested.

Suppose also that that end of the string far distant from the $y$-axis is held fixed, $y \to 0$ as $x \to \infty$, but that at the $y$-axis end the string is moved up and down according to some prescribed law, $y \to F(t)$ as $x \to 0^+$, with $F(t)$ known. Figure 64 shows the position of the string at some $t > 0$.

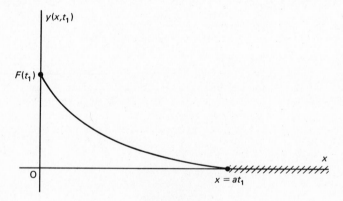

FIGURE 64

The problem of determining the transverse displacement $y$ in terms of $x$ and $t$ is that of solving the boundary value problem:

$$\frac{\partial^2 y}{\partial t^2} = a^2 \frac{\partial^2 y}{\partial x^2}, \qquad \text{for } t > 0, x > 0; \tag{1}$$

$$t \to 0^+, y \to 0, \qquad \text{for } x \geqq 0; \tag{2}$$

$$t \to 0^+, \frac{\partial y}{\partial t} \to 0, \qquad \text{for } x > 0; \tag{3}$$

$$x \to 0^+, y \to F(t), \qquad \text{for } t \geqq 0; \tag{4}$$

$$\lim_{x \to \infty} y(x, t) = 0, \qquad \text{for all } t \geqq 0. \tag{5}$$

The prescribed function $F(t)$ must vanish at $t = 0$ to retain continuity of the string.

This problem satisfies the criteria, page 504, that suggest the use of the Laplace transform. Let

$$L\{y(x, t)\} = u(x, s), \qquad L\{F(t)\} = f(s). \tag{6}$$

Note that $F(t)$ must be continuous because of its physical meaning here. The operator $L$ converts the problem (1) through (5) into the new problem

$$s^2 u = a^2 \frac{d^2 u}{dx^2}, \qquad \text{for } x > 0; \tag{7}$$

$$x \to 0^+, u \to f(s); \tag{8}$$

$$\lim_{x \to \infty} u(x, s) = 0. \tag{9}$$

From (7) we write at once the general solution

$$u(x, s) = c_1(s) \exp\left(-\frac{sx}{a}\right) + c_2(s) \exp\left(\frac{sx}{a}\right). \tag{10}$$

With $s > 0$, $x > 0$, the condition (9) requires

$$c_2(s) \equiv 0. \tag{11}$$

Thus (10) becomes

$$u(x, s) = c_1(s) \exp\left(-\frac{sx}{a}\right), \tag{12}$$

and (8) requires that

$$f(s) = c_1(s).$$

We therefore have

$$u(x, s) = f(s) \exp\left(-\frac{sx}{a}\right), \qquad x > 0, s > 0. \tag{13}$$

Equation (13) yields the desired solution

$$y(x, t) = F\left(t - \frac{x}{a}\right)\alpha\left(t - \frac{x}{a}\right), \qquad x > 0, t > 0, \tag{14}$$

in which we assume that $F(t)$ is defined in some manner for negative argument so that Theorem 22, page 208, can be used.

Verification of the solution (14) is a simple matter. Note that

$$\frac{\partial y}{\partial t} = F'\left(t - \frac{x}{a}\right)\alpha\left(t - \frac{x}{a}\right), \qquad \frac{\partial y}{\partial x} = -\frac{1}{a}F'\left(t - \frac{x}{a}\right)\alpha\left(t - \frac{x}{a}\right)$$

and

$$\frac{\partial^2 y}{\partial t^2} = F''\left(t - \frac{x}{a}\right)\alpha\left(t - \frac{x}{a}\right), \qquad \frac{\partial^2 y}{\partial x^2} = \frac{1}{a^2}F''\left(t - \frac{x}{a}\right)\alpha\left(t - \frac{x}{a}\right).$$

We are forced to assume existence of two derivatives of the prescribed function $F(t)$. It is particularly convenient to choose $F(t)$ so that $F'(0)$ and $F''(0)$ vanish along with $F(0)$, so that the continuity of $y$ and its derivatives is not interrupted along the line $x = at$. Completion of the verification of the solution is left to the student.

In Section 157 we studied the transverse displacement of a string of finite length held fixed at both ends. Fourier series methods seem superior to Laplace transform techniques for such problems. Try, for instance, transform methods on exercise 1, p. 482.

### Exercise

1. Interpret and solve the problem:

$$\frac{\partial^2 y}{\partial t^2} = \frac{\partial^2 y}{\partial x^2}, \qquad \text{for } t > 0, 0 < x < 1;$$

$$t \to 0^+, y \to x - x^2, \qquad \text{for } 0 < x < 1;$$

$$t \to 0^+, \frac{\partial y}{\partial t} \to 0, \qquad \text{for } 0 < x < 1;$$

$$x \to 0^+, y \to 0, \qquad \text{for } t > 0;$$

$$x \to 1^-, y \to 0, \qquad \text{for } t > 0.$$

Verify your solution directly.

ANS. $y = x - x^2 - t^2 + \displaystyle\sum_{n=0}^{\infty} (-1)^n[(t - n - x)^2\alpha(t - n - x)$

$$+ (t - n - 1 + x)^2\alpha(t - n - 1 - x)].$$

## 165. Diffusion in a semi-infinite solid

Consider the solid defined by $x \geq 0$, occupying one half of three-dimensional space. If the initial temperature within the solid and the conditions at the surface $x = 0$ are independent of the coordinates $y$ and $z$, the temperature $u$ will be independent of $y$ and $z$ for all $t > 0$. We may visualize, for example, a huge flat slab of concrete with an initial temperature distribution dependent only upon the distance from the plane surface of the slab. If the temperature at that surface is thereafter ($t > 0$) maintained at some specified function of $t$, or if the surface is insulated, the problem of finding the temperature for all positive $x$ and $t$ is one involving the simple heat equation (2) of Section 153.

EXAMPLE: Consider a semi-infinite slab $x \geqq 0$, initially at a fixed temperature $u = A$ and thereafter subjected to a surface temperature $(x \to 0^+)$ which is $u = B$ for $0 < t < t_0$ and then $u = 0$ for $t \geqq t_0$. Find the temperature within the solid for $x > 0, t > 0$.

The boundary value problem to be solved is:

$$\frac{\partial u}{\partial t} = h^2 \frac{\partial^2 u}{\partial x^2}, \qquad \text{for } x > 0, t > 0; \tag{1}$$

$$t \to 0^+, u \to A, \qquad \text{for } x > 0; \tag{2}$$

$$x \to 0^+, u \to B, \qquad \text{for } 0 < t < t_0, \tag{3}$$

$$u \to 0, \qquad \text{for } t > t_0; $$

$$\lim_{x \to \infty} u(x, t) \text{ exists}, \qquad \text{for each fixed } t > 0. \tag{4}$$

In this problem $A$, $B$, and $h^2$ are constants. We use the $\alpha$ function to reword the boundary condition (3) in the form

$$x \to 0^+, \qquad u \to B[1 - \alpha(t - t_0)], \qquad \text{for } t > 0. \tag{5}$$

Note also that the physical problem dictates that the value of the limit in (4) is to be $A$. This furnishes us with an additional check on our work.

The problem satisfies the criteria, page 504, that suggest the use of the Laplace transform. Let

$$L\{u(x, t)\} = w(x, s), \qquad x > 0, s > 0. \tag{6}$$

The equation (1) with condition (2) is transformed into

$$sw - A = h^2 \frac{d^2w}{dx^2}, \qquad x > 0,$$

or

$$\frac{d^2w}{dx^2} - \frac{s}{h^2} w \doteqdot -\frac{A}{h^2}, \qquad x > 0. \tag{7}$$

Conditions (4) and (5) become

$$\lim_{x \to \infty} w(x, s) \text{ exists,} \quad \text{for fixed } s > 0, \tag{8}$$

and

$$x \to 0^+, \quad w \to \frac{B}{s}[1 - \exp(-t_0 s)]. \tag{9}$$

The differential equation (7) has the general solution

$$w = c_1 \exp\left(-\frac{x\sqrt{s}}{h}\right) + c_2 \exp\left(\frac{x\sqrt{s}}{h}\right) + \frac{A}{s}, \quad x > 0, s > 0, \tag{10}$$

in which $c_1$ and $c_2$ may be functions of $s$, but not of $x$. As $x \to \infty$, the $w$ of (10) will approach a limit if, and only if, $c_2 = 0$. Hence condition (8) yields the result

$$c_2 = 0 \tag{11}$$

and the $w$ of (10) becomes

$$w = c_1 \exp\left(-\frac{x\sqrt{s}}{h}\right) + \frac{A}{s}. \tag{12}$$

By letting $x \to 0^+$ and using (9), we obtain,

$$\frac{B}{s}[1 - \exp(-t_0 s)] = c_1 + \frac{A}{s}. \tag{13}$$

Therefore the solution of the problem (7) through (9) is

$$w(x, s) = \frac{A}{s}\left[1 - \exp\left(-\frac{x\sqrt{s}}{h}\right)\right] + \frac{B}{s}\exp\left(-\frac{x\sqrt{s}}{h}\right)[1 - \exp(-t_0 s)]. \tag{14}$$

We know that

$$L^{-1}\left\{\frac{1}{s}\exp\left(-\frac{x\sqrt{s}}{h}\right)\right\} = \operatorname{erfc}\left(\frac{x}{2h\sqrt{t}}\right), \quad x > 0. \tag{15}$$

Hence we may write

$$L^{-1}\left\{\frac{1}{s}\exp\left(-\frac{x\sqrt{s}}{h}\right)\exp(-t_0 s)\right\} = \operatorname{erfc}\left(\frac{x}{2h|t - t_0|^{1/2}}\right)\alpha(t - t_0), \tag{16}$$

where absolute value signs have been inserted to permit $t$ to be used in the range 0 to $t_0$ in which range the $\alpha$ function will force the right member of (16) to be zero.

We are now in a position to write the inverse transform of the $w$ of equation (14). For $x > 0$ and $t > 0$,

$$u(x, t) = A\left[1 - \operatorname{erfc}\left(\frac{x}{2h\sqrt{t}}\right)\right]$$

$$+ B\left[\operatorname{erfc}\left(\frac{x}{2h\sqrt{t}}\right) - \operatorname{erfc}\left(\frac{x}{2h|t - t_0|^{1/2}}\right)\alpha(t - t_0)\right], \quad (17)$$

or

$$u(x, t) = A\operatorname{erf}\left(\frac{x}{2h\sqrt{t}}\right) + B\left[\operatorname{erfc}\left(\frac{x}{2h\sqrt{t}}\right) - \operatorname{erfc}\left(\frac{x}{2h|t - t_0|^{1/2}}\right)\alpha(t - t_0)\right].$$

$$(18)$$

The $u$ of (17), or of (18), is the desired solution.

It is a matter of direct substitution to show that each term of (18) is a solution of the one-dimensional heat equation. That the conditions (2), (3), and (4) are also satisfied follows rapidly from the properties

$$\lim_{z \to 0} \operatorname{erf} z = 0, \qquad \lim_{z \to \infty} \operatorname{erf} z = 1$$

and the corresponding properties of the erfc function. Indeed, for the $u$ of (18),

As $x \to 0^+$, $u \to A \cdot 0 + B[1 - \alpha(t - t_0)] = B[1 - \alpha(t - t_0)]$,     for $t > 0$;

As $t \to 0^+$, $u \to A \cdot 1 + B(0 - 0) = A$,     for $x > 0$;

As $x \to \infty$, $u \to A \cdot 1 + B \cdot 0 = A$,     for $0 < t < t_0$;

As $x \to \infty$, $u \to A \cdot 1 + B(0 - 0) = A$,     for $t > t_0$.

## 166. Canonical variables

As we attack problems of increasing complexity, it becomes important that we simplify our work by the introduction of what are called *canonical variables*. These variables are dimensionless combinations of the physical variables and parameters of the original problem. We now illustrate a method for selecting such variables.

In Section 167 we shall solve a diffusion problem that can be expressed in the following way:

$$\frac{\partial u}{\partial t} = h^2 \frac{\partial^2 u}{\partial x^2}, \qquad \text{for } t > 0, 0 < x < c; \quad (1)$$

$$t \to 0^+, u \to A, \qquad \text{for } 0 < x < c; \tag{2}$$

$$x \to 0^+, u \to 0, \qquad \text{for } t > 0; \tag{3}$$

$$x \to c^-, u \to 0, \qquad \text{for } t > 0. \tag{4}$$

A consistent set of units for the measure of the various constants (parameters) and variables in this problem is

$$u = \text{temperature (°F)},$$

$$t = \text{time (hr)},$$

$$x = \text{space coordinate (ft)}$$

$$h^2 = \text{thermal diffusivity (ft}^2/\text{hr)},$$

$$c = \text{length (ft)},$$

$$A = \text{initial temperature (°F)}.$$

We seek dimensionless new variables $\zeta$, $\tau$, $\psi$, proportional to the physical variables $x$, $t$, $u$. For the moment let

$$x = \beta\zeta, \qquad t = \gamma\tau, \qquad u = \delta\psi, \tag{5}$$

in which $\beta$, $\gamma$, $\delta$ are positive constants to be so determined that the new variables will each be of dimension zero. The changes of variable (5) transform (1) through (4) into

$$\frac{\delta}{\gamma} \frac{\partial \psi}{\partial \tau} = \frac{h^2 \delta}{\beta^2} \frac{\partial^2 \psi}{\partial \zeta^2}, \qquad \text{for } \tau > 0, 0 < \beta\zeta < c; \tag{6}$$

$$\tau \to 0^+, \delta\psi \to A, \qquad \text{for } 0 < \beta\zeta < c; \tag{7}$$

$$\zeta \to 0^+, \psi \to 0, \qquad \text{for } \tau > 0; \tag{8}$$

$$\beta\zeta \to c^-, \psi \to 0, \qquad \text{for } \tau > 0. \tag{9}$$

Because of (7), we choose $\delta = A$ and $\beta = c$. Because of (6), we choose

$$\frac{1}{\gamma} = \frac{h^2}{\beta^2},$$

from which

$$\gamma = \frac{c^2}{h^2}.$$

We thus find that the introduction of the new variables

$$\zeta = \frac{x}{c}, \qquad \tau = \frac{h^2 t}{c^2}, \qquad \psi = \frac{u}{A}, \tag{10}$$

transforms the problem (1) through (4) into the canonical form

$$\frac{\partial \psi}{\partial \tau} = \frac{\partial^2 \psi}{\partial \zeta^2}, \qquad \text{for } \tau > 0, 0 < \zeta < 1; \tag{11}$$

$$\tau \to 0^+, \psi \to 1, \qquad \text{for } 0 < \zeta < 1; \tag{12}$$

$$\zeta \to 0^+, \psi \to 0, \qquad \text{for } \tau > 0; \tag{13}$$

$$\zeta \to 1^-, \psi \to 0, \qquad \text{for } \tau > 0. \tag{14}$$

Note that the canonical variables in (10) are of dimension zero; $\zeta$ has dimension feet over feet, and so on.

The solution of (11) through (14) is independent of the parameters $h^2$, $c$, and $A$ of the original problem, a fact of great importance in applications. The solution of the original problem (1) through (4) is a function of two variables and three parameters,

$$u = f(x, t, c, h, A). \tag{15}$$

The solution of (11) through (14), for which see Section 167, is a function of two variables

$$\psi = F(\zeta, \tau), \tag{16}$$

so (15) actually takes the form

$$u = AF\left(\frac{x}{c}, \frac{h^2 t}{c^2}\right). \tag{17}$$

The function $F$, of two variables, can be tabulated and it thus yields the solution of the original problem no matter what the values of $c$, $A$, and $h^2$.

There are problems, such as in the study of temperatures in a concrete dam, in which it is important to know the mean value with respect to $x$ of the temperature $u$ of (15) over the range $0 < x < c$. That mean value may be computed by using (16), and the result is a function of the one variable $\tau$. Thus a single curve can be drawn in the $\psi \tau$ plane to give the pertinent mean temperature for all problems (1) through (4).

## 167. Diffusion in a slab of finite width

We shall now solve by transform methods the slab problem of Section 139 for the special case $f(x) = A$. Let the thickness of the slab be $c$ units of length. Let the coordinate $x$ denote distance from one face of the slab and assume that the slab extends very far in the $y$ and $z$ directions. Assume that the initial temperature of the slab is a constant $A$ and that the surfaces $x = 0$, $x = c$ are maintained at zero temperature for all $t > 0$. If the slab is considered infinite in the $y$ and $z$ directions or, more specifically, if we treat only cross sections nearby (far from the distant surfaces of the slab), then the temperature $u$ at any time $t$ and position $x$ is determined by the boundary value problem:

$$\frac{\partial u}{\partial t} = h^2 \frac{\partial^2 u}{\partial x^2}, \qquad \text{for } t > 0, 0 < x < c; \tag{1}$$

$$t \to 0^+, u \to A, \qquad \text{for } 0 < x < c; \tag{2}$$

$$x \to 0^+, u \to 0, \qquad \text{for } t > 0; \tag{3}$$

$$x \to c^-, u \to 0, \qquad \text{for } t > 0. \tag{4}$$

We shall solve the corresponding problem in canonical variables. That is, in (1) through (4) we put

$$\zeta = \frac{x}{c}, \qquad \tau = \frac{h^2 t}{c^2}, \qquad \psi = \frac{u}{A}. \tag{5}$$

In the new variables $\zeta, \tau, \psi$, the problem to be solved is

$$\frac{\partial \psi}{\partial \tau} = \frac{\partial^2 \psi}{\partial \zeta^2}, \qquad \text{for } \tau > 0, 0 < \zeta < 1; \tag{6}$$

$$\tau \to 0^+, \psi \to 1, \qquad \text{for } 0 < \zeta < 1; \tag{7}$$

$$\zeta \to 0^+, \psi \to 0, \qquad \text{for } \tau > 0; \tag{8}$$

$$\zeta \to 1^-, \psi \to 0, \qquad \text{for } \tau > 0. \tag{9}$$

Let

$$L\{\psi(\zeta, \tau)\} = w(\zeta, s) = \int_0^\infty e^{-s\tau} \psi(\zeta, \tau)\, d\tau. \tag{10}$$

Application of the Laplace operator transforms the problem (6) through (9) into

$$sw - 1 = \frac{d^2 w}{d\zeta^2}, \qquad \text{for } 0 < \zeta < 1; \tag{11}$$

$$\zeta \to 0^+, w \to 0; \tag{12}$$

$$\zeta \to 1^-, w \to 0. \tag{13}$$

The general solution of (11) may be written

$$w = c_1 \sinh(\zeta\sqrt{s}) + c_2 \cosh(\zeta\sqrt{s}) + \frac{1}{s}. \tag{14}$$

From (12) it follows that

$$0 = c_2 + \frac{1}{s} \tag{15}$$

and (13) yields

$$0 = c_1 \sinh \sqrt{s} + c_2 \cosh \sqrt{s} + \frac{1}{s}. \tag{16}$$

By solving (15) and (16) we obtain

$$c_2 = -\frac{1}{s}, \qquad c_1 = \frac{\cosh \sqrt{s} - 1}{s \sinh \sqrt{s}}, \tag{17}$$

from which we see that

$$w = \frac{1}{s} + \frac{(\cosh \sqrt{s} - 1) \sinh (\zeta\sqrt{s}) - \sinh \sqrt{s} \cosh (\zeta\sqrt{s})}{s \sinh \sqrt{s}}. \tag{18}$$

Since

$$\sinh B_1 \cosh B_2 - \cosh B_1 \sinh B_2 = \sinh (B_1 - B_2),$$

the $w$ of (18) may be written in the form

$$w(\zeta, s) = \frac{1}{s}\left[ 1 - \frac{\sinh (\zeta\sqrt{s})}{\sinh \sqrt{s}} - \frac{\sinh \{(1 - \zeta)\sqrt{s}\}}{\sinh \sqrt{s}} \right]. \tag{19}$$

The desired solution $\psi(\zeta, \tau)$ is the inverse of the $w(\zeta, s)$ of (19), with $\zeta$ on the range $0 < \zeta < 1$.

We already know from equation (29), page 497, that for $0 < x < 1$,

$$L^{-1}\left\{ \frac{\sinh (x\sqrt{s})}{s \sinh \sqrt{s}} \right\} = \sum_{n=0}^{\infty} \left[ \operatorname{erfc}\left( \frac{1 - x + 2n}{2\sqrt{t}} \right) - \operatorname{erfc}\left( \frac{1 + x + 2n}{2\sqrt{t}} \right) \right]. \tag{20}$$

Applying (20) twice, once with $\zeta$ and once with $(1 - \zeta)$ replacing $x$, we obtain from (19) the desired solution

$$\psi(\zeta, \tau) = 1 - \sum_{n=0}^{\infty} \left[ \operatorname{erfc}\left( \frac{1 - \zeta + 2n}{2\sqrt{\tau}} \right) - \operatorname{erfc}\left( \frac{1 + \zeta + 2n}{2\sqrt{\tau}} \right) \right]$$
$$- \sum_{n=0}^{\infty} \left[ \operatorname{erfc}\left( \frac{\zeta + 2n}{2\sqrt{\tau}} \right) - \operatorname{erfc}\left( \frac{2 - \zeta + 2n}{2\sqrt{\tau}} \right) \right]. \tag{21}$$

The complementary error functions in (21) may be replaced by error functions, since

$$\operatorname{erfc} z = 1 - \operatorname{erf} z. \tag{22}$$

With the aid of the properties

$$\text{erfc } 0 = 1, \qquad \lim_{z \to \infty} \text{erfc } z = 0, \tag{23}$$

the solution (21) is easily verified, assuming that the summation sign and the pertinent limits may be interchanged. With the theorems of advanced calculus the assumption can be shown to be valid.

From (21) we get, as $\zeta \to 0^+$,

$$\psi \to 1 - \sum_{n=0}^{\infty} \left[ \text{erfc} \left( \frac{2n+1}{2\sqrt{\tau}} \right) - \text{erfc} \left( \frac{2n+1}{2\sqrt{\tau}} \right) \right]$$

$$- \sum_{n=0}^{\infty} \left[ \text{erfc} \left( \frac{n}{\sqrt{\tau}} \right) - \text{erfc} \left( \frac{n+1}{\sqrt{\tau}} \right) \right].$$

In the first series each term is zero. The second series telescopes; in it we replace the series by the limit of the partial sums to get

$$\psi \to 1 - \lim_{n \to \infty} \sum_{k=0}^{n} \left[ \text{erfc} \left( \frac{k}{\sqrt{\tau}} \right) - \text{erfc} \left( \frac{k+1}{\sqrt{\tau}} \right) \right],$$

or

$$\psi \to 1 - \lim_{n \to \infty} \left[ \text{erfc } 0 - \text{erfc} \left( \frac{n+1}{\sqrt{\tau}} \right) \right].$$

For fixed $\tau > 0$, $(n+1)/\sqrt{\tau} \to \infty$ as $n \to \infty$. Hence, by (23),

$$\psi \to 1 - 1 + 0 = 0, \qquad \text{as } \zeta \to 0^+. \tag{24}$$

The solution (21) is unchanged when $\zeta$ is replaced by $(1 - \zeta)$, because the two series merely change places. Therefore, because of (24),

$$\psi \to 0, \qquad \text{as } \zeta \to 1^-. \tag{25}$$

For any $\zeta$ in the range $0 < \zeta < 1$, the argument of each erfc in (21) is positive and approaches infinity as $\tau \to 0^+$. Hence each erfc $\to 0$ and each term of the two series $\to 0$. Thus, because the order of limit and summation can be interchanged,

$$\psi \to 1, \quad \text{as } \tau \to 0^+, \qquad \text{for } 0 < \zeta < 1. \tag{26}$$

Perhaps the most valuable single fact about the solution (21) is the fact that the series converge very rapidly for small $\tau$ because the arguments of the various erfc functions are then very large. By the methods of separation of variables and Fourier series, the problem (6) through (9) at the start of this section can be shown to have the solution

$$\psi(\zeta, \tau) = \frac{4}{\pi} \sum_{k=0}^{\infty} \frac{\exp\left[-\pi^2 (2k+1)^2 \tau\right] \sin\left[(2k+1)\pi\zeta\right]}{2k+1}. \tag{27}$$

The solutions given by (21) and (27) are identical, though the uniqueness of such solutions is not proved here.

The series in (27) converges rapidly for large $\tau$ and slowly for small $\tau$. The series in (21) converge rapidly for small $\tau$ and slowly for large $\tau$. The two forms of solution complement each other neatly.

The solution of the original problem (1) through (4) may be obtained from (21) or (27) by making the substitutions in (5).

### Exercise

**1.** Interpret and solve the following problem.

$$\frac{\partial u}{\partial t} = \frac{\partial^2 u}{\partial x^2}, \qquad \text{for } t > 0, 0 < x < 1;$$

$$t \to 0^+, u \to 1, \qquad \text{for } 0 < x < 1;$$

$$x \to 0^+, u \to 0, \qquad \text{for } t > 0;$$

$$x \to 1^-, \frac{\partial u}{\partial x} \to 0, \qquad \text{for } t > 0.$$

ANS.  $u = 1 - \displaystyle\sum_{n=0}^{\infty} (-1)^n \left[ \operatorname{erfc} \left( \frac{2n + x}{2\sqrt{t}} \right) + \operatorname{erfc} \left( \frac{2n + 2 - x}{2\sqrt{t}} \right) \right].$

## 168.  Diffusion in a quarter-infinite solid

As a final application let us study the temperatures near a square corner of a huge slab initially at a constant temperature and having its surfaces thereafter held at a constant temperature different from the initial interior temperature. We assume that all temperatures are independent of one rectangular space coordinate. By introducing canonical variables, we may express the mathematical problem as follows:

$$\frac{\partial u}{\partial t} = \frac{\partial^2 u}{\partial x^2} + \frac{\partial^2 u}{\partial y^2}, \qquad \text{for } t > 0, x > 0, y > 0; \qquad (1)$$

$$t \to 0^+, u \to 1, \qquad \text{for } x > 0, y > 0; \qquad (2)$$

$$x \to 0^+, u \to 0, \qquad \text{for } t > 0, y > 0; \qquad (3)$$

$$y \to 0^+, u \to 0, \qquad \text{for } t > 0, x > 0; \qquad (4)$$

$$\lim_{x \to \infty} u(x, y, t) \text{ exists}, \qquad \text{for fixed positive } t \text{ and } y; \qquad (5)$$

$$\lim_{y \to \infty} u(x, y, t) \text{ exists}, \qquad \text{for fixed positive } t \text{ and } x. \qquad (6)$$

The solution of the problem (1) through (6) will be accomplished by combining separation of variables with the Laplace transform technique. First we separate the function $u$ of the three variables $x$, $y$, $t$ into the product of a function of $x$ and $t$ alone by a function of $y$ and $t$ alone. This separation is possible only because of the peculiar simplicity of the boundary value problem. Let

$$u(x, y, t) = v(x, t)w(y, t). \tag{7}$$

From (7) it follows that

$$\frac{\partial u}{\partial t} = v\frac{\partial w}{\partial t} + w\frac{\partial v}{\partial t},$$

$$\frac{\partial^2 u}{\partial x^2} = w\frac{\partial^2 v}{\partial x^2},$$

$$\frac{\partial^2 u}{\partial y^2} = v\frac{\partial^2 w}{\partial y^2}.$$

Hence equation (1) yields

$$v\frac{\partial w}{\partial t} + w\frac{\partial v}{\partial t} = w\frac{\partial^2 v}{\partial x^2} + v\frac{\partial^2 w}{\partial y^2}, \tag{8}$$

which will be satisfied if both

$$\frac{\partial v}{\partial t} = \frac{\partial^2 v}{\partial x^2}, \qquad \text{for } t > 0,\ x > 0, \tag{9}$$

and

$$\frac{\partial w}{\partial t} = \frac{\partial^2 w}{\partial y^2}, \qquad \text{for } t > 0,\ y > 0, \tag{10}$$

are satisfied.

If we impose the conditions

$$t \to 0^+,\ v \to 1, \qquad \text{for } x > 0, \tag{11}$$

$$t \to 0^+,\ w \to 1, \qquad \text{for } y > 0, \tag{12}$$

condition (2) will be satisfied.

From condition (3) we get

$$x \to 0^+,\ v \to 0, \qquad \text{for } t > 0, \tag{13}$$

and from (4),

$$y \to 0^+,\ w \to 0, \qquad \text{for } t > 0. \tag{14}$$

Conditions (5) and (6) will be satisfied if

$$\lim_{x \to \infty} v(x, t) \text{ exists,} \qquad \text{for fixed positive } t, \tag{15}$$

and

$$\lim_{y \to \infty} w(y, t) \text{ exists,} \qquad \text{for fixed positive } t. \tag{16}$$

We must now find $v$ from

$$\frac{\partial v}{\partial t} = \frac{\partial^2 v}{\partial x^2}, \qquad \text{for } t > 0, x > 0; \tag{9}$$

$$t \to 0^+, v \to 1, \qquad \text{for } x > 0; \tag{11}$$

$$x \to 0^+, v \to 0, \qquad \text{for } t > 0; \tag{13}$$

$$\lim_{x \to \infty} v(x, t) \text{ exists,} \qquad \text{for fixed positive } t. \tag{15}$$

The function $w$ must satisfy (10), (12), (14), and (16); it is therefore the same function as $v$ except that $y$ replaces $x$.

To obtain $v$ we use the Laplace transform. Let

$$L\{v(x, t)\} = g(x, s) = \int_0^\infty e^{-st} v(x, t) \, dt. \tag{17}$$

Then (9) and (11) yield

$$sg - 1 = \frac{d^2 g}{dx^2}, \tag{18}$$

for which the general solution is easily written by inspection because of our experience in handling equations with constant coefficients. We thus get

$$g = \frac{1}{s} + c_1(s) \exp\left(-x\sqrt{s}\right) + c_2(s) \exp\left(x\sqrt{s}\right). \tag{19}$$

The function $g$ must, because of (13) and (15), satisfy the conditions

$$x \to 0^+, \qquad g \to 0, \tag{20}$$

$$\lim_{x \to \infty} g(x, s) \text{ exists.} \tag{21}$$

Because of (21), $c_2(s) = 0$. Because of (20),

$$0 = \frac{1}{s} + c_1(s).$$

Therefore

$$g(x, s) = \frac{1}{s} - \frac{1}{s} \exp\left(-x\sqrt{s}\right). \tag{22}$$

The function $v(x, t)$ is an inverse transform of $g(x, s)$:

$$v(x, t) = 1 - \operatorname{erfc}\left(\frac{x}{2\sqrt{t}}\right). \tag{23}$$

But $1 - \operatorname{erfc} z = \operatorname{erf} z$. Hence

$$v(x, t) = \operatorname{erf}\left(\frac{x}{2\sqrt{t}}\right). \tag{24}$$

Therefore the solution of our original problem (1) through (6) is

$$u = \operatorname{erf}\left(\frac{x}{2\sqrt{t}}\right)\operatorname{erf}\left(\frac{y}{2\sqrt{t}}\right). \tag{25}$$

The student should verify that the $u$ of (25) satisfies all the conditions of the boundary value problem (1) through (6) introduced at the beginning of this section.

## Exercises

1. Show that for the $u$ of (25), $0 < u < 1$, for all $x, y, t > 0$.
2. Let the point with coordinates $(x, y, t)$ be in the first octant of the rectangular $x, y, t$ space. Let that point approach the origin along a curve

$$x^2 = 4a^2 t,$$

$$y^2 = 4a^2 t,$$

in which $a$ is positive but otherwise arbitrary. Show that as $x, y, t \to 0^+$ in the manner described above, $u$ may be made to approach any desired number between zero and unity.

# Index